Natural Products and Nano-Formulations in Cancer Chemoprevention

This book covers various aspects of cancer chemoprevention, including an overview of chemoprevention in the process of tumorigenesis; the roles of various phytochemicals, functional foods, and dietary interventions in disease prevention; and techniques such as cancer stem cell targeting, nano-formulations, and so forth.

The nutrigenomic and epigenetic effects of natural products at the molecular and genetic levels are also covered alongside their potential for additive and synergistic effect, as well as overcoming drug resistance. The key selling features of the book are as follows:

- Discusses holistic and comprehensive areas of chemoprevention
- Includes diverse techniques, such as cancer stem cell targeting, nano-formulations, and nanotechnology-based drug delivery systems
- Introduces various mechanisms involved in prevention of the diseases, including targeting cancer stem cells
- Reviews various aspects which can reduce the toxicity and cost of treatment of diseases by alternative medicine
- Explores various sources, mechanisms, and ways to develop cancer chemopreventive agents with minimal toxicity compared to traditional cancer therapy drugs

This book is focused on researchers and graduate students in drug delivery and formulation, nanobiotechnology, cancer chemoprevention, prevention, and therapeutics.

Advances in Bionanotechnology

Series Editors
Ravindra Pratap Singh
Department of Biotechnology, Indira Gandhi National Tribal University,
Anuppur, Madhya Pradesh, India
Jay Singh
Department of Chemistry, Institute of Science, Banaras Hindu University,
Varanasi, Uttar Pradesh, India
Charles Oluwaseun Adetunji
Department of Microbiology, Edo State University Uzairue, Iyamho, Edo State, Nigeria

Bionanotechnology is a multi-disciplinary field that shows immense applicability in different domains, namely chemistry, physics, material sciences, biomedical, agriculture, environment, robotics, aeronautics, energy, electronics and so forth. This book series will explore the enormous utility of bionanotechnology for biomedical, agricultural, environmental, food technology, space industry, and many other fields. It aims to highlight all the spheres of bionanotechnological applications and its safety and regulations for using biogenic nanomaterials that are a key focus of the researchers globally.

Bionanotechnology Towards Sustainable Management of Environmental Pollution
Edited by Naveen Dwivedi and Shubha Dwivedi

Natural Products and Nano-Formulations in Cancer Chemoprevention
Edited by Shiv Kumar Dubey

For more information about this series, please visit: www.routledge.com/Advances-in-Bionano technology/book-series/CRCBIONAN

Natural Products and Nano-Formulations in Cancer Chemoprevention

Edited by Shiv Kumar Dubey

CRC Press
Taylor & Francis Group
Boca Raton London New York

CRC Press is an imprint of the
Taylor & Francis Group, an **informa** business

Designed cover image: © Shutterstock

First edition published 2023
by CRC Press
6000 Broken Sound Parkway NW, Suite 300, Boca Raton, FL 33487–2742

and by CRC Press
4 Park Square, Milton Park, Abingdon, Oxon, OX14 4RN

CRC Press is an imprint of Taylor & Francis Group, LLC

ISBN: 978-1-032-31303-0 (hbk)
ISBN: 978-1-032-43818-4 (pbk)
ISBN: 978-1-003-36896-0 (ebk)

DOI: 10.1201/b23311

Contents

Editor Biography...vii
List of Contributors...ix
Preface... xiii

Chapter 1   Overview of Natural Products and Nano-Formulations in Cancer
Chemoprevention... 1

*Ananya Bahuguna, Shiv Kumar Dubey, Kanchan Gairola,
and Rohit Pujari*

Chapter 2   Polyphenols in Cancer Chemoprevention 21

*Pushpendra Koli, Dinesh K. Yadav, Mahesh K. Samota,
and Sandeep Kumar*

Chapter 3   Fat-Soluble Vitamins in Cancer Chemoprevention....................... 39

*Neetika Naudiyal, Gaurav Singh Rana, Ankur Adhikari,
and Shiv Kumar Dubey*

Chapter 4   Targeting Cancer Stem Cells by Natural Products for Chemoprevention 57

Kamana Singh, Prabha Arya, and Ram Sunil Kumar L.

Chapter 5   Functional Foods in Cancer Chemoprevention and Therapeutics 75

Shreya Joshi, Pranshu Sharma, and Atul Kumar Joshi

Chapter 6   Phytochemicals: Potential Source of Interventions in Cancer Prevention
and Treatment..97

*Sonu Kumar Mahawer, Himani, Sushila Arya, Tanuja Kabdal,
Ravendra Kumar, and Om Prakash*

Chapter 7   Nano-Formulation-Based Approaches for Chemoprevention................................. 109

Dipanjana Ghosh, Wean Sin Cheow, and Rohit Saluja

Chapter 8   Potential of Natural-Product-Based Nano-Formulations as Chemopreventive
Agents... 129

*Sneh Gautam, Pushpa Lohani, Shiv Dutt Purohit, Sonu Ambwani,
and Poonam Maan*

Chapter 9   Nano-Formulations of Chemotherapeutic Drugs: Recent Outlook, Advances,
and Challenges in Cancer Management... 143

Deepa Bisht, Shilpa Maddheshiya, Seema Nara, and Manisha Sachan

Chapter 10 Nutrigenomics in Cancer Chemoprevention 165

  Kanchan Gairola, Shiv Kumar Dubey, Ananya Bahuguna, and Shruti

Chapter 11 Synergistic Effects of Natural Products in Cancer Treatment 183

  Ritu Saini, Chitra Gupta, Shalini Pundir, Meenakshi Bajpai,
  Akansha Agrwal, and Sharad Visht

Chapter 12 Medicinal Chemistry, Pharmacodynamics, and Pharmacokinetics
  of Camptothecin and Its Derivatives in Clinical Chemotherapeutics 207

  Thadiyan Parambil Ijinu, Parameswaran Sasikumar, Vandhanam Aparna,
  Sreejith Pongillyathundiyil Sasidharan, Vipin Mohan Dan,
  Farkhodjon Tukhtaev, Varughese George, and Palpu Pushpangadan

Chapter 13 *Brassica* Phytochemicals: A Potential Source of Cancer Prevention
  and Treatment 227

  Himanshu Punetha and Shivanshu Garg

Chapter 14 Opportunities with Nano-Formulations in Cancer Chemoprevention 243

  Akansha Agrwal and Sheetal Mittal

Chapter 15 Obstacles in Utilizing Nanomedicine for Cancer Management 259

  Kamana Singh, Vineeta Kashyap, and Addanki P Kumar

Chapter 16 Mangroves as an Alternative Source of Anticancer Drug Leads:
  Current Evidence and Future Prospects 275

  Vasantha Kavunkal Hridya, Kokkuvayil Vasu Radhakrishnan,
  Nainarpandian Chandrasekar, and Thadiyan Parambil Ijinu

Index 285

Editor Biography

Dr. Shiv Kumar Dubey is Assistant Professor in the Biochemistry Department at the College of Basic Sciences and Humanities, G. B. Pant University of Agriculture and Technology in Pantnagar, India, since 2015. He completed his master's and bachelor's degrees at the University of Allahabad in Uttar Pradesh, India, where he also accomplished his PhD degree. He pursued his postdoctoral research at the Institute of Pharmacology and Structural Biology, CNRS, Toulouse, France, and at the University of Michigan Cancer Comprehensive Center, Ann Arbor, Michigan, USA.

His numerous fellowships, including the NET-JRF (National Eligibility Test for Junior Research Fellowship), given by the UGC-CSIR in India, have been awarded to Dr. Shiv Kumar Dubey. More than 25 research papers, book chapters, and reviews by Dr. Dueby have been published in reputable national and international journals. The results of his research have also been presented in numerous national and international forums, including conferences, seminars, and symposiums. Dr. Dubey has created a number of e-content modules and organized webinars and e-training sessions for the benefit of students. To his credit, he has supervised many student theses, filed patents, and developed many courses for undergraduate and postgraduate students. He is actively engaged in the areas of advanced biochemistry, chemoprevention, and natural product research. His work has considerably impacted the study of appropriately constructed curcumin bioconjugates, the validation of analytical methods, the effectiveness of nutraceuticals, and the development of their synthetic analogs for the treatment of cancer. His pursuit of bioprospecting of novel natural compounds for cancer chemoprevention is ongoing.

Contributors

Himani
Govind Ballabh Pant University of Agriculture
and Technology
Pantnagar

Shruti
Govind Ballabh Pant University of Agriculture
and Technology
Pantnagar

Ankur Adhikari
Govind Ballabh Pant University of Agriculture
and Technology
Pantnagar

Akansha Agrwal
KIET Group of Institutions
Ghaziabad

Sonu Ambwani
Govind Ballabh Pant University of Agriculture
and Technology
Pantnagar

Vandhanam Aparna
Mahatma Gandhi University
Kottayam, Kerala

Prabha Arya
Deshbandhu College, University of Delhi
New Delhi

Sushila Arya
Govind Ballabh Pant University of Agriculture
and Technology
Pantnagar

Ananya Bahuguna
Govind Ballabh Pant University of Agriculture
and Technology
Pantnagar

Meenakshi Bajpai
Institute of Pharmaceutical Research, GLA
University
Chaumuhan, Mathura

Deepa Bisht
Motilal Nehru National Institute of Technology
Allahabad

Nainarpandian Chandrasekar
Centre for Geotechnology, Manonmaniam
Sundaranar University
Tirunelveli, Tamil Nadu

Wean Sin Cheow
Singapore Institute of Technology
Singapore

Vipin Mohan Dan
KSCSTE-Jawaharlal Nehru Tropical Botanic
Garden and Research Institute
Palode, Thiruvananthapuram, Kerala

Shiv Kumar Dubey
Govind Ballabh Pant University of Agriculture
and Technology
Pantnagar

Kanchan Gairola
Govind Ballabh Pant University of Agriculture
and Technology
Pantnagar

Shivanshu Garg
Govind Ballabh Pant University of Agriculture
and Technology
Pantnagar

Sneh Gautam
Govind Ballabh Pant University of Agriculture
and Technology
Pantnagar

Varughese George
Amity Institute for Herbal and Biotech
Products Development
Thiruvananthapuram, Kerala

Dipanjana Ghosh
People's University
Bhanpur, Bhopal

Chitra Gupta
Smt. Tarawati Institute of Biomedical and
    Allied Sciences
Roorkee

Vasantha Kavunkal Hridya
CSIR-National Institute for Interdisciplinary
    Science and Technology
Thiruvananthapuram

Thadiyan Parambil Ijinu
Amity Institute for Herbal and Biotech
    Products Development
Thiruvananthapuram, Kerala

Atul Joshi
Patanjali Ayurveda Limited
Haridwar

Shreya Joshi
Community Health Centre
Agastmuni

Tanuja Kabdal
Govind Ballabh Pant University of Agriculture
    and Technology
Pantnagar

Vineeta Kashyap
Deshbandhu College, University of Delhi
New Delhi

Pushpendra Koli
Murdoch University
Perth

Addanki Pratap Kumar
University of Texas Health, San Antonio
Texas

Ravendra Kumar
Govind Ballabh Pant University of Agriculture
    and Technology
Pantnagar

Sandeep Kumar
ICAR-Indian Agriculture Research Institute
New Delhi

Ram Sunil Kumar L
Kirori Mal College, University of Delhi
New Delhi

Pushpa Lohani
Govind Ballabh Pant University of Agriculture
    and Technology
Pantnagar

Poonam Maan
Sardar Vallabhbhai Patel University of
    Agriculture and Technology
Meerut

Shilpa Maddheshiya
Motilal Nehru National Institute of Technology
Allahabad

Sonu Kumar Mahawer
Govind Ballabh Pant University of Agriculture
    and Technology
Pantnagar

Sheetal Mittal
KIET Group of Institutions
Ghaziabad

Seema Nara
Motilal Nehru National Institute of Technology
Allahabad

Neetika Naudiyal
Govind Ballabh Pant University of Agriculture
    and Technology
Pantnagar

Om Prakash
Govind Ballabh Pant University of Agriculture
    and Technology
Pantnagar

Rohit Pujari
Govind Ballabh Pant University of Agriculture
    and Technology
Pantnagar

Shalini Pundir
Smt. Tarawati Institute of Biomedical and
    Allied Sciences
Roorkee

Himanshu Punetha
Govind Ballabh Pant University of Agriculture
    and Technology
Pantnagar

Shiv Dutt Purohit
Yeungnam University
Gyeongsan, Korea

Palpu Pushpangadan
Amity Institute for Herbal and Biotech
    Products Development
Thiruvananthapuram, Kerala

Kokkuvayil Vasu Radhakrishnan
CSIR-National Institute for Interdisciplinary
    Science and Technology
Thiruvananthapuram, Kerala

Gaurav Singh Rana
Govind Ballabh Pant University of Agriculture
    and Technology
Pantnagar

Manisha Sachan
Motilal Nehru National Institute
    of Technology
Allahabad

Ritu Saini
Institute of Pharmaceutical Research, GLA
    University
Chaumuhan, Mathura

Rohit Saluja
All India Institute of Medical Sciences AIIMS
Bibinagar, Hyderabad

Mahesh K Samota
ICAR-Central Institute of Post-Harvest
    Engineering & Technology
Abhor, Punjab

Sreejith Pongillyathundiyil Sasidharan
Government Medical College
Thiruvananthapuram, Kerala

Parameswaran Sasikumar
Government Ayurveda College
Thiruvananthapuram, Kerala

Pranshu Sharma
Sidney Kinmmel Medical College
Philadelphia

Kamana Singh
Deshbandhu College, University of Delhi
New Delhi

Farkhodjon Tukhtaev
Tashkent Research Institute of Vaccines and
    Serum
Uzbekistan

Sharad Visht
Tishk International University
Erbil, Iraq

Dinesh K. Yadav
ICAR-Indian Institute of Soil Science
Bhopal

Preface

In the current book, *Natural Products and Nano-Formulations in Cancer Chemoprevention*, the various domains of cancer chemoprevention by natural products and nano-formulations are covered in detail. Most of the chemotherapeutic strategies used in contemporary cancer therapy regimens, which combine a variety of synthetic medications, have not been very effective in curing the disease. Price affordability, economic burden, accessibility and availability of standard rugs, toxicity along with poor prognosis, survival rate, and recurrence of the disease are a few major concerns and domains of disease treatment that need to be effectively addressed while the process of elimination and cure of diseases. Many diseases like smallpox and polio are virtually completely eradicated from the world by prevention mechanisms justifying the paradigm "prevention is better than treatment," and cancer chemoprevention appears to be a feasible and workable approach in dealing with this chronic noncommunicable disease.

The reports of the World health organization (WHO) predict around 10 million cancer-related deaths by the end of this decade and around 30 million cancer-related deaths globally by end of the next two decades. However, about one-third of these deaths could be avoided by various chemopreventive agents and nutritional interventions that are closely associated with cancer chemoprevention. The emergence of nutrigenomics has revealed the correlation of natural products, functional food, and nutraceuticals to the molecular and genetic levels. The evidence-based studies and many clinical and preclinical trials have proved that natural products are known to influence and modulate the process of tumorigenesis either by delaying/reversing or regulating all the stages of neoplastic transformations, including the initiation, progression, and recurrence of the disease. Natural products are known to potentiate efficacy and have demonstrated synergistic effects with cancer-curing drugs.

The limitations of using natural products, including their poor solubility, low oral bioavailability, and ineffective systemic distribution, can be overcome by the developing science of nanotechnology. Additionally, nano-formulation-based approaches in chemopreventive mechanisms and nanotechnology-based drug delivery systems have also been documented to increase the efficacy and minimize the side effects. The incorporation of different drugs or active ingredients into nano-formulations is a wonderful way to ensure that they are delivered efficiently to the site of action without compromising their activity. Thus, it is crucial to comprehend the structure and purpose of nano-formulations in the context of cancer treatment and prevention. The clinical viewpoint of nanotherapeutics in cancer has also been discussed in this book, along with the prospects for the future in this field.

Chapter 1, authored by the editor and his associates, provides an overview of various strategies and types of natural products used in cancer chemoprevention. It also discusses obstacles to using natural products in cancer chemoprevention and treatment. The authors also highlight the development of chemoprevention methods based on nanotechnology, including nanoliposomes, polymeric micelles, dendrimers, quantum dots, nanofibers, polymer-drug conjugates, and nanoemulsions, in addition to protein-, peptide-, and carbohydrate-based systems.

The function of polyphenols in cancer chemoprevention is covered in detail in Chapter 2, by Pushpendra Koli and colleagues. The authors explain the nature, grouping, function, and mechanism of polyphenols in the treatment of cancer. The chapter goes into detail about the function of dietary phytochemicals and natural polyphenols in the chemoprevention of illness.

The significance of fat-soluble vitamins in cancer chemotherapy is discussed in Chapter 3, by Ankur Adhikari and colleagues. The importance of vitamins A and K in cancer chemotherapy is also covered in the chapter. The chapter describes the mechanism, including the effect of vitamin D in TGF mutations and the role of vitamin D in regulating the DNA repair mechanism, cell cycle arrest, induction of autophagy, and apoptosis by tocotrienols and their effect on cancer stem cells.

Targeting cancer stem cells with natural products for chemoprevention is the topic of Chapter 4, by Prof. Kamana Singh and colleagues. The authors present a detailed cancer stem model and identify many cell signaling pathways as prospective cancer chemoprevention targets. The authors also go through a number of natural and dietary supplements that target cancer stem cells and give a brief overview of the role of various antioxidants in the prevention of various cancers.

Chapter 5, by Shreya Joshi and colleagues, covers a variety of rules and laws, technological difficulties, market trends, and health advantages of using functional foods in cancer chemoprevention and therapeutics. The authors cover the pathogenesis of cancer and provide a thorough and all-encompassing overview of numerous functional foods that may be candidates for use in cancer chemoprevention and therapy.

The topic of phytochemicals as a potential source of intervention in the prevention and treatment of cancer sickness is covered in Chapter 6, by Dr. Ravendra Kumar and colleagues. The authors describes how numerous phytochemicals that are now the subject of preclinical and clinical research, including as curcumin, lycopene, isothiocyanates, genistein, and resveratrol, affect the process of carcinogenesis.

Approaches for chemoprevention based on nano-formulation are described by Dr. Rohit Saluja and colleagues in Chapter 7. The authors go into extensive detail about the structure, physico-chemical characteristics, and advantages of nano-formulations compared to traditional medicinal formulations. This chapter discusses with clinical context the active and passive targeting of cancer tissue by nano-formulations, as well as nano-formulation-based chemoprevention in a variety of malignancies, including pancreatic, breast, colorectal, lung, and bronchial cancers.

Sneh Gautam and colleagues highlight the potential of nano-formulations based on natural products as chemopreventive drugs in Chapter 8. The authors list the natural compounds' anticancer activities and any associated impedances. This chapter covers a variety of nano-formulations, including liposomes, nanoparticles, dendrimers, carbon nanotubes, and nanomicelles, in greater detail.

Chapter 9 is written by Dr. Manisha Sachan and colleagues. The authors go into detail about the function of nano-formulations as a means of delivering chemotherapy medicines. Additionally, current developments in nano-co-delivery for the treatment of cancer and clinical uses of various nano-formulated chemotherapeutic medicines and their study of cellular pathways are presented. The chapter also includes an overview of the rules, obstacles, and toxicity concerns related to medications with nano-formulated dosage forms.

Chapter 10, written by the editor and his team, describes the use of nutrigenomics in cancer chemoprevention. The chapter on nutrigenomics decoding the genomic puzzle of the disease emphasizes the genetic relationship between food patterns and cancer. The authors go into detail about natural products that affect miRNA, as well as epigenetic alterations and critical genes implicated in carcinogenesis.

The synergistic effects of natural items in the treatment of cancer are discussed in Chapter 11 by Dr. Ritu Saini and colleagues. The chapter addresses several herbal techniques to treating cancer, possible interactions between natural items and synthetic medications, the state of these approaches currently, and their potential for the future.

The medicinal chemistry, pharmacodynamics, and pharmacokinetics of camptothecin and its derivatives in clinical chemotherapeutics are discussed in Chapter 12, by Dr. Ijinu and colleagues. The potential for camptothecin to be used as a chemotherapeutic agent has also been addressed. The authors cover the biosynthesis, chemistry, and synthesis of camptothecin, as well as the synthesis of camptothecin derivatives.

Chapter 13, by Dr. Himanshu Punetha and Shivanshu Garg, discusses the role of *Brassica* in cancer chemoprevention. The authors explore the chemistry, diversity, and method of action of indole-3-carbinol in the treatment of cancer, as well as emphasize how widespread and distinctive organosulfur compounds are.

Akansha Agrawal and colleagues in Chapter 14 examine the opportunities of nano-formulations in cancer chemoprevention. The chapter discusses the opportunities and problems that come with using nano-formulations for cancer chemoprevention. The authors discuss the use of carbon-based nano-formulations for cancer treatment and drug delivery.

The challenges of using nanomedicine for cancer treatment are described in Chapter 15, by Prof. Kamana Singh and her colleagues. The authors discuss different obstructions and obstacles that prevent natural products from reaching the tumor location. The authors go into detail about the difficulties involved in producing nanomedicines, government rules and laws governing their production and commercialization, and different difficulties in moving nanomaterial from the lab to the clinic.

Dr. Hridya and her colleagues in Chapter 16 describe the distribution and diversity of the mangrove ecosystem, the therapeutic potential of mangrove species, potential anticancer leads from mangroves, and the potential of mangroves to overcome drug resistance.

1 Overview of Natural Products and Nano-Formulations in Cancer Chemoprevention

*Ananya Bahuguna, Shiv Kumar Dubey,
Kanchan Gairola, and Rohit Pujari*

CONTENTS

1.1 Introduction 2
1.2 Cancer Chemoprevention 2
1.3 Natural Products for Cancer Chemoprevention...3
1.4 Types of Natural Products for Cancer Chemoprevention..3
    1.4.1   Polyphenols..3
        1.4.1.1   Epigallocatechin-3-Gallate (EGCG) 3
        1.4.1.2   Resveratrol and Trans-resveratrol (3,5,4′-Trihydroxy-trans-stilbene)...........4
        1.4.1.3   Curcumin (bis-α, β-unsaturated β-diketone).....................................4
    1.4.2   Flavones 4
        1.4.2.1   Quercetin...4
        1.4.2.2   Carotenoids 4
        1.4.2.3   Lycopene 5
    1.4.3   Monoterpene and Triterpenoids..5
        1.4.3.1   Thymoquinone 5
    1.4.4   Sulfur Compounds..5
        1.4.4.1   Sulforaphane 6
    1.4.5   Genistein...6
    1.4.6   Dihydroartemisinin (DHA) 6
    1.4.7   Indole-3-carbinol 7
1.5 Shortcomings of Natural Products 7
1.6 The Emergence of Nanotechnology for Chemoprevention 7
1.7 Nano-Formulations...7
1.8 Types of Nano-Formulations 8
    1.8.1   Nanoliposomes...8
    1.8.2   Polymeric Micelles 8
    1.8.3   Dendrimers 9
    1.8.4   Quantum Dots..9
    1.8.5   Nanofibers...9
    1.8.6   Polymer-Drug Conjugates...9
    1.8.7   Nanoemulsions...9
    1.8.8   Protein-Based Systems 10
        1.8.8.1   Albumin-Based Nanoparticles..10
    1.8.9   Polypeptide Nanoparticles 10
    1.8.10   Carbohydrate-Based System..10
        1.8.10.1   Cyclodextrins 10
    1.8.11   Multi-biopolymer Systems...11

DOI: 10.1201/b23311-1

1.9   Challenges in Producing Anticancer Nano-Formulations Based on Natural Products.......... 11
1.10 Conclusion and Future Perspective.. 12
1.11 References... 13

1.1 INTRODUCTION

Due to its high occurrence, cancer causes a significant disease burden worldwide, resulting in disability and premature death in people [1]. Cancer alone claimed the lives of 8.8 million people in 2015. Over the next two decades, this number is predicted to climb by almost 70%. Men are most usually harmed by prostate, lung, bronchial, and colorectal cancers, while women are most frequently harmed by breast, lung, and bronchial cancers. Twenty-nine percent of all new cancer diagnoses in women will be breast cancer [2]. There is a great deal of scientific interest in discovering more anticancer medications from natural sources because of the high morbidity and mortality rates linked to cancer. Many cancers are thought to be caused by environmental and/or lifestyle factors, with inborn genetic abnormalities accounting for just a small percentage (up to 10%) [3, 4]. Numerous studies have found that reactive oxygen species (ROS), in excess, contribute to the development of cancer. A multitude of unfavorable metabolic pathways that are crucial in the conversion of healthy cells into malignant ones can be activated by ROS-induced DNA instability. Extremely reactive free radicals and their byproducts frequently form DNA cross-links and adducts, altering the base sequence and triggering the development of cancer. Cancer progresses via a series of steps involving a variety of signaling pathways, all of which can be slowed or stopped at an early stage [5].

Cancer development is assumed to be influenced by oxidative stress-induced DNA damage, liver damage brought on by xenobiotics and carcinogens, tumor suppressor gene (TSG) alterations, chronic inflammation, dysregulated apoptosis, and unchecked cellular proliferation. Plant-based chemopreventive medications have been discovered to either slow down or stop carcinogenesis to varying degrees. Plant-based medicines have been used in the Ayurvedic system of medicine as traditional cures for hundreds of years [6, 7]. Numerous dietary phytochemicals are helpful in cancer chemoprevention because they can block one or more cellular pathways and inhibit carcinogenic processes [8, 9].

1.2 CANCER CHEMOPREVENTION

Chemoprevention is the application of varied strategic processes by natural or synthetic agents to halt, slow down, or reverse the course of the carcinogenic process at various stages [10, 11]. Chemoprevention is the proactive use of ostensibly nontoxic chemicals, such as organic, synthetic, or biological substances, to prevent cancer. There have been many reports of issues involving conventional cancer therapy methods [12, 13]. For instance, adverse effects of synthetic chemotherapeutic drugs include continuing side effects, high cost, and the progression of synthetic drug resistance in cancer cells. In addition, a number of phytochemicals have a direct effect on cancer cells. These might have an impact on cancer cells or cancer stem cells (CSCs) that have already weathered stress, such as significantly elevated ROS levels brought on by excessive metabolic activities that are strictly controlled in normal cells [14, 15]. The first stage of carcinogenesis or the development of premalignant cells into cancer are all potential targets for chemoprevention medications, as are its reversal, inhibition, prevention, and delay. The vast majority of known occurrences are focused on the modification of recognized tumorigenic pathways, including EGFR signaling pathways, the p53 cascade, inflammation, and oxidative stress. Nowadays, chemopreventive chemicals can be found in nature, dietary supplements, and synthesized molecules that are either repurposed pharmaceuticals or known chemotherapeutic agents. A deeper comprehension of the connection between host carcinogenesis and the microbiota brings up new possibilities for cancer prevention.

When administered in small doses, chemopreventive phytochemicals tend to target just cancer cells and CSCs, elevating cellular stress and ultimately killing them. On the other hand, normal cells may withstand phytochemical-induced changes without having their physiological response suffer [16]. As shown by several scientific, epidemiological, and pharmacological studies, several strong natural compounds from medicinal plants have attracted a lot of attention in recent years. This is because it possesses antioxidative, antimutagenic/antigenotoxic, and anticarcinogenic qualities, in addition to other bioactive and chemopreventive ones. These qualities have made them desirable candidates for contemporary therapeutic methods, as they have led to their inclusion in traditional medicines for the management and treatment of a range of illnesses, including cancer [17].

The large number of papers on chemoprevention discovered through PubMed bibliographic searches over the preceding ten years illustrates the significance of phytochemicals and the chemopreventive properties of medicinal plants. Plants create phytochemicals, which are non-nutritive substances, as part of their defense mechanism against harmful environmental stresses and a variety of illnesses. Terpenoids, polyphenols, and alkaloids are currently used to categorize a variety of phytochemicals that contribute to the medicinal properties of plants and are also utilized by people for nutraceutical purposes [18].

1.3   NATURAL PRODUCTS FOR CANCER CHEMOPREVENTION

Natural products have long been used to treat human illnesses. Just like penicillin's famous discovery revolutionized worldwide existence, similarly, salicylates and willow are well-known instances, as is quinine's link with cinchona [19]. The majority of today's medications are made from natural ingredients, particularly in the domains of cancer and antibiotic therapy. The leading cause of death worldwide remains cancer, notwithstanding recent breakthroughs. Human cancer can never be treated effectively; prevention is always preferable. Another major method for tackling this serious public health issue is cancer chemoprevention, which involves using artificial or natural substances to inhibit, slow down, or reverse the process of carcinogenesis.

1.4   TYPES OF NATURAL PRODUCTS FOR CANCER CHEMOPREVENTION

The use of herbal medicines, diet, and spices as natural cancer preventatives has progressively increased since they have been quite safe for people to consume for thousands of years,

1.4.1   POLYPHENOLS

Natural phytochemicals known as phenolic compounds, which are abundant in food and nutraceuticals, are part of the family of secondary metabolites. They are largely produced from phenylalanine and, to a lesser extent, from tyrosine [20]. The most prevalent and frequent substances in tea are the polyphenols catechins, which possess (−)-epicatechin (EC), (−)-epigallocatechin (EGC), (−)-epicatechin-3-gallate (ECG), and (−)-epigallocatechin-3-gallate (EGCG) (*Camellia sinensis*, Theaceae) [21, 22].

1.4.1.1   Epigallocatechin-3-Gallate (EGCG)

Tea is one of the most popular drinks in the world. Green tea, black tea, oolong tea, and other varieties of tea are only a few of the many varieties that exist. The most popular form of tea consumed worldwide is green tea. Green tea has been the subject of extensive research because of its high polyphenol content. Epicatechin, epicatechin-3-gallate, epigallocatechin, and epigallocatechin-3-gallate are just a few of the significant polyphenols found in green tea (EGCG) [23]. EGCG, which makes up 50% to 80% of the catechin content and has been discovered as one of the most powerful epigenetic modifiers for cancer treatment and chemoprevention, has gotten a lot of attention [24]. In human-xenograft mice, EGCG blocks the intrinsic route of CSCs and eradicates cancer stem cells. According to a study, EGCG interferes with CSCs' ability to self-renew by affecting the transcription and translation of genes that code for stemness markers. Despite being slightly less effective than parental cells at inhibiting CSC stemness, EGCG's activity rises when combined with anticancer medications [25].

1.4.1.2   Resveratrol and Trans-resveratrol (3,5,4′-Trihydroxy-trans-stilbene)

A phytoalexin shown to be produced by nearly 70 plant species, including *Polygonum cuspidatum*, red grapes, berries, peanuts, and pines [26, 27]. These phytoalexins are formed as protection against physical trauma, UV radiation, and fungi. A growing body of research suggests that resveratrol may have anticancer capabilities against a range of tumor types, impacting various stages of tumorigenesis and growth. Oral ingestion and intravenous administration are the two most typical ways of receiving resveratrol, with oral ingestion being more widespread. Resveratrol is mostly metabolized by the enterocytes and hepatocytes after oral administration; hence, gastrointestinal and liver cancers were frequently treated with it [28].

1.4.1.3   Curcumin (bis-α, β-unsaturated β-diketone)

Many different diseases have been treated with curcumin for a very long time in different cultures [29, 30]. Diverse malignancies, including those of the breast, leukemia, and lymphoma, have been proven to be successfully treated with curcumin, also known chemically as diferuloylmethane [31]. Recent decades have seen an increase in clinical research aimed at understanding whether curcumin has a significant effect in colorectal cancer models because of its preferred accumulation in colonic mucosa. Studies have demonstrated that curcumin can considerably slow down the spontaneous incidence and tumor formation in ovarian cancer when used as chemoprophylaxis for the disease [32]. Curcumin has also been discovered to inhibit B[a]P, a procarcinogen that induces lung carcinogenesis and is present in both the environment and cigarette smoke [33].

1.4.2   FLAVONES

1.4.2.1   Quercetin

A bioflavonoid that often develops spontaneously as a glycoside and is found in apples, onions, tomatoes, broccoli, and citrus fruits [34]. Due to its important functioning in modulating both tumor-associated signaling pathways, it can be seen as a potential chemopreventive formulation. Quercetin can be a potential agent for the chemoprevention of oral squamous cell carcinoma (OSCC) since it can enhance tumor cell apoptosis by altering the NF-κB signaling system and its target genes Bcl-2 and Bax [35, 36]. Quercetin may also be utilized as an analgesic for discomfort associated with cancer, according to another study. The tumor-growth-inhibiting effects of different chemopreventive drugs, including green tea, were increased by increasing quercetin bioavailability. The study reports that after four weeks of tumor inoculation, a combined diet of green tea and quercetin demonstrated greater suppression of tumor growth in tumor-inoculated mice than green tea alone, and decreased tumor growth by 47% when compared to control. Cotreatment increased tissue concentrations of total green tea polyphenols and nonmethylated EGCG by 1.5 and 1.8 times, respectively [37, 38]. According to earlier studies on both animals and humans, quercetin's bioavailability diminishes after just one oral intake since it is mostly dependent on macronutrients for absorption. Contrarily, quercetin enters phase II metabolism after being absorbed by the digestive tract, changing the makeup of the intestinal microbiota and safeguarding the intestinal lining. It also has a wide range of intestinal first-pass metabolism [39, 40].

1.4.2.2   Carotenoids

They belong to the class of isoprenoid polyenes, which includes the naturally occurring lipid-soluble pigments that give plants and animals their bright hues. Regulating carotenoid anabolism in plants is a challenging process that is affected by a number of variables [37]. Today, there are about 750 different varieties of carotenoids that have been found in nature [41]. There are four categories of carotenoids:

1. Precursors to vitamin A, such as beta-carotene
2. Pigments that only partially activate vitamin A, such as cryptoxanthin
3. Precursors that are not vitamin A, like violaxanthin and neoxanthin
4. Precursors that are not vitamin A, like lutein and zeaxanthin

Recent research has shown that carotenoids are important for human health; eating foods rich in carotenoids reduces the risk of heart disease. Most significantly, α-carotene prevented colon, liver, skin, and lung cancer more potently than β-carotene. According to recent studies, beta-carotene prevents hepatocellular carcinoma brought on by diethylnitrosamine (DEN) and is increased by phenobarbital more effectively than retinoic acid (RA) (PB) [42].

1.4.2.3 Lycopene

Lycopene commonly occurs in fruits and vegetables, particularly in tomatoes. By boosting antioxidants including glutathione-S-transferase-omega-1 (GSTO-1), superoxide dismutase-1 (SOD-1), and ERO-1, lycopene reduces intercellular reactive oxygen species (ROS). Lycopene has been demonstrated to significantly lower the risk of colon, lung, breast, and prostate cancer by preventing the cellular growth of these diseases. It has also been demonstrated to lessen the growth of ovarian tumors. Additionally, radiation- and cisplatin-induced nephropathy can be treated with lycopene.

1.4.3 MONOTERPENE AND TRITERPENOIDS

An extensive study on plant-based therapeutic chemicals over the last two decades has suggested a possible function for triterpenoids in prevention.

1.4.3.1 Thymoquinone

Nigella sativa L. seed oil contains the monoterpene thymoquinone (2-methyl-5-isopropyl-1,4-benzoquinone) (family Ranunculaceae). In numerous experimental studies, thymoquinone has been found to reduce the risk of cancer in different organs and its mechanism may be related to the regulation of the RTK signaling pathways. According to experimental investigations, it has been reported that the addition of thymoquinone to water can reduce the risk of developing stomach, lung, and liver cancer, as well as hepatic carcinogenesis, brought on by benzo(a)pyrene, 20-methyclonathrene diethylnitrosamine (DENA), and diethylnitrosamine, respectively [43, 44].

Triterpenoids are abundantly available in its free form, triterpene glycosides (saponins), and their precursors having anticancer activities. Examples of these compounds include betulinic acid, lupeol, oleanolic acid, and cucurbitacin [45].

1.4.4 SULFUR COMPOUNDS

Certain foods like garlic and other *Allium* plants have an anticarcinogenic effect because they contain organosulfur compounds, such as allicin, which have a strong protective effect against cancer in animal models [46]. The substance that gives garlic its characteristic smell, allicin, has been shown to have a number of health advantages, including the ability to prevent cancer. Allicin decreases lymphangiogenesis by blocking the activation of the vascular endothelial growth factor (VEGF) receptor, a critical biological mechanism in tumor spreading [47]. Allicin can also be used as adjuvant therapy for thyroid cancer since it induces autophagic cell death, which slows the disease's growth [48].

1.4.4.1 Sulforaphane

Is the most well-known isothiocyanate (ITC) substance, which is predominantly present in significant quantities in broccoli and is considered to be a potentially promising agent in the protection of breast cancer. Numerous studies have confirmed that cruciferous vegetables, including brussels sprouts, cauliflower, broccoli, and cabbage, are effective at preventing chemotherapy. It is believed that the active lead isothiocyanates found in cruciferous vegetables are what give them their cancer-fighting properties. Sulforaphane is one of the principal isothiocyanates present in cruciferous vegetables, especially broccoli and broccoli sprouts. Sulforaphane not only prevents cancer from developing, but it also reduces the growth of existing cancers [49]. It absorbs quickly and reaches a peak level in under an hour [50]. Breast cancer stem cells can be selectively eliminated by sulforaphane by inhibiting the NF-κB p65 subunit's translocation, downregulating p52 and its downstream transcriptional activity (CSC). Sulforaphane and docetaxel significantly reduce primary tumor volume and future tumor development when compared to treatment alone [46]. However, sulforaphane outperforms glucoraphanin in its capacity to change the phase I and phase II enzymes involved in carcinogen metabolism in vitro [51]. Sulforaphane is also effective in treating breast cancer because it shrinks primary and secondary mammospheres and prevents taxane-induced enrichment of aldehyde dehydrogenase-positive (ALDH+) cells [46].

1.4.5 GENISTEIN

Breast and prostate cancer are thought to be potentially prevented by genistein, an isoflavanone found in soyabeans. It prevents PMA from activating AP1 and ERK while also promoting the expression of c-FOS in human mammary cell lines [52]. Genistein, predominantly found in soybeans, is an isoflavone that regulates a number of signaling pathways in process of tumorigenesis. By blocking the NF-kB and Akt signaling pathways, this bioactive chemical promotes G2/M and G0/G1 arrest in several cancer cell lines, altering the cell cycle and apoptotic process in prostate, breast, and lung cancer cell lines. Genistein inhibits the expression of the anti-apoptotic proteins BCL-XL and BCL-2 while promoting the expression of the pro-apoptotic proteins Bax, Bak, and Bad. Genistein affects both healthy and leukemic stem cells. According to a study, the CSC-inhibiting factor is present in the blood following genistein consumption since animals fed with soy protein isolate at various concentrations had sera that limit the growth of human breast CSC spheroids in cell culture [53]. Docetaxel, gemcitabine, and cisplatin are examples of chemotherapy medicines that activate NF-kB in cancer cells, which may result in treatment resistance. When compared to chemotherapeutic treatments alone in in vivo and in vitro models, pretreatment with genistein decreased cell proliferation and enhanced apoptosis.

1.4.6 DIHYDROARTEMISININ (DHA)

Artemisinin, a chemical derived from the daisy-like wormwood plant *Artemisia annua*, was once employed by traditional Chinese herbalists to treat fevers. DHA is a derivative of this molecule. In RPMI18226 multiple myeloma cells, it reduces VEGF expression and inhibits HUVE cell proliferation, motility, and tube formation. It kills C6 glioma cells and prevents the activation of hypoxia-inducible factor-1 [54]. In rat C6 glioma cells, DHA synergizes with temozolomide to produce cytotoxic effects [55]. It was shown that it accelerated the degradation and ubiquitination of human fortilin (an anti-apoptotic protein overexpressed in many malignancies). Research suggests that DHA may have a molecular target in fortilin. It was created as an antimalarial drug and has undergone in vivo testing on both humans and animals. It has been reported to increase the susceptibility of human ovarian cancer cells to carboplatin therapy [56]. Jiao et al. discovered that dihydroartemisinin therapy alone was effective against ovarian cancer cells and that the cancer cell lines were five to ten times more susceptible than healthy ovarian cell lines [57].

1.4.7 INDOLE-3-CARBINOL

The phytochemical in cruciferous plants includes broccoli, cabbage, and cauliflower called indole-3-carbinol (I3C) retards the growth of cancer. It has been shown, in a number of cancer cell lines, including prostate, melanoma, and breast cancer, to produce G1 cell cycle arrest, activation of apoptosis, and interference with signal transduction pathways [58, 59]. I3C protects the cell from oxidative injury brought on by oxygen radicals (ROS). Another I3C component, 3′-diindolylmethane, inhibits angiogenesis, causes apoptosis, and lowers activation of the Akt/NF-κB signaling pathway in the MDA-MB-231 breast cancer cells. It has strong qualities in protecting the liver from a number of mutagens. Due to its wide range of activities and low lethality, I3C has been lauded as a potential anticancer medication [60].

1.5 SHORTCOMINGS OF NATURAL PRODUCTS

Natural products provide a wide range of health advantages, but their effectiveness is still constrained by their low levels of water solubility, absorption, and bioavailability, as well as their brief retention times in biological systems. Natural substances are subjected to a variety of chemical and physical problems and opportunities after administration. These ailments cause modifications to their natural structure, which affects their anticancer efficacy. To deliver the most potent chemoprophylaxis and chemotherapeutical actions, novel approaches have been devised, including creative formulations and delivery systems.

1.6 THE EMERGENCE OF NANOTECHNOLOGY FOR CHEMOPREVENTION

Nanotechnology-based products broadly refer to nano-formulations containing particles smaller than 100 nanometers and, from a literature standpoint, those smaller than 1,000 nanometers [61]. This nanoscale size confers superior features on these drug carrier systems in terms of absorption, targeting, and safety. In a nutshell, a medicine's overall efficacy and safety are determined by its intrinsic potency and variables such as drug pharmacokinetics, toxicity, targeted delivery, and stability. Medicinal carriers based on nanotechnology have been shown to improve these variable qualities and so increase drug efficacy. Furthermore, in light of chemotherapy-related side effects, the use of nanotechnology-based formulations in cancer prevention and therapy becomes increasingly relevant, as they can allow for possible dose reduction and drug targeting, hence improving medication safety.

1.7 NANO-FORMULATIONS

In recent years, a lot of research teams have become interested in nano-formulations as potential medication carriers. With a high surface-to-mass ratio, nanosized particles with a surface area of 10 nm to 100 nm can easily penetrate and get absorbed via the plasma membrane. These characteristics are essential to address the problem of poor medicinal intake of phytonutrients as novel synthetic units and drugs. Numerous nano-formulations have been used in drug delivery studies to enhance drug confinement, bioaccessibility, water solubility, and anticipated drug administration, as well as to reduce side effects like toxicity concerns. Environmental concerns, such as pH fluctuations, enzyme attacks, and possible biochemical breakdown, are all protected by phytochemicals contained in the nano-formulation. Nanoencapsulation allows for tailored distribution of phytochemicals while preserving their original structure. The majority of nano-based cancer therapeutic formulations are still in the early stages of development. In vivo investigations must be completed first, followed by clinical trials, before they may be employed.

By modifying their size, shape, and other properties, nanoparticles can be made to target particular cells. To target tumor cells, they may employ passive targeting or aggressive targeting.

Anticancer drugs, which include nanostructures, are designed in such a way that they actively target tumor cells through ligand-receptor or antigen-antibody recognition. When cancer cells are passively targeted, nanoparticles can also interact with them. However, a distinct pathway or reason for selective nanoparticle accumulation may be that in a solid tumor, abnormal angiogenesis leads to amplified porosity and reservation, and the nanosized particle can subsequently become particular to the therapy of the tumor. Additionally, customized administration lowers medication concentrations in other organs and body parts while increasing drug bioaccessibility at the location of the activity, minimizing formulation toxicity.

1.8 TYPES OF NANO-FORMULATIONS

Excipient composition can be used to classify nano-formulations, discussed in the following sections (Figure 1.1).

1.8.1 NANOLIPOSOMES

Traditional liposomes are spherical amphiphilic phospholipid vesicles with a diameter of 25–2,500 nm that form a closed bilayer around hydrophobic or hydrophilic molecules to protect them from aqueous or non-aqueous surroundings [62]. Both hydrophobic and hydrophilic chemotherapy can be delivered to precise areas using this capability. With regard to composition and distribution methods, there are five classes of liposomes that can be employed to maximize the possibility that a product will have the desired benefits. The various types of liposomes include cationic, conventional, long-circulating, and immuno-liposomes [63, 64]. Nanoliposomes are lipid vesicles with a diameter of less than a micron that has been used to transport medicinal, nutraceutical, and cosmetic substances [65, 66]. Nanoliposomes have proved to offer great potential for improving medication stability, bioavailability, and targeting, as well as preventing unwanted interactions [65, 66]. Mohan et al. (2014) investigated the efficacy of utilizing resveratrol and 5-fluorouracil in PEGylated lipid nanoliposomes to treat head and neck squamous cell carcinoma. Triglycerides are dissolved in chloroform to produce nanoliposomes, which are then dried in a vacuum to eliminate the solvent [67]. A polycarbonate membrane with a pore size of 200 nm is used to extrude the resulting liposomal suspension [67]. Depending on the medication's solubility in water, it is either dissolved in PBS or a triglyceride solution. This procedure resulted in nanoliposomes with encapsulation effectiveness of 3–54% and a size range of 200–250 nm [67]. By altering triglyceride content, drug-to-lipid mass ratio, and PEGylation level, drug loading capacity and efficiency can be changed [67].

1.8.2 POLYMERIC MICELLES

Amphiphilic block copolymers known as polymeric micelles (PMs) with average dimensions of 10–100 nm self-assemble into nanoscale core-shell structures [68]. A dense hydrophobic region makes up the core, whereas hydrophilic copolymers make up the shell. PMs can be modified in terms of blood stability and drug release rate by using different chemical linkages to their exterior components, such as esters [69]. Often researched copolymers are polyesters, poly L-amino acid, and polypropylene oxide [70]. They have a hydrophilic surface that shields them from indefinite absorption, making them suitable for the fundamental administration of hydrophobic chemotherapeutic drugs [71]. Extended blood circulation is made possible by the hydrophilic shell and small size of PMs, which inhibit the mechanical clearance of the micelles by renal filtration, reticuloendothelial system (RES) absorption, and the spleen [72]. An example of an anticancer drug with a PM formulation is Genexol-PM, which is paclitaxel in a PM form [73]. Ovarian cancer, breast cancer, non-small-cell lung cancer, and Kaposi's sarcomas linked to AIDS are all treated with the chemotherapy medication paclitaxel, which has low water solubility. Due to its capacity to prevent hypersensitive reaction, Genexol-PM has been granted a license in Korea for the cure of progressive lung cancer and metastatic breast tumor [73].

1.8.3 DENDRIMERS

The shape and size of dendrimers, which have a branched structure and a tree-like structure, may be easily changed and regulated through the use of polymerization methods. With the dendrimer ends available for molecule attachment and conjugation, branching nanostructures develop around a spherical core [74, 75]. This structure enables great entrapment efficiency, molecular size and weight selectivity, and medication and gene delivery (DNA/RNA). Gene therapy and immune system activation are a few applications of dendrimers [75]. By encapsulating water-insoluble molecules in their interior cavities or adhering them to their surface via electrostatic or hydrophobic interactions, dendrimers, like liposomes, can escalate the solubility and bioaccessibility of such substances [76]. Nucleic-acid-based chemotherapies that involve challenging passage of hydrophilic molecules via cell membrane can also be transported via dendrimers [76].

1.8.4 QUANTUM DOTS

Semiconducting particles of the size of nanocrystals that have an organic shell over an inorganic core [77, 78]. Fluorescence can be produced by QDs when they are excited by light. Because of a special characteristic, QDs are an effective instrument for seeing and monitoring intracellular processes [79, 80]. Since QDs can build up in tumor tissue, tumors can be found and seen noninvasively [80]. QDs demonstrated promising outcomes when evaluated in vitro for the quick localization of HER-2 receptors, targeted chemotherapy, and imaging-guided therapy [81].

1.8.5 NANOFIBERS

In recent years, the creation of nanofibers for the transport of chemopreventive chemicals has gotten a lot of interest [82, 83]. Nanofibers are usually made using an electrospinning process [82]. This process is not only easy and inexpensive, but it can also produce nanofibers of various materials in a variety of fibrous assemblages [82, 83]. To create the nanofibers, enough PVA (6–10% w/v PVA) is dissolved in double distilled water. Double-distilled water has been thoroughly blended with all additional components (drug and excipients). After being subjected to a syringe, the resulting solution is electrospun [84]. The resulting PVA nanofibers had superior shape, durability, drug envelopment efficiency, easy drug delivery, mucosal penetration, and mucoadhesive power [84].

1.8.6 POLYMER-DRUG CONJUGATES

For cancer therapy, polymer-drug conjugates with a size of 5–100 nm have been investigated [85]. In this method, an anticancer medication is linked to a water-soluble polymer's functional group either directly or indirectly [86]. Numerous anticancer medications, including doxorubicin, paclitaxel, and camptothecin, have been successfully conjugated to the polymers [87]. After intravenous injection, the polymer-drug conjugates provide superior tumor-targeted medicine delivery [88]. When scheming and establishing polymer-drug conjugates for intraoral site-specific chemoprevention, it is important to take into account the drug's molecular weight, the polymer's physicochemical properties, the polymer's ability to carry sufficient drug payloads, the kind and uniformity of the linker, and the method of intracellular drug release [89].

1.8.7 NANOEMULSIONS

These are thermodynamically stable colloidal dispersions with a droplet size of about 10–1,000 nm [90]. Making nanoemulsions requires an appropriate oil, an emulsifying agent (such as a surfactant or cosurfactant), and water [90, 91]. The system benefits from the unique traits of the nanoscale droplets, such as their large surface area per unit volume, improved durability, optical transparency, and controllable rheology [91]. Many methods, including forceful amalgamation, ultrasonication, phase inversion temperature, emulsion inversion point, and bubble bursting, can be targeted to create nanoemulsions [90, 91]. These transporters find importance in a variety of products, including food, cosmetics, and drug delivery [92].

1.8.8 PROTEIN-BASED SYSTEMS

Proteins can be regarded as the best material for nanoparticle production due to its amphiphilic nature. Biodegradable protein nanosized particles have a surface that is adaptable for the attachment of ligands and drugs [93]. In an effort to address a number of drug delivery issues, including (1) the insufficient solubility of antitumor drugs like doxorubicin and cisplatin, (2) the ineffectiveness of many medications in causing side effects when they target particular sites, and (3) the interim livelihood of numerous drugs already in use, nanoparticles are currently receiving a lot of attention as drug delivery agents [94, 95]. Two processes are used to create protein nanoparticles: one requires the crosslinking of a native protein with functional groups, and the other employs derivative groups that have been modified on the surface of the protein molecules [96, 97].

1.8.8.1 Albumin-Based Nanoparticles

Albumin has a molar mass of 66.5 kDa and is the major component of plasma proteins. It is a great substance for drug delivery systems due to its strong stability over a broad pH spectrum and elevated temperatures, preferential uptake by tumor and swollen tissue, environment-friendly, and decreased lethality, among others [98].

Albumin particles are produced primarily through coacervation, controlled heat gelation, emulsion formation, dissolution, and self-assembly [98]. In a study, the cancer drug tamoxifen was loaded onto albumin nanoparticles by coacervation [99]. An albumin-particle formulation containing paclitaxel has been approved for use by patients with metastatic breast tumor.

1.8.9 POLYPEPTIDE NANOPARTICLES

Polypeptide-based nanoparticles are more biocompatible and have a clearly defined chemical makeup. Elastin-like polypeptides (ELPs) are the most frequently produced polypeptide nanosized particles, which alternate the hydrophobic blocks with cross-linking domains. ELPs exhibit a unique phase behavior that encourages protein purification, and recombinant expression [98]. These nanosized particles were created in a study by the self-assembly of ELPs, and they released loaded dexamethasone phosphate for about 30 days [100]. This, therefore, forms the basis of these polymeric nanoparticles being extensively used in drug delivery systems [101].

1.8.10 CARBOHYDRATE-BASED SYSTEM

Polysaccharides are naturally occurring substances having potential uses in the delivery of medication and theranostics. They can be used to conjugate diagnostic sensing components, nanoparticles, and other tiny molecules, thanks to their flexible functional groups. Additionally, polysaccharides are available in a range of sizes and have minimal in vivo toxicity. More and more imaging and medical delivery systems are using polysaccharide-based nanoparticles [102, 103].

1.8.10.1 Cyclodextrins

The enzymatic breakdown of starch results in the formation of cyclodextrins, which are hydrophobic macromolecules that are water-soluble. Cyclodextrins have a truncated cone form as a result of the presence of glucopyranose. Cyclodextrins come in two different varieties: naturally occurring and synthetically manufactured (for instance, through starch degradation) [104]. Cyclodextrins are important for building an efficient medication delivery system because of their unique features. To

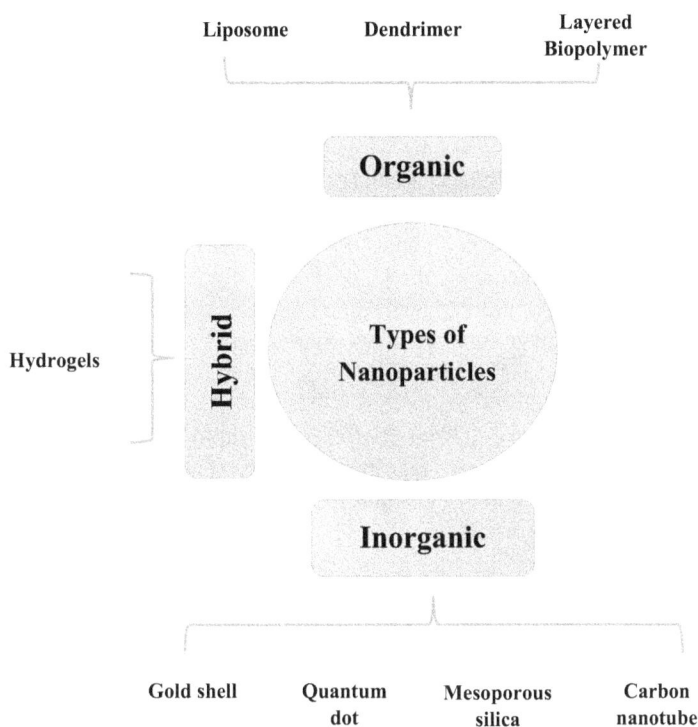

FIGURE 1.1   Different types of nano-formulations.

improve medication pharmacokinetics, several anticancer drugs have been incorporated into cyclo-dextrin [105]. The blood-brain barrier became more permeable when doxorubicin and cyclodextrin were combined. The solubility of chlorambucil was improved by the folic acid combination of cyclodextrin and polyethylene glycol (CD-PEG). 9-nitrocamptothecin's antitumor efficacy and tox-icity were both markedly improved by the addition of hydroxypropyl-cyclodextrin (HP-CD) [105].

1.8.11   Multi-biopolymer Systems

Applications of biopolymer-functionalized nanoparticles in catalysis, optics, electronics, antibacte-rial action, and drug transport have been effective. Nanoparticles have recently played a key char-acter as medication and vaccine carriers in target cells or tissues because they may bind with many chemical compounds [106]. Self-assembly of biocompatible materials poly-glutamic acid and chito-san led to nanoparticles of 80–150 nm that suppressed tumor growth better than free nanoparticles when loaded with doxorubicin [106]. Nanoparticles utilized in drug delivery systems as anticancer agents have been produced using a variety of polymeric polymers, such as polylactic acid and poly-ethylene glycol [107].

The encapsulated medicines were released slowly and had cytotoxic effects on the cells.

1.9   CHALLENGES IN PRODUCING ANTICANCER NANO-FORMULATIONS BASED ON NATURAL PRODUCTS

Even though a variety of in vitro preclinical settings have shown that natural compounds have a wide spectrum of anticancer actions, there have been difficulties in converting these positive find-ings to in vivo systems, falling short of expectations.

Aside from the effective targeting strategy, nano-formulations encounter a slew of challenges in terms of construction, distribution, biological system interaction, and toxicity [108]. The most essential property of nano-formulation is particle size and size distribution, which can have a major impact on pharmacokinetics, biodistribution, and safety. The mononuclear phagocytic system consumes larger particles (greater than 200 nm), while the excretory system removes smaller nanoparticles (20–30 nm) quickly [108, 109]. The surface properties of nanoparticles, such as charge, hydrophobicity, and functional groups, are also significant regulators of nanoparticle stability [110]. Nano-formulation is seriously concerned about the immunogenicity of nanoparticles because they have the potential to act as antigenic agents [108]. Potentially lethal allergy and hypersensitivity reactions, humoral and cellular immunological responses and anaphylactic shock can all result from the activation of the complement system when nanoparticles are opsonized by plasma proteins and recognized as foreign particles [111]. Also, hematologic safety issues such as thrombogenicity and hemolysis have been linked to nano-formulations [108].

These problems have posed significant challenges in the development and safety of nano-formulation-based therapy. The features and complicated systems of nanoparticle production outlined above can result in severe toxicities. Rigid reconciliation of all dimensions, along with the distribution of size and the inclusion of additional components, are necessary throughout development in order to enhance the safety profiles of natural-product-based nano-formulations.

1.10   CONCLUSION AND FUTURE PERSPECTIVE

It is now possible to get beyond the drawbacks of traditional therapy regimens for cancer, which have been hampered by problems including vulnerability, low aqueous solubility, high lethality, inadequate absorption, low selectivity, and multidrug resistance. In studies conducted around the world, the use of nanotechnology to treat cancer has been continuously investigated. It has been demonstrated to be a viable strategy for diagnostics, imaging, and treatments. Regarding nanocarriers/nano delivery systems, numerous nanotechnology-based structures are currently undergoing clinical or preclinical testing, some of which have already received FDA approval and are available for purchase. In the near future, it is anticipated that a cancer cure will be available, and nanotechnology will grow to be a $1 trillion business with a significant concentration on cancer therapy [112]. Chemotherapeutics' nonspecificity has long destroyed patients' normal proliferating tissues, leaving them immunodeficient and causing long-term negative effects. Nanotechnology has provided outstanding treatment strategies for effective and less damaging medication delivery systems because of its sensitivity to target tumor cells. This has been proven by accounts of patients who received nanoparticle therapy having an enhanced treatment response and long-term survival. The idea of nano-chemoprevention may help to reduce some of the typical worries associated with nanotechnology, such as long-term toxicity, degradation, and metabolism of nanotechnology agents. A low-cost, palatable, and successful cancer control and management technique called nano-chemoprevention (distribution of bioactive food components via nanotechnology-based carriers) could be created [113]. This method is quite effective [114] since natural agents have never produced any issues at the levels used in typical research and NPs rarely cause harm in normal cells. Due to their biodegradability, these NPs are likewise regarded as safe. As a result, the cancer therapies of the nanomedicine era have the ability to successfully address the problems that standard therapeutic regimens encounter. Creating patient-specific drug delivery systems will help with the tailoring and administration of treatment based on the patient's clinical profile as nanotherapeutics become more complicated. Through early tumor diagnosis and improved patient care, nanomedicine has the potential to alter the treatment front in the fight against cancer. According to this theory, it may be possible to enhance the biodistribution and pharmacokinetics of encapsulated compounds and selectively kill tumor cells without damaging healthy cells [115]. Increases in patient survival and quality of life may be associated with these core traits. With the help of this discovery, it may be simpler to obtain supraphysiological levels of bioactive food substances, which are challenging to do when ingested as part of a diet. A low-cost, palatable, and simple-to-implement cancer control and management strategy might be created using nano-chemoprevention.

1.11 REFERENCES

1. Shih, Alan H, Omar Abdel-Wahab, Jay P Patel and Ross L Levine. 2012. "The role of mutations in epigenetic regulators in myeloid malignancies." *Nature Reviews Cancer* 12 (9): 599–612. https://doi.org/10.1038/nrc3343
2. Miller, Kimberly D, Rebecca L Siegel, Chun Chieh Lin, Angela B Mariotto, Joan L Kramer, Julia H Rowland, Kevin D Stein, Rick Alteri and Ahmedin Jemal. 2016. "Cancer treatment and survivorship statistics." *CA: A Cancer Journal for Clinicians* 66 (4): 271–289. https://doi.org/10.3322/caac.21349
3. Weiderpass, Elisabete. 2010. "Lifestyle and cancer risk." *Journal of Preventive Medicine and Public Health* 43 (6): 459–471. https://doi.org/10.3961/jpmph.2010.43.6.459
4. Heikkilä, K, ST Nyberg, T Theorell, EI Fransson, L Alfredsson, JB Bjorner, S Bonenfant, M Borritz, K Bouillon, H Burr and N Dragano. 2013. "Work stress and risk of cancer: Meta-analysis of 5700 incident cancer events in 116000 European men and women." *British Medical Journal* 346: 165–175. https://doi.org/10.1136/bmj.f165
5. Martincorena, Iñigo and Peter J Campbell. 2015. "Somatic mutation in cancer and normal cells." *Science* 349 (6255): 1483–1489. https://doi.org/10.1126/science.aab4082
6. Mukherjee, Pulok K, P Venkatesh and S Ponnusankar. 2010. "Ethnopharmacology and integrative medicine-let the history tell the future." *Journal of Ayurveda and Integrative Medicine* 1 (2): 100–109. https://doi.org/10.4103/0975-9476.65077
7. Jaiswal, Yogini S and Leonard L Williams. 2017. "A glimpse of Ayurveda–The forgotten history and principles of Indian traditional medicine." *Journal of Traditional and Complementary Medicine* 7 (1): 50–53. https://doi.org/10.1016%2Fj.jtcme.2016.02.002
8. Surh, Young-Joon. 2003. "Cancer chemoprevention with dietary phytochemicals." *Nature Reviews Cancer* 3 (10): 768–780. https://doi.org/10.1038/nrc1189
9. Weng, Chia-Jui and Gow-Chin Yen. 2012. "Chemopreventive effects of dietary phytochemicals against cancer invasion and metastasis: Phenolic acids monophenol, polyphenol, and their derivatives." *Cancer Treatment Reviews* 38 (1): 76–87. https://doi.org/10.1016/j.ctrv.2011.03.001
10. Sporn, MB and N Suh. 2000. "Chemoprevention of cancer." *Carcinogenesis* 21 (3): 525–530. https://doi.org/10.1093/carcin/21.3.525
11. Steward, WP and K Brown. 2013. "Cancer chemoprevention: A rapidly evolving field." *British Journal of Cancer* 109 (1): 1–7. https://doi.org/10.1038/bjc.2013.280
12. Yazbeck, Victor Y, Liza Villaruz, Marsha Haley and Mark A Socinski. 2013. "Management of normal tissue toxicity associated with chemoradiation (primary skin esophagus, and lung)." *The Cancer Journal* 19 (3): 231–237. https://doi.org/10.1097%2FPPO.0b013e31829453fb
13. Wong, Hai Ming. 2014. "Oral complications and management strategies for patients undergoing cancer therapy." *The Scientific World Journal* 2014 (4): 1–14. http://dx.doi.org/10.1155/2014/ 581795.
14. Raffaghello, Lizzia, Changhan Lee, Fernando M Safdie, Min Wei, Federica Madia, Giovanna Bianchi and Valter D Longo. 2008. "Starvation-dependent differential stress resistance protects normal but not cancer cells against high-dose chemotherapy." *Proceedings of the National Academy of Sciences of the United States of America* (24): 8215–8220. https://doi.org/10.1073/pnas.0708100105105
15. Zhou, Daohong, Lijian Shao and Douglas R Spitz. 2014. "Reactive oxygen species in normal and tumor stem cells." *Advances in Cancer Research* 122: 1–67. https://doi.org/10.1016%2FB978-0-12-420117-0.00001-3
16. Tanveer, S, E Fathi and F Guy. 2011. "Towards new anticancer strategies by targeting cancer stem cells with phytochemical compounds." In: Stanley Shostak (Ed.), *Cancer Stem Cells-The Cutting Edge*, London: IntechOpen, pp. 431–456. http://dx.doi.org/10.5772/18695
17. Yin, Shu-Yi, Wen-Chi Wei, Feng-Yin Jian and Ning-Sun Yang. 2013. "Therapeutic applications of herbal medicines for cancer patients." *Evidence Based Complementary and Alternative Medicine* 2013: 1–15. http://dx.doi.org/10.1155/2013/302426.
18. Jimenez-Garcia, SN, MA Vazquez-Cruz, Ramon G Guevara-González, I Torres-Pacheco, A Cruz-Hernandez and AA Feregrino-Perez. 2013. "Current approaches for enhanced expression of secondary metabolites as bioactive compounds in plants for agronomic and human health purposes–a review." *Polish Journal of Food and Nutrition Sciences* 63 (2): 67–78. https://doi.org/10.2478/v10222-012-0072-6

19. Cragg, Gordon M and John M Pezzuto. 2016. "Natural products as a vital source for the discovery of cancer chemotherapeutic and chemopreventive agents." *Medical Principles and Practice* 25 (2): 41–59. doi: 10.1159/000443404.

20. Shankar, Sharmila and Rakesh K. Srivastava. 2012. "Curcumin: Structure, biology and clinical applications." In: *Nutrition, Diet and Cancer*, 413–457. Springer: Dordrecht, The Netherlands. https://doi. org/10.1007/978-94-007-2923-0

21. Isemura, M, K Saeki, T Kimura, S Hayakawa, T Minami and M Sazuka. 2000. "Tea catechins and related polyphenols as anti-cancer agents." *Biofactors* 13 (1–4): 81–85. https://doi.org/10.1002/biof.5520130114

22. Shukla, Yogeshwar. 2007. "Tea and cancer chemoprevention: A comprehensive review." *Asian Pacific Journal of Cancer Prevention* 8 (2): 155–166.

23. Li, Yanyan and Tao Zhang. 2013. "Targeting cancer stem cells with sulforaphane, a dietary component from broccoli and broccoli sprouts." *Future Oncology* 9 (8): 1097–1103. https://doi.org/10.2217/fon.13.108

24. Naujokat, Cord and Dwight McKee. 2020. "The "Big Five" phytochemicals targeting cancer stem cells: Curcumin, EGCG, sulforaphane, resveratrol, and genistein." *Current Medicinal Chemistry* 27 (22): 1–21. doi.10.2174/0929867327666200228110738.

25. Fujiki, Hirota, Eisaburo Sueoka, Anchalee Rawangkan and Masami Suganuma. 2017. "Human cancer stem cells are a target for cancer prevention using epigallocatechin gallate." *Journal of Cancer Research and Clinical Oncology* 143 (12): 2401–2412. https://doi.org/10.1007/s00432-017-2515-2

26. Ogas, Talysa, Tamara P Kondratyuk and John M Pezzuto. 2013. "Resveratrol analogs: Promising chemopreventive agents." *Annals of the New York Academy of Sciences* 1290: 21–29. https://doi.org/10.1111/nyas.12196

27. Leipert, Jenny, Franziska Kässner, Susanne Schuster, Norman Händel, Antje Körner, Wieland Kiess and Antje Garten. 2016. "Resveratrol potentiates growth inhibitory effects of rapamycin in PTEN-deficient lipoma cells by suppressing p70S6 kinase activity." *Nutrition and Cancer* 68 (2): 342–349. https://doi.org/10.1080/01635581.2016.1145244

28. Ma, Li, Mengmeng Zhang, Rong Zhao, Dan Wang, Yuerong Ma and Li Ai. 2021. "Plant natural products: Promising resources for cancer chemoprevention" *Molecules* 26 (4): 933. https://doi.org/10.3390/molecules26040933.

29. Shanmugam, Muthu K, Grishma Rane, Madhu Mathi Kanchi, Frank Arfuso, Arunachalam Chinnathambi, ME Zayed, Sulaiman Ali Alharbi, Benny KH Tan, Alan Prem Kumar and Gautam Sethi. 2015. "The multifaceted role of curcumin in cancer prevention and treatment." *Molecules* 20 (2): 2728–2769. https://doi.org/10.3390%2Fmolecules20022728

30. Park, Wungki, ARM Ruhul Amin, Zhuo Georgia Chen and Dong M Shin. 2013. "New perspectives of curcumin in cancer prevention." *Cancer Prevention Research* 6 (5): 387–400. https://doi.org/10.1158/1940-6207.capr-12-0410

31. Noureddin, Sawsan A, Reda M El-Shishtawy and Khalid O Al-Footy. 2019. "Curcumin analogues and their hybrid molecules as multifunctional drugs." *European Journal of Medicinal Chemistry* 182: 111631. https://doi.org/10.1016/j.ejmech.2019.111631

32. Kazim, Sahin, Cemal Orhan, Mehmet Tuzcu, Nurhan Sahin, Hakkı Tastan, Ibrahim Hanifi Ozercan, Osman Guler, Nermin Kahraman, Omer Kucuk and Bulent Ozpolat. 2017. "Chemopreventive and antitumor efficacy of curcumin in a spontaneously developing hen ovarian cancer model." *Cancer Prevention Research* 11 (1): 59–67. https://doi.org/10.1158/1940–6207.CAPR-16–0289

33. Puliyappadamba, Vineshkumar T, Arun Kumar T Thulasidasan, Vinod Vijayakurup, Jayesh Antony, Smitha V Bava, Shabna Anwar, Sankar Sundaram and Ruby John Anto. 2015. "Curcumin inhibits B a]PDE-induced procarcinogenic signals in lung cancer cells, and curbs B a]P-induced mutagenesis and lung carcinogenesis." *Biofactors* 41 (6): 431–442. https://doi.org/10.1002/biof.1244

34. Gibellini, Lara, Marcello Pinti, Milena Nasi, Jonas P Montagna, Sara De Biasi, Erika Roat, Linda Bertoncelli, Edwin L Cooper and Andrea Cossarizza. 2011. "Quercetin and cancer chemoprevention." *Evidence Based Complementary and Alternative Medicine* 2011: 591356. https://doi.org/10.1093/ecam/neq053

35. Zhang, Wen, Gang Yin, Jianguo Dai, YU Sun, Robert M Hoffman, Zhijian Yang and Yuan Fan. 2017. "Chemoprevention by quercetin of oral squamous cell carcinoma by suppression of the NF-κb signalling pathway in dmba-treated hamsters." *Anticancer Research* 37 (8): 4041–4050. https://doi.org/10.21873/anticanres.11789

36. Wang, Piwen, Jaydutt V Vadgama, Jonathan W Said, Clara E Magyar, Ngan Doan, David Heber and Susanne M Henning. 2014. "Enhanced inhibition of prostate cancer xenograft tumor growth by

combining quercetin and green tea." *The Journal of Nutritional Biochemistry* 25 (1): 73–80. https://doi.org/10.1016/j.jnutbio.2013.09.005

37. Wang, Piwen, David Heber and Susanne M Henning. 2012. "Quercetin increased bioavailability and decreased methylation of green tea polyphenols in vitro and in vivo." *Food and Function* 3 (6): 635–642. https://doi.org/10.1039/c2fo10254d

38. Rich, Gillian T, Maria Buchweitz, Mark S Winterbone, Paul A Kroon and Peter J Wilde. 2017. "Towards an understanding of the low bioavailability of quercetin: A study of its interaction with intestinal lipids." *Nutrients* 9 (2): 111. https://doi.org/10.3390%2Fnu9020111

39. Ader, P, A Wessmann and S Wolffram. 2000. "Bioavailability and metabolism of the flavonol quercetin in the pig." *Free Radical Biology and Medicine* 28 (7): 1056–1067. https://doi.org/10.1016/s0891-5849(00)00195-7

40. Wu, Yuanyuan, Yufeng Yu and Yihui Wang. 2020. "Research progress in anabolic control mechanisms of plant carotenoids." *Botanical Research* 9 (3): 217–225. https://doi.org/10.12677/br.2020.93026

41. Tanaka, Takuji, Masahito Shnimizu and Hisataka Moriwaki. 2012. "Cancer chemoprevention by carotenoids." *Molecules* 17 (3): 3202–3242. https://doi.org/10.3390/molecules17033202

42. Bishayee, Anupam, Alok Sarkar and Malay Chatterjee. 2000. "Further evidence for chemopreventive potential of beta-carotene against experimental carcinogenesis: Diethylnitrosamine-initiated and phenobarbital-promoted hepatocarcinogenesis is prevented more effectively by beta-carotene than by retinoic acid." *Nutrition and Cancer* 37 (1): 89–98. https://doi.org/10.1207/S15327914NC3701_12

43. Yi, Tingfang, Sung-Gook Cho, Zhengfang Yi, Xiufeng Pang, Melissa Rodriguez, Ying Wang, Gautam Sethi, Bharat B Aggarwal and Mingyao Liu. 2008. "Thymoquinone inhibits tumor angiogenesis and tumor growth through suppressing AKT and ERK signalling pathways." *Molecular Cancer Therapeutics* 7 (7): 1789–1796. https://doi.org/10.1158%2F1535-7163.MCT-08-0124

44. Aziza, Samy Hussein, SA Abdel-Aal and Khalaf A Heba. 2014. "Chemopreventive effect of thymoquinone on benzo(a)pyrene-induced lung cancer in male swiss albino mice." *Benha Veterinary Medical Journal* 27 (2): 330–340. www.bvmj.bu.edu.eg/

45. Patlolla, Jagan MR and Chinthalapally V Rao. 2012. "Triterpenoids for cancer prevention and treatment: Current status and future prospects." *Current Pharmaceutical Biotechnology* 13 (1): 47–155. https://doi.org/10.2174/138920112798868719

46. Herman-Antosiewicz, Anna, Anna A Powolny and Shivendra V Singh. 2007. "Molecular targets of cancer chemoprevention by garlic-derived organosulfides." *Acta Pharmacologica Sinica* 28 (9): 1355–1364. https://doi.org/10.1111/j.1745-7254.2007.00682.x

47. Wang, Weicang, Elvira Sukamtoh, Hang Xiao and Guodong Zhang. 2015. "Curcumin inhibits lymphangiogenesis in vitro and in vivo." *Molecular Nutrition & Food Research* 59 (12): 2345–2354. https://doi.org/10.1002/mnfr.201500399

48. Xiang, Yangfeng, Jianqiang Zhao, Ming Zhao and Kejing Wang. 2018. "Allicin activates autophagic cell death to alleviate the malignant development of thyroid cancer." *Experimental and Therapeutic Medicine* 15 (4): 3537–3543. https://doi.org/10.3892/etm.2018.5828

49. Gairola, Kanchan, Shriya Gururani, Ananya Bahuguna, Vaishali Garia, Rohit Pujari and Shiv Kumar Dubey. 2021. "Natural products targeting cancer stem cells: Implications for cancer chemoprevention and therapeutics." *Journal of Food Biochemistry* 45 (7): 1–14. https://doi.org/10.1111/jfbc.13772

50. Ong, Chau and Fawzy Elbarbry. 2016. "A new validated HPLC method for the determination of sulforaphane: Application to study pharmacokinetics of sulforaphane in rats." *Biomedical Chromatography* 30 (7): 1016–1021. https://doi.org/10.1002/bmc.3644

51. Burnett, Joseph P, Gi Lim, Yanyan Li, Ronak B Shah, Rebekah Lim, Hayley J Paholak, Sean P McDermott, Lichao Sun, Yasuhiro Tsume, Shuhua Bai, Max S Wicha, Duxin Sun and Tao Zhang. 2017. "Sulforaphane enhances the anticancer activity of taxanes against triple negative breast cancer by killing cancer stem cells." *Cancer Letters* 394: 52–64. https://doi.org/10.1016/j.canlet.2017.02.023

52. Razis, Ahmad Faizal Abdull and Noramaliza Mohd Noor. 2013. "Sulforaphane is superior to glucoraphanin in modulating carcinogen-metabolising enzymes in Hep G2 cells." *Asian Pacific Journal of Cancer Prevention* 14 (7): 4235–4238. https://doi.org/10.7314/apjcp.2013.14.7.4235

53. Montales, Maria Theresa E, Omar M Rahal, Hajime Nakatani, Tsukasa Matsuda and Rosalia CM Simmen. 2013. "Repression of mammary adipogenesis by genistein limits mammosphere formation of human MCF-7 cells." *Journal Endocrinology* 218 (1): 135–149. https://doi.org/10.1530/joe-12-0520

54. Huang, Xiao-Jia, Zhen-Qiu Ma, Wei-Ping Zhang, Yun-Bi Lu and Er-Qing Wei. 2007. "Dihydroartemisinin exerts cytotoxic effects and inhibits hypoxia inducible factor-1α activation in C6 glioma cells." *Journal of Pharmacy and Pharmacology* 59 (6): 849–856. https://doi.org/10.1211/jpp.59.6.0011

55. Huang, Xiao-Jia Huang, Cheng-Tan Li, Wei-Ping Zhang, Yun-Bi Lu, San-Hua Fang and Er-Qing Wei. 2008. "Dihydroartemisinin potentiates the cytotoxic effect of temozolomide in rat c6 glioma cells." *Pharmacology* 82 (1): 1–9. https://doi.org/10.1159/000125673

56. Chen, Tao, Mian Li, Ruiwen Zhang and Hui Wang. 2009. "Dihydroartemisinin induces apoptosis and sensitizes human ovarian cancer cells to carboplatin therapy." *Journal of Cellular and Molecular Medicine* 13 (7): 1358–1370. https://doi.org/10.1111/j.1582-4934.2008.00360.x

57. Jiao, Yang, Chun-min Ge, Qing-hui Meng, Jian-ping Cao, Jian Tong and Sai-jun Fan. 2007. "Dihydroartemisinin is an inhibitor of ovarian cancer cell growth." *Acta Pharmacologica Sinica* 28 (7): 1045–1056. https://doi.org/10.1111/j.1745-7254.2007.00612.x

58. Kim, D-S, Y-M Jeong, S-I Moon, S-Y Kim, S-B Kwon, E-S Park, S-W Youn and K-C Park. 2006. "Indole-3-carbinol enhances ultraviolet B-induced apoptosis by sensitizing human melanoma cells." *Cellular and Molecular Life Sciences CMLS* 63 (22): 2661–2668. https://doi.org/10.1007/s00018-006-6306-1

59. Kim, Young S and JA Milner. 2005. "Targets for indole-3-carbinol in cancer prevention." *The Journal of Nutritional Biochemistry* 16 (2): 65–73. https://doi.org/10.1016/j.jnutbio.2004.10.007

60. Weng, Jing-Ru, Chen-Hsun Tsai, Samuel K Kulp and Ching-Shih Chen. 2008. "Indole-3-carbinol as a chemopreventive and anti-cancer agent." *Cancer Letters* 262 (2): 153–163. https://doi.org/10.1016/j.canlet.2008.01.033

61. Jeevanandam, Jaison, Ahmed Barhoum, Yen S Chan, Alain Dufresne and Michael K Danquah. 2018. "Review on nanoparticles and nanostructured materials: History, sources, toxicity and regulations." *Beilstein Journal of Nanotechnology* 9: 1050–1074. https://doi.org/10.3762/bjnano.9.98

62. Akbarzadeh, Abolfazl, Rogaie Rezaei-Sadabady, Soodabeh Davaran, Sang Woo Joo, Nosratollah Zarghami, Younes Hanifehpour, Mohammad Samiei, Mohammad Kouhi and Kazem Nejati-Koshki. 2013. "Liposome: Classification, preparation, and applications." *Nanoscale Research Letters* 8: 102. https://doi.org/10.1186/1556-276X-8-102

63. Perez-Herrero, Edgar and Alberto Fernández-Medarde. 2015. "Advanced targeted therapies in cancer: Drug nanocarriers, the future of chemotherapy." *European Journal of Pharmaceutics and Biopharmaceutics* 93: 52–79. https://doi.org/10.1016/j.ejpb.2015.03.018

64. Paliwal, Shivani Rai, Rishi Paliwal and Suresh P Vyas. 2015. "A review of mechanistic insight and application of pH-sensitive liposomes in drug delivery." *Drug Delivery* 22: 231–242. https://doi.org/10.3109/10717544.2014.882469

65. Mozafari, MR. 2010. "Nanoliposomes: Preparation and analysis." *Methods in Molecular Biology* 605: 29–50. http://dx.doi.org/10.1007/978-1-60327-360-2_2

66. Abreu, Ana S, Elisabete Ms Castanheira, Maria-João Rp Queiroz, Paula Mt Ferreira, Luís A Vale-Silva and Eugénia Pinto. 2011. "Nanoliposomes for encapsulation and delivery of the potential antitumoral methyl 6-methoxy-3-(4-methoxyphenyl)-1Hindole-2-carboxylate." *Nanoscale Research Letters* 6 (1):1–6. https://doi.org/10.1186/1556-276x-6-482

67. Mohan, Aarti, Shridhar Narayanan, Swaminathan Sethuraman and Uma Maheswari Krishnan. 2014. "Novel resveratrol and 5-fluorouracil coencapsulated in PEGylated nanoliposomes improve chemotherapeutic efficacy of combination against head and neck squamous cell carcinoma." *Biomed Research International* 2014: 1–14. Article ID 424239. doi: 10.1155/2014/424239.

68. Deshantri, Anil K, Aida Varela Moreira, Veronika Ecker, Sanjay N Mandhane, Raymond M Schiffelers, Maike Buchner and Marcel HAM Fens. 2018. "Nanomedicines for the treatment of hematological malignancies." *Journal of Controlled Release* 287: 194–215. https://doi.org/10.1016/j.jconrel.2018.08.034

69. Shin, Dae Hwan, Yu Tong Tam and Glen S Kwon. 2016. "Polymeric micelle nanocarriers in cancer research." *Frontiers of Chemical Science and Engineering* 10: 348–359. https://doi.org/10.1007/s11705-016-1582-2

70. Croy, SR and GS Kwon. 2006. "Polymeric micelles for drug delivery." *Current Pharmaceutical Design* 12 (36): 4669–4684. https://doi.org/10.2174/138161206779026245

71. Cho, Hyunah, Tsz Chung Lai, Keishiro Tomoda and Glen S Kwon. 2015. "Polymeric micelles for multi-drug delivery in cancer." *AAPS PharmSciTech* 16 (1): 10–20. https://doi.org/10.1208/s12249-014-0251-3

72. Kulthe, S Sushant, Yogesh M Choudhari, Nazma N Inamdar and Vishnukant Mourya. 2012. "Polymeric micelles: Authoritative aspects for drug delivery." *Designed Monomers and Polymers* 15 (5): 465–521. https://doi.org/10.1080/1385772X.2012.688328

73. Ventola, C Lee. 2017. "Progress in nanomedicine: Approved and investigational nanodrugs." *P & T: A Peer-reviewed Journal for Formulary Management* 42 (12): 742–755.

74. Castro, I Ricardo, Oscar Forero-Doria and Luis Guzmán. 2018. "Perspectives of dendrimer-based nanoparticles in cancer therapy." *Anais da Academia Brasileira de Ciencias* 90 (2): 2331–2346. https://doi.org/10.1590/0001-3765201820170387

75. Abbasi, Elham, Sedigheh Fekri Aval, Abolfazl Akbarzadeh, Morteza Milani, Hamid Tayefi Nasrabadi, Sang Woo Joo, Younes Hanifehpour, Kazem Nejati-Koshki and Roghiyeh Pashaei-Asl. 2014. "Dendrimers: Synthesis, applications, and properties." *Nanoscale Research Letters* 9 (1): 247. https://doi.org/10.1186%2F1556-276X-9-247

76. Mendes, Livia Palmerston, Jiayi Pan and Vladimir P Torchilin. 2017. "Dendrimers as nanocarriers for nucleic acid and drug delivery in cancer therapy." *Molecules* 22 (9): 1401. https://doi.org/10.3390/molecules22091401

77. Cuenca, G Alex, Huabei Jiang, Steven N Hochwald, Matthew Delano, William G Cance and Stephen R Grobmyer. 2006. "Emerging implications of nanotechnology on cancer diagnostics and therapeutics." *Cancer* 107 (3): 459–466. https://doi.org/10.1002/cncr.22035

78. Fang, Min, Chun-wei Peng, Dai-Wen Pang and Yan Li. 2012. "Quantum dots for cancer research: Current status, remaining issues, and future perspectives." *Cancer Biology and Medicine* 9 (3): 151–163. https://doi.org/10.7497%2Fj.issn.2095-3941.2012.03.001

79. Pisanic, TR, Y Zhang and TH Wang. 2014. "Quantum dots in diagnostics and detection: Principles and paradigms." *Analyst* 139 (12): 2968–2981. https://doi.org/10.1039/C4AN00294F

80. Zdobnova, TA, EN Lebedenko and SM Deyev. 2011. "Quantum dots for molecular diagnostics of tumors." *Acta Naturae* 3 (1): 29–47. http://dx.doi.org/10.32607/20758251-2011-3-1-29-47

81. Rizvi, Sarwat B, Sepideh Rouhi, Shohei Taniguchi, Shi Yu Yang, Mark Green, Mo Keshtgar, and Alexander M Seifalian. 2014. "Near-infrared quantum dots for HER2 localization and imaging of cancer cells." *International Journal of Nanomedicine* 9: 1323–1337. https://doi.org/10.2147%2FIJN.S51535

82. Hu, Xiuli, Shi Liu, Guangyuan Zhou, Yubin Huang, Zhigang Xie and Xiabin Jing. 2014. "Electrospinning of polymeric nanofibers for drug delivery applications." *Journal of Controlled Release* 185: 12–21. https://doi.org/10.1016/j.jconrel.2014.04.018

83. Sharma, Rahul, Harmanpreet Singh, Munish Joshi and Abhinandan Sharma. 2014. "Recent advances in polymeric electrospun nanofibers for drug delivery." *Critical Reviews in Therapeutic Drug Carrier Systems* 31 (3): 187–217. http://dx.doi.org/10.1615/CritRevTherDrugCarrierSyst.2014008193

84. Singh, Harmanpreet, Rahul Sharma, Munish Joshi, Tarun Garg, Amit Kumar Goyal and Goutam Rath. 2015. "Transmucosal delivery of docetaxel by mucoadhesive polymeric nanofibers." *Artificial Cells, Nanomedicine and Biotechnology* 43 (4): 263–269. https://doi.org/10.3109/21691401.2014.885442

85. Khandare, Jayant and Tamara Minko. 2006. "Polymer-drug conjugates: Progress in polymeric prodrugs." *Progress in Polymer Science* 31 (4): 359–397. https://doi.org/10.1016/J.PROGPOLYMSCI.2005.09.004

86. Li, Chun and Sidney Wallace. 2008. "Polymer-drug conjugates: Recent development in clinical oncology." *Advanced Drug Delivery Reviews* 60 (8): 886–898. https://doi.org/10.1016/j.addr.2007.11.009

87. Vicent, María J and Ruth Duncan. 2006. "Polymer conjugates: Nanosized medicines for treating cancer." *Trends in Biotechnology* 24 (1): 39–47. https://doi.org/10.1016/j.tibtech.2005.11.006

88. Wadhwa, Saurabh and Russell J Mumper. 2015. "Polymer-drug conjugates for anticancer drug delivery." *Critical Reviews in Therapeutic Drug Carrier Systems* 32 (3): 215–245. https://doi.org/10.1615/critrevtherdrugcarriersyst.2015010174

89. Duncan, Ruth. 2006. "Polymer conjugates as anticancer nanomedicines." *Nature Reviews Cancer* 6 (9): 688–701. https://doi.org/10.1038/nrc1958

90. Jaiswal, Manjit, Rupesh Dudhe and PK Sharma. 2015. "Nanoemulsion: An advanced mode of drug delivery system." *3 Biotech* 5 (2):123–127. https://doi.org/10.1007/s13205-014-0214-0

91. Gupta, Ankur, H Burak Eral, T Alan Hatton and Patrick S Doyle. 2016. "Nanoemulsions: Formation, properties and applications." *Soft Matter* 12 (11): 2826–2841. https://doi.org/10.1039/C5SM02958A

92. Gavin, Amy, Jimmy Th Pham, Dawei Wang, Bill Brownlow and Tamer A Elbayoumi. 2015. "Layered nanoemulsions as mucoadhesive buccal systems for controlled delivery of oral cancer therapeutics." *International Journal of Nanomedicine* 10: 1569–1584. https://doi.org/10.2147/ijn.s75474

93. Lohcharoenkal, Warangkana, Liying Wang, Yi Charlie Chen and Yon Rojanasakul. 2014. "Protein nanoparticles as drug delivery carriers for cancer therapy." *BioMed Research International* 2014: 1–12. https://doi.org/10.1155/2014/180549.

94. Soppimath, Kumaresh S, Tejraj M Aminabhavi, Anandrao R Kulkarni and Walter E Rudzinski. 2001. "Biodegradable polymeric nanoparticles as drug delivery devices." *Journal of Controlled Release* 70 (1–2): 1–20. https://doi.org/10.1016/s0168-3659(00)00339-4.

95. Jahanshahi, M, Z Zhang and A Lyddiatt. 2005. "Subtractive chromatography for purification and recovery of nano-bioproducts." *IEE Proceedings Nanobiotechnology* 152 (3): 121–126, IET Digital Library. https://doi.org/10.1049/ip-nbt:20045004

96. Cheng, Dan, Xueqing Yong, Tianwen Zhu, Yining Qiu, Jun Wang, Hui Zhu, Baoliang Ma and Jinbing Xie. 2016. "Synthesis of protein nanoparticles for drug delivery." *European Journal of BioMedical Research* 2 (2): 8–11. https://doi.org/10.18088/ejbmr.2.2.2016

97. Andrew Mackay, J and Ashutosh Chilkoti. 2008. "Temperature sensitive peptides: Engineering hyperthermia-directed therapeutics." *International Journal of Hyperthermia* 24 (6): 483–495. https://doi.org/10.1080/02656730802149570.

98. Nitta, Kaihara Sachiko and Keiji Numata. 2013. "Biopolymer-based nanoparticles for drug/gene delivery and tissue engineering." *International Journal of Molecular Sciences* 14 (1): 1629–1654. https://doi.org/10.3390/ijms14011629

99. Martínez, A, M Benito-Miguel, I, Iglesias, JM Teijón and MD Blanco. 2012. "Tamoxifenloaded thiolated alginate-albumin nanoparticles as antitumoral drug delivery systems." *Journal of Biomedical Materials Research* Part A 100 (6): 1467–1476. https://doi.org/10.1002/jbm.a.34051.

100. Herrero-Vanrell, R, AC Rincon, M Alonso, V Reboto, IT Molina-Martinez and JC Rodriguez-Cabello. 2005. "Self-assembled particles of an elastin-like polymer as vehicles for controlled drug release." *Journal of Controlled Release* 102 (1): 113–122. https://doi.org/10.1016/j.jconrel.2004.10.001.

101. Yasmin, Rehana, Mohsin Shah, Saeed Ahmad Khan and Roshan Ali. 2017. "Gelatin nanoparticles: A potential candidate for medical applications." *Nanotechnology Reviews* 6 (2): 191–207. https://doi.org/10.1515/ntrev-2016-0009.

102. Gidwani, Bina and Amber Vyas. 2015. "A comprehensive review on cyclodextrin-based carriers for delivery of chemotherapeutic cytotoxic anticancer drugs." *BioMedical Research International* 2015 (3): 1–15. https://doi.org/10.1155/2015/198268.

103. Ahmad, MZ, S Akhter, I Ahmad, M Rahman, M Anwar, GK Jain, FJ Ahmad and RK Khar. 2011. "Development of polysaccharide-based colon targeted drug delivery system: Design and evaluation of Assam Bora rice starch-based matrix tablet." *Current Drug Delivery* 8 (5): 575–581. https://doi.org/10.1517/17425247.2012.633507

104. Merisko-Liversidge, E, GG Liversidge and ER Cooper. 2003. "Nanosizing: A formulation approach for poorly-water-soluble compounds." *European Journal of Pharmaceutical Sciences* 18 (2): 113–120. https://doi.org/10.1016/s0928-0987(02)00251-8

105. Jiang, Ye, Xinyi Jiang, Kitki Law, Yanzuo Chen, Jijin Gu, Wei Zhang, Hongliang Xin, Xianyi Sha and Xiaoling Fang. 2011. "Enhanced anti-tumor effect of 9-nitro-camptothecin complexed by hydroxypropylβ-cyclodextrin and safety evaluation." *International Journal of Pharmaceutics* 415 (1–2): 252–258. http://dx.doi.org/10.1016/j.ijpharm.2011.05.056

106. Gopi, Sreeraj, Augustine Amalraj and Sabu Thomas. 2016. "Effective drug delivery system of biopolymers based on nanomaterials and hydrogels-a review." *Drug Design* 5 (129): 2169–0138. https://doi.org/10.4172/2169-0138.1000129.

107. Akhter, Sohail, Mohammad Ahmad Zaki, Anjali Singh, Iqbal Ahmad, Mahfoozur Rahman, Mohammad Anwar, Gaurav Kumar Jain, Farhan Ahmad Jalees and Roop Khar Krishen. 2011. "Cancer targeted metallic nanoparticle: Targeting overview, recent advancement and toxicity concern." *Current Pharmacological Design* 17 (18): 1834–1850. https://doi.org/10.2174/138161211796391001.

108. Desai, Neil. 2012. "Challenges in development of nanoparticle-based therapeutics." *The AAPS Journal* 14 (2): 282–295. https://doi.org/10.1208/s12248-012-9339-4.

109. Moghimi, Seyed Moein, A Christy Hunter and J Clifford Murray. 2001. "Long-circulating and target-specific nanoparticles: Theory to practice." *Pharmacological Reviews* 53 (2): 283–318. http://pharmrev.aspetjournals.org.

110. Moghimi, SM and J Szebeni. 2003. "Stealth liposomes and long circulating nanoparticles: Critical issues in pharmacokinetics, opsonization and protein-binding properties." *Progress in Lipid Research* 42 (6): 463–478. https://doi.org/10.1016/S0163-7827(03)00033-X.

111. Zolnik, Banu S, África González-Fernández, Nakissa Sadrieh and Marina A Dobrovolskaia. 2010. "Nanoparticles and the immune system." *Endocrinology* 151 (2): 458–465. https://doi.org/10.1210/en.2009-1082.

112. Kawasaki, Ernest S and Audrey Player. 2005. "Nanotechnology, nanomedicine, and the development of new, effective therapies for cancer." *Nanomedicine* 1 (2): 101–109. https://doi.org/10.1016/j.nano.2005.03.002

113. Zhang, L, FX Gu, JM Chan, AZ Wang, RS Langer and OC Farokhzad. 2008. "Nanoparticles in medi-cine: Therapeutic applications and developments." *Clinical Pharmacology and Therapeutics* 83 (5): 761–769. https://doi.org/10.1038/sj.clpt.6100400

114. Gref, R, Y Minamitake, MT Peracchia, V Trubetskoy, V Torchilin and R Langer. 1994. "Biodegradable long-circulating polymeric nanospheres." *Science* 263 (5153): 1600–1603. https://doi.org/10.1126/science.8128245

115. Siddiqui, Imtiaz A and Vanna Sanna. 2016. "Impact of nanotechnology on the delivery of natural prod-ucts for cancer prevention and therapy." *Molecular Nutrition & Food Research* 60 (6): 1330–1341. https://doi.org/10.1002/mnfr.201600035.

2 Polyphenols in Cancer Chemoprevention

Pushpendra Koli, Dinesh K. Yadav, Mahesh K. Samota, and Sandeep Kumar

CONTENTS

2.1   Introduction 21
2.2   Nature of Polyphenols and Their Classifications..22
2.3   Challenges in Cancer Prevention and Treatment...22
      2.3.1   Growing Inequality and Societal Costs of Cancer Prevention 22
      2.3.2   Western Lifestyle and Limited Access 23
2.4   Role and Mechanism of Polyphenols in Cancer Chemoprevention.............24
      2.4.1   Role of Reactive Oxygen Species in the Development of Carcinogenesis 24
      2.4.2   Polyphenols as Pro-oxidants..24
      2.4.3   Phytochemicals Modulating Signaling Molecules 25
2.5   Naturally Occurring Polyphenols for Cancer Prevention and Treatment.............26
      2.5.1   Flavonoids...26
      2.5.2   Phenolic Acids 27
2.6   Dietary Phytochemicals as Chemopreventives...27
      2.6.1   Common Examples of Dietary Phytochemicals......................................29
      2.6.2   Use of Medicinal Plants in Chemoprevention of Cancer 29
2.7   References...30

2.1 INTRODUCTION

Cancer is a greater concern of human health across the globe, and it is increasing becoming worse without authentic therapies or controlled treatment. It is still the prominent cause of human death worldwide; it is assumed that one in six deaths is due to cancer only [1]. Polyphenol-derived foods are considered because of their safety and therapeutic potential. Polyphenols are a group of natural compounds that are produced biosynthetically in the plant system and have strong antioxidant properties. These are heterogeneous groups of secondary metabolites, and the presence of a phenolic group in their chemical structure gives them antioxidant, anti-inflammatory, and cytotoxic properties [2]. In addition to this, these are responsible for color development in the fruits and other parts of the plants [3]. These molecules are widely present in almost every plant system but are mainly found in fruits, vegetables, coffee, tea, and grains [2, 4]. Due to their presence in almost every plant system, dietary polyphenols contribute a significant role in human nutrition. The antioxidant nature of polyphenols accelerates their use in food, and therefore, they are more frequently used to prevent diseases like cancer and cardiovascular over the last 30 years [5]. Several reports have been published on polyphenols that show anticancer effects by involving several mechanisms like change in signaling pathways, initiation of apoptosis, and eventually blockage of the cell cycle [6]. In general, polyphenol interferes with enzymes that are the key factors for tumor cell proliferation. In recent studies, other mechanisms that are antiangiogenic and

DOI: 10.1201/b23311-2

antimetastatic and the interference and interaction with DNA were also reported in the prevention of cancer by polyphenols [7, 8].

In this chapter, we discuss recent findings on various types of polyphenols, their role and difficulties in cancer prevention, and also, briefly summarized a few current advancements in polyphenol-based cancer drug development.

2.2 NATURE OF POLYPHENOLS AND THEIR CLASSIFICATIONS

Polyphenols are a group of natural compounds having a phenolic functional group. Chemically, this group is very diverse, and it consists of several subgroups of phenolic compounds. They can be subclassified based on the number of phenol units per molecule and the linkage type between the two groups. The interesting finding is that till today around 8,000 phenolic structures have been identified and among them over 4,000 flavonoids have been identified [9]. Therefore, to understand polyphenols, the basic structure of flavonoids is considered as the reference (Figure 2.1).

The flavonoids show the basic structure of diphenyl propane (C6-C3-C6). Here, the right phenolic ring (ring A) and the left phenolic ring (ring B) are connected through a heterocyclic ring (in middle, ring C) [10]. The various hydroxylation patterns and different oxidation states of oxygen in the heterocyclic ring result in the diversity of these molecules and because of this diversity several molecules arise, for example, flavanols, anthocyanidins, anthocyanins, isoflavones, flavones, flavonols, flavanones, and flavanonols (Figure 2.2). Flavones, flavonols, flavanones, and flavanonols mainly represent the polyphenolic group [11]. In the majority of flavonoids structure, C2 of the C-ring is attached to the B-ring, but C3 and C4 attachments are also found. However, chalcones also represent the flavonoid family without a C-ring.

2.3 CHALLENGES IN CANCER PREVENTION AND TREATMENT

2.3.1 GROWING INEQUALITY AND SOCIETAL COSTS OF CANCER PREVENTION

According to the report, around 19 million people were diagnosed with cancer and 10 million people have died. It is expected that by 2040 there will be about 28 million cancer patients and 16 million deaths [11, 12]. The treatment of cancer is so expensive, it is believed that in a year, the total cost comes to approximately $1.2 trillion annually, which was about 2% of global GDP in 2019 [13]. There are many challenges in cancer care, like health policies, lack of trained professionals, lack of infrastructure, providers having less faith in care, and so on. Many countries have not enforced the WHO framework on the control of tobacco use, which causes cancer and is intensified marketing by tobacco companies [14, 15]. There are huge disparities in HPV vaccine in many countries, and the coverage of dose is <40% globally [16]. In Asian countries and African countries, only 40% and 31% of HPV programmes were introduced as of 2020 [17].

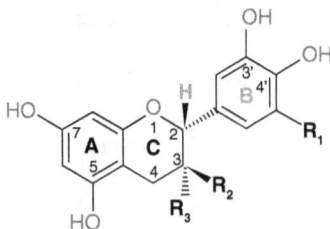

FIGURE 2.1 The basic structure of a flavonoid.

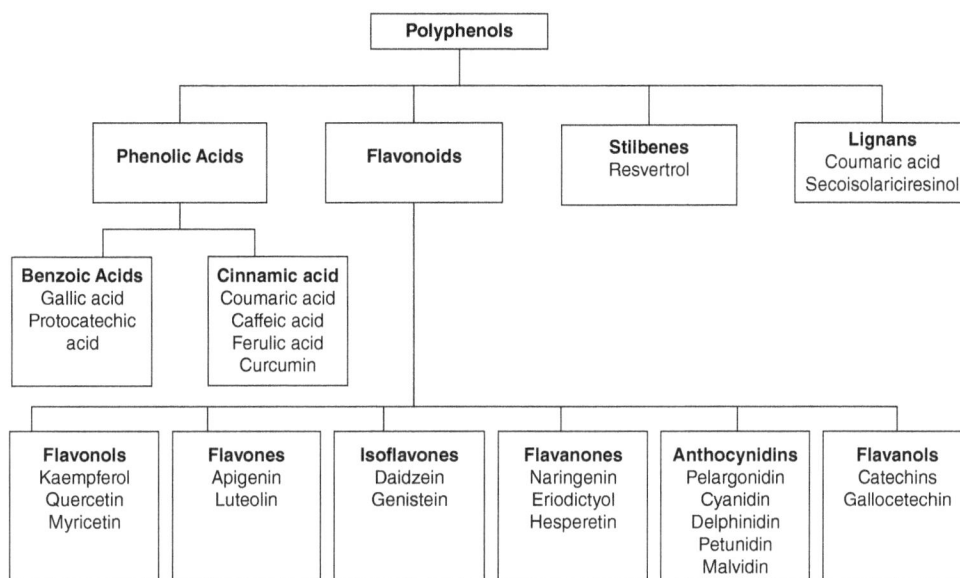

FIGURE 2.2 Polyphenol classifications.

2.3.2 WESTERN LIFESTYLE AND LIMITED ACCESS

An undesirable lifestyle trend is the additional cause of incidence and challenges in responding to premature death caused by it. The cancer incidence increases observed in Western lifestyle countries than in their home countries [18–20]. Breast cancer rates are low in China, Brazil, and India but now a day's metropolitan cities of these countries have the same rate as western countries because of changes in breastfeeding practices and delays in pregnancy as well as changes in lifestyle [21, 22]. For example, in Brazil, around 50% population is underweight due to the unavailability of sufficient calories and that resulted in an adaptation to consuming fast and processed foods that are low in nutrition and another reason is the lack of physical activities due to the modern lifestyle [23].

There are many disparities in cancer treatment globally, like fragmented care and lack of resources. Novel cancer therapies like immunotherapies and targeted treatment are not available for the majority of the population. Africa, rural Latin America, and rural Asia have limited access to effective and more recent cancer treatments [24]. The radiotherapy machines per a million people in low-income countries was almost nonexistent, at only a range of 0.0 to 0.4 compared, while high-income countries have a range of 0.4 to 11.6 [25]. Recently, a study reported that there is a 25-fold difference between low-income countries and high-income countries in the survival of women diagnosed with breast cancer. Further, authors estimated that in low-income countries, implementing a set of treatments like surgery, imaging, medical oncology, radiation, and screening improves the quality of care and increases the survival rate from 3.5% to 55.3% [26]. Another important reason is that there is less priority on health policies and health insurance coverage that suppress the treatment of such a fatal disease as cancer [27]. In lower-income countries, it is very hard to bear the medical expenses by several patients which causes the dominance of healthcare financing in these countries [28]. In sub-Saharan countries, 50% of patients consult a local and traditional healer due to high medical costs, uneasy access, and several myths bestowed with western treatments [29]. Another important factor is lack of education, which hampers the proper and timely diagnosis and eventually poor outcomes.

The increase in cancer survivors globally is due to growth of the aging population. This makes it difficult to count the exact figure of the cancer-suffering population. There is a large gap in progress in the supportive care of survivors and addressing symptoms such as pain, insomnia, anxiety, neuropathy, fatigue, and cognitive dysfunction [30–32]. Further, there are more challenges in the healthcare system for proper and sufficient resources and infrastructures [33]. According to the WHO, the early integration of palliative care is considered an essential component of healthcare to improve life quality [34]; however, it is not accessible easily. For example, only 10% of patients are allowed to access oral morphine to treat pain relief in low-income countries [35]. The national policy for palliative care is equally important, but some countries like China and India do not have such policies, and only a few African countries have palliative care in their national policy [36, 37].

2.4 ROLE AND MECHANISM OF POLYPHENOLS IN CANCER CHEMOPREVENTION

2.4.1 ROLE OF REACTIVE OXYGEN SPECIES IN THE DEVELOPMENT OF CARCINOGENESIS

Redox imbalance means increased reactive oxygen species is one of the most important reasons to develop cancer in human cells. These reactive oxygen species have been generated from intrinsic sources, such as mitochondria, several enzymatic cellular complexes, and inflammatory cells, or from extrinsic sources, such as radiation, alcohol, smoking, drugs, and pro-oxidant toxins [38, 39]. These free radicals can cause damage to lipids, proteins, and nucleic acids that lead to damaging different tissues of humans and downregulates the activity of antioxidants enzymes that cause malignant transformation via molecular targets like NF-B and Nrf-2 [40, 41]. The ROS cause oxidative stress, which leads the toxicity, and these ROS interact with reactive nitrogen species, which disturbed the signaling pathways, such as phosphatases, protein kinases, and transduction mechanisms [42]. Oxidative stress is attributed to pro-oxidants changing disulfide/thiol redox state, which leads to cancer, diabetes mellitus, chronic inflammation, and atherosclerosis [43].

2.4.2 POLYPHENOLS AS PRO-OXIDANTS

The polyphenols show pro-oxidant activity because of transition metals that contribute to redox cycling and accelerate the formation of hydroxyl radicals through reactions such as the Fenton reaction. Since the deployment of endogenous metal ions like iron and copper and polyphenols exhibit cytotoxicity against cancer cells, the phenolic compounds, such as chlorogenic, caffeic, and ferulic acid, are more effective against DNA cleavage in human promyelocytic leukemia HL-60 cells in the presence of Cu(II) ions [44]. On the other hand, grapes and rhubarb phenols, like piceatannol and stilbene, combined with oxidized Cu ions, boost the Haber-Weiss and Fenton reactions, which help DNA cleavage, and eventually, the piceatannol is converted into *ortho*-quinone and forms adducts with DNA [45]. Some other relevant reports do explain the importance of transition metals in the pro-oxidant activity of polyphenols because of the Warburg effect. Cancer cells demonstrate a higher rate of glycolysis, which results in lowering the pH, which affects the DNA structure and exposes the chromatin-bound Cu ions to attack of pro-oxidant agents such as resveratrol [46]. There is some disagreement regarding polyphenol-induced ROS as a reason for apoptosis. Some studies showed that ROS accumulated in early steps in the pathway towards polyphenol-induced apoptosis, and these pro-apoptotic effects were reduced effectively by pre-incubation with antioxidant enzymes, NAC, or Triton [47, 48]. It was also listed that NAC inactivates caspase-3, which results in the decline in apoptosis [49].

The pro-oxidant activity of polyphenols has been controlled by different pathways to the cell cycle arrest and induction of apoptosis. Anthocyanins show pro-apoptotic activity in cancer cells due to the formation of higher ROS [47]. Delphinidin, an anthocyanin, causes G2/M phase cell cycle arrest and apoptosis via the downregulation of cyclin B and cdc2 protein expression and the upregulation of the protein p53 tumor suppressor and its downstream target p21 in HCT116 human colon cells [50]. Emodin, an anthraquinone, through the generation of ROS, induces apoptosis and

DNA damage in H460 human carcinoma lung cells [51]. Emodin, also through the ROS mechanism, downregulates the MRP 1 (multidrug-resistance-related protein 1) expression in human ovarian carcinoma cells, and these effects are reversed by the NAC antioxidant [52]. Gingerol reveals the pro-oxidant action to sanitize the human glioblastoma multiforme cells to TNF-related apoptosis-inducing ligand (TRAIL; TNF means "tumor necrosis factor") stimulated apoptosis [53]. Gingerol fosters the generation of ROS, upregulation of p53, and inhibition of Bcl-2 and surviving proteins that show anti-apoptotic activity. Chalcones derive some extra mechanisms for redox-sensitive apoptosis in the cells of cervix cancer HeLa [54]. Chalcones also induce the formation of ROS via disruption of the mitochondrial function and activation of caspase-3 and caspase-9, eventually resulting in the caspase-12-dependent apoptosis [55].

Coumarins, such as esculetin, generate the activities against the proliferation of hepatocellular carcinoma, both in vivo and in vitro. The esculetin shows the caspase-9- and caspase-3-mediated apoptosis with a reduction of the expression of Bcl-2 [56]. In addition, esculetin induced the release of cytochrome C into the cytosol, the formation of apoptotic bodies, and the activation of caspases that exerted a cytotoxic effect on cervical cancer HeLa cells [57]. Curcumin activated the ASK 1 (apoptosis signal regulating kinase 1) signaling cascade, which displays apoptosis in gastric BGC-823 cells [48]. Green tea flavanols, like catechin, display antimutagenic, pro-oxidant, and cytotoxic activities against cancer cells [58]. Catechin-induced high ROS production accompanies human DNA damage in leukemia HL-60 cells and increases the level of 8-hydroxyguanine [59]. A flava-none called hesperidin is responsible for apoptosis induction in breast MCF-7 cells [60]. Apigenin, luteolin, quercetin, and kaempferol show pro-apoptotic activity in human against hepatocarcinoma HepG2 cells and increased the expression of the PIG3 gene (p53-inducible gene 3), pro-oxidant effect, and accumulation of ROS [61]. Isoflavones isolated from *Glycine max* are also associated with a lower incidence of cancer [62]. Daidzein, an isoflavone, can form ROS and cause apoptosis in breast cancer MCF-7 cells [63]. Proanthocyanidins extracted from cranberry induce ROS genera-tion, which is responsible for pro-apoptotic activity in ovarian adenocarcinoma SKOV-3 cells and downregulate the p-Akt pathway resulting in decreased cell viability [64].

2.4.3 PHYTOCHEMICALS MODULATING SIGNALING MOLECULES

The epigenetic machinery of human cells is affected by nutritional and environmental factors. Polyphenols and dietary compounds are involved in the histone modification, DNA methylation, and post-transcriptional regulation of genes to shape the epigenome [65]. Recently, many studies focused on identifying new therapies for cancer, and to achieve this, phytochemicals have been tested extensively [66]. Most phytochemicals have functional roles, such as having chemopreven-tive, antioxidant, and anti-inflammatory activities [67]. For example, the polyphenols extracted from green tea show pro-apoptotic, antioxidant, and antiangiogenic activities to modulate cancer [68, 69]. Some phytochemicals display cytotoxic effects that are because of unsuitable combina-tions, improper use, and high dose [70]. Cytotoxic effects, such as goitrogenic effects, and hemolytic effects can be caused by *Brassica* plants and fava beans, respectively [71]. Tamoxifen drug inhibits tumor proliferation by stopping the hormone-triggered signaling pathways from binding to the ERs of cancer cells [72, 73].

Phytochemicals are naturally occurring compounds that can modulate the tumor genetic expres-sions and change the sensitivity to antihormonal therapies. Their molecular mechanism is linked to the expression and function of genes or ncRNA. Polyphenols can regain the ER phenotype from ER negative cancer [74] influencing the reading of genetic code and affecting the signaling pathway in cancer cells can be the result of histone modifications like acetylations and deacetylations [75, 76]. The sensitivity of ER tumors increased with the treatment of demethylating agents, which involve in gene silencing. DNA methyl transferase (DNMT) is a demethylating agent, which slows down the mRNA and reduces the protein expressions by affecting the hypermethylation of the promoter of ER [77]. Nowadays, the application of dietary intervention has been increased for cancer prevention for example dietary phenols have important potential to hinder oncogenic signaling pathways and play a role as chemopreventive and therapeutic agents. The phytochemicals have low bioavailability, which helps these compounds to reach the target site in ample quantity [78]. For example, curcumin limitation has been adjusted by the analog synthesis, which enhances efficiency [79]. However, these phytochemicals need further assessments before being used as chemopreventive and chemo-therapeutic agents.

Phytochemicals maintain the normal signal pathways by modulating the expression of cod-ing and noncoding RNA [80]. Many studies reported that the phytochemicals in chemotherapy enhanced the efficiency and decrease the side effects, such as reducing drug resistance and induc-ing apoptosis in cancer cells [81, 82]. Micro RNA (miRNA) expression is twisted in diseases like cancer where they suppress the growth of tumors [83, 84]. Therefore, the manipulation of miRNA as molecular targets can be encouraged in cancer treatment [85]. Phytochemicals modulate the miRNA expression and suppress or activate different miRNA levels in tumor cells as suppressor roles restore normal expression levels [86]. Several studies showed the effects of phytochemicals to modify miRNA expression and their mRNA targets, by changing angiogenesis, apoptosis cell pro-liferation and differentiation, metastasis, and assimilation of drug resistance in cancer [87, 88]. The let-7 family of miRNA decreases tumor growth when its expression is increased [89–93]. miR-16, miR-15a, and miR-34a have constructive results in chronic lymphocytic leukemia and pancreatic cells, respectively, in the form of increased apoptosis [94–96]. miR-203 overexpression inhibits the tumor growth in a mouse and pancreatic cell proliferation [97, 98]. miR-210 is an important bio-marker, and the poor prognosis in glioblastoma and acute leukemia is linked to its overexpression [99, 100].

The tea polyphenol administration in lung cancer cells increases the expression of miR-210, reduces the proliferation activity of cancer cells via stabilization of hypoxia-inducible factor-1 [101], and downregulates the miRNA-30b in the HepG2 liver cancer cells [102]. Polyphenols isolated from tea can upregulate the expression of miR-34a/E2F3/Sirt1 in colon cancer cells [103]. Recently, one study shows that tea polyphenols increase the expression of tumor suppressor miRNAs, such as miR-7-1, miR-34a, and miR-99a, and decrease the expression of oncogenic miRNAs, such as miR-92, miR-93, and miR-106b [104]. They also downregulate 5-miRNA in HepG2 hepatocellular carcinoma cells [105] and upregulate 13-miRNA and downregulate 48-miRNA in HCC cells [106]

2.5 NATURALLY OCCURRING POLYPHENOLS FOR CANCER PREVENTION AND TREATMENT

Cancer is recognized to be among the most serious issues confronting humanity on a global scale, and following cardiovascular disease, it is the second top cause of death. According to the WHO, there will be approximately 27 million cases of cancer by 2050, with an annual fatality rate of 17.5 million people [107]. Polyphenols, such as phenolic acids, flavonoids, stilbenes, and lignans, are abundant in fruits, vegetables, and spices. Flavonoids and phenolic acids are the well-known classes, which account for roughly 60% and 30% of all natural polyphenolic compounds, respectively. Such compounds are plant secondary metabolites that are present in the regular diet and provide sig-nificant benefits to the human body [108]. Natural polyphenols' anticancer attributes have become a major topic in many laboratories over the last two decades. Natural polyphenols show anticancer activity with antioxidative and anti-inflammatory activities and their ability to modify molecular tar-gets and signal transduction pathways involved in cell existence/multiplication/segmentation, tumor growth, hormonal processes, detoxifying enzymes, immunity, and so on [109, 110].

2.5.1 Flavonoids

Flavonoids are a class of compounds that includes anthocyanins, flavanols, flavanones, flavones, flavonols, and isoflavonoids. Fruits and vegetables contain anthocyanins, which are pigments that vary in color. Among the anthocyanins, delphinidin has potent anticancer properties. Several studies have found that delphinidin causes apoptosis and cell cycle arrest in a variety of cancers [50]. Anthocyanins found in black rice, such as peonidin and cyanidin, have been shown to reduce tumorigenesis and cell cycle arrest in breast cancer [111]. After extensive research, it was discovered that a combination of anthocyanins may be more effective than a single anthocyanin in cancer treatment. Xanthohumol isolated from hops (*Humulus lupulus*) proved anticancer activity with apoptosis and cell cycle arrest. In transgenic mice, it was discovered that xanthohumol promotes ROS and inhibits tumor growth in prostate cancer [112].

Flavanols, which have the most complicated structures of any flavonoid subclass and are found in foods. The anticancer activity of epigallocatechin gallate (EGCG) was reported in terms of cell cycle arrest, tumorigenesis decrease, and antiproliferation. Its anticancer properties may involve hormone activity modulation [113, 114]. In breast cancer, EGCG can reduce cell proliferation, while in prostate cancer, it can antagonize androgen [115, 116].

Citrus fruits, particularly the solid parts of the fruit, are rich in flavanones. The most important flavanones are naringenin and hesperidin. Hesperidin inhibits cell proliferation and increases intracellular ROS in gastric cancer cells [117]. The flavone luteolin induces cell cycle arrest and apoptosis in lung cancer cells, resulting in significant cytotoxicity [118, 119]. Flavonols seem to be the most common and widespread flavonoids in foods. Quercetin inhibits the growth of lung cancer cells by causing cell cycle arrest and apoptosis. Quercetin showed a significant reduction in tumor growth during feeding in an experiment [118]. The most important isoflavonoids—genistein and daidzein—induced apoptosis in liver cancer cells. In soy and soy products, the isoflavonoid genistein is the most abundant, and it regulates cell growth, which is followed by apoptosis and cell cycle arrest in lung cancer [120].

2.5.2 PHENOLIC ACIDS

There are two types of phenolic acids: hydroxybenzoic and hydroxycinnamic acid. Ellagic acid is a flavonoid found in a variety of fruits. Ellagic acid has been shown to induce apoptosis and decrease tumorigenesis in colon and liver cancers, respectively [121, 122]. Gallic acid is found in a variety of plant-based foods. Cereal grains are the primary dietary sources of ferulic acid. Because of its therapeutic properties against a variety of diseases, this polyphenol has received considerable interest. Ferulic acid was found to be effective in causing apoptosis and cell cycle arrest. At high concentrations, it was also found to be a pro-oxidant [123].

2.6 DIETARY PHYTOCHEMICALS AS CHEMOPREVENTIVES

Phytochemicals are biologically active substances found in a variety of fruits, vegetables, plants, and spices. They have a variety of biological properties, such as anti-inflammatory and antioxidant properties, and are particularly well-known for inhibiting the growth of cancer cells. Evidently, a consistent inverse relationship between increased fruit and vegetable consumption and a lower incidence of oral cancer has been reported [140]. Phytochemicals can exert a chemopreventive effect and reverse precancerous conditions by blocking important events in the onset and promotion of tumors. These phytochemicals may also help to prevent tumor formation by reducing the growth of cancer cells, which is one of the foremost reasons for death across the world. It was also classified based on the second most important factor in reducing death. Even though it has been going on for the past ten years, it continues to be a global lethal cause of death. In epidemiological research, aside from factors that cannot be changed, we have shown strong evidence of this. The main risk factors for developing cancer are age, heredity, diet, and lifestyle. Diets high in lean and processed meats and excessively cooked foods may cause cancer [141]. Much research has been done recently to understand cancer, and most cases of cancer are known to be caused by certain lifestyle [142] This leads to well-known new ways to prevent cancer to suppress, prevent, or delay the development of

TABLE 2.1
Natural Polyphenol's Anticancer Properties

Nature	Type of Cancer	Role and Mechanism	References
		Flavonoids	
Anthocyanin	Lung cancer	Inducing apoptosis in HepG-2 cells	[124]
Epigallocatechin gallate (EGCG)	Breast cancer	Modulating microRNA, cell cycle arrest, decreasing tumorigenesis, VEGF-D expression	[115] [125]
Quercetin, luteolin	Lung cancer	Inhibiting tumorigenesis	[119]
Genistein	Lung cancer	Inhibiting cell growth, followed by apoptosis	[120]
EGCG	Gastric cancer	Causing apoptosis by inhibiting the catenin signaling pathway	[113] [114]
Quercetin	Gastric cancer	Causing apoptosis	[126]
Kaempferol	Gastric cancer	Inhibiting tumor growth	[127]
Delphinidin	Colorectal cancer	Apoptosis, cell cycle arrest, and oxidative stress	[50]
Genistein	Colon cancer	Preventing angiogenesis and metastasis	[128]
Quercetin	Liver cancer	Reducing tumorigenesis	[129]
Kaempferol	Liver cancer	Cell cycle arrest and autophagy inducer	[130]
Anthocyanin	Breast cancer	Preventing tumor growth	[111]
Genistein	Breast cancer	Inducing apoptosis and arresting the cell cycle	[131]
EGCG	Prostate cancer	Inhibiting tumor growth	[116]
Delphinidin	Prostate cancer	Causing apoptosis and arresting the cell cycle	[132]
Quercetin	Cervical cancer	Inducing apoptosis	[117]
Hesperetin	Cervical cancer	Arresting the cell cycle	[133]
		Phenolic acids	
Gallic acid	Gastric cancer	Inhibiting tumor growth	[134]
Ellagic acid	Colon cancer	Inducing apoptosis	[122]
	Liver cancer	Reducing tumorigenesis	[121]
Ferulic acid	Prostate cancer	Inducing apoptosis and arrest cell cycle	[123]
		Stilbenoids	
Resveratrol	Lung cancer	Downregulating PD-1 expression on pulmonary T cells	[135]
	Gastric cancer	Reducing tumorigenesis, inducing apoptosis, and producing ROS	[136] [81]
	Colon cancer	Inducing apoptosis and reducing tumorigenesis	[137]
		Lignans	
Sesamin	Liver cancer	Inducing apoptosis and cell cycle arrest	[138]
	Breast cancer	Suppressing tumor growth	[139]

carcinogenesis [143, 144]. Current natural products derived from plants are becoming more important due to their role as raw materials. Research is being conducted on anticancer drugs and their chemopreventive effects against cancer.

The first type of chemotherapy agents, such as flavonoids, indoles, and isothiocyanates, work by inhibiting the metabolic activation of the carcinogen or tumor promoter by enhancing the detoxification system. The second type of agents, such as vitamin D, vitamin A, retinoids, monoterpenes, and calcium, prevents the progression of neoplastic processes in cells that have the potential to become malignant [145]. Now researchers are moving toward using natural products as chemical preventatives to improve cancer survival. We investigated the chemical preventive effect of cancer using a variety of drugs or dietary agents. Biologically active plant-based compounds are classified based on their chemical structure [146]. Nutrition is the entire process involved in the absorption and utilization of nutrients that achieve growth, repair, and maintenance of the body (health guide to better health). Chemoprophylaxis is the long-term use of synthetic, natural, or biological agents to prevent or postpone the development of malignant tumors.

2.6.1 COMMON EXAMPLES OF DIETARY PHYTOCHEMICALS

Carotenoids, polyphenolic compounds, protease inhibitors, and terpenoids are found in almost every fruit and vegetable. Phytochemicals contained in food can help prevent cancer. Dietary phytochemicals are quercetin, capsaicin, epigallocatechin-3-gallate, resveratrol, sulforaphane, gingerol, lycopene, inositol hexaphosphate, curcumin, folic acid, organosulfur compounds, and genistein have a chemopreventive effect on cancer.

Some scientists assume that you can lessen the risk for cancer by consuming extra vegetables and different plant meals that have positive phytochemicals in them. Research has proven that, additionally, a few phytochemicals might assist in forestalling the capacity of most cancer-inflicting substances (carcinogens) and assist cells in forestalling and wiping out most cancer-like adjustments. The most useful phytochemicals are polyphenols. Broccoli, mustard greens, turnips, and cauliflower contain isothiocyanates. Flavonoids are found in masses of grains and vegetables. The flavonoids in pulses may also act similarly to estrogen, reducing the risk of breast cancers. Phytoestrogens are the estrogen-like compounds found in that vegetation. However, most phytoestrogens have very vulnerable estrogen-like activity. However, phytoestrogens are only found in trace amounts in those foods. Ellagic acid, found in pomegranate, has antiproliferative properties and promotes apoptosis in cancer cells [147]. Genistein, an isoflavone found primarily in soybeans, regulates multiple signaling targets in cancer cells. Genistein inhibits anti-apoptotic proteins while increasing pro-apoptotic proteins [148].

Antioxidants can be found in a range of vegetables, leafy vegetables, nuts, fish, seeds, grains, and green tea. On average, dark-colored fruits and vegetables contain more antioxidants than other fruits and vegetables. Carotenoids can be found in various foods, including carrots, yams, squash, and apricots. Anthocyanins have been shown in the lab to have anti-inflammatory and antitumor properties. Garlic and its organosulfur ingredients exert anticancer results with the aid of using improving carcinogen detoxing and immunity, inducing apoptosis, and inhibiting angiogenesis and inflammation [149]. Resveratrol has anticancer, anti-inflammatory, and other biological properties [150]. Tea polyphenols are flavonols known as catechins. Green tea has effectively been used to hinder the development of cancers of the bladder, breast, ovary, cervix, and prostate [151]. Sulforaphane is a nutritional phytochemical acquired from broccoli. It shows cycle arrest and apoptosis in cancer cells [150]. Sulfides, discovered in garlic and onions, might also additionally toughen the immune system. Much of the evidence for the efficacy of phytochemicals come from studying people who consume primarily plant-based diets. These humans seem to have markedly decreased risks for cancer and coronary heart disease.

2.6.2 USE OF MEDICINAL PLANTS IN CHEMOPREVENTION OF CANCER

Tablets derived from plants that have undergone successful scientific trials are well-known for their scientific advancement. Many species have been studied in the treatment of various diseases. There is a great demand for medicinal plants in developing nations, putting excessive pressure on plant populations. Bark, bulbs, and tubers are examples of medicinal plant parts that are harvested [152]. Some of the medicinal plants—*Asparagus racemosa*, *Boswellia serrata*, *Erthyrina suberosa*, *Euphorbia hirta*, *Gynandropis pentaphylla*, *Peaderia foetida*, and *Picrorrhiza kurroa*—has proven anticancer properties [153]. Medicinal foods include cruciferous vegetables and berries. Industrial byproducts can be used to extract anticancer agents from sources that contain these substances.

2.7 REFERENCES

1. WHO (World Health Organization), 2022. Retrieved from www.who.int/news-room/fact-sheets/detail/cancer#:~:text=Cancer%20is%20a%20leading%20cause,and%20rectum%20and%20prostate%20cancers.

2. Singla, R.K., Dubey, A.K., Garg, A., Sharma, R.K., Fiorino, M., Ameen, S.M., Haddad, M.A. and Al-Hiary, M., 2019. Natural polyphenols: Chemical classification, definition of classes, subcategories, and structures. *Journal of AOAC International, 102*(5), pp. 1397–1400.

3. Erdman Jr, J.W., Balentine, D., Arab, L., Beecher, G., Dwyer, J.T., Folts, J., Harnly, J., Hollman, P., Keen, C.L., Mazza, G. and Messina, M., 2007. Flavonoids and heart health: Proceedings of the ILSI North America flavonoids workshop, May 31–June 1, 2005, Washington, DC. *The Journal of Nutrition, 137*(3), pp. 718S–737S.

4. Fukushima, Y., Tashiro, T., Kumagai, A., Ohyanagi, H., Horiuchi, T., Takizawa, K., Sugihara, N., Kishimoto, Y., Taguchi, C., Tani, M. and Kondo, K., 2014. Coffee and beverages are the major contributors to polyphenol consumption from food and beverages in Japanese middle-aged women. *Journal of Nutritional Science, 3*.

5. Scalbert, A. and Williamson, G., 2000. Dietary intake and bioavailability of polyphenols. *The Journal of Nutrition,* 130(8), pp. 2073S-2085S.

6. Bhosale, P.B., Ha, S.E., Vetrivel, P., Kim, H.H., Kim, S.M. and Kim, G.S., 2020. Functions of polyphenols and its anticancer properties in biomedical research: A narrative review. *Translational Cancer Research,* 9(12), p. 7619.

7. Abbas, M., Saeed, F., Anjum, F.M., Afzaal, M., Tufail, T., Bashir, M.S., Ishtiaq, A., Hussain, S. and Suleria, H.A.R., 2017. Natural polyphenols: An overview. *International Journal of Food Properties, 20*(8), pp. 1689–1699.

8. Spatafora, C. and Tringali, C., 2012. Natural-derived polyphenols as potential anticancer agents. *Anti-Cancer Agents in Medicinal Chemistry (Formerly Current Medicinal Chemistry-Anti-Cancer Agents), 12*(8), pp. 902–918.

9. Tsao, R., 2010. Chemistry and biochemistry of dietary polyphenols. *Nutrients, 2*(12), pp. 1231–1246.

10. Khoddami, A., Wilkes, M.A. and Roberts, T.H., 2013. Techniques for analysis of plant phenolic compounds. *Molecules, 18*(2), pp. 2328–2375.

11. Sung, H., Ferlay, J., Siegel, R.L., Laversanne, M., Soerjomataram, I., Jemal, A. and Bray, F., 2021. Global cancer statistics 2020: GLOBOCAN estimates of incidence and mortality worldwide for 36 cancers in 185 countries. *CA: A Cancer Journal for Clinicians, 71*(3), pp. 209–249.

12. American Cancer Society, 2018. *Cancer facts & figures 2018.* Atlanta, GA: American Cancer Society.

13. Wild, C.P., Weiderpass, E. and Stewart, B.W., eds., 2020. *World cancer report: Cancer research for cancer prevention.* International Agency for Research on Cancer. Accessed May 12, 2021.publi cations.iarc.fr/586

14. World Health Organization (WHO). *WHO Reports Progress in the Fight against Tobacco Epidemic.* Accessed October 21, 2021. who.int/news/item/27-07-2021-who-reports-progress-in-the-fight-against-tobacco-epidemic.

15. GBD 2015 Tobacco Collaborators, 2017. Smoking prevalence and attributable disease burden in 195 countries and territories, 1990–2015: A systematic analysis from the Global Burden of Disease Study 2015. *Lancet, 389*, pp. 1885–1906.

16. Asiki, G., Newton, R., Marions, L., Seeley, J., Kamali, A. and Smedman, L., 2016. The impact of maternal factors on mortality rates among children under the age of five years in a rural Ugandan population between 2002 and 2012. *Acta Paediatrica, 105*(2), pp. 191–199.

17. Bruni, L., Saura-Lázaro, A., Montoliu, A., Brotons, M., Alemany, L., Diallo, M.S., Afsar, O.Z., LaMontagne, D.S., Mosina, L., Contreras, M. and Velandia-González, M., 2021. HPV vaccination introduction worldwide and WHO and UNICEF estimates of national HPV immunization coverage 2010–2019. *Preventive Medicine, 144*, p. 106399.

18. Defo, B.K., 2014. Demographic, epidemiological, and health transitions: Are they relevant to population health patterns in Africa? *Global Health Action, 7*(1), p. 22443.

19. Holmboe-Ottesen, G. and Wandel, M., 2012. Changes in dietary habits after migration and consequences for health: A focus on South Asians in Europe. *Food & Nutrition Research, 56*(1), p. 18891. doi:10.3402/fnr.v56i0.18891

20. Shah, S.C., Kayamba, V., Peek Jr, R.M. and Heimburger, D., 2019. Cancer control in low-and middle-income countries: Is it time to consider screening? *Journal of Global Oncology, 5*, pp. 1–8.

21. Mathur, P., Sathishkumar, K., Chaturvedi, M., Das, P., Sudarshan, K.L., Santhappan, S., Nallasamy, V., John, A., Narasimhan, S., Roselind, F.S. and ICMR-NCDIR-NCRP Investigator Group, 2020. Cancer statistics, 2020: Report from national cancer registry programme, India. *JCO Global Oncology*, 6, pp. 1063–1075.

22. Lee, B.L., Liedke, P.E., Barrios, C.H., Simon, S.D., Finkelstein, D.M. and Goss, P.E., 2012. Breast cancer in Brazil: Present status and future goals. *The Lancet Oncology*, *13*(3), pp. e95-e102.

23. Rtveladze, K., Marsh, T., Webber, L., Kilpi, F., Levy, D., Conde, W., McPherson, K. and Brown, M., 2013. Health and economic burden of obesity in Brazil. *PloS One*, *8*(7), p. e68785.

24. Gopal, S. and Sharpless, N.E., 2021. Cancer as a global health priority. *JAMA*, *326*(9), pp. 809–810.

25. DIRAC (International Atomic Energy Agency. Directory of Radiotherapy Centres). Accessed March 30, 2020. dirac.iaea.org/

26. Ward, Z.J., Atun, R., Hricak, H., Asante, K., McGinty, G., Sutton, E.J., Norton, L., Scott, A.M. and Shulman, L.N., 2021. The impact of scaling up access to treatment and imaging modalities on global disparities in breast cancer survival: A simulation-based analysis. *The Lancet Oncology*, *22*(9), pp. 1301–1311.

27. McIntyre, D., 2007, July. *Learning from experience: Health care financing in low-and middle-income countries*. Geneva: Global forum for health research.

28. Ataguba, J.E.O., 2012. Reassessing catastrophic health-care payments with a Nigerian case study. *Health Economics, Policy and Law*, *7*(3), pp. 309–326.

29. Afungchwi, G.M., Hesseling, P.B. and Ladas, E.J., 2017. The role of traditional healers in the diagnosis and management of Burkitt lymphoma in Cameroon: Understanding the challenges and moving forward. *BMC Complementary and Alternative Medicine*, *17*(1), pp. 1–7.

30. WHO, World Report on Aging and Health, 2015. Accesses on March 3, 2021. https://apps.who.int/iris/bitst%20ream/handl%20e/10665/%20186463/97892%2040694%20811_eng.pdf

31. Gallicchio, L., Tonorezos, E., de Moor, J.S., Elena, J., Farrell, M., Green, P., Mitchell, S.A., Mollica, M.A., Perna, F., Saiontz, N.G. and Zhu, L., 2021. Evidence gaps in cancer survivorship care: A report from the 2019 National Cancer Institute Cancer Survivorship Workshop. *JNCI: Journal of the National Cancer Institute*, *113*(9), pp. 1136–1142.

32. Nekhlyudov, L., Mollica, M.A., Jacobsen, P.B., Mayer, D.K., Shulman, L.N. and Geiger, A.M., 2019. Developing a quality of cancer survivorship care framework: Implications for clinical care, research, and policy. *JNCI: Journal of the National Cancer Institute*, *111*(11), pp. 1120–1130.

33. Geerse, O.P., Lakin, J.R., Berendsen, A.J., Alfano, C.M. and Nekhlyudov, L., 2018. Cancer survivorship and palliative care: Shared progress, challenges, and opportunities. *Cancer*, *124*(23), pp. 4435–4441.

34. WHO. Sixty-Seventh World Health Assembly, 2014. *Strengthening of palliative care as a component of comprehensive care throughout the life course*. Accessed October5, 2021. apps.who.int/gb/ebwha/pdf_files/ wha67/ a67_r19-en.pdf

35. International Narcotics Control Board (INCB). *Narcotic drugs—Technical report: Estimated world requirements for 2020-statistics for 2018*. Accessed March 30, 2020. incb.org/incb/en/narcotic-drugs/Techn ical_Repor ts/narco tic_drugs_repor ts.html

36. Kumar, S., 2013. Models of delivering palliative and end-of-life care in India. *Current Opinion in Supportive and Palliative Care*, *7*(2), pp. 216–222.

37. Li, J., Davis, M.P. and Gamier, P., 2011. Palliative medicine: Barriers and developments in mainland China. *Current Oncology Reports*, *13*(4), pp. 290–294.

38. Saikolappan, S., Kumar, B., Shishodia, G., Koul, S. and Koul, H.K., 2019. Reactive oxygen species and cancer: A complex interaction. *Cancer Letters*, *452*, pp. 132–143.

39. Zhang, J., Ahn, K.S., Kim, C., Shanmugam, M.K., Siveen, K.S., Arfuso, F., Samym, R.P., Deivasigamanim, A., Lim, L.H.K., Wang, L. and Goh, B.C., 2016. Nimbolide-induced oxidative stress abrogates STAT3 signaling cascade and inhibits tumor growth in transgenic adenocarcinoma of mouse prostate model. *Antioxidants & Redox Signaling*, *24*(11), pp. 575–589.

40. Morgan, M.J. and Liu, Z.G., 2011. Crosstalk of reactive oxygen species and NF-κB signaling. *Cell Research*, *21*(1), pp. 103–115.

41. Puar, Y.R., Shanmugam, M.K., Fan, L., Arfuso, F., Sethi, G. and Tergaonkar, V., 2018. Evidence for the involvement of the master transcription factor NF-κB in cancer initiation and progression. *Biomedicines*, *6*(3), p. 82.

42. Franco, R., Schoneveld, O., Georgakilas, A.G. and Panayiotidis, M.I., 2008. Oxidative stress, DNA methylation and carcinogenesis. *Cancer Letters*, *266*(1), pp. 6–11.

43. Valko, M., Leibfritz, D., Moncol, J., Cronin, M.T., Mazur, M. and Telser, J., 2007. Free radicals and antioxidants in normal physiological functions and human disease. *The International Journal of Biochemistry & Cell Biology*, 39(1), pp. 44–84.

44. Fan, G.J., Jin, X.L., Qian, Y.P., Wang, Q., Yang, R.T., Dai, F., Tang, J.J., Shang, Y.J., Cheng, L.X., Yang, J. and Zhou, B., 2009. Hydroxycinnamic acids as DNA-cleaving agents in the presence of CuII ions: Mechanism, structure–activity relationship, and biological implications. *Chemistry–A European Journal*, 15(46), pp. 12889–12899.

45. Li, Z., Yang, X., Dong, S. and Li, X., 2012. DNA breakage induced by piceatannol and copper (II): Mechanism and anticancer properties. *Oncology Letters*, 3(5), pp. 1087–1094.

46. Shamim, U., Hanif, S., Albanyan, A., Beck, F.W., Bao, B., Wang, Z., Banerjee, S., Sarkar, F.H., Mohammad, R.M., Hadi, S.M. and Azmi, A.S., 2012. Resveratrol-induced apoptosis is enhanced in low pH environments associated with cancer. *Journal of Cellular Physiology*, 227(4), pp. 1493–1500.

47. Alhosin, M., León-González, A.J., Dandache, I., Lelay, A., Rashid, S.K., Kevers, C., Pincemail, J., Fornecker, L.M., Mauvieux, L., Herbrecht, R. and Schini-Kerth, V.B., 2015. Bilberry extract (Antho 50) selectively induces redox-sensitive caspase 3-related apoptosis in chronic lymphocytic leukemia cells by targeting the Bcl-2/Bad pathway. *Scientific Reports*, 5(1), pp. 1–10.

48. Liang, T., Zhang, X., Xue, W., Zhao, S., Zhang, X. and Pei, J., 2014. Curcumin induced human gastric cancer BGC-823 cells apoptosis by ROS-mediated ASK1-MKK4-JNK stress signaling pathway. *International Journal of Molecular Sciences*, 15(9), pp. 15754–15765.

49. Khan, M.A., Gahlot, S. and Majumdar, S., 2012. Oxidative stress induced by curcumin promotes the death of cutaneous T-cell lymphoma (HuT-78) by disrupting the function of several molecular targets. *Molecular Cancer Therapeutics*, 11(9), pp. 1873–1883.

50. Yun, J.M., Afaq, F., Khan, N. and Mukhtar, H., 2009. Delphinidin, an anthocyanidin in pigmented fruits and vegetables, induces apoptosis and cell cycle arrest in human colon cancer HCT116 cells. *Molecular Carcinogenesis: Published in Cooperation with the University of Texas MD Anderson Cancer Center*, 48(3), pp. 260–270.

51. Lee, H.Z., Lin, C.J., Yang, W.H., Leung, W.C. and Chang, S.P., 2006. Aloe-emodin induced DNA damage through generation of reactive oxygen species in human lung carcinoma cells. *Cancer Letters*, 239(1), pp. 55–63.

52. Ma, J., Yang, J., Wang, C., Zhang, N., Dong, Y., Wang, C., Wang, Y. and Lin, X., 2014. Emodin augments cisplatin cytotoxicity in platinum-resistant ovarian cancer cells via ROS-dependent MRP1 downregulation. *BioMed Research International*, 2014, pp. 1–8. doi:10.1155/2014/107671

53. Lee, D.H., Kim, D.W., Jung, C.H., Lee, Y.J. and Park, D., 2014. Gingerol sensitizes TRAIL-induced apoptotic cell death of glioblastoma cells. *Toxicology and Applied Pharmacology*, 279(3), pp. 253–265.

54. Karthikeyan, C., SH Narayana Moorthy, N., Ramasamy, S., Vanam, U., Manivannan, E., Karunagaran, D. and Trivedi, P., 2015. Advances in chalcones with anticancer activities. *Recent Patents on Anticancer Drug Discovery*, 10(1), pp. 97–115.

55. Xuan, Y.U.A.N., Zhang, B., Lu, G.A.N., Wang, Z.H., Yu, B.C., Liu, L.L., Zheng, Q.S. and Wang, Z.P., 2013. Involvement of the mitochondrion-dependent and the endoplasmic reticulum stress-signaling pathways in isoliquiritigenin-induced apoptosis of HeLa cell. *Biomedical and Environmental Sciences*, 26(4), pp. 268–276.

56. Wang, J., Lu, M.L., Dai, H.L., Zhang, S.P., Wang, H.X. and Wei, N., 2014. Esculetin, a coumarin derivative, exerts in vitro and in vivo antiproliferative activity against hepatocellular carcinoma by initiating a mitochondrial-dependent apoptosis pathway. *Brazilian Journal of Medical and Biological Research*, 48, pp. 245–253.

57. Yang, J., Xiao, Y.L., He, X.R., Qiu, G.F. and Hu, X.M., 2010. Aesculetin-induced apoptosis through a ROS-mediated mitochondrial dysfunction pathway in human cervical cancer cells. *Journal of Asian Natural Products Research*, 12(3), pp. 185–193.

58. Lambert, J.D. and Elias, R.J., 2010. The antioxidant and pro-oxidant activities of green tea polyphenols: A role in cancer prevention. *Archives of Biochemistry and Biophysics*, 501(1), pp. 65–72.

59. Oikawa, S., Furukawa, A., Asada, H., Hirakawa, K. and Kawanishi, S., 2003. Catechins induce oxidative damage to cellular and isolated DNA through the generation of reactive oxygen species. *Free Radical Research*, 37(8), pp. 881–890.

60. Palit, S., Kar, S., Sharma, G. and Das, P.K., 2015. Hesperetin induces apoptosis in breast carcinoma by triggering accumulation of ROS and activation of ASK1/JNK pathway. *Journal of Cellular Physiology*, 230(8), pp. 1729–1739.

61. Zhang, S., Lai, N., Liao, K., Sun, J. and Lin, Y., 2015. MicroRNA-210 regulates cell proliferation and apoptosis by targeting regulator of differentiation 1 in glioblastoma cells. *Folia Neuropathologica*, *53*(3), pp. 236–244.

62. Iwasaki, M., Inoue, M., Otani, T., Sasazuki, S., Kurahashi, N., Miura, T., Yamamoto, S. and Tsugane, S., 2008. Plasma isoflavone level and subsequent risk of breast cancer among Japanese women: A nested case-control study from the Japan Public Health Center-based prospective study group. *Journal of Clinical Oncology*, *26*(10), pp. 1677–1683.

63. Jin, S., Zhang, Q.Y., Kang, X.M., Wang, J.X. and Zhao, W.H., 2010. Daidzein induces MCF-7 breast cancer cell apoptosis via the mitochondrial pathway. *Annals of Oncology*, *21*(2), pp. 263–268.

64. Kim, D.S., Jeon, B.K., Lee, Y.E., Woo, W.H. and Mun, Y.J., 2012. Diosgenin induces apoptosis in HepG2 cells through generation of reactive oxygen species and mitochondrial pathway. *Evidence-Based Complementary and Alternative Medicine*, *2012*, pp. 1–8. doi:10.1155/2012/981675

65. Ducasse, M. and Brown, M.A., 2006. Epigenetic aberrations and cancer. *Molecular Cancer*, *5*(1), pp. 1–10.

66. Chen, H.Y., Yang, Y.M., Stevens, B.M. and Noble, M., 2013. Inhibition of redox/Fyn/c-Cbl pathway function by Cdc42 controls tumour initiation capacity and tamoxifen sensitivity in basal-like breast cancer cells. *EMBO Molecular Medicine*, *5*(5), pp. 723–736.

67. Bunea, A., Rugină, D., Sconţa, Z., Pop, R.M., Pintea, A., Socaciu, C., Tăbăran, F., Grootaert, C., Struijs, K. and VanCamp, J., 2013. Anthocyanin determination in blueberry extracts from various cultivars and their antiproliferative and apoptotic properties in B16-F10 metastatic murine melanoma cells. *Phytochemistry*, *95*, pp. 436–444.

68. Yang, C.S., Wang, X., Lu, G. and Picinich, S.C., 2009. Cancer prevention by tea: Animal studies, molecular mechanisms, and human relevance. *Nature Reviews Cancer*, *9*(6), pp. 429–439.

69. Khan, N. and Mukhtar, H., 2010. Cancer and metastasis: Prevention and treatment by green tea. *Cancer and Metastasis Reviews*, *29*(3), pp. 435–445.

70. Menéndez Menendez, J.A., Quirantes Piné, R., Rodríguez Gallego, E., Cufí González, S., Corominas Faja, B., Cuyàs, E., Bosch Barrera, J., Martin Castillo, B., Segura Carretero, A. and Joven, J., 2014. Oncobiguanides: Paracelsus' law and nonconventional routes for administering diabetobiguanides for cancer treatment. *Oncotarget*, *5*(9), pp. 2344–2348.

71. Shibamoto, T. and Bjeldanes, L.F., 2009. *Introduction to food toxicology*. Amsterdam, the Netherlands: Elsevier Science.

72. Murphy, L.C., Seekallu, S.V. and Watson, P.H., 2011. Clinical significance of estrogen receptor phosphorylation. *Endocrine-related Cancer*, *18*(1), p. R1.

73. Gu, Y., Chen, T., López, E., Wu, W., Wang, X., Cao, J. and Teng, L., 2014. The therapeutic target of estrogen receptor-alpha36 in estrogen-dependent tumors. *Journal of Translational Medicine*, *12*(1), pp. 1–12.

74. Li, G.X., Chen, Y.K., Hou, Z., Xiao, H., Jin, H., Lu, G., Lee, M.J., Liu, B., Guan, F., Yang, Z. and Yu, A., 2010. Pro-oxidative activities and dose–response relationship of (–)-epigallocatechin-3-gallate in the inhibition of lung cancer cell growth: A comparative study in vivo and in vitro. *Carcinogenesis*, *31*(5), pp. 902–910.

75. Sharma, D., Blum, J., Yang, X., Beaulieu, N., Macleod, A.R. and Davidson, N.E., 2005. Release of methyl CpG binding proteins and histone deacetylase 1 from the estrogen receptor alpha (ER) promoter upon reactivation in ER-negative human breast cancer cells. *Molecular Endocrinology*, *19*(7), pp. 1740–1751.

76. Macaluso, M., Montanari, M., Noto, P.B., Gregorio, V., Bronner, C. and Giordano, A., 2007. Epigenetic modulation of estrogen receptor-α by pRb family proteins: A novel mechanism in breast cancer. *Cancer Research*, *67*(16), pp. 7731–7737.

77. Oh, A.S., Lorant, L.A., Holloway, J.N., Miller, D.L., Kern, F.G. and El-Ashry, D., 2001. Hyperactivation of MAPK induces loss of ERα expression in breast cancer cells. *Molecular Endocrinology*, *15*(8), pp. 1344–1359.

78. Aqil, F., Munagala, R., Jeyabalan, J. and Vadhanam, M.V., 2013. Bioavailability of phytochemicals and its enhancement by drug delivery systems. *Cancer Letters*, *334*(1), pp. 133–141.

79. Sarkar, H. F., Li, Y., Wang, Z. and Padhye, S., 2010. Lesson learned from nature for the development of novel anti-cancer agents: Implication of isoflavone, curcumin, and their synthetic analogs. *Current Pharmaceutical Design*, *16*(16), pp. 1801–1812.

80. Wang, L. and Chen, C., 2013. Emerging applications of metabolomics in studying chemopreventive phytochemicals. *The AAPS Journal*, *15*(4), pp. 941–950.

81. Wang, Z., Li, W., Meng, X. and Jia, B., 2012. Resveratrol induces gastric cancer cell apoptosis via reactive oxygen species, but independent of sirtuin1. *Clinical and Experimental Pharmacology and Physiology*, *39*(3), pp. 227–232.

82. Pandey, K.B. and Rizvi, S.I., 2009. Plant polyphenols as dietary antioxidants in human health and disease. *Oxidative Medicine and Cellular Longevity*, *2*(5), pp. 270–278.

83. Srivastava, S.K., Arora, S., Averett, C., Singh, S. and Singh, A.P., 2015. Modulation of microRNAs by phytochemicals in cancer: Underlying mechanisms and translational significance. *BioMed Research International*, *2015*, pp. 1–9. doi:10.1155/2015/848710

84. Thakur, V.S., Deb, G., Babcook, M.A. and Gupta, S., 2014. Plant phytochemicals as epigenetic modulators: Role in cancer chemoprevention. *The AAPS Journal*, *16*(1), pp. 151–163.

85. Masika, J., Zhao, Y., Hescheler, J. and Liang, H., 2016. Modulation of miRNAs by Natural Agents: Nature's way of dealing with cancer. *RNA & DISEASE*, *3*.

86. Shankar, S., Kumar, D. and Srivastava, R.K., 2013. Epigenetic modifications by dietary phytochemicals: Implications for personalized nutrition. *Pharmacology & Therapeutics*, *138*(1), pp. 1–17.

87. Naselli, F., Belshaw, N.J., Gentile, C., Tutone, M., Tesoriere, L., Livrea, M.A. and Caradonna, F., 2015. Phytochemical indicaxanthin inhibits colon cancer cell growth and affects the DNA methylation status by influencing epigenetically modifying enzyme expression and activity. *Lifestyle Genomics*, *8*(3), pp. 114–127.

88. Petric, R.C., Braicu, C., Raduly, L., Zanoaga, O., Dragos, N., Monroig, P., Dumitrascu, D. and Berindan-Neagoe, I., 2015. Phytochemicals modulate carcinogenic signaling pathways in breast and hormone-related cancers. *OncoTargets and Therapy*, *8*, p. 2053.

89. Barh, D., Malhotra, R., Ravi, B. and Sindhurani, P., 2010. MicroRNA let-7: An emerging next-generation cancer therapeutic. *Current Oncology*, *17*(1), pp. 70–80.

90. Takamizawa, J., Konishi, H., Yanagisawa, K., Tomida, S., Osada, H., Endoh, H., Harano, T., Yatabe, Y., Nagino, M., Nimura, Y. and Mitsudomi, T., 2004. Reduced expression of the let-7 microRNAs in human lung cancers in association with shortened postoperative survival. *Cancer Research*, *64*(11), pp. 3753–3756.

91. Brueckner, B., Stresemann, C., Kuner, R., Mund, C., Musch, T., Meister, M., Sültmann, H. and Lyko, F., 2007. The human let-7a-3 locus contains an epigenetically regulated microRNA gene with oncogenic function. *Cancer Research*, *67*(4), pp. 1419–1423.

92. Xia, X.M., Jin, W.Y., Shi, R.Z., Zhang, Y.F. and Chen, J., 2010. Clinical significance and the correlation of expression between Let-7 and K-ras in non-small cell lung cancer. *Oncology Letters*, *1*(6), pp. 1045–1048.

93. Zhao, Y., Deng, C., Wang, J., Xiao, J., Gatalica, Z., Recker, R.R. and Xiao, G.G., 2011. Let-7 family miRNAs regulate estrogen receptor alpha signaling in estrogen receptor positive breast cancer. *Breast Cancer Research and Treatment*, *127*(1), pp. 69–80.

94. Cimmino, A., Calin, G.A., Fabbri, M., Iorio, M.V., Ferracin, M., Shimizu, M., Wojcik, S.E., Aqeilan, R.I., Zupo, S., Dono, M. and Rassenti, L., 2005. miR-15 and miR-16 induce apoptosis by targeting BCL2. *Proceedings of the National Academy of Sciences*, *102*(39), pp. 13944–13949.

95. Aqeilan, R.I., Calin, G.A. and Croce, C.M., 2010. miR-15a and miR-16-1 in cancer: Discovery, function and future perspectives. *Cell Death & Differentiation*, *17*(2), pp. 215–220.

96. Chang, T.C., Wentzel, E.A., Kent, O.A., Ramachandran, K., Mullendore, M., Lee, K.H., Feldmann, G., Yamakuchi, M., Ferlito, M., Lowenstein, C.J. and Arking, D.E., 2007. Transactivation of miR-34a by p53 broadly influences gene expression and promotes apoptosis. *Molecular Cell*, *26*(5), pp. 745–752.

97. Xu, D., Wang, Q., An, Y. and Xu, L., 2013. miR-203 regulates the proliferation, apoptosis and cell cycle progression of pancreatic cancer cells by targeting Survivin. *Molecular Medicine Reports*, *8*(2), pp. 379–384.

98. Sonkoly, E., Lovén, J., Xu, N., Meisgen, F., Wei, T., Brodin, P., Jaks, V., Kasper, M., Shimokawa, T., Harada, M. and Heilborn, J., 2012. MicroRNA-203 functions as a tumor suppressor in basal cell carcinoma. *Oncogenesis*, *1*(3), pp.e3-e3.

99. Tang, R., Liang, L., Luo, D., Feng, Z., Huang, Q., He, R., Gan, T., Yang, L. and Chen, G., 2015. Downregulation of MiR-30a is associated with poor prognosis in lung cancer. *Medical Science Monitor: International Medical Journal of Experimental and Clinical Research*, *21*, p. 2514.

100. Zhang, S., Lai, N., Liao, K., Sun, J. and Lin, Y., 2015. MicroRNA-210 regulates cell proliferation and apoptosis by targeting regulator of differentiation 1 in glioblastoma cells. *Folia Neuropathologica*, *53*(3), pp. 236–244.

101. Thomas, R. and Kim, M.H., 2005. Epigallocatechin gallate inhibits HIF-1α degradation in prostate cancer cells. *Biochemical and Biophysical Research Communications*, *334*(2), pp. 543–548.

102. Arola-Arnal, A. and Blade, C., 2011. Proanthocyanidins modulate microRNA expression in human HepG2 cells. *PLoS One*, *6*(10), p.e25982.

103. Kumazaki, M., Noguchi, S., Yasui, Y., Iwasaki, J., Shinohara, H., Yamada, N. and Akao, Y., 2013. Anti-cancer effects of naturally occurring compounds through modulation of signal transduction and miRNA expression in human colon cancer cells. *The Journal of Nutritional Biochemistry*, *24*(11), pp. 1849–1858.

104. Chakrabarti, M., Khandkar, M., Banik, N.L. and Ray, S.K., 2012. Alterations in expression of specific microRNAs by combination of 4-HPR and EGCG inhibited growth of human malignant neuroblastoma cells. *Brain Research*, *1454*, pp. 1–13.

105. Farooqi, A.A., Gadaleta, C.D., Ranieri, G., Fayyaz, S. and Marech, I., 2016. New frontiers in promoting trail-mediated cell death: Focus on natural sensitizers, mirnas, and nanotechnological advancements. *Cell Biochemistry and Biophysics*, *74*(1), pp. 3–10.

106. Tsang, W.P. and Kwok, T.T., 2010. Epigallocatechin gallate up-regulation of miR-16 and induction of apoptosis in human cancer cells. *The Journal of Nutritional Biochemistry*, *21*(2), pp. 140–146.

107. Sharma, A., Kaur, M., Katnoria, J.K. and Nagpal, A.K., 2018. Polyphenols in food: Cancer prevention and apoptosis induction. *Current Medicinal Chemistry*, *25*(36), pp. 4740–4757.

108. Montenegro-Landivar, M.F., Tapia-Quiros, P., Vecino, X., Reig, M., Valderrama, C., Granados, M., Cortina, J.L. and Saurina, J., 2021. Polyphenols and their potential role to fight viral diseases: An overview. *Science of the Total Environment*, *801*, p. 149719.

109. Li, F., Li, S., Li, H.B., Deng, G.F., Ling, W.H. and Xu, X.R., 2013. Antiproliferative activities of tea and herbal infusions. *Food & Function*, *4*(4), pp. 530–538.

110. Law, B.Y.K., Mok, S.W.F., Wu, A.G., Lam, C.W.K., Yu, M.X.Y. and Wong, V.K.W., 2016. New potential pharmacological functions of Chinese herbal medicines via regulation of autophagy. *Molecules*, *21*(3), p. 359.

111. Liu, W., Xu, J., Wu, S., Liu, Y., Yu, X., Chen, J., Tang, X., Wang, Z., Zhu, X. and Li, X., 2013. Selective anti-proliferation of HER2-positive breast cancer cells by anthocyanins identified by high-throughput screening. *PLoS One*, *8*(12), p. e81586.

112. Vene, R., Benelli, R., Minghelli, S., Astigiano, S., Tosetti, F. and Ferrari, N., 2012. Xanthohumol impairs human prostate cancer cell growth and invasion and diminishes the incidence and progression of advanced tumors in TRAMP mice. *Molecular Medicine*, *18*(9), pp. 1292–1302.

113. Onoda, C., Kuribayashi, K., Nirasawa, S., Tsuji, N., Tanaka, M., Kobayashi, D. and Watanabe, N., 2011. (-)-Epigallocatechin-3-gallate induces apoptosis in gastric cancer cell lines by down-regulating survivin expression. *International Journal of Oncology*, 38(5), pp. 1403–1408.

114. Tanaka, T., Ishii, T., Mizuno, D., Mori, T., Yamaji, R., Nakamura, Y., Kumazawa, S., Nakayama, T. and Akagawa, M., 2011. (–)-Epigallocatechin-3-gallate suppresses growth of AZ521 human gastric cancer cells by targeting the DEAD-box RNA helicase p68. *Free Radical Biology and Medicine*, *50*(10), pp. 1324–1335.

115. Mocanu, M.M., Ganea, C., Georgescu, L., Varadi, T., Shrestha, D., Baran, I., Katona, E., Nagy, P. and Szöllősi, J., 2014. Epigallocatechin 3-O-gallate induces 67 kDa laminin receptor-mediated cell death accompanied by downregulation of ErbB proteins and altered lipid raft clustering in mammary and epidermoid carcinoma cells. *Journal of Natural Products*, *77*(2), pp. 250–257.

116. Siddiqui, I.A., Asim, M., Hafeez, B.B., Adhami, V.M., Tarapore, R.S. and Mukhtar, H., 2011. Green tea polyphenol EGCG blunts androgen receptor function in prostate cancer. *The FASEB Journal*, *25*(4), pp. 1198–1207.

117. Alshatwi, A.A., Ramesh, E., Periasamy, V.S. and Subash-Babu, P., 2013. The apoptotic effect of hesperetin on human cervical cancer cells is mediated through cell cycle arrest, death receptor, and mitochondrial pathways. *Fundamental & Clinical Pharmacology*, *27*(6), pp. 581–592.

118. Lu, J., Li, G., He, K., Jiang, W., Xu, C., Li, Z., Wang, H., Wang, W., Wang, H., Teng, X. and Teng, L., 2015. Luteolin exerts a marked antitumor effect in cMet-overexpressing patient-derived tumor xenograft models of gastric cancer. *Journal of Translational Medicine*, *13*(1), pp. 1–11.

119. Zheng, S.Y., Li, Y., Jiang, D., Zhao, J. and Ge, J.F., 2012. Anticancer effect and apoptosis induction by quercetin in the human lung cancer cell line A-549. *Molecular Medicine Reports*, *5*(3), pp. 822–826.

120. Tian, T., Li, J., Li, B., Wang, Y., Li, M., Ma, D. and Wang, X., 2014. Genistein exhibits anti-cancer effects via down-regulating FoxM1 in H446 small-cell lung cancer cells. *Tumor Biology*, *35*(5), pp. 4137–4145.

121. Srigopalram, S., Jayraaj, I.A., Kaleeswaran, B., Balamurugan, K., Ranjithkumar, M., Kumar, T.S., Park, J.I. and Nou, I.S., 2014. Ellagic acid normalizes mitochondrial outer membrane permeabilization and attenuates inflammation-mediated cell proliferation in experimental liver cancer. *Applied Biochemistry and Biotechnology*, *173*(8), pp. 2254–2266.

122. Yousef, A.I., El-Masry, O.S. and Abdel Mohsen, M.A., 2016. Impact of cellular genetic make-up on colorectal cancer cell lines response to ellagic acid: Implications of small interfering RNA. *Asian Pacific Journal of Cancer Prevention*, 17(2), pp. 743–748.

123. Eroglu, C., Seçme, M., Bagcı, G. and Dodurga, Y., 2015. Assessment of the anticancer mechanism of ferulic acid via cell cycle and apoptotic pathways in human prostate cancer cell lines. *Tumor Biology*, 36(12), pp. 9437–9446.

124. Zhou, F., Wang, T., Zhang, B. and Zhao, H., 2018. Addition of sucrose during the blueberry heating process is good or bad? Evaluating the changes of anthocyanins/anthocyanidins and the anticancer ability in HepG-2 cells. *Food Research International*, 107, pp. 509–517.

125. Mineva, N.D., Paulson, K.E., Naber, S.P., Yee, A.S. and Sonenshein, G.E., 2013. Epigallocatechin-3-gallate inhibits stem-like inflammatory breast cancer cells. *PloS One*, 8(9), p. e73464

126. Wang, K., Liu, R., Li, J., Mao, J., Lei, Y., Wu, J., Zeng, J., Zhang, T., Wu, H., Chen, L. and Huang, C., 2011. Quercetin induces protective autophagy in gastric cancer cells: Involvement of Akt-mTOR-and hypoxia-induced factor 1α-mediated signaling. *Autophagy*, 7(9), pp. 966–978.

127. Song, H., Bao, J., Wei, Y., Chen, Y., Mao, X., Li, J., Yang, Z. and Xue, Y., 2015. Kaempferol inhibits gastric cancer tumor growth: An in vitro and in vivo study. *Oncology Reports*, 33(2), pp. 868–874.

128. Qin, J., Teng, J., Zhu, Z., Chen, J. and Huang, W.J., 2016. Genistein induces activation of the mitochondrial apoptosis pathway by inhibiting phosphorylation of Akt in colorectal cancer cells. *Pharmaceutical Biology*, 54(1), pp. 74–79.

129. Dai, W., Gao, Q., Qiu, J., Yuan, J., Wu, G. and Shen, G., 2016. Quercetin induces apoptosis and enhances 5-FU therapeutic efficacy in hepatocellular carcinoma. *Tumor Biology*, 37(5), pp. 6307–6313.

130. Huang, W.W., Tsai, S.C., Peng, S.F., Lin, M.W., Chiang, J.H., Chiu, Y.J., Fushiya, S., Tseng, M.T. and Yang, J.S., 2013. Kaempferol induces autophagy through AMPK and AKT signaling molecules and causes G2/M arrest via downregulation of CDK1/cyclin B in SK-HEP-1 human hepatic cancer cells. *International Journal of Oncology*, 42(6), pp. 2069–2077.

131. Rigalli, J.P., Tocchetti, G.N., Arana, M.R., Villanueva, S.S.M., Catania, V.A., Theile, D., Ruiz, M.L. and Weiss, J., 2016. The phytoestrogen genistein enhances multidrug resistance in breast cancer cell lines by translational regulation of ABC transporters. *Cancer Letters*, 376(1), pp. 165–172.

132. Bin Hafeez, B., Asim, M., Siddiqui, I.A., Adhami, V.M., Murtaza, I. and Mukhtar, H., 2008. Delphinidin, a dietary anthocyanidin in pigmented fruits and vegetables: A new weapon to blunt prostate cancer growth. *Cell Cycle*, 7(21), pp. 3320–3326.

133. Bishayee, K., Ghosh, S., Mukherjee, A., Sadhukhan, R., Mondal, J. and Khuda-Bukhsh, A.R., 2013. Quercetin induces cytochrome-c release and ROS accumulation to promote apoptosis and arrest the cell cycle in G2/M, in cervical carcinoma: Signal cascade and drug-DNA interaction. *Cell Proliferation*, 46(2), pp. 153–163.

134. Ho, H.H., Chang, C.S., Ho, W.C., Liao, S.Y., Lin, W.L. and Wang, C.J., 2013. Gallic acid inhibits gastric cancer cells metastasis and invasive growth via increased expression of RhoB, downregulation of AKT/small GTPase signals and inhibition of NF-κB activity. *Toxicology and Applied Pharmacology*, 266(1), pp. 76–85.

135. Han, X., Zhao, N., Zhu, W., Wang, J., Liu, B. and Teng, Y., 2021. Resveratrol attenuates TNBC lung metastasis by down-regulating PD-1 expression on pulmonary t cells and converting macrophages to M1 phenotype in a murine tumor model. *Cellular Immunology*, 368, p. 104423.

136. Yang, Q., Wang, B., Zang, W., Wang, X., Liu, Z., Li, W. and Jia, J., 2013. Resveratrol inhibits the growth of gastric cancer by inducing G1 phase arrest and senescence in a Sirt1-dependent manner. *PloS One*, 8(11), p. e70627.

137. Miki, H., Uehara, N., Kimura, A., Sasaki, T., Yuri, T., Yoshizawa, K. and Tsubura, A., 2012. Resveratrol induces apoptosis via ROS-triggered autophagy in human colon cancer cells. *International Journal of Oncology*, 40(4), pp. 1020–1028.

138. Deng, P., Wang, C., Chen, L., Wang, C., Du, Y., Yan, X., Chen, M., Yang, G. and He, G., 2013. Sesamin induces cell cycle arrest and apoptosis through the inhibition of signal transducer and activator of transcription 3 signalling in human hepatocellular carcinoma cell line HepG2. *Biological and Pharmaceutical Bulletin*, 36(10), pp. 1540–1548.

139. Xu, P., Cai, F., Liu, X. and Guo, L., 2015. Sesamin inhibits lipopolysaccharide-induced proliferation and invasion through the p38-MAPK and NF-κB signaling pathways in prostate cancer cells. *Oncology Reports*, 33(6), pp. 3117–3123.

140. Steinmetz, K.A. and Potter, J.D., 1996. Vegetables, fruit, and cancer prevention: A review. *Journal of the American Dietetic Association*, 96(10), p. 1027–1039.

141. Jaganathan, S.K., Vellayappan, M.V., Narasimhan, G., Supriyanto, E., Dewi, D.E.O., Narayanan, A.L.T., Balaji, A., Subramanian, A.P. and Yusof, M., 2014. Chemopreventive effect of apple and berry fruits against colon cancer. *World Journal of Gastroenterology: WJG*, *20*(45), p. 17029.
142. Li, Y.H., Niu, Y.B., Sun, Y., Zhang, F., Liu, C.X., Fan, L. and Mei, Q.B., 2015. Role of phytochemicals in colorectal cancer prevention. *World Journal of Gastroenterology: WJG*, *21*(31), p. 9262.
143. Pericleous, M., Mandair, D. and Caplin, M.E., 2013. Diet and supplements and their impact on colorectal cancer. *Journal of Gastrointestinal Oncology*, *4*(4), p. 409.
144. Steward, W.P. and Brown, K., 2013. Cancer chemoprevention: A rapidly evolving field. *British Journal of Cancer*, *109*(1), pp. 1–7.
145. Wattenberg, L.W., 1996. Chemoprevention of cancer. *Preventive Medicine*. 1, p. 44.
146. Kotecha, R., Takami, A. and Espinoza, J.L., 2016. Dietary phytochemicals and cancer chemoprevention: A review of the clinical evidence. *Oncotarget*, *7*(32), p. 52517.
147. Bell, C. and Hawthorne, S., 2008. Ellagic acid, pomegranate and prostate cancer—a mini review. *Journal of Pharmacy and Pharmacology*, *60*(2), pp. 139–144.
148. Banerjee, S., Li, Y., Wang, Z. and Sarkar, F.H., 2008. Multi-targeted therapy of cancer by genistein. *Cancer Letters*, *269*(2), pp. 226–242.
149. Nagini, S., 2008. Cancer chemoprevention by garlic and its organosulfur compounds-panacea or promise? *Anti-Cancer Agents in Medicinal Chemistry (Formerly Current Medicinal Chemistry-Anti-Cancer Agents)*, *8*(3), pp. 313–321.
150. Aly, M. S. and Mahmoud, A. A. E. (2013). *Cancer chemoprevention by dietary polyphenols, carcinogenesis*. Dr. Kathryn Tonissen (Ed.) IntechOpen.
151. Mandel, S., Weinreb, O., Amit, T. and Youdim, M.B., 2004. Cell signaling pathways in the neuroprotective actions of the green tea polyphenol (-) -epigallocatechin-3-gallate: Implications for neurodegenerative diseases. *Journal of Neurochemistry*, *88*(6), pp. 1555–1569.
152. Parveen, S., Jan, U. and Kamili, A., 2013. Importance of Himalayan medicinal plants and their conservation strategies. *Australian Journal of Herbal Medicine*, *25*(2), pp. 63–67.
153. Dhar, M.L., Dhar, M.M., Dhawan, B.N., Mehrotra, B.N. and Ray, C., 1968. Screening of Indian plants for biological activity: Part I. *Indian Journal of Experimental Biology*, *6*(4) pp. 232–247.

3 Fat-Soluble Vitamins in Cancer Chemoprevention

*Neetika Naudiyal, Gaurav Singh Rana, Ankur Adhikari,
and Shiv Kumar Dubey*

CONTENTS

3.1   Introduction 39
3.2   Current Scenario of Cancer 40
3.3   Mechanism of Vitamins in Cancer Prevention..40
3.4   Vitamins Exploited for Cancer Prevention 43
    3.4.1   Vitamin A 43
        3.4.1.1   Vitamin A and Cancer 44
    3.4.2   Vitamin D 45
        3.4.2.1   Tumor Progression and Vitamin D 45
        3.4.2.2   Effect of Vitamin D on TGF- β 46
        3.4.2.3   Vitamin D in Regulating DNA Repair Proteins 46
    3.4.3   Vitamin E—Tocotrienols (TTs) and Tocopherols (TOCs) 46
        3.4.3.1   Cell Cycle Arrest Caused by Tocotrienols...................................46
        3.4.3.2   Induction of Autophagy and Apoptosis by TTs 47
        3.4.3.3   Inhibition of Invasion, Metastasis, and Angiogenesis by Tocotrienols 47
        3.4.3.4   The Effect of Tocotrienols on Cancer Stem Cells 47
    3.4.4   Vitamin K 48
3.5   Conclusion 49
3.6   References..49

3.1   INTRODUCTION

Cancer is a multistage process designated by an accumulation of tumor suppressor gene inactivation, molecular aberration oncogene activation, and epigenetic plasticity that disrupt intracellular signaling pathways and promote cancer initiation and progression (**1**). Due to paramount levels of reactive oxygen species (ROS), metabolic reprogramming, and oncogenic transformation, cancer cells are subjected to high levels of oxidative stress (**2**). Oxidative stress arises when there is an imbalance between the generation of free radicals (O^{2-}, NO, H_2O_2, OH^-, and others), and antioxidant defense mechanisms (glutathione peroxidase, catalase, superoxide dismutase, and others) eliminate them, resulting in lipid-peroxidation-induced cell damage, which causes a change in cell membrane integrity of the cell membrane, loss of function, and derangement, along with DNA damage, which promotes cell proliferation and genomic instability, thereby expanding the neoplastic transformation and somatic mutations (**3**). Chemoprevention is the employment of a variety of natural, biological, or synthetic substances to prevent cancer from starting or progressing (**4**). Chemoprevention is the introduction of particular substances to inhibit tumor growth or reverse carcinogenesis. Hormones, medicines, vitamins, minerals, and vaccines are examples of chemopreventive agents (**4**). Epidemiological studies have found a clear linkage associating the ingestion of vegetables and fruits with the chances of not getting cancer (**5**). According to World Health Organization (WHO), consuming fruits and vegetables can help prevent cancer because they include elements such as vitamins, minerals, and fiber (**6**).

DOI: 10.1201/b23311-3

Enzyme detoxification, antioxidant activity, change of hormone metabolism, prevention of nitrosa-mine development, and the potential to control carcinogenic cellular processes are all examples of phytochemical ingredients that have complementary or overlapping modes of action (7).

As per INCA (Instituto Nacional de Câncer José Alencar Gomes da Silva, 2011), antioxidant-containing edibles are high in vitamins A, C, and E, selenium, and flavonoids and are recommended due to their antagonistic action, as they can aid in impeding cancer by inhibiting free radical pro-duction, as well as eliminating oxidative stress and even hindering carcinogenesis (8). Vitamin sup-plementation can support cancer patients to gain more benefits and promote wellness. This chapter examines the cancer chemoprevention and therapy by vitamins.

3.2   CURRENT SCENARIO OF CANCER

Cancer is a severe health issue that is regarded as one of the leading causes of death around the world (9). Furthermore, among cancers, breast cancer is the most prevalent and the second leading cause of cancer death in women globally (10, 11). In 2018, cancer was the foremost cause of illness worldwide among all diseases, with 18.1 million new cases and 9.6 million deaths (12). It was evalu-ated by the NIH (National Institutes of Health), USA, that in 2020, there were 1.8 million cancer patients, with 606,520 deaths in the USA (13). According to the Brazilian National Cancer Institute, there were 14.1 million cancer cases worldwide in 2012, with 8.2 million deaths due to the disease. The international cancer burden is predicted to rise by 21.4 million new cases and 13.2 million deaths in 2030, owing to population expansion and aging (14). Over 10 million deaths worldwide, or nearly one in every six people, was attributed to cancer in 2020, making it the leading cause of death. The most prominent form of cancer is lung (1.80 million deaths), rectal and colon (916,000 deaths), breast (685,000 deaths), liver (830,000 deaths), and prostate cancers (1.41 million cases). There are a number of factors that can cause cancer, but the ones that cause about one-third of can-cer fatalities are high levels of tobacco consumption, a high body mass index, alcohol consumption, a diet low in fruits and vegetables, and a lack of physical activity. If detected early and given the right care, some cancers are curable.

3.3   MECHANISM OF VITAMINS IN CANCER PREVENTION

Scientists have shown that vitamin D has a wide range of biological consequences besides its usual participation in phosphate and calcium homeostasis (15). The discovery of vitamin D receptor (VDR) in practically all tissues of the body and the cloning of the VDR in 1987 ignited widespread interest in its physiological activities (16). Various studies reported that vitamin-D-regulated cell proliferation, differentiation, and growth. VDR can be found in a variety of body tissues, including pancreatic cells, bone, parathyroid gland, kidneys, skin, breast, liver, prostate, brain, heart, testes, intestines, lungs, adipose cells, skeletal muscle tissues, and immune cells like activated T- and B-cells, dendritic cells, and macrophages (17, 18). The VDR alters the DNA in these tissues by attaching to regulatory vitamin D response elements (VDREs) found in the promoter regions of target genes. The most common VDRE motif comprises two half-sites divided by three additional nucleotides from any sequence, each with the six-nucleotide consensus sequence GGTCCA (19). Vitamin D can also work through non-genomic ways that do not require increased protein syn-thesis or transcriptional effects (Figure 3.1). Vitamin D may impede cell development by repress-ing several vital molecules engaged in cell cycle control (Figure 3.2). Vitamin D represses c-Myc and c-Fos (a proto-oncogene) of MCF-7 breast cancer cell line or colorectal tumors and promotes MAD1/MXD1 (mitotic arrest deficiency 1/max dimerization protein 1; transcriptional repressor), an antagonist, but its analog, MART-10 (19Modified-nor-2α-(3-hydroxypropyl)-1α,25-dihydroxy vitamin D3), to arrest cell cycle at the G_0/G_1 transition step (20, 21). Vitamin D inhibits the activity of Skp1-cullin-F-box protein/Skp2 (S-phase kinase-associated protein-2) ubiquitin ligase and cyclin E/cyclin-dependent kinase 2 to stop the G1 and G0/G1 phase of the cell cycle in ovarian cancer cells

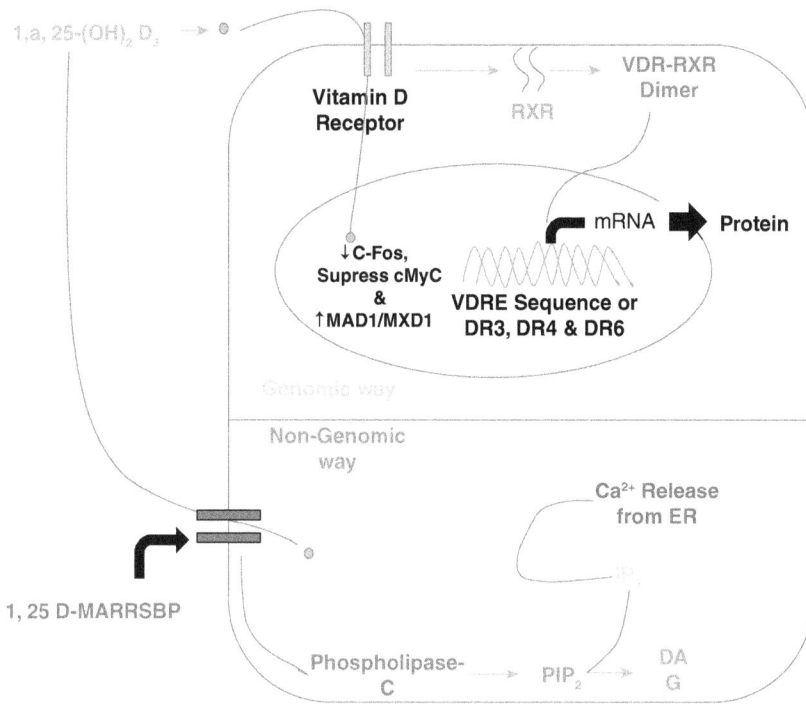

FIGURE 3.1 Molecular mechanism of vitamin D.

DR = direct repeat; 1, 25 D-MARRSBP = 1,25-$(OH)_2$D membrane-associated rapid-response steroid-binding protein, or known as ER stress protein 57; RXR = retinoic X receptor; PIP_2 = phosphatidylinositol 4,5-bisphosphate; DAG = diacylglycerol; IP_3 = inositol trisphosphate; ER = endoplasmic reticulum.

and squamous cell carcinoma (SCC25) cells (22, 23). According to one study, vitamin D and its analogues are therapeutically beneficial in the treatment of numerous cancer cells, including those that cause colon, breast, and prostate cancer (24, 25).

Reports suggest that reduced levels of active vitamin D are linked to increased prostate cancer incidence and death (26). As a result, numerous lines of data from *in vitro* research indicate vitamin D's role in triggering cell cycle arrest and apoptosis in malignantly transformed cells. Vitamin B9 (folic acid) may be useful in the chemoprevention of gastrointestinal carcinogenesis. After the administration of folic acid, the tumor suppressor p53 expression in the stomach mucosa increased considerably, but the expression of the oncogene protein Bcl-2 plummeted (27).

In several experimental systems, retinoids have been demonstrated to induce cancer cell differentiation and cell death (28). The most well-studied case of retinoid receptor deficiency is acute promyelocyte leukemia (APL), in which a reciprocal chromosomal t(15;17) translocation between the RAR-α (retinoic acid receptor alpha) and the promyelocyte leukemia (PML) protein alters signaling through both proteins. The resulting PML/RXRa (retinoid X receptor alpha) fusion protein suppresses transcription and is the sole cause of APL. Preclinical studies demonstrated that atRA (all-trans-retinoic acid) triggers terminal differentiation in leukemic cells by driving fusion protein breakdown via the UBE1L ubiquitin-activating enzyme (29). UBP43 (ubiquitin-specific proteases), a UBE1L antagonist, has been discovered as an anticancer target, as its inhibition destabilizes the t(15;17) PML/RARa fusion protein and causes death in APL cells (30). The discovery of a single molecular flaw as the causal agent cleared the path for differentiation-based therapy, as a result, atRA and 13-cis-retinoic acid were successfully used in clinical trials to treat APL. (31). In APL patients, the therapy of retinoids changed the course of the disease for the significant majority of

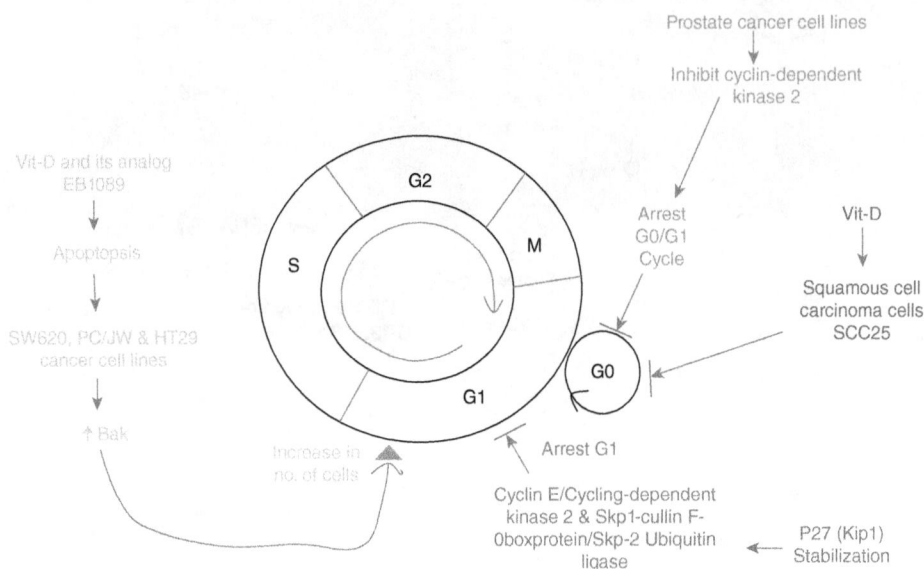

FIGURE 3.2   Cell cycle arrest at various checkpoints.

individuals. The clinical trial found that the anticancer mechanism of action reduced tumor size when treated with ascorbic acid. Antioxidants, such as tocopherols, carotenoids, and ascorbic acid, have been shown in human, animal, and *in vitro* studies to inhibit the progress of neoplastic cells, induce apoptosis, boost cell differentiation, and inhibit the activity of protein kinase C and adenylyl cyclase, asserting their antitumor effect and confirming that individuals can benefit from high-dose therapy by enhancing their prognosis and therapeutic efficacy (**32**). According to studies, ascorbic acid's pro-oxidant mechanisms include the ability to reduce metal ions such as Fe^{3+} and Cu^{3+}, a procedure that produces free radicals like OH^- radicals, which connect with DNA, causing breaks in the phosphodiester bonds and changes made in the bases, going to result in induced cytotoxicity (**33**). Another anticancer strategy is the expansion of natural killer cells without impairing their normal functioning, according to some studies. These cells can "kill" tumor cells without the requirement for directed sensitization, and vitamin C stimulates its multiplication (**34**). According to Tsao and colleagues, high ascorbic acid intake can change the levels of some amino acids in body fluids and may reduce the bioavailability of lysine, glutamine, and cysteine, which are needed for rapidly expanding tumors. Ascorbic acid inhibits tumor proliferation through a variety of mechanisms (**35**). Furthermore, according to Uetaki and colleagues, ascorbic acid impeded energy metabolism by depleting NAD, resulting in cancer cell death (**36**).

Vitamin K is a vital micronutrient. It is involved in the coagulation cascade, regulates bone metabolism, and involves gamma-carboxylation (**37**). It also exhibited anticancer efficacy *in vitro* against human cancers such as leukemia, hepatic, breast, intestinal, oral, bladder and lung cancers, and lungs (**38**). Vitamin K anticancer action was initially identified about six decades ago (**39**). The majority of anticancer activity of vitamin K1 (phylloquinone) and vitamin K2 (menaquinone) is interceded by non-oxidative processes, most likely via transcription factors (TF), although vitamin K3 (menadione) operates in at higher doses by lowering oxidative stress and arylation (**40**). Tamori *et al.* claim that patients with hepatic cirrhosis who took vitamin K2 had less hepatocarcinogenesis (**41**). The intrinsic apoptotic pathway and NF-B activation inhibition are two examples of signaling pathways that can be blocked or activated. After a partial hepatectomy, Zhong noted that an analogue of vitamin K2 was found to decrease the development of secondary tumors and increase overall survival in HCC patients (**42**). Menadione, a synthetic vitamin K derivative, was administered intravenously to patients with inoperable bronchial cancer to boost their survival rate (**43**). In particular, when paired

TABLE 3.1

Vitamins' Ability to Act on a Variety of Cancerous Cells and Different Organs in the Human Body

Vitamin	Target	Effects	References
Vitamin A	Basal cell skin cancer	Acts upon RAR-alpha, RAR-gamma, RAR-beta	(49, 31)
	Metastatic breast cancer	Acts upon RAR-gamma, RXR (metobolites), RXR-alpha	(50, 51)
	Neuroblastoma	Acts upon RAR-gamma, RXR (metobolites)	(52, 53)
Vitamin D	Duodenum	Increases calbindin D28k and intestinal Ca^{2+} absorption	(54)
	Osteoblasts and chondrocytes	Increases RANKL for osteoblasts to activate osteoclasts Increase bone mineralization and matrix formation	(55)
	Cells of various types and cancerous cells	Increases differentiation (p21, p27); decreases cell growth (c-Fos, c-Myc); increases apoptosis (decrease Bcl-2)	(56)
	Ovarian cancer cells	Decreases (hTERT)	(57, 58)
	MCF-7 (human breast cancer)	Blocks cells in G1/S transition; minimal effects on the expression levels of mRNA coding for p21	(59)
	Prostate cancer	Blocks cells in the G1/S transition	(20)
	LNCaP	Increases in the p21 gene	(60)
	Human neuroblastoma cells	Deacetylation and the dephosphorylation of FoxO; the arrest of cell cycle progression	(61)
Vitamin E	Breast cancer cells	Reduces ROS and p53 expression in mice	(62)

TRPV6 = transient receptor potential cation channel subfamily V member 6; SPP-1 = secreted phosphoprotein 1; RANKL = receptor activator of nuclear factor kappa-B ligand; hTERT = human telomerase reverse transcriptase; SCC = squamous cell carcinoma; RAR = retinoic acid receptors; LNCaP = human prostate cancer cell line; HNSCC = human head and neck SCC; CRP = C-reactive protein; MMP-2,9 = matrix metalloproteins-2,9; VEGF = vascular endothelial growth factor; RXR-α = retinoid X receptor alpha

with vitamin C, menadione synergistically inhibits the growth of cancer cells by increasing oxidative stress and reducing cellular thiols (44). Phylloquinone and menaquinone have antiproliferative properties by targeting proto-oncogene TF, such as c-jun, c-fos, and c-myc, causing cell cycle arrest and death (45). Phylloquinone and menaquinone have anticancer effects in various types of cancers, such as breast, stomach, and liver (46). Menadione has also been demonstrated to have significant cytotoxicity against malignancies of the mouth, prostate, kidney, and breast (47, 48).

3.4 VITAMINS EXPLOITED FOR CANCER PREVENTION

3.4.1 VITAMIN A

The term "vitamin A" refers to a group of chemicals that are all linked. Vitamin A precursors are typically referred to as retinal (alcohol) and retinol (aldehyde). Retinoids are chemicals like retinal, retinol, and retinoic acid (RA). Retinol is formed by provitamin-A carotenoid, which is a beta-carotene (63).

Vitamin A and its derivatives are necessary for many tasks during a lifetime, notably fertilization, development, cognition, metabolism, cell division and proliferation, epithelial cell integrity preservation, and physiological functions (64–66). Vitamin A's hormone-like actions and major retinoic acid metabolite are connected to the ability to detect and impact various approaches and its carefully controlled synthesis (67). RA's mode of action is very similar to that of steroid hormones, including progestins, estrogens, thyroid hormones, and glucocorticoids, all of which work by activating the gene. Those types of hormones, such as retinoic acid (RA), are soluble in plasma cell membranes and are delivered in circulation together with the hormone-binding protein. They have intracellular receptors that function on gene expression rather than translation (68).

3.4.1.1 Vitamin A and Cancer

Vitamin A was one of the first vitamins researched about carcinogenesis since cancer is a disease that disrupts normal tissue development and division. Vitamin A deficiency in animal trials has been shown in animal trials to increase susceptibility to specific kinds of carcinogenesis. The effects of retinoids on carcinogenesis have been extensively studied (**69**). The activity in the cell nucleus, which involves the expression of genetic information that governs cell division, is assumed to be involved in these pathways. Certain retinol- and retinoic-acid-binding proteins are thought to be important for retinol and retinoic acid transit within the cell and across the cell membrane, which boosts hormone-like levels to control cell division. Retinol also has several impacts on cell membranes, including altered glycoprotein production and modifications in hormone receptor membranes, including those that control c-AMP. Cell activation, cell adhesion, and cell membrane invasion can all be affected by activity on these receptors. In addition, animal studies have indicated that retinol improves both humoral and cell-mediated immune responses, suggesting that it may help the immune system fight tumors. Retinoids tend to affect exclusively epigenetic alterations, implying that they only affect carcinogenesis' promoting phases. In animal studies, the antagonistic impact of retinoids on plant promoters is frequently described (**69**). Early epidemiological studies have found a link between vitamin A use and the incidence of cancer in the lungs, bladder, and gastrointestinal system. However, early studies did not distinguish between preformed vitamin A and β-carotene, and more current research has found that protective benefits are only evident with β-carotene intake, not with earlier vitamin A ingestion (**70**). Expected serum research on serum vitamin A concentration and later cancer risk yielded reliable results, indicating that the connections might be attributable to cancer in the early stages (**71**). Furthermore, because serum retinol levels are under homeostatic regulation and do not rise with rising ingestion, these results are tough to describe. As a result, the findings appear to suggest that vitamin A counterparts play a key role in cellular processes linked to carcinogenesis.

Concurrent alterations in gene expression caused by genetic and epigenetic abnormalities promote the formation of oncogenes and pro-metastatic genes, or genetic inhibition of tumor suppression, as well as genome mutation and instability, which drive cancer initiation and progression (**72**). Because of their recurrence, epigenetic alterations have gotten a lot of interest in the screening and treatment of a variety of illnesses, including metastatic cancer. Isolation has been the most frequent clinical therapy in cancer patients, and isolating drugs, such as RA, are perhaps the most frequent medical treatment (**73**). External retinoids, for example, have been found to reduce malignant tumors and prevent the occurrence of subsequent malignancy (**74**). The RA sign is a potential molecular candidate for neuroblastic tumors because of its capacity to demonstrate differentiation, which is a significant aspect of pathologic divergence and prognosis. Neuroblastic tumors respond to RA with remarkable neural maturity due to the genetic regulation of genes engaged mostly in the differentiation process (**75**). The tyrosine kinase receptor RET (rearranged during transfection) is induced by RA in neuroblastoma cells, according to several studies, and plays an important role in proliferation and differentiation (**76**). Angrisano with colleagues explored key epigenetic alterations in the RET site in the neuroblastoma cell line in response to RA stimulation, finding that the complexity of a sequence of biological processes, comprising mutations in both chromatin and DNA methylation status, is connected to retinoic acid-mediated RET function. Retinoic-acid-induced RET stimulation involves an elevation in the methylation level of H3K4me3 (H3 histone is trimethylated at fourth lysine) in the promoter region of RET, but a large reduction in H3K27me3 indicated the tyrosine kinase receptor RET gene activity in retinoic acid stimulation was found in the RET enhancing area. Furthermore, in both the promoter and developer areas, a systemic rise in histone H3 acetylation was discovered (**77**). In mouse P19 embryonal cancer cells, further histone alterations in response to all-trans retinoic acid (ATRA) were discovered (**78**). ATRA therapy had little effect on the acetylation of H3 and H4 histone protein, but the transcriptional activity of the mRARb2 promoter is upregulated by H3 histone phosphorylation. HeLa nuclear extracts contain HDAC, whose activity is inhibited by ATRA through protein acetylation processes (**79**). Stefanska's colleagues demonstrated that epigenetic techniques can enhance the anticancer impact of RA in breast cancer cells in *in vitro* tests. ATRA prevented methylation of the RARb2 promoter, a gene that reduces tumors that are generally repressed during carcinogenesis owing to the methylation promoter, while also enhancing RARb2 production (**80**). This ATRA impact was only observed in nonbinding MCF-7 cells. The influence of ATRA on methylation and accessibility to phosphatase and tensin homologue (PTEN), a gene involved in gastric bypass, was studied in another research (**81**). The researchers discovered that ATRA has substantial potential in lowering PTEN promoter methylation not in greatly invasive MDA-MB-231 cells but in noninvasive MCF-7 cells, which may improve their appearance in the early phase of carcinogenesis (**81**). Treatment with RA increases the emergence of acute promyelocytic leukemia (APL) in a non-leukemic phenotype and lowers the expression level of DNMT (DNA methyltransferases) (**82**). RA treatment resulted in inappropriate and pervasive exhaustion of DNMT1, DNMT3a, and expression/activity of DNMT3b, which is linked to methylation discharge of the promoter/exon-1 receptors of its aim RARb2 type in the APL reflex. Di Corce proposed the following approach: characteristics of oncogenic transcription wrongly appeal to DNMTs to address promoters. The qualities of HDAC (histone deacetylases) and DNMT react with the properties of the raw methylated CpGs, which produce metabolic protein binding spots. The interaction between DNMT and HDAC1 may improve the gathered qualities even further. If the initial stage of hiring is not stopped, hypermethylation may extend to neighboring areas, resulting in the formation of stable chromatin (**83**). Retinoid interaction with RAR/RXR receptors triggers a series of chromatin modifications that should promote differentiation and produce constant epigenetic alterations (**84**). Those mutations can have the ability to enhance neoplastic stagnation diversity in cancer cells. Retinoids do this by boosting stem cell proliferation and modifying the gene expression profile in animal cells, making them more susceptible to other treatments. As a result, retinoids may be used in a variety of cancer therapies in the coming future (**85**).

3.4.2 Vitamin D

Vitamin D is made up of a set of fat-soluble prohormones that help in maintaining Ca^{2+} and PO_4^{3-} homeostasis, as well as bone and muscle integrity. Vitamin D2 (ergocalciferol) is produced by plants, yeasts, and fungi, while Vitamin D3 (cholecalciferol) is produced by animals. Vitamin D2 is made from ergosterol, which itself is transformed into viosterol and then to ergocalciferol by UV light. This type of vitamin D2 can be gained by food and supplementation. When skin is exposed to UV-B radiation, 7-dehydrocholesterol is converted to pre-vitamin-D3 (1,25-dihydroxycholecalciferol or calcitriol), which also isomerizes to vitamin D3 (**54, 86**). Vit. D has been shown to have a range of impacts on cancerous cells boosting cellular differentiation, preventing tumor blood vessel formation, delaying cancer cell proliferation, inducing cell death (apoptosis), and stimulating cellular differentiation (angiogenesis) and tumors in mice, suggesting that it may help to reduce or prevent the development of cancer. (**87, 88**).

3.4.2.1 Tumor Progression and Vitamin D

Colston was the first to demonstrate that 1, 25(OH)2D3 treatment of melanoma cells decreased cell proliferation and that this effect was dose-dependent (**89**). Later, it was discovered that this chemical inhibited the growth of several tumor cell lines, including breast, prostate, and colon cancer cells (**90**). On the other hand, cell growth arrest is potentiated by overexpression of VDR. Unexpectedly, it has been found that 1, 25(OH)2D3 prevents the spread of ovarian cancer by restricting the amount of hTERT (human telomerase reverse transcriptase) mRNA by using a short noncoding RNA. In ovarian tumor and cancer cell lines, microRNA-498 (miR-498) generated by 1, 25(OH)2D3 decreased hTERT mRNA expression, enhanced cell death, and suppressed tumor growth. In cell lines as well as tumor-bearing animals, however, the ability of 1, 25(OH)$_2$D3 to lower hTERT mRNA and inhibit ovarian cancer development was diminished in the absence of miR-498 (**58**).

3.4.2.2 Effect of Vitamin D on TGF-β

A member of an unidentified family of growth factors, transforming growth factor (TGF-) controls a number of biological processes, including cell division, adhesion, motility, adhesion, proliferation, and systemic cell death. (91). The effects of vitamin D and TGF-β share similar effects on cell growth and differentiation (92). According to numerous research, breast cancer cells that have been exposed briefly to 1,25 (OH)2D3 or its homologue EB1089 have higher levels of TGF- and/ or TGF- receptors (93). Surprisingly, 1,25(OH)2D3 has long-lasting therapeutic benefits at mRNA levels encoding several TGF-family members in other tumor cells, including colorectal cancer cells and SCC (94).

3.4.2.3 Vitamin D in Regulating DNA Repair Proteins

In vivo experiments show that mice without VDR are more prone to skin cancer caused by carcinogens such as 7,12-dimethylbenzanthracene (DMBA) or progressive UVR. These findings suggest that 1,25(OH)$_2$D3 may protect the skin from malignant tumors by regulating keratinocyte proliferation and dissociation, resulting in DNA repair, and reduced hedgehog stimulation in response to UV B exposure (95). Vitamin D treatment protects BPH-1 prostate epithelial cells from cancer-induced genotoxic stress by regulating the ATM and RAD50 gene regulation through VDER-mediated transcriptional upregulation, which allows for the correction of DNA of double strands (96). Other findings have shown that vitamin D inhibits the degradation of DNA 53BP1 protein-derived protein cysteine proteinases Cathepsin L (97), a lysosomal endopeptidase which promotes tumor cell proliferation in BRCA1 breast cancer cells (98).

3.4.3 VITAMIN E—TOCOTRIENOLS (TTs) AND TOCOPHEROLS (TOCs)

Vitamin E is shown to be an antioxidant fat-soluble molecule that regulates female reproductive function. Tocotrienols (α-TT, δ-TT, γ-TT, and β-TT) and tocopherols (α-TOC, δ-TOC, γ-TOC, and β-TOC) are lipid-soluble hydrophobic compounds made up of eight natural isoforms (99). Animals and humans are just unable to manufacture vitamin E; thus, they must obtain it from plant sources. Rice bran oil, palm oil, annatto oil, and cereals, including wheat, oats, and barley, contain large levels of tocopherols, whereas tocotrienols are found in barley (100).

Tocotrienols are capable of detecting cancer cells while inflicting no damage to healthy cells. In a study published in 2006, Srivastava and Gupta discovered that tocotrienols might lead to death in prostate cancer cells but never in malignant prostate epithelial cells (101). According to previous studies, tocotrienols cause apoptotic death in prostate cancer cells but never in non-breast cells or prostate epithelial cells (102). According to various research over the last several years, tocotrienols have been shown to restrict the growth or cause death in a range of cancer, including colon, cervix, breast, hepatic, lungs, and pancreas (99, 103). In contrast, TTs interact with a variety of intracellular pathways involved in invasion, apoptosis, cell proliferation paraptosis, autophagy, and metastasis (100).

3.4.3.1 Cell Cycle Arrest Caused by Tocotrienols

Tocotrienols have been studied in recent years for their impact on cancer cell line cell-cycle block entrance. Both estrogen-dependent (MCF-7) and estrogen-independent (MCF-7) cells experienced a G1/S cell cycle arrest after exposure to γ-tocotrienol (MDA-MB-231) cancerous breast cell lines (104). Furthermore, γ-tocotrienol-induced G1/S cell cycle arrest was found to be associated with rising amounts of p27 as well as phosphorylated Rb expression in full form of SA$^+$ mammary tumor cells and decreased levels of CDK6, CDK4, CDK2, and cyclin D1 (105).

Tocotrienol-rich fraction (TRF) couples the G1/S cell cycle in prostate cancer cell lines LNCaP (lymph node carcinoma of the prostate), DU-145, and PC-3; however, tocotrienols inhibit cell proliferation in prostate cancer cell line VCaP (101). In this investigation, p27 and p21 control were used to achieve both goals; however, δ-tocotrienol was more successful than γ-tocotrienol in minimizing

(106). Infusing δ-tocotrienol with cell cycle binding reduced the susceptibility of murine melanoma cells to cyclin-dependent kinase-4 (CDK4) in the G1 phase (107).

3.4.3.2 Induction of Autophagy and Apoptosis by TTs

The endoplasmic stress response is crucial in the triggering of apoptosis by γ-tocotrienol. γ-tocotrienol stimulation of SA+ mammary tumor cells resulting to PERK/eIF2a/ATF-4 excitation, an ER stress response marker, raised C/EBP homologous protein (CHOP) levels, a vital part of endoplasmic reticulum stress-mediated apoptosis, as well as cleavage of poly-ADP-ribose polymerase (PARP) (108). Park and colleagues discovered that γ-tocotrienol activated JNK, p38, MAP kinase, and C/EBP homologous protein (CHOP) in MDA-MB-231 cells, as well as caused caspase-9, caspase-8, and PARP cleavage (endoplasmic reticulum's stress marker) (109).

Autophagy is a tightly regulated lysosomal digestion process that may help cells survive or die throughout the body. γ-tocotrienol was found to stimulate autophagy in breast cancer cell lines in studies (110). The γ-tocotrienol treatment decreased mTOR/Akt/PI3K expression and increased the Bcl-2/Bax proportion and caspase-3 activity while decreasing mTOR/Akt/PI3K expression. As a result, this investigation's findings showed that autophagy and caspase-mediated death are connected (autophagy may lead to apoptosis) (111). The same team then examined the relationship between autophagy stimulation and the ER response to stress in human breast cancer cells MDA-MB-231 and MCF-7. An increase in the initial (LC3B-II, Beclin-1) and delayed (LAMP-1 and cathepsin-D) autophagy manifestations after tocotrienol therapy was observed in comparison to prior therapy with autophagy blockers to prevent these effects. Tocotrienols increase the effectiveness of an endoplasmic reticulum response to stress, which causes cervical cancer cells to undergo apoptosis. (112).

In breast cancer cells, tocotrienols employ caspase-based apoptosis (112) and concentrate in breast cancer cells at elevated amounts than tocopherols (113). Additionally, δ-tocotrienol was employed to induce apoptosis in both estrogen-dependent MCF-7 and estrogen-independent MDA-MB-435 breast cancer cell lines, with the latter being linked to the modulation of TGF- receptor II and TGF, Fas, and JNK signaling, as well as their efficacy (114). In human cancer cell lines, DNA fragmentation, PARP separation, and NF-κB suppression were all connected to the triggering of apoptosis by tocotrienols (115).

3.4.3.3 Inhibition of Invasion, Metastasis, and Angiogenesis by Tocotrienols

γ-tocotrienol has been shown to restrict the attack of prostate cancer cells LNCaP and PC-3 by hindering mesenchymal signals and boosting the utterance of E-cadherin and β-catenin. In comparison to the control group, cell adhesion inside the gastric cell SGC-7901 was reduced in γ-tocotrienol-treated groups (116). The Matrigel invasion of these cells was considerably reduced when the expression levels of matrix metallopeptidase (2 and 9) and the tissue blocker of metalloproteinase (TIMP-2 and TIMP-1) were controlled. The results of this investigation suggest a potential mode of action for γ-tocotrienol-mediated antitumor metastasis.

δ-tocotrienol inhibits angiogenesis and eukaryotic polymerase lambda (λ) activity in a separate investigation (117). In breast cancer patients, the anti-angiogenic effect of δ-tocotrienol has been shown to reduce pro-angiogenic indications, such as vascular endothelial growth factor and IL-8 (118). In conclusion, tocotrienols have been shown to reduce angiogenesis in cancer via reducing matrix metallopeptidase-9, tumor necrosis factor, angiopoietin-1, epidermal growth factor, matrix metallopeptidase-2, and vascular endothelial growth factor, as well as the regulation of metalloproteinases (TIMP-2 and TIMP-1) (119).

3.4.3.4 The Effect of Tocotrienols on Cancer Stem Cells

Various studies have reported the use of tocotrienols in reducing the proliferation of the cancer stem cells (CSCs) (120). Treatment with γ -tocotrienol prevents the growth of prostate stem cell cancer and triggers docetaxel-induced apoptosis in the affected cells (121). They also found that γ-tocotrienol affects the transcription of CSCs' prostate signal (CD-133 and CD-44) in prostate cancer androgens (PC-3 and DU-145). Additionally, pretreating PC-3 cells with γ-tocotrienol prevent tumorigenic cell proliferation. These findings imply that γ-tocotrienol may be a useful drug for controlling prostate CSCs, which might explain its anticancer and chemo-sensitizing properties found in earlier research. Inhibiting a mevalonate pathway and then using the de novo ceramide synthesis technique eliminated breast CSCs, as well as silenced exposure of facilitators displaying STAT-3 by blocking the mevalonate process, using the de-novo ceramide synthesis technique, respectively (122). According to studies, simvastatin and γ-tocotrienol, alone or in conjunction, can destroy CSCs from drug-resistant cancerous cells. γ-tocotrienol alone and in conjunction with docosahexaenoic acids (DHA) caused three times higher apoptotic cell death triple human ALDH cancerous cells (123).

3.4.4 Vitamin K

Vitamin K is available in three different natural and synthetic forms: vitamins K1, K2, and K3. Phylloquinone (vitamin K1) is a naturally occurring form present mainly in leafy green vegetables. Menaquinone (vitamin K2), also a naturally occurring form, is synthesized by the gut flora. Anticancer activity of vitamin K was first investigated in 1947. Anticancer action of vitamin K1 has been discovered in several cell lines, like nasopharynx, lung, breast, liver, oral epidermoid, stomach, colon cancer, and leukemia (124). *In vitro* and *in vivo* investigations demonstrated the anticancer properties of vitamin K2. Vitamin K2 inhibited glioma cell growth in human and rat cell types by cell cycle arrest and death in a dose-dependent manner (125). Takami conducted a trial on a patient suffering from myelodysplastic. An oral dose of 45 mg/day of vitamin K2 was administered to her. After 14 months, her pancytopenia improved, and she no longer required transfusions (126). Vitamin K2 promoted apoptosis in glioma, leukemia, and hepatoma cell lines in a dose-dependent manner and also causes cell cycle arrest at G0 and G1 transition (127).

Vitamin K2, including K1, triggers apoptosis by a non-oxidative process that involves transcription factors owing to their incompetence to use 1-e⁻ redox cycling (128). The first and prior mechanism involves ROS generation via the 1-e⁻ cycling of quinone. Increasing the redox cycle of menadione and ROS production within a cell can increase the cell's oxidative capacity and further lead to cell death. Other mechanisms involved are apoptosis along with cell cycle arrest triggered by transcription factor modification (37).

Vitamin K3 has two separate modes of action. At high concentrations, it causes oxidative stress, necrosis, and autoschizis. At lower concentrations, it induces apoptosis by a non-oxidative mechanism (129). Vitamin K3 exhibits anticancer activity in cell lines from leukemia, stomach, colon, lung, nasopharynx, liver, breast, cervix, and lymphoma (124, 130). Several studies reported that vitamin C and K co-administration increases their potential toxicity. Dose, exposure time, and subsequent level of oxidative stress, vitamins K3 and C either cause cell necrosis or activate apoptosis (131).

When the two vitamins are combined, 1-e⁻ reduction of vitamin K3 is stimulated, and redox cycling of quinones is accelerated (132). At concentrations lower than vitamin K alone, the combination showed specific anticancer action against human breast, endometrial, and oral epidermoid carcinoma cell lines. An efficient combination of vitamins K3 and C as radio-sensitizer and chemo-sensitizer in another investigation, with no systemic or major organ damage (133). Vitamins C and K have synergistic anticancer efficacy against two androgen-independent human carcinoma cells, as well as other urinary tumor cell lines (134).

Noto and colleagues noted that 1.38 g/mL of K3 increased 74% inhibition and 104 μmol/L of vitamin C and 105 nmol/L of K3 resulted in a 93% inhibition (133). Vitamin K3 showed differential activation of alkaline DNA in whole malignant tumor cells, in comparison to reactivation of acid DNase by vitamin C, which further showed suppression in both malignant tumors and non-necrotic cells in males and in laboratory animals and also during the initial experimental carcinogenesis stage (**135**). When vitamin K3 was coupled with 5-fluorouracil, a standard chemotherapeutic drug, Waxman and Bruckner found a synergistic impact *in vivo* and *in vitro* trials. This combination increased the effectiveness of the treatment against hepatoma cells (**136**). The enhancing effects of vitamin K3 on the deleterious effect of several clinically relevant anticancer drugs, including activity against MDR (multidrug-resistant) human cancer cell lines with fewer side effects (**130**). Vitamin K shows an antitumor effect against multidrug-resistant human cancer cell lines, having fewer side effects (**130**).

3.5 CONCLUSION

Cancer is among the leading causes of death around the globe. Chemotherapy and radiotherapy are the two most common cancer therapies today; however, both have significant adverse effects. The ultimate goal is to pick out alternative methods for the prevention of cancer, aiming to reduce the side effects. In addition, new techniques should be researched and developed to reduce the risk of cancer. Vitamins and other various antioxidants have been reported to help reduce cancer. Vitamins are found in fruits, vegetables, herbs, eggs, and other meaty products that are not harmful to one's health. In various studies, it has been reported that vitamins are helpful in chemoprevention. There are numerous reports in the literature that vitamins, at different concentrations/doses, act as an antioxidant by various mechanisms, such as reduction in apoptosis, inhibition of cytotoxicity, protection of neoplastic cells against lipid peroxidation, and decline in tumor growth. And as anticarcinogenic agents, vitamins act by inhibiting the cell progression and tumor cells' proliferative activity, among many others. Thus, additional research is required to fully comprehend dose-response variations, including its target-specific mechanism of action, both as an antioxidant and antitumor agent, to aid in cancer control and diagnosis, with the goal of enhancing the quality of life of the population.

3.6 REFERENCES

1. Flavahan, W.A., Gaskell, E. and Bernstein, B.E., 2017. Epigenetic plasticity and the hallmarks of cancer. *Science*. 357(6348): eaal2380.
2. Akladios, F.N., Andrew, S.D. and Parkinson, C.J., 2015. Selective induction of oxidative stress in cancer cells via synergistic combinations of agents targeting redox homeostasis. *Bioorganic & Medicinal Chemistry*. 23(13): 3097–3104.
3. Rajakumar, T., Pugalendhi, P. and Thilagavathi, S., 2015. Dose response chemopreventive potential of allyl isothiocyanate against 7, 12-dimethylbenz (a) anthracene induced mammary carcinogenesis in female Sprague-Dawley rats. *Chemico-Biological Interactions*. 231: 35–43.
4. Benetou, V., Lagiou, A. and Lagiou, P., 2015. Chemoprevention of cancer: Current evidence and future prospects. *F1000Research*. 4 (F1000 Faculty Rev). Faculty of 1000 Ltd. doi:10.12688/F1000RESEARCH.6684.1.
5. Tavsan, Z. and Kayali, H.A., 2019. Flavonoids showed anticancer effects on the ovarian cancer cells: Involvement of reactive oxygen species, apoptosis, cell cycle and invasion. *Biomedicine & Pharmacotherapy*. 116: 109004.
6. Xiao, H.J., Liang, H., Wang, J.B., *et al.*, 2011. Attributable causes of cancer in China: Fruit and vegetable. *Chinese Journal of Cancer Research*. 23(3): 171–176.
7. Hamilton, Z. and Parsons, J.K., 2016. Prostate cancer prevention: Concepts and clinical trials. *Current Urology Reports*. 17(4): 1–9.
8. INCA. Instituto Nacional de Câncer José Alencar Gomes Silva (INCA). Consenso Nacional de Nutrição Oncológica. v. 2; 2011. Available from: http://www1.inca.gov.br/inca/Arquivos/consenso_nutricao_vol2.pdf.
9. Brozmanová, J., Mániková, D., Vlčková, V. *et al.*, 2010. Selenium: A double-edged sword for defense and offence in cancer. *Archives of Toxicology*. 84(12): 919–938.
10. Park, B., Shin, A., Jung-Choi, K., *et al.*, 2014. Correlation of breast cancer incidence with the number of motor vehicles and consumption of gasoline in Korea. *Asian Pacific Journal of Cancer Prevention*. 15(7): 2959–2964.

11. Seifabadi, S., Vaseghi, G., Javanmard, S.H., *et al.*, 2017. The cytotoxic effect of memantine and its effect on cytoskeletal proteins expression in metastatic breast cancer cell line. *Iranian Journal of Basic Medical Sciences.* 20(1): 41.

12. Bray, F., Ferlay, J., Soerjomataram, I., *et al.*, 2018. Global cancer statistics 2018: GLOBOCAN estimates of incidence and mortality worldwide for 36 cancers in 185 countries. *CA: A Cancer Journal for Clinicians.* 68(6): 394–424.

13. Siegel, R. L., Miller, K. D., and Jemal, A., 2020. Cancer statistics, 2020. *CA Cancer Journal of Clinicians.* 70(1): 7–30.

14. INCA. Instituto Nacional de Câncer José Alencar Gomes da Silva (INCA). Incidência de Câncer no Brasil. Estimativa 2014. Available from: www.inca.gov.br/estimativa/2014/estimativa-24042014.pdf.

15. Moukayed, M. and Grant, W.B., 2013. Molecular link between vitamin D and cancer prevention. *Nutrients.* 5(10): 3993–4021.

16. Baker, A.R., McDonnell, D.P., Hughes, M., *et al.*, 1988. Cloning and expression of full-length cDNA encoding human vitamin D receptor. *Proceedings of the National Academy of Sciences.* 85(10): 3294–3298.

17. Welsh, J., 2012. Cellular and molecular effects of vitamin D on carcinogenesis. *Archives of Biochemistry and Biophysics.* 523(1): 107–114.

18. Zinser, G.M. and Welsh, J., 2004. Accelerated mammary gland development during pregnancy and delayed postlactational involution in vitamin D3 receptor null mice. *Molecular Endocrinology.* 18(9): 2208–2223.

19. Haussler, M.R., Whitfield, G.K., Haussler, C.A., *et al.*, 1998. The nuclear vitamin D receptor: Biological and molecular regulatory properties revealed. *Journal of Bone and Mineral Research.* 13(3): 325–349.

20. Jensen, S.S., Madsen, M.W., Lukas, J., *et al.*, 2001. Inhibitory effects of 1α, 25-dihydroxyvitamin D3 on the G1–S phase-controlling machinery. *Molecular Endocrinology.* 15(8): 1370–1380.

21. Salehi-Tabar, R., Nguyen-Yamamoto, L., Tavera-Mendoza, L.E., *et al.*, 2012. Vitamin D receptor as a master regulator of the c-MYC/MXD1 network. *Proceedings of the National Academy of Sciences.* 109(46): 18827–18832.

22. Li, P., Li, C., Zhao, X., *et al.*, 2004. p27Kip1 stabilization and G1 arrest by 1, 25-dihydroxyvitamin D3 in ovarian cancer cells mediated through down-regulation of cyclin E/cyclin-dependent kinase 2 and Skp1-Cullin-F-box protein/Skp2 ubiquitin ligase. *Journal of Biological Chemistry.* 279(24): 25260–25267.

23. Bao, B.Y., Hu, Y.C., Ting, H.J. and Lee, Y.F., 2004. Androgen signaling is required for the vitamin D-mediated growth inhibition in human prostate cancer cells. *Oncogene.* 23(19): 3350–3360.

24. Krishnan, A.V., Swami, S. and Feldman, D., 2012. The potential therapeutic benefits of vitamin D in the treatment of estrogen receptor positive breast cancer. *Steroids.* 77(11): 1107–1112.

25. Leyssens, C., Verlinden, L. and Verstuyf, A., 2013. Antineoplastic effects of 1, 25 (OH) 2D3 and its analogs in breast, prostate and colorectal cancer. *Endocrine-related Cancer.* 20(2): R31–R47.

26. Garland, C.F., Garland, F.C., Gorham, E.D., *et al.*, 2006. The role of vitamin D in cancer prevention. *American Journal of Public Health.* 96(2): 252–261.

27. Cao, D.Z., Sun, W.H., Ou, X.L., *et al.*, 2005. Effects of folic acid on epithelial apoptosis and expression of Bcl-2 and p53 in premalignant gastric lesions. *World Journal of Gastroenterology: WJG.* 11(11): 1571.

28. Idres, N., Benoît, G., Flexor, M.A., *et al.*, 2001. Granulocytic differentiation of human NB4 promyelocytic leukemia cells induced by all-trans retinoic acid metabolites. *Cancer Research.* 61(2): 700–705.

29. Kitareewan, Sutisak, Pitha-Rowe, I., Sekula, D. *et al.*, 2002. UBE1L is a retinoid target that triggers PML/RARα degradation and apoptosis in acute promyelocytic leukemia. *Proceedings of the National Academy of Sciences of the United States of America.* 99(6): 3806–3811. doi:10.1073/PNAS.052011299.

30. Guo, Yongli, Dolinko, A.v., Chinyengetere, F. *et al.*, 2010. Blockade of the ubiquitin protease UBP43 destabilizes transcription factor PML/RARa and inhibits the growth of acute promyelocytic leukemia. *Cancer Research.* 70(23). American Association for Cancer Research: 9875–9885. doi:10.1158/0008-5472.CAN-10-1100/649410/AM/BLOCKADE-OF-THE-UBIQUITIN-PROTEASE-UBP43

31. Petrie, K., Zelent, A. and Waxman, S., 2009. Differentiation therapy of acute myeloid leukemia: Past, present and future. *Current Opinion in Hematology.* 16(2): 84–91. doi:10.1097/MOH.0B013E3283257AEE.

32. Gröber, U., 2009. Antioxidants and other micronutrients in complementary oncology. *Breast Care.* 4(1). Karger Publishers: 13–20. doi:10.1159/000194972.

33. Putchala, M.C., Ramani, P., Sherlin, H.J., *et al.*, 2013. Ascorbic acid and its pro-oxidant activity as a therapy for tumours of oral cavity–a systematic review. *Archives of Oral Biology.* 58(6): 563–574.

34. Huijskens, M.J., Walczak, M., Sarkar, S., *et al.*, 2015. Ascorbic acid promotes proliferation of natural killer cell populations in culture systems applicable for natural killer cell therapy. *Cytotherapy.* 17(5): 613–620.

35. Tsao, C.S. and Miyashita, K., 1985. Effect of large intake of ascorbic-acid on the urinary-excretion of amino-acids and related-compounds. *IRCS Medical Science-Biochemistry*. 13(9): 855–856.

36. Uetaki, M., Tabata, S., Nakasuka, F. *et al.*, 2015. Metabolomic alterations in human cancer cells by vitamin C-induced oxidative stress. *Scientific Reports*. 5(1). Nature Publishing Group: 1–9. doi:10.1038/srep13896.

37. Plaza, S.M., 2003. The anticancer effects of vitamin K. *Alternative Medicine Review*. 8(3): 303–318.

38. Ishibashi, M., Arai, M., Tanaka, S., *et al.*, 2012. Antiproliferative and apoptosis-inducing effects of lipophilic vitamins on human melanoma A375 cells *in vitro*. *Biological and Pharmaceutical Bulletin*. 35(1): 10–17.

39. Mitchell, J.S. and Simon-Reuss, I., 1947. Combination of some effects of x-radiation and a synthetic vitamin K substitute. *Nature*. 160(4055): 98–99.

40. Wang, Z., Wang, M., Finn, F. *et al.*, 1995. The growth inhibitory effects of vitamins K and their actions on gene expression. *Hepatology*. 22(3): 876–882.

41. Tamori, A., Habu, D., Shiomi, S. *et al.*, 2007. Potential role of vitamin K2 as a chemopreventive agent against hepatocellular carcinoma. *Hepatology Research*. 37(s2 Fourth JSH S). John Wiley & Sons, Ltd: S303–307. doi:10.1111/J.1872-034X.2007.00202.X.

42. Zhong, J.H., Mo, X.S., Xiang, B. de *et al.*, 2013. Postoperative use of the chemopreventive vitamin K2 analog in patients with hepatocellular carcinoma. *PLOS ONE*. 8(3). Public Library of Science: e58082. doi:10.1371/JOURNAL.PONE.0058082.

43. Plaza, S.M., 2003. The anticancer effects of vitamin K. *Alternative Medicine Review*. 8(3): 303–318.

44. Verrax, J., Cadrobbi, J., Delvaux, M., *et al.*, 2003. The association of vitamins C and K3 kills cancer cells mainly by autoschizis, a novel form of cell death. Basis for their potential use as coadjuvants in anticancer therapy. *European Journal of Medicinal Chemistry*. 38(5): 451–457.

45. Tokita, H., Tsuchida, A., Miyazawa, K., *et al.*, 2006. Vitamin K2-induced antitumor effects via cell-cycle arrest and apoptosis in gastric cancer cell lines. *International Journal of Molecular Medicine*. 17(2): 235–243.

46. Otsuka, M., Kato, N., Shao, R.X., *et al.*, 2004. Vitamin K2 inhibits the growth and invasiveness of hepatocellular carcinoma cells via protein kinase A activation. *Hepatology*. 40(1): 243–251.

47. Degen, M., Alexander, B., Choudhury, M., *et al.*, 2013. Alternative therapeutic approach to renal-cell carcinoma: Induction of apoptosis with combination of vitamin K3 and D-fraction. *Journal of Endourology*. 27(12): 1499–1503.

48. Yang, C.R., Liao, W.S., Wu, Y.H., *et al.*, 2013. CR108, a novel vitamin K3 derivative induces apoptosis and breast tumor inhibition by reactive oxygen species and mitochondrial dysfunction. *Toxicology and Applied Pharmacology*. 273(3): 611–622.

49. Lo-Coco, F., Avvisati, G., Vignetti, M., *et al.*, 2013. Retinoic acid and arsenic trioxide for acute promyelocytic leukemia. *New England Journal of Medicine*. 369(2). Massachusetts Medical Society: 111–121. doi:10.1056/NEJMOA1300874/SUPPL_FILE/NEJMOA1300874_DISCLOSURES.PDF.

50. Recchia, F., Sica, G., Candeloro, G., *et al.*, 2009. Beta-interferon, retinoids and tamoxifen in metastatic breast cancer: Long-term follow-up of a phase II study. *Oncology Reports*. 21(4). Spandidos Publications: 1011–1016. doi:10.3892/OR_00000317/HTML.

51. Chiesa, M.D., Passalacqua, R., Michiara, M., *et al.*, 2007. Tamoxifen vs tamoxifen plus 13-Cis-retinoic acid vs tamoxifen plus interferon alpha-2a as first-line endocrine treatments in advanced breast cancer: Updated results of a phase II, prospective, randomised multicentre trial. *Acta Bio-Medica: Atenei Parmensis*. 78(3): 204–209. https://europepmc.org/article/med/18330080.

52. Matthay, Katherine K., Patrick Reynolds, C., Seeger, R.C., *et al.*, 2009. Long-term results for children with high-risk neuroblastoma treated on a randomized trial of myeloablative therapy followed by 13-Cis-retinoic acid: A children's oncology group study. *Journal of Clinical Oncology*. 27(7). American Society of Clinical Oncology: 1007. doi:10.1200/JCO.2007.13.8925.

53. Maurer, B.J., Kang, M.H., Villablanca, J.G., *et al.*, 2013. Phase I trial of fenretinide delivered orally in a novel organized lipid complex in patients with relapsed/refractory neuroblastoma: A report from the new approaches to neuroblastoma therapy (NANT) consortium. *Pediatric Blood & Cancer*. 60(11). John Wiley & Sons, Ltd: 1801–1808. doi:10.1002/PBC.24643.

54. Bouillon, R., Carmeliet, G., Verlinden, L., *et al.*, 2008. Vitamin D and human health: Lessons from vitamin D receptor null mice. *Endocrine Reviews*. 29(6): 726–776.

55. Rosen, C.J., Adams, J.S., Bikle, D.D., *et al.*, 2012. The nonskeletal effects of vitamin D: An Endocrine Society scientific statement. *Endocrine Reviews*. 33(3): 456–492.

56. Moukayed, M. and Grant, W.B., 2013. Molecular link between vitamin D and cancer prevention. *Nutrients*. 5(10): 3993–4021.

57. Ikeda, N., Uemura, H., Ishiguro, H., *et al.*, 2003. Combination treatment with 1α, 25-dihydroxyvitamin D3 and 9-cis-retinoic acid directly inhibits human telomerase reverse transcriptase transcription in prostate cancer cells. *Molecular Cancer Therapeutics*. 2(8): 739–746.

58. Kasiappan, R., Shen, Z., Anfernee, K.W., *et al.* 2012. 1,25-dihydroxyvitamin D3 Suppresses Telomerase Expression And Human Cancer Growth Through MicroRNA-498 *. *Journal of Biological Chemistry*. 287(49). Elsevier: 41297–309. doi:10.1074/JBC.M112.407189.

59. Istfan, N.W., Person, K.S., Holick, M.F. *et al.*, 2007. 1α, 25-Dihydroxyvitamin D and fish oil synergistically inhibit G1/S-phase transition in prostate cancer cells. *The Journal of Steroid Biochemistry and Molecular Biology*. 103(3–5):726–730.

60. Flores, O., Wang, Z., Knudsen, K.E. *et al.*, 2010. Nuclear targeting of cyclin-dependent kinase 2 reveals essential roles of cyclin-dependent kinase 2 localization and cyclin E in vitamin D-mediated growth inhibition. *Endocrinology*. 151(3):896–908.

61. An, B.S., Tavera-Mendoza, L.E., Dimitrov, V., *et al.*, 2010. Stimulation of Sirt1-regulated FoxO protein function by the ligand-bound vitamin D receptor. *Molecular and Cellular Biology*. 30(20): 4890–4900.

62. Diao, Q.X., Zhang, J.Z., Zhao, T., Xue, F., *et al.*, 2016. Vitamin E promotes breast cancer cell proliferation by reducing ROS production and p53 expression. *European Review for Medical and Pharmacological Science*. 20(12): 2710–2717.

63. Theodosiou, M., Laudet, V. and Schubert, M., 2010. From carrot to clinic: An overview of the retinoic acid signaling pathway. *Cellular and Molecular Life Sciences*. 67(9): 1423–1445.

64. Mark, M., Ghyselinck, N.B. and Chambon, P., 2006. Function of retinoid nuclear receptors: Lessons from genetic and pharmacological dissections of the retinoic acid signaling pathway during mouse embryogenesis. *Annual Review of Pharmacology and Toxicology*. 46: 451–480.

65. Stephensen, C.B., 2001. Vitamin A, infection, and immune function. *Annual Review of Nutrition*. 21(1): 167–192.

66. Duester, G., 2008. Retinoic acid synthesis and signaling during early organogenesis. *Cell*. 134(6): 921–931.

67. Ross, A.C. and Ternus, M.E., 1993. Vitamin A as a hormone: Recent advances in understanding the actions of retinol, retinoic acid, and beta carotene. *Journal of the American Dietetic Association*. 93(11): 1285–1290.

68. Brinkmann, A.O., 1994. Steroid hormone receptors: Activators of gene transcription. *Journal of Pediatric Endocrinology and Metabolism*. 7(4): 275–282.

69. DiGiovanni, J., 1990. Inhibition of chemical carcinogenesis. In *Chemical carcinogenesis and mutagenesis II*. Springer, Berlin, Heidelberg, pp. 159–223.

70. Ziegler, R.G., 1989. A review of epidemiologic evidence that carotenoids reduce the risk of cancer. *The Journal of Nutrition*. 119(1): 116–122.

71. Knekt, P., Aromaa, A., Maatela, J., Aaran, R.K., Nikkari, T., Hakama, M., Hakulinen, T., Peto, R. and Teppo, L., 1990. Serum vitamin A and subsequent risk of cancer: Cancer incidence follow-up of the Finnish Mobile Clinic Health Examination Survey. *American Journal of Epidemiology*. 132(5): 857–870.

72. Stefanska, B., Salamé, P., Bednarek, A., *et al.*, 2012. Comparative effects of retinoic acid, vitamin D and resveratrol alone and in combination with adenosine analogues on methylation and expression of phosphatase and tensin homologue tumour suppressor gene in breast cancer cells. *British Journal of Nutrition*. 107(6): 781–790.

73. Mongan, N.P. and Gudas, L.J., 2007. Diverse actions of retinoid receptors in cancer prevention and treatment. *Differentiation*. 75(9): 853–870.

74. Pastorino, U., Infante, M., Maioli, M., *et al.*, 1993. Adjuvant treatment of stage I lung cancer with high-dose vitamin A. *Journal of Clinical Oncology*. 11(7): 1216–1222.

75. Oppenheimer, O., Cheung, N.K. and Gerald, W.L., 2007. The RET oncogene is a critical component of transcriptional programs associated with retinoic acid–induced differentiation in neuroblastoma. *Molecular Cancer Therapeutics*. 6(4): 1300–1309.

76. Cerchia, L., d'Alessio, A., Amabile, G., *et al.*, 2006. An autocrine loop involving ret and glial cell–derived neurotrophic factor mediates retinoic acid–induced neuroblastoma cell differentiation. *Molecular Cancer Research*. 4(7): 481–488.

77. Angrisano, T., Sacchetti, S., Natale, F., *et al.*, 2011. Chromatin and DNA methylation dynamics during retinoic acid-induced RET gene transcriptional activation in neuroblastoma cells. *Nucleic Acids Research*. 39(6): 1993–2006.

78. Lefebvre, B., Ozato, K. and Lefebvre, P., 2002. Phosphorylation of histone H3 is functionally linked to retinoic acid receptor β promoter activation. *EMBO Reports*. 3(4): 335–340.

79. Rahim, R. and Strobl, J.S., 2009. Hydroxychloroquine, chloroquine, and all-trans retinoic acid regulate growth, survival, and histone acetylation in breast cancer cells. *Anti-cancer Drugs*. 20(8): 736–745.
80. Stefanska, B., Rudnicka, K., Bednarek, A. *et al.*, 2010. Hypomethylation and induction of retinoic acid receptor beta 2 by concurrent action of adenosine analogues and natural compounds in breast cancer cells. *European Journal of Pharmacology*. 638(1–3). 47–53.
81. Stefanska, B., Salamé, P., Bednarek, A., *et al.*, 2012. Comparative effects of retinoic acid, vitamin D and resveratrol alone and in combination with adenosine analogues on methylation and expression of phosphatase and tensin homologue tumour suppressor gene in breast cancer cells. *British Journal of Nutrition*. 107(6): 781–790.
82. Fazi, F., Travaglini, L., Carotti, D., *et al.*, 2005. Retinoic acid targets DNA-methyltransferases and histone deacetylases during APL blast differentiation in vitro and in vivo. *Oncogene*. 24(11): 1820–1830.
83. Di Croce, L., Raker, V.A., Corsaro, M., *et al.*, 2002. Methyltransferase recruitment and DNA hypermethylation of target promoters by an oncogenic transcription factor. *Science*. 295(5557): 1079–1082.
84. Zuchegna, C., Aceto, F., Bertoni, A., *et al.*, 2014. Mechanism of retinoic acid-induced transcription: Histone code, DNA oxidation and formation of chromatin loops. *Nucleic Acids Research*. 42(17): 11040–11055.
85. Gudas, L.J., 2012. Emerging roles for retinoids in regeneration and differentiation in normal and disease states. *Biochimica et Biophysica Acta (BBA)-Molecular and Cell Biology of Lipids*. 1821(1): 213–221.
86. Jäpelt, R.B. and Jakobsen, J., 2013. Vitamin D in plants: A review of occurrence, analysis, and biosynthesis. *Frontiers in Plant Science*. 4: 136.
87. Thorne, J. and Campbell, M.J., 2008. The vitamin D receptor in cancer: Symposium on 'Diet and cancer.' *Proceedings of the Nutrition Society*. 67(2): 115–127.
88. Deeb, K.K., Trump, D.L. and Johnson, C.S., 2007. Vitamin D signalling pathways in cancer: Potential for anticancer therapeutics. *Nature Reviews Cancer*. 7(9): 684–700.
89. Colston, K.A.Y., COLSTON, M.J. and FELDMAN, D., 1981.1, 25-dihydroxyvitamin D3 and malignant melanoma: The presence of receptors and inhibition of cell growth in culture. *Endocrinology*. 108(3): 1083–1086.
90. Welsh, J., 2004. Vitamin D and breast cancer: Insights from animal models. *The American Journal of Clinical Nutrition*. 80(6):1721S–1724S.
91. Massagué, J., 2008. TGFβ in cancer. *Cell*. 134(2): 215–230.
92. Daniel, C., Schaub, K., Amann, K., *et al.*, 2007. Thrombospondin-1 is an endogenous activator of TGF-β in experimental diabetic nephropathy in vivo. *Diabetes*. 56(12): 2982–2989.
93. Yang, L., Yang, J., Venkateswarlu, S., *et al.*, 2001. Autocrine TGFβ signaling mediates vitamin D3 analog-induced growth inhibition in breast cells. *Journal of Cellular Physiology*. 188(3): 383–393.
94. Pálmer, H.G., Larriba, M.J., García, J.M., *et al.*, 2004. The transcription factor SNAIL represses vitamin D receptor expression and responsiveness in human colon cancer. *Nature Medicine*. 10(9): 917–919.
95. Bikle, D.D., Xie, Z. and Tu, C.L., 2012. Calcium regulation of keratinocyte differentiation. *Expert Review of Endocrinology & Metabolism*. 7(4): 461–472.
96. Ting, H.J., Yasmin-Karim, S., Yan, S.J., *et al.*, 2012. A positive feedback signaling loop between ATM and the vitamin D receptor is critical for cancer chemoprevention by vitamin D. *Cancer Research*. 72(4): 958–968.
97. Gonzalo, S., 2014. Novel roles of 1α, 25 (OH) 2D3 on DNA repair provide new strategies for breast cancer treatment. *The Journal of Steroid Biochemistry and Molecular Biology*. 144: 59–64.
98. Lankelma, J.M., Voorend, D.M., Barwari, T., *et al.*, 2010. Cathepsin L, target in cancer treatment? *Life Sciences*. 86(7–8): 225–233.
99. Peh, H.Y., Tan, W.D., Liao, W. *et al.*, 2016. Vitamin E therapy beyond cancer: Tocopherol versus tocotrienol. *Pharmacology & Therapeutics*. 162: 152–169.
100. Sailo, B.L., Banik, K., Padmavathi, G., *et al.*, 2018. Tocotrienols: The promising analogues of vitamin E for cancer therapeutics. *Pharmacological Research*. 130: 259–272.
101. Srivastava, J.K. and Gupta, S., 2006. Tocotrienol-rich fraction of palm oil induces cell cycle arrest and apoptosis selectively in human prostate cancer cells. *Biochemical and Biophysical Research Communications*. 346(2): 447–453.
102. Yap, W.N., Chang, P.N., Han, H.Y., *et al.*, 2008. γ-Tocotrienol suppresses prostate cancer cell proliferation and invasion through multiple-signalling pathways. *British Journal of Cancer*. 99(11): 1832–1841.
103. Cardenas, E. and Ghosh, R., 2013. Vitamin E: A dark horse at the crossroad of cancer management. *Biochemical Pharmacology*. 86(7): 845–852.
104. Prasad, S., Yadav, V.R., Sung, B., *et al.*, 2012. Ursolic acid inhibits growth and metastasis of human colorectal cancer in an orthotopic nude mouse model by targeting multiple cell signaling pathways: Chemosensitization with capecitabine. *Clinical Cancer Research*. 18(18): 4942–4953.

105. Montagnani Marelli, M., Marzagalli, M., Fontana, F., *et al.*, 2019. Anticancer properties of tocotrienols: A review of cellular mechanisms and molecular targets. *Journal of Cellular Physiology.* 234(2): 1147–1164.
106. Huang, Y., Wu, R., Su, Z.Y., *et al.*, 2017. A naturally occurring mixture of tocotrienols inhibits the growth of human prostate tumor, associated with epigenetic modifications of cyclin-dependent kinase inhibitors p21 and p27. *The Journal of Nutritional Biochemistry.* 40: 155–163.
107. Fernandes, N.V., Guntipalli, P.K. and Mo, H., 2010. d-δ-Tocotrienol-mediated cell cycle arrest and apoptosis in human melanoma cells. *Anticancer Research.* 30(12): 4937–4944.
108. Wali, V.B., Bachawal, S.V. and Sylvester, P.W., 2009. Endoplasmic reticulum stress mediates γ-tocotrienol-induced apoptosis in mammary tumor cells. *Apoptosis.* 14(11): 1366–1377.
109. Park, S.K., Sanders, B.G. and Kline, K., 2010. Tocotrienols induce apoptosis in breast cancer cell lines via an endoplasmic reticulum stress-dependent increase in extrinsic death receptor signaling. *Breast Cancer Research and Treatment.* 124(2): 361–375.
110. Patacsil, D., Tran, A.T., Cho, Y.S., *et al.*, 2012. Gamma-tocotrienol induced apoptosis is associated with unfolded protein response in human breast cancer cells. *The Journal of Nutritional Biochemistry.* 23(1): 93–100.
111. Tiwari, R.V., Parajuli, P. and Sylvester, P.W., 2014. γ-Tocotrienol-induced autophagy in malignant mammary cancer cells. *Experimental Biology and Medicine.* 239(1): 33–44.
112. Comitato, R., Guantario, B., Leoni, G., *et al.*, 2016. Tocotrienols induce endoplasmic reticulum stress and apoptosis in cervical cancer cells. *Genes & Nutrition.* 11(1): 1–15.
113. McIntyre, B.S., Briski, K.P., Gapor, A. *et al.*, 2000. Antiproliferative and apoptotic effects of tocopherols and tocotrienols on preneoplastic and neoplastic mouse mammary epithelial cells (44544). *Proceedings of the Society for Experimental Biology and Medicine.* 224(4): 292–301.
114. Shun, M.C., Yu, W., Gapor, A., *et al.*, 2004. Pro-apoptotic mechanisms of action of a novel vitamin E analog (α-TEA) and a naturally occurring form of vitamin E (δ-tocotrienol) in MDA-MB-435 human breast cancer cells. *Nutrition and Cancer.* 48(1): 95–105.
115. Loganathan, R., Selvaduray, K.R., Nesaretnam, K. *et al.*, 2013. Tocotrienols promote apoptosis in human breast cancer cells by inducing poly (ADP-ribose) polymerase cleavage and inhibiting nuclear factor kappa-B activity. *Cell Proliferation.* 46(2): 203–213.
116. Liu, H.K., Wang, Q., Li, Y., *et al.*, 2010. Inhibitory effects of γ-tocotrienol on invasion and metastasis of human gastric adenocarcinoma SGC-7901 cells. *The Journal of Nutritional Biochemistry.* 21(3): 206–213.
117. Mizushina, Y., Nakagawa, K., Shibata, A., *et al.*, 2006. Inhibitory effect of tocotrienol on eukaryotic DNA polymerase λ and angiogenesis. *Biochemical and Biophysical Research Communications.* 339(3): 949–955.
118. Selvaduray, K.R., Radhakrishnan, A.K., Kutty, M.K. *et al.*, 2012. Palm tocotrienols decrease levels of pro-angiogenic markers in human umbilical vein endothelial cells (HUVEC) and murine mammary cancer cells. *Genes & Nutrition.* 7(1): 53–61.
119. Abdullah, A. and Atia, A., 2013. Tocotrienols: Molecular aspects beyond its antioxidant activity. *Journal of Medical Research and Practice, North America.* 2: 246–250.
120. Kaneko, S., Sato, C., Shiozawa, N., *et al.*, 2018. Suppressive effect of delta-tocotrienol on hypoxia adaptation of prostate cancer stem-like cells. *Anticancer Research.* 38(3): 1391–1399.
121. Luk, S.U., Yap, W.N., Chiu, Y.T., *et al.*, 2011. Gamma-tocotrienol as an effective agent in targeting prostate cancer stem cell-like population. *International Journal of Cancer.* 128(9): 2182–2191.
122. Gopalan, A., Yu, W., Sanders, B.G. *et al.*, 2013. Eliminating drug resistant breast cancer stem-like cells with combination of simvastatin and gamma-tocotrienol. *Cancer Letters.* 328(2): 285–296.
123. Xiong, A., Yu, W., Liu, Y., *et al.*, 2016. Elimination of ALDH+ breast tumor initiating cells by docosahexanoic acid and/or gamma tocotrienol through SHP-1 inhibition of Stat3 signaling. *Molecular Carcinogenesis.* 55(5): 420–430.
124. Wu, F.Y.H., Liao, W.C. and Chang, H.M., 1993. Comparison of antitumor activity of vitamins K1, K2 and K3 on human tumor cells by two (MTT and SRB) cell viability assays. *Life Sciences.* 52(22): 1797–1804.
125. Sun, L., Yoshii, Y., Miyagi, K. *et al.*, 1999. Proliferation inhibition of glioma cells by vitamin K2. *No Shinkei geka. Neurological Surgery.* 27(2): 119–125.
126. Takami, A., Nakao, S., Ontachi, Y., *et al.*, 1999. Successful therapy of myelodysplastic syndrome with menatetrenone, a vitamin K2 analog. *International Journal of Hematology.* 69(1): 24–26.
127. Miyazawa, K., Yaguchi, M., Funato, K *et al.*, 2001. Apoptosis/differentiation-inducing effects of vitamin K2 on HL-60 cells: Dichotomous nature of vitamin K2 in leukemia cells. *Leukemia.* 15(7): 1111–1117.

128. Cantoni, O., Fiorani, M., Cattabeni, F. *et al.*, 1991. DNA breakage caused by hydrogen peroxide produced during the metabolism of 2-methyl-1, 4-naphthoquinone (menadione) does not contribute to the cytotoxic action of the quinone. *Biochemical Pharmacology.* 42: S220–S222.

129. Sata, N., Klonowski-Stumpe, H., Han, B., *et al.*, 1997. Menadione induces both necrosis and apoptosis in rat pancreatic acinar AR4–2J cells. *Free Radical Biology and Medicine.* 23(6): 844–850.

130. Nutter, L.M., Ann-Lii, C., Hsiao-Ling, H., *et al.*, 1991. Menadione: Spectrum of anticancer activity and effects on nucleotide metabolism in human neoplastic cell lines. *Biochemical Pharmacology. 41*(9): 1283–1292.

131. Juan, C.C. and Wu, F.Y., 1993. Vitamin K3 inhibits growth of human hepatoma HepG2 cells by decreasing activities of both p34cdc2 kinase and phosphatase. *Biochemical and Biophysical Research Communications.* 190(3): 907–913.

132. Jarabak, R. and Jarabak, J., 1995. Effect of ascorbate on the DT-diaphorase-mediated redox cycling of 2-methyl-1, 4-naphthoquinone. *Archives of Biochemistry and Biophysics.* 318(2): 418–423.

133. Noto, V., Taper, H.S., Yi-Hua, J., *et al.*, 1989. Effects of sodium ascorbate (vitamin C) and 2-methyl-1, 4-naphthoquinone (vitamin K3) treatment on human tumor cell growth *in vitro.* I. Synergism of combined vitamin C and K3 action. *Cancer.* 63(5): 901–906.

134. Taper, H.S., De Gerlache, J., Lans, M. *et al.*, 1987. Non-toxic potentiation of cancer chemotherapy by combined C and K3 vitamin pre-treatment. *International Journal of Cancer.* 40(4): 575–579.

135. Taper, H.S. and Bannasch, P., 1976. Histochemical correlation between glycogen, nucleic acids and nucleases in pre-neoplastic and neoplastic lesions of rat liver after short-term administration of N-nitro somorpholine. *Zeitschrift für Krebsforschung und Klinische Onkologie.* 87(1): 53–65.

136. Waxman, S. and Bruckner, H., 1982. The enhancement of 5-fluorouracil antimetabolic activity by leucovorin, menadione and α-tocopherol. *European Journal of Cancer and Clinical Oncology.* 18(7): 685–692.

4 Targeting Cancer Stem Cells by Natural Products for Chemoprevention

Kamana Singh, Prabha Arya, and Ram Sunil Kumar L.

CONTENTS

4.1 Introduction 58
4.2 Cancer Overview 58
4.3 The CSC Model 59
4.4 Cell Signaling Pathways That May Lead to the Blossoming of Different Types
of Cancers and Are Potential Targets for Cancer Prevention 59
    4.4.1 Wingless/Integrated (Wnt) Pathway 60
    4.4.2 Notch Pathway 60
    4.4.3 Hh Signaling Pathway 60
    4.4.4 PI3K/AKT/mTOR Pathways 62
    4.4.5 NF-κB Signaling Pathway 62
    4.4.6 JAK-STAT and FAK Signaling Pathway 63
    4.4.7 TGF/SMAD Signaling Pathway in CSCs 63
    4.4.8 PPAR Signaling Pathways in CSCs 63
    4.4.9 Regulation and Crosstalk Between Different Signaling Pathways 64
4.5 Chemical and Radiation Treatment and Their Harmful Effects 64
4.6 Natural Products Which Are Used in the Prevention of Different Diseases 64
4.7 Natural Products Which Are Used in the Prevention of Different Cancers
by Acting on CSCs 65
    4.7.1 Antioxidants Can Help in the Prevention of Different Types of Cancer 65
        4.7.1.1 Vitamin C 65
        4.7.1.2 Vitamin D3 65
        4.7.1.3 Vitamin E 65
        4.7.1.4 Selenium 66
    4.7.2 Studies on Food Products as Prospective Preventive Treatment for Cancer 66
        4.7.2.1 Curcumin in Different Types of Cancers 66
        4.7.2.2 Black Cumin 66
        4.7.2.3 Ginger 66
        4.7.2.4 Garlic 66
        4.7.2.5 Piperine in Pepper 67
        4.7.2.6 Saffron 67
        4.7.2.7 Chili Pepper 67
        4.7.2.8 Flavonoids in Mint 67
        4.7.2.9 Fenugreek 67
        4.7.2.10 Quinines 68
        4.7.2.11 Cloves 68
        4.7.2.12 Koenimbin from Curry Leaves 68
        4.7.2.13 Resveratrol (Found in Grapes, Berries, and Peanuts) 68

DOI: 10.1201/b23311-4

4.7.2.14 Green Tea..69
4.7.2.15 α-Pinene from Bay Leaf 69
4.7.2.16 Lycopene from Tomato 69
4.7.2.17 Soybean..69
4.7.2.18 Use of Probiotics for Prevention of Cancer.................................69
4.8 Conclusion and Prospects 70
4.9 References..70

4.1 INTRODUCTION

In recent times, cancer has been considered a lifestyle disorder; maybe it is more prevalent or maybe there are simply clear diagnoses and improvements in diagnostic technologies. The number of diagnosed cancer cases is growing worldwide with time. Around 26 million new cancer cases will be found every year with 17 million deaths by cancer annually by 2030 expectedly [1]. Though no doubt, the number of cases has increased due to changes in food habits of people globally as they are going away from natural products and their natural forms. Also, there have been many adulterants added to the food which may act as carcinogens [2, 3]. Apart from that, industrialization has increased the level of pollutants and a reduction in natural cover has accentuated the woes, making it more difficult to tackle the pollutants. These carcinogens can cause mutations in genes that can lead to cancer of different types, depending on the encounter of the organ with the carcinogen. Natural products have been considered for chemoprevention for different lifestyle disorders, be it hypertension, diabetes, or metabolic syndrome.

As all cells have programming to differentiate into different types of cells, CSCs are also destined to become cancerous once they find an appropriate environment. CSCs are known to have resistance to treatment by chemotherapy or radiotherapy. As stem cells are the cause of recurrence of cancer, there has been keen interest in bringing the lifestyle toward a more organic way to reduce the occurrence of cancer and it can be stopped at the stem cell level and can be prevented from spreading to different areas.

To maximize the utility and long-term effects of the treatments, recent trials have been in the area of combining different types of therapies like chemotherapy, radiation therapy, and the usage of compounds from natural sources. For this purpose, natural compounds from different sources have been tested for their efficacy in cancer treatment [4]. There are various types of natural compounds with prospective therapeutic purposes that have been screened from plants, animals, marine organisms, and microorganisms. Some plant-derived molecules are already in use for the treatment of cancer, like paclitaxel from *Taxus brevifolia* [5], irinotecan from *Camptotheca acuminata* [6], vincristine from *Catharanthus roseus* [7], and etoposide from *Podophyllum peltatum* [8]. Also, some of the antibiotics produced by microorganisms have been applied for the treatment of cancer like dactinomycin, bleomycin, and doxorubicin. Cytarabine, isolated from a marine source, the Caribbean sponge *Cryptotheca crypta*, is also used for the treatment of cancer. Other compounds isolated are aplidine, bryostatin-1, dolastatin, and ET-,74,3, among others, from marine sources, which are under trial. First-in-class life-saving chemotherapeutic agent, camptothecin, which was initially isolated and identified from *Camptotheca acuminata* tree, can poison topoisomerase-I and kill the cells. On the other hand, paclitaxel from *Taxus brevifolia* is discovered to inhibit cancer cell growth by stabilizing the microtubules and has been used in the treatment of ovarian cancer, breast cancer, non-small-cell lung cancer, and Kaposi's sarcoma [9].

4.2 CANCER OVERVIEW

Different types of cancers have been found based on their location in the body, like skin cancer, breast cancer, prostate cancer, lung cancer, oral cancer, glioma, leukemia, and many others. What they have in common is a tendency to divide uncontrollably. There has been active research in finding causes of cancer, such as possible carcinogens, mutations, radiations, and changes in the microenvironment. These effects are prominent in various genes involved like proto-oncogenes and tumor suppressor genes.

There are many morphological, physiological, genetic, and epigenetic changes that occur in cancer cells. For example, cancer cells have the unique feature of maintaining telomere length, which enables them to escape the apoptotic route, which happens in the case of normal cells after a few cell divisions. On the other hand, there are other proteins involved in the signaling pathways, and changes in these signaling pathways can give rise to uncontrolled cell division of cancer cells. A normal cell requires certain conditions in which they flourish, but cancer cells acquire this characteristic to divide without the requirement of the things for growth, like growth factors.

Another feature that cancer cells show is their ability to escape the check at the cell cycle points. Various proteins are involved in cell cycle regulation and are categorized into oncogenes and tumor suppressor genes. These genes are mutated in cancer cells, which remove the regulation and unstoppable cell division as the result. Also, cancer cells have different requirements for the growth factors and other molecules or can produce them, which makes them independent. Cancer cells have other morphological properties, like insensitivity to cell density and anchorage-independent growth. Apart from this, cancer cells also affect the rate of apoptosis, which helps in making a heap of live cells in tumors, which is also a reason for tumorigenesis in many cancers.

4.3   THE CSC MODEL

According to the CSC (cancer stem cell) model, tumor formation starts with a small group of particular cells with stem-like characteristics, which is why they are termed CSCs. Normal stem cells own three important properties—uncontrolled cell division, apoptosis, and the capability to differentiate into different kinds of cells [10]. CSCs can renew themselves and also inhibit the apoptotic pathways, which gives tumors resistance to chemotherapeutic drugs, hormone replacement therapy, and radiotherapy [11], as well as epithelial to mesenchymal transitions (EMT) and metastasis. It is important to eradicate CSCs during the treatment of cancer; failure to do so leads to tumor recurrence and metastasis. Recently, the focus has been on understanding the signaling pathways that underlie self-renewal and drug resistance and the characteristic properties of CSCs from different tumor types. Understanding the signaling and other pathways that the CSCs follow to escape cause and causing drug resistance can give insight into the probable target of CSCs.

The CSCs have characteristic cell surface markers that are utilized for their isolation from CSCs that are involved in different types of cancers—for example, CD29+ in breast cancer (involved in cell adhesion); CD117+ receptor tyrosine kinase, also known as c-kit, in prostate cancer; CD90+, an N-glycosylated glycophosphatidylinositol anchored cell surface protein, in brain cancer; CD133+, a prominent-1, in stomach cancer; CD200+, a membrane glycoprotein, in colorectal cancer; CD24+, a glycosylated mucin-like cell surface protein, in the liver and ovarian cancer; CD34+, a receptor for chemokine, in AML; and CD20+, a glycosylated phosphoprotein, in bladder cancer; among others. These markers are specific proteins involved in a variety of functions depending on the type of tumor in which they are present on CSCs and in many cancers more than one type of marker is found to be present upon testing.

CSCs have also evolved the mechanism to overpower the stress of chemicals given for chemotherapy. They have increased expression of aldehyde dehydrogenase, which can neutralize the chemical used for treatment. Aldehyde dehydrogenase is also utilized as a marker to distinguish CSCs from normal cells [12].

4.4   CELL SIGNALING PATHWAYS THAT MAY LEAD TO THE BLOSSOMING OF DIFFERENT TYPES OF CANCERS AND ARE POTENTIAL TARGETS FOR CANCER PREVENTION

Different pathways are found to help CSCs to develop into a tumor. The CSCs undergo self-renewal and remain silent or less aggressive with the help of wingless/integrated (Wnt), PI3K/AKT/mTOR,

Hedgehog (Hh), and Notch signaling pathways and MDR-ABC transporters. Signaling pathways also help CSCs in various functions, like differentiation, multidrug resistance, and metastasis. There have been constant efforts to find the ways by which these signaling pathways can be altered, and CSCs can be prevented from turning into cancer cells with the help of natural products that have been shown to reduce malignancy rates.

4.4.1 WINGLESS/INTEGRATED (WNT) PATHWAY

The Wnt signaling system is associated with many cancers, like invasive ductal breast carcinomas, colorectal cancer, colorectal cancer, esophageal cancer, and papillary thyroid cancer [13]. This pathway functions via ubiquitylation and degradation, is involved in embryonal growth, and is responsible for the development of the central nervous system, cardiovascular system, and limbs, as well as the differentiation of CSCs. The pathway has two types: canonical and non-canonical. In canonical pathways, a destruction box is formed by APC protein, Axin, and glycogen synthase kinase-3 (GSK3). In the absence of a Wnt ligand, GSK3 is responsible for the phosphorylation of protein β-catenin. Tagged by phosphorylation β-catenin is degraded by proteasomal by ubiquitylation. In the presence of Wnt protein, Axin is recruited to bind to the receptor and GSK3 is unable to phosphorylate the β-catenin. Unphosphorylated β-catenin then enters the nucleus and binds to tcf, which is a transcription factor, to activate the gene expression responsible for CD44, myc, MMP7, SMYD3, PMP22, and so on (Figure 4.1).

β-catenin is not involved in non-canonical Wnt signaling pathways. In non-canonical pathways, different sub-pathways—namely, frizzled (FZD) or ROR1/ROR2/RYK receptors to the Wnt/planer cell polarity (PCP), Wnt/receptor tyrosine kinase, and Wnt/Ca^{2+} signaling cascades—are followed [14]. In this signaling pathway, cytoskeletal rearrangement occurs by the transcriptional outcome. There is an adaptor protein called disheveled adaptor proteins (Dvl) to which two proteins, Wnt signaling and ROR frizzled proteins, bind to activate it. Activated Dvl then inhibits the binding of DAAM1, which is a cytoplasmic protein and GTPase Rho.

CSCs alter the signaling pathway in such a manner that their differentiation into mature cancer cells starts. There are multiple points down the Wnt pathway which are manipulated in the CSCs. Apart from this, there are many ways by which apoptosis is inhibited by the Wnt pathway [13].

4.4.2 NOTCH PATHWAY

The Notch pathway is found to be involved in many tumors, like those of glial cells in brain or spinal cord, also called glioblastoma, those of white blood cells (leukemia), and those of the breast, pancreas, colon, lung, and many others [13]. The Notch regulatory pathway is a unique pathway that is involved in straight cell-cell interactions. This pathway comprises a receptor, a ligand, the transcription factor CSL (CBF-1, Suppressor of Hairless, Lag), a DNA binding protein, effectors, and other Notch regulatory molecules. The ligand Delta binds to the Notch receptor and causes the Notch domain is cleaved proteolytically by γ-secretase, which occurs at the inside of the cell membrane. The cleaved domain from Notch receptor moves in the nucleus, where it leads to activation of gene expression by binding to the CSL, which is a transcription factor, activating the genes of the transcription inhibitor family. There are subtypes of Notch receptor and ligand expressed in different types of tumors, which can act as oncogenes and tumor suppressor genes. The local environment surrounding the tumor have a crucial role in the tumor development involving the Notch signaling pathway.

4.4.3 HH SIGNALING PATHWAY

The Hh pathway is reportedly related to CSCs of human cancers, like lung cancer, breast cancer, bladder cancer, pancreatic cancer, and many others. As there are variations in activation of Hh

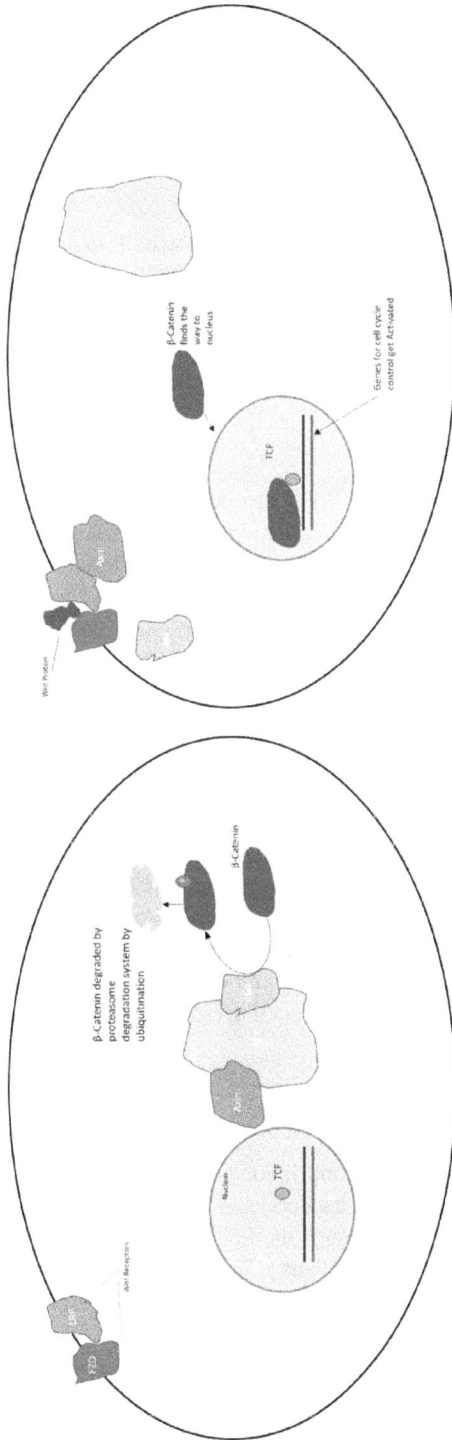

FIGURE 4.1 Wnt pathway. In this pathway, when (a) Wnt protein is not there, a destruction complex forms with three proteins—Axin, APC, and GSK3—which phosphorylates and destroys the β-catenin. When Wnt protein is present, Axin is recruited to the inner surface of the membrane, and β-catenin enters the nucleus in unphosphorylated form and binds with transcription factor TCF and control the expression the genes.

signaling, there are differences in trigger points in different tumors. The Hh signaling pathway is a complex pathway that consists of number of ligands and receptors and is involved in embryonic development and the maturation of the nervous system, as well as the organs, like the heart, lungs, skeleton, limbs, and gut, of the embryo [15].

Some of the Hh ligands are the PTCH (patched; which is a transmembrane protein receptor), the transmembrane protein SMO (smoothened), intermediate transduction molecules, and the transcription factor Gli [16]. PTCH and SMO have the opposite effect as SMO is involved in positive regulation while PTCH regulates negatively. Another protein is Gli, and some subtypes of Gli are Gli1, Gli2, and Gli3. Gli1 is involved in activating transcription, Gli3 inhibits transcription, and Gli2 has a dual role, depending on the circumstances.

PTCH binds to SMO when there is no signal present at the cell surface in the form of ligand and prevents it from activating Gli. When Hh signal is present, PTCH is bound to it and its conformation changes, which prevents PTCH binding to SMO, and SMO then activates Gli, and Gli then enters the nucleus where it activates the genes involved in regulation of cell growth, proliferation, and differentiation [13].

4.4.4 PI3K/AKT/mTOR Pathways

The PI3K/Akt/mTOR signaling pathway is found to be a causative factor in some tumors, such as breast cancer, prostate cancer, non-small-cell lung cancer (NSCLC), esophageal adenocarcinoma, Burkitt lymphoma, and colorectal cancer, by suppressing the apoptosis by phosphorylation and alleviating the inhibition on cell cycle [13]. Through its capability to catalyze the phosphorylation of several key target proteins, Akt suppresses apoptosis and inhibits cell cycle arrest. Phosphatidylinositol-3-kinase (PI3K) is an intracellular phosphatidylinositol kinase that catalyzes the phosphorylation at the 3'-OH at PIP2 (phosphatidylinositol-4,5 bisphosphate) and converts it into phosphatidylinositol-3,4,5 bisphosphate). PIP3 then binds to specific serine/threonine kinases known as Akt and take it to inner surface of the plasma membrane, where it is activated by phosphorylation by mTOR complex (mammalian target of rapamycin), and pyruvate dehydrogenase kinases (PDK). Activated Akt then phosphorylates a number of target proteins that regulate cell survival, such as foxo (member of the Forkhead family), and a transcription factor, when phosphorylated, forms a site for a specific chaperon and gets inactivated. In the absence of activated Akt, unphosphorylated foxo gets into the nucleus and causes expression of protein involved in apoptosis.

Phosphate and tensin homologue (PTEN) is a phosphatase that removes the phosphate group from PIP2 and prevents the activation of Akt [14] and deletion of PTEN gene in neural stem cells leads to occurrence of cancer cell properties, like cell division, cell enlargement, inhibition of apoptosis, and metastasis [13].

4.4.5 NF-κB Signaling Pathway

The NF-κB signaling pathway has been found involved in cancers of genital and urinary tracts, head and neck cancers, breast cancer, blood cancers, and multiple myeloma [13]. As the cancer cells have a component of inflammatory components, the first regulation of NF-κB occurs at the level of translocation of transcription factors to the nucleus [15]. NF-κB contains family of five transcription factors that play major roles in immune response and regulation and survival of proliferation of different types of cells. The NF-κB pathway has two types: canonical pathway and non-canonical pathway (like Wnt pathway). The canonical pathway, involved in inflammatory response, is activated by inflammatory molecules like bacterial cell components, such as lipopolysaccharides (LPS), while the non-canonical pathway is stimulated by activated NIK, which induces the phosphorylation of IKK1, which leads to phosphorylation and generation of p52 from p100 [13]. NF-κB signaling is activated by different molecules produced by the tumor, like cytokines, growth factors, proteases, and angiogenic factors [17].

4.4.6 JAK-STAT and FAK Signaling Pathway

The Janus kinase/signal transducers and activators of transcription (JAK-STAT) signaling pathway is also stimulated by cytokines. Mutation in JAK2 and abnormal activation of STAT3 have been found in many tumors and have been reported in cancers like thrombocythemia, myeloproliferative neoplasms, polycythemia vera, and myelofibrosis.

The JAK-STAT pathway is of non-receptor type of kinase. These have a receptor with an extracellular domain to bind with ligand (cytokine mostly) and a transmembrane helix followed by cytosolic domain. The cytosolic domain does not contain any enzyme activity, but it binds and activates the other effectors, like cytosolic tyrosine kinases. In this pathway ligand binds to the extracellular domain of JAK which leads to dimerization of receptors and cross phosphorylation of tyrosine kinases which are attached with JAK noncovalently.

Signal transducer and activator of transcription (STAT) plays a major role in signal transduction and transcriptional activation. Numerous cytokines and growth factors send signals through the JAK-STAT signaling pathway, which includes interleukin (IL-2–7), granulocyte/macrophage colony-stimulating factor (GCSF), growth hormone (GF), epidermal growth factor (EGF), platelet-derived growth factor (PDGF), and interferon, as they have respective receptors on the cell membrane. When the signal reaches at the outer segment of receptor molecule, JAK present in cytosol is brought to activate the receptor, which results in receptor phosphorylation at the tyrosine residue. The STAT protein has the SH2 site, which binds with the phosphorylated tyrosine on the receptor molecule, leading to tyrosine phosphorylation of STAT. STAT is a transcription factor after dimerization, and upon entering the nucleus, STAT dimer directly regulates related genes, which results in proliferation or differentiation of target cells.

Integrin is another molecule having related mechanism of action and involved in cell matric interaction. This molecule binds with the receptor, which is associated with non-receptor kinase FAK (focal adhesion kinase), which now gets autophosphorylated and forms a binding site for Src protein as it contains SH2 domains. FAK pathway has been found associated in breast cancer solid tumors [11].

4.4.7 TGF/SMAD Signaling Pathway in CSCs

The transforming growth factor (TGF-β) pathway seems to have modulated prostate cancer, lung cancer, liver cancer, colorectal cancer, glioma stem cells, and breast CSCs. The TGF signaling pathway is involved in regulation of various cellular events, like organism and embryo development, cell differentiation, cell proliferation, homeostasis, and apoptosis. There are two types of receptors in the TGF-β signaling pathway: type I and type II. These receptors are also non-receptor kinases like JAK-STAT, but the phosphate group is added most commonly at serine/threonine residue on the surface of enzyme. In a series of events, TGF-β binds to a type II receptor, which then phosphorylates the type I receptor, which in turn phosphorylates the transcription factor Smad, and Smad then binds Co-smad (common pathway Smad). Smad/co-Smad forms a complex and function as a transcription factor, which regulates the expression of target genes in nucleus. Few examples of the TGF-β superfamily ligands are growth and differentiation factors, anti-Mullerian hormone, and so on [18].

4.4.8 PPAR Signaling Pathways in CSCs

Peroxisome proliferator-activated receptors, originally recognized in peroxisomes, have been involved in many types of cancer cells, like breast cancer, prostate cancer, liver cancer, and bladder cancer. The ligand for PPAR are fats or alike molecules. There are different subtypes of PPAR: PPARα, PPARδ, and PPARγ. These receptors bind with another receptor, RXR, in cytosol and then move to nucleus, where they activate the genes responsible for uptake and oxidation of fatty acids and the enzymes involved in formation of ketone bodies during fasting. Although the function of PPAR in CSCs is not clear, these are expressed in different organs, like kidneys, heart, liver, and brown adipose tissues (BAT). Different subtypes of PPAR have different roles. PPARγ functions as tumor suppressor by binding to canonical pathway response element in the miR-15a gene in CSCs to reduce the mesenchymal stem cells, which inhibit angiogenesis. The PPARγ/NF-κB pathway prevents cell death in ovarian CSCs by activating M2 polarization of macrophages [13].

4.4.9 REGULATION AND CROSSTALK BETWEEN DIFFERENT SIGNALING PATHWAYS

So far, we have discussed signaling pathways as if they are existing independent of each other, but actually there are more than one pathway present in particular types of CSCs. These pathways get regulated by feedback, and they have interconnection among themselves. This interaction can be positive regulatory, leading to the potentiation of each other's effect, or they can be negative regulatory, which culminates in the modulation of function; for example, PPARγ activation promotes the expression of its target gene PTEN, which inhibits PI3K/Akt/mTOR signaling to inhibit self-renewal, tumor development, and metastasis in CSCs of glioblastoma, liver cancer, and so on.

4.5 CHEMICAL AND RADIATION TREATMENT AND THEIR HARMFUL EFFECTS

Many drugs are given for the treatment of cancer, which target particular enzyme molecules; for example, methotrexate targets dihydrofolate reductase, but these drugs are nonspecific and affect normal cells too. Apart from that, multidrug resistance (MDR) has been another concern for the researchers for the treatment using chemotherapy, for example, by methotrexate. This resistance is pertaining to the ABC (ATP-binding cassettes) transport involved in the transport of molecules from outside to inside the cells or from inside to outside the cells. There are several strategies a CSC follows, to reduce the concentration: (1) expulsion of the drugs outside and production of enzymes to detoxify the drug remaining in the cell, (2) decreased intake of drugs, (3) activation of DNA repair system, and (4) inhibition of apoptosis. There are several subtypes of transporters that can be addressed for the treatment of cancer to hold the drug or combination of drugs inside the cell in effective concentration, as well as blocking the enzymes to prevent the drug detoxification [19].

Radiation therapy is also not selective. Hence, there is a requirement for selectivity in the prevention of stem cells from dividing and treatment, which will be helping in protecting the useful cells from the harmful effect of chemicals and radiation. In a pool of cancer stem cells, those cells that have drug resistance via clonal selection survive and develop into cells that have drug resistance, such as lung resistance proteins (LRPs) and ABC transporter proteins. Natural products may have inhibitory action by inhibiting drug efflux, inhibiting the gene expression of these transporters.

Another type of MDR transporter is P-glycoprotein, a product of ABCB1 gene expression. It has been found that there are many natural products that have inherent ABCB1 inhibiting properties. This is a molecular target for the treatment of many human cancers. It is a 72-kDa protein that has been involved in the development of drug resistance, which suggests involvement of CSCs and its regulation by various factors, like growth factors like EGFR and transcription factors like NF-κB.

4.6 NATURAL PRODUCTS WHICH ARE USED IN THE PREVENTION OF DIFFERENT DISEASES

There has been some interest in different types of products obtained from natural and least-processed sources that can be used as a treatment for this purpose. India is land with different types of natural products that have been under investigation for the prevention of cancer. Spices, like turmeric, clove, aniseed, and fenugreek, have been shown to have cancer-preventing properties. Many of the compounds have properties that have benefits in preventing and treating different disorders, like cardiovascular diseases, diabetes, skin disorders, and other metabolic disorders. Most of the components of these natural products are organic compounds that are widely present in different concentrations in different food items and have been tested for their benefits in specific disorders.

4.7 NATURAL PRODUCTS WHICH ARE USED IN THE PREVENTION OF DIFFERENT CANCERS BY ACTING ON CSCS

There is a wide variety of natural compounds that have been tested for their efficacies in the prevention of CSCs from growing by different mechanisms and combination of mechanisms, like preventing the CSCs from dividing and differentiating and causing the apoptosis of cancer cells [20]. Some of the food products and components of food products are discussed here.

4.7.1 ANTIOXIDANTS CAN HELP IN THE PREVENTION OF DIFFERENT TYPES OF CANCER

4.7.1.1 Vitamin C

Vitamin C, which is an anti-scurvy vitamin and an electron donor in different oxidation-reduction reactions, has a role in treating cancer. Vitamin C is responsible for making collagen, and that is the reason that those who have a suboptimal intake of vitamin C suffer from gum bleeding as lack of collagen weakens the tissue. In the case of cancer cells, vitamin C has been found to improve the treatment of the patients in a dose-dependent manner by blocking the glycolysis in cancer cells and nitroso synthesis [21]. Vitamin C is also an antioxidant, and antioxidant vitamins are useful for clearing the free radical load, which can lead to inflammation and alteration in signaling pathways, which lead to cancerous growth [22].

4.7.1.2 Vitamin D3

Vitamin D3, also called the sunshine vitamin, is required for bone health and is also considered a hormone. Accumulating evidence has shown the inhibitory effects of vitamin D and its analogs on the CSC signaling pathways, suggesting that vitamin D is a potential preventive/therapeutic agent against CSCs. Vitamin D3 is one of the first molecule to demonstrate crosstalk with Notch signaling pathways. In one study it is found that Vitamin D3 affects mRNA level of Notch, and other effectors of pathways in prostate epithelial cells and breast cancer cells while failing to inhibit Notch signaling in brain cancer cell line. Other pathways, like Hh, Wnt, and TGF-β, were also found to be modulated with the help of vitamin D3 [18].

4.7.1.3 Vitamin E

Vitamin E is also known as an antioxidant, anti-sterility, and anti-inflammatory vitamin and is reported to have anticancer properties, especially in case of prostate cancer. It has been tested on other types of cancer incidences. In other cancers, there is not much improvement, but in prostate cancer, a statistically significant reduction is observed. In a sense, vitamin E can be used as preventive medicine to protect men at high risk of prostate cancer. However, it is difficult to estimate the concentration of vitamin E and decide the concentration required for the treatment of cancer because of its solubility in oil in comparison to vitamin C, which is water-soluble.

Vitamin E is involved in the modulation of signaling pathways, like NF-κB, JAK-STAT, and Akt/mTOR, by changing the properties of molecules involved in cancer cell proliferation, reduction in apoptosis and, metastasis. There are different molecular forms of vitamin E, such as γ-tocopherol (γT), δ-tocopherol (δT), γ-tocotrienol (γTE), and δ-tocotrienol (δTE), which have been found to be effective in preventing the progression of various types of cancer in preclinical animal models.

These vitamin E forms have been tested on different types of cancer cells, and it is found that they are able to inhibit the CSC proliferation effects via modulating various signaling pathways. These vitamin E molecular forms vary in their effectiveness in inhibiting inflammation and preventing CSCs to grow; for example, αT has the least activity [23].

4.7.1.4   Selenium

There are several mechanisms proposed for anticancer effects of selenium by anti-inflammatory response, immune response, stimulation of apoptosis, inhibition of cell division and differentiation, and so on. As selenium is involved in anti-inflammatory response by helping in scavenging the free radicals, this activates the P53/ATM/FOXO3a pathway involved in regulation of apoptosis [24, 25].

4.7.2   STUDIES ON FOOD PRODUCTS AS PROSPECTIVE PREVENTIVE TREATMENT FOR CANCER

4.7.2.1   Curcumin in Different Types of Cancers

Turmeric is a well-known Indian spice and is widely used for imparting yellow color to the curries. Curcumin is a compound present in turmeric and some other natural sources, like ginger, and has anti-inflammatory and antioxidant properties [26]. Also, curcumin has been found to have numerous cytotoxic effects on CSCs. This is due to its suppression of the release of cytokines, particularly interleukin (IL)-6 [27], IL-8 [28], and IL-1 [29], which stimulate CSCs, and it also has effects on multiple sites along CSC pathways, such as Wnt, Notch, Hh, and FAK pathways. The Wnt/β-catenin pathway is inhibited by curcumin, which induces the breakdown of β-catenin by caspase-3 [30].

4.7.2.2   Black Cumin

Black cumin, or *Nigella sativa* L., possesses bioactive compounds like thymoquinone (TQ), thymol (THY), and α-hederin, which have antioxidant effects and protect liver inflammatory damage by different scavenging mechanisms [31]. These extracts also have been found effective against different types of cells, like mouse kidney cells, normal human lung fibroblasts, and normal human intestinal cells [32]. TQ is observed to induce the apoptosis in CSCs by generating ROS, causing DNA damage, activating caspases, and altering the pathways like p53, Wnt, STAT, and so on, along with inhibiting the metastasis by inhibiting the NF-κB pathway, making TQ as promising therapeutic ingredient to be involved in combinatorial therapy.

4.7.2.3   Ginger

Ginger (*Zingiber officinale*) is essential ingredient of Indian cooking and has been known for its therapeutic properties since ages. Apart from its properties to enhance the test, ginger is also used for treating small ailments, like cough, flatulence, nausea, heartburn, diarrhea, loss of appetite, infections, cough, and so on. There are active compounds in ginger like 6-gingerol and 6-shogaol, which are found to exert anticancer activities particularly against GI cancer. Further studies have revealed that these molecules are able to affect many signaling pathways, like NF-κB, STAT3, MAPK, PI3K, ERK1/2, Akt, TNF-α, COX-2, cyclin D1, cdk, MMP-9, survivin, cIAP-1, XIAP, Bcl-2, caspases, and other cell growth regulatory proteins [33, 34].

4.7.2.4   Garlic

Another spice widely used in different cuisines is garlic (*Allium sativum*), which is also known for its medicinal value since ancient times. It has been used as ingredient in food and also used for treatment of different types of disorders, like cold and osteoarthritis, and it is also a blood thinner, so it helps in reducing hypertension by reducing the stiffness of blood vessels [35]. Some of the compounds found in garlic are allicin, diallyl sulfide, diallyl disulfide, diallyl trisulfide, E/Z-ajoene, S-allyl-cysteine, and S-allyl-cysteine sulfoxide (alliin) [36]. Allicin was tested on mice and human cancer cells and checked by the MTT assay, and it was found that cell cycle arrest occurred at specific stages, depending on the type of cell and whether it is a CD44+ or CD44– marker. Also, apoptosis was increased as found out by increase caspase-3 expression. This is also suggested that combining the allicin with other natural compound like MSM can be a treatment of choice [37].

Diallyl trisulfide (DATS) is known the inhibit breast CSCs. It is found that DATS targets the breast CSCs in a dose-dependent manner, and the level of another marker, aldehyde dehydrogenase-1, is found to be increased [38].

4.7.2.5   Piperine in Pepper

An alkaloid by nature, piperine has anticancer properties as it represses the growth of the CD44+/CD133+ CSCs from HepG2 cells and arrests the cells at the G1/G0 phase. In molecular docking studies, it is found that piperine is effective against the epithelial-mesenchymal (EMT) induced by the transcription growth factor-β (TGF-β) in hepatocarcinogenesis [39].

4.7.2.6   Saffron

Saffron (*Crocus sativus* L.) is used as a spice and a food colorant. In different cultures, like Indian, Chinese, and Arabian cultures, the dry stigmas of the plant have been used as an herbal treatment for different ailments. Crocetin is an important carotenoid and is present in saffron, which has been tested for its antitumor properties in different systems. Crocetin inhibits the nucleic acid synthesis, elicits antioxidative properties, induces apoptosis, and modulates growth factor signaling pathways [40]. Saffron and its derivatives, particularly crocetin, have demonstrated significant anticancer activity in breast, lung, pancreatic, and leukemic cells. It has a promising role in the prevention of the division of cells by different mechanisms. For example, experiments on cell line in breast cancer and pancreatic cancer has resulted in reduced proliferation and increased apoptosis; in cervical cancer, DNA, RNA, and protein synthesis is reduced by a different mechanism, which reduces proliferation in colorectal cancer by mismatch repair [40]; in leukemia, it reduces proliferation and increases apoptosis; in liver and lung cancer cells, the production of antioxidants is reduced [41].

4.7.2.7   Chili Pepper

There are many varieties of chili papers that are used globally. It is used in different cuisines as well it is also a traditional medicine. Chilies have different nutrients, like vitamins and minerals, that impart it medicinal properties and have been used for the treatment of different types of ailments, like bronchitis, cough, headache, rheumatism, stiff joints, arthritis, and arrhythmia. Chili pepper is found to have ingredients that are tested for their action against cancer cells and microbes. One component of chili is quercetin, which is found to have an inhibitory effect on pancreatic CSCs. Aldehyde dehydrogenase activity was reduced by the treatment of quercetin in CSCs and enhanced the apoptosis, as analyzed by substrate assays, FACS, and western blot analysis [42].

4.7.2.8   Flavonoids in Mint

Mint is an herb used mostly in Indian cuisine for taste enhancement. It has compounds that can be used for different treatments of disorders; cancer is one of them. The chemical found in it has been found to have anticancer properties as it prevents CSC proliferation. Flavonoids present in mint, and other natural products have a vast variety of mechanisms to prevent cancer, like affecting the metabolism of CSCs, invasion and metastasis, decreasing the drug resistance, epigenetic changes, cell cycle arrest at the G0 level, and apoptosis, and involve many signaling pathways, like EGFR, MAPK, IGF, PI3K/Akt/mTOR, NF-κB, and Wnt β-catenin/Hh pathway [10].

4.7.2.9   Fenugreek

Fenugreek (*Trigonella foenum-graecum*) is a dietary plant. Its leaves and seeds are used in different cuisines. Apart from that, it is a potential candidate for the treatment of cancer because of its pro-apoptotic properties [43]. Diosgenin, a steroidal saponin, is one of the compounds produced by fenugreek. Diosgenin modulates the Wnt/β-catenin signaling via the Wnt-antagonist-secreted frizzled-related protein-4 [44].

4.7.2.10 Quinines

Cinchona, from the Rubiaceae family, is well known for its antimalarial activity. Cinchona has many alkaloids like quinine, chichonine, quinidine, and cinchonidine in different quantities in various species. Quinine is found to increase apoptosis in cancer cells and inhibits cell division and differentiation in a dose-dependent manner. Currently, quinine can be used as a combination therapy as it also possesses anticancer properties and can be used along with bleomycin, cisplatin, anthracyclines, and radiation therapy. The mechanism of action of quinine includes increase in ROS production, which induces many morphological changes, such as changed apoptotic signals, which culminates into loss of apoptotic bodies and causes adhesion, cell contraction, membrane vesication, condensation of nuclear material, and finally formation of nuclear fragments. Because of these effects, quinine has a role in increasing apoptosis in cancer cells, and there is a possibility of using it as effective anticancer drug in the future [45]. In another study, it was found that tumor cell proliferation is inhibited with the help of quinine, which interacts with receptor-associated factor 6-AKT [46]. While another type of natural triterpenoid quinine compound, pristimerin, which is isolated from Celastraceae and Hippocrateacea plants, inhibits the PI3K/Akt/FoxO3a signaling pathway [47].

4.7.2.11 Cloves

Clove bud (*Syzygium aromaticum*) is used as pain reliever in toothache, and its oil has been an active component of many ayurvedic medicinal preparations. Cloves possess antiseptic, antimicrobial, and anticancer properties [48]. The active fraction of clove (AFC), which is extracted from dried buds of cloves, has been studied for its anticancer properties in which it is found that AFC induces the cellular apoptosis in human colorectal cancer HCT-116 cells [49]. Eugenol, which is also present in bay leaf (11%) [50], is major components of cloves (76.8%) targets the β-catenin in lung cancer and has shown anticancer activity both in vivo and in vitro [51]. β- caryophyllene (BCP) and β-caryophyllene oxide (BCPO) components of clove and other plants, like guava [52], are also found to modulate the receptors in glioblastoma cells to inhibit the cell proliferation [53]. BCPO follows different pathways other than binding to a receptor, by altering the pathways for cancer development, like MAPK, PI3K/Akt/mTOR/S6K1, and STAT3 pathways. Apart from this, BCPO has shown anti-apoptotic activity [54].

4.7.2.12 Koenimbin from Curry Leaves

Koenimbin, a natural dietary compound from *Murraya koenigii* L Spreng has antiproliferative activity. It has been tested on PC-3 cells and found to activate the apoptosis of cells by increasing caspase activity, cytochrome-C release, decreasing the level of anti-apoptotic proteins like Bcl-2 and HSP70 and inhibiting the translocation of NF-κB from the cytoplasm to nucleus [55]. In another in vitro study, koenimbin inhibited the breast-cancer-cell-derived MCF7 cells and targeted them for apoptosis as in PC-3 cells [56]. These studies can be relevant for the prevention and treatment of prostate cancer and breast cancer.

4.7.2.13 Resveratrol (Found in Grapes, Berries, and Peanuts)

Resveratrol, which is present in many natural sources, has been tested in different types of CSCs for their efficacy. Breast CSCs are inhibited by resveratrol by many routes. In mice, it was found that treatment of resveratrol decreased the size of xenograft, while apoptosis was promoted via Wnt/β-catenin and Hh signaling pathways. Derivatives of resveratrol, like pterostilbene, are also tested, and it is found that they involve an increased level of tumor-suppressive microRNA. In leukemia, resveratrol treatment depicted decreased resistance to drugs, as well as immunomodulation and apoptosis. A similar result was found in colorectal CSCs in which the Wnt/β-catenin pathway seems to play a critical role. Other types of cancer like glioblastoma and other CSCs are also tested for their behavior under the treatment by resveratrol [57].

4.7.2.14   Green Tea

Polyphenolic catechins present in green tea have chemoprevention activity. The most widely studied epigallocatechin-3-gallate (EGCG), present in green tea, inhibits NF-κB activity, MAPK activity, and many signaling pathways [58]. EGCG also blocks the Wnt pathway by stabilizing HBP1 and causes a reduction in proliferation and metastasis of breast cancer. Macha green tea, as a whole solution, has also been tested on human breast cancer cells and has shown reduced mitochondrial metabolism and glycolytic flux [59].

4.7.2.15   α-Pinene from Bay Leaf

α-pinene is a terpenoid that is present in many plants [60], like a bay leaf, which is used in Indian curries for increasing the taste and aroma, and has been tested in animal cells. It is seen that α-pinene can inhibit cancerous growth by affecting the ERK/AKT pathway [61].

4.7.2.16   Lycopene from Tomato

Lycopene, which is responsible for the red color of tomato, is also investigated for its anticancer properties, and it is found that lycopene, in combination with β-carotene supplementation, suppressed the growth of lung cancer or prostate cancer cell line by inhibiting angiogenesis. Lycopene has a dual role in creating ROS and scavenging them, which depends on its concentration. Therefore, a lower concentration of lycopene prevents oxidative damage of DNA in the HT29 cell line, which is increased by an increased concentration of lycopene. Epigenetic changes are also important for changes in CSC behavior, and lycopene can show epigenetic changes by partial methylation of the GSTP1 gene in MDA-MB-468 breast cancer cell lines at the promotor region [62].

4.7.2.17   Soybean

Soybean and soy products have been in focus for their health benefits in terms of cardiovascular health. As soybean protein has less sulfur content in comparison to other animal proteins, it is found to inhibit the development of tumors in animals by different means, like antiprotease activity, to prevent cancer. A peptide found in soybean is also shown to have antimitotic activity [63]. Isoflavones like aglycones daidxein, genistein, and glycitein have estrogen-modulating activity. Genistin is able to inhibit the growth of estrogen-dependent and independent tumors, as it has been tested on the MCF-7 breast cancer cells and causes apoptosis [64]. In another study, genistein is found to decrease breast CSCs both in vitro and in vivo through the downregulation of the Hedgehog-Gli-signaling pathway [63].

Saponins in soybean, like dammarane, oleanane, ursane, lupine, and their conjugated forms, have anticancer properties. They prevent cancer by forming antioxidants, and they are also able to interfere with the enzymes involved in cancer formation. Saponins are also reported to have capabilities to modulate estrogen-dependent cancer, like breast cancer, as different saponins have shown different effects on the development of tumors [64].

4.7.2.18   Use of Probiotics for Prevention of Cancer

The benefits of the gut microbiome are of considerable understanding in maintaining homeostasis. They are known to have role in immune responses and also possess anticancer properties. At first stance, bacterial strains can help combat the carcinogens by modulating the secretion of cytokines and by stimulating the phagocytosis of the nascent cancer cells. Gut microbiota have been found useful in the prevention of colon and colorectal cancer and in the treatment [65]. Kefir is a type of probiotic, which is produced by the combination of lactic acid bacteria and yeast added to fresh milk. Components of kefir are involved in the synthesis of different types of molecules, which have a role as anticancer agents and in cancer treatment like different peptides, polysaccharides, and sphingolipids [66].

Bacteriocins are peptides that are generated by bacteria, whether gram-positive or gram-negative, and located on either plasmid or chromosomes. These peptides possess charges, and cationic charged bacteriocins can bind to the negatively charged membranes of cancer cells, which is due to the different compositions of membrane lipids than normal cells, and kill them by making pores in the membrane [67]. There are a few bacteriocins that have been investigated for their anticancer properties, like nisin from *Lactococcus lactis*, plantaricin A from *Lactobacillus plantarum*, azurin and pyocins from *Pseudomonas aeruginosa*, colicins from *Escherichia coli*, micorcins from the members of the family Enterobacteriaceae, pediocins from the members of *Pediococcus* and other lactic acid-producing genera, bovicin from *Streptococcus bovis*, and lacterosporulins from *Brevibacillus* sp. [68].

4.8   CONCLUSION AND PROSPECTS

Natural products have always been considered a safe cure for different types of illnesses, with the least side effects. There are plenty of natural compounds available in our surroundings; many of them are to be explored. For different types of cancers, researchers are trying to find out the compounds that can arrest the duplication of CSCs, which can differentiate into different types of cells and need to be stopped at the initial level from dividing. As rapid cell growth culminating in cancer involves signals and their transduction via different pathways, compounds that are helping in the modulation of pathways that lead to cell division can help prevent the cancer development along with helping in managing multidrug resistance. Many compounds are routinely used in our food, like curcumin in turmeric and nigella in black cumin and mint, and their use in appropriate quantities and ways can help arrest the cancers at the stem cell level. Apart from discussed compounds in this chapter, there are many compounds present in different natural sources, which gives the possibility of exploring the arrest of cancers of different types at the various levels, and the mechanisms involved in their action can be explored.

4.9   REFERENCES

1. M. J. Thun, J. O. DeLancey, M. M. Center, A. Jemal, and E. M. Ward, "The global burden of cancer: Priorities for prevention," *Carcinogenesis*, vol. 31, no. 1, pp. 100–110, Jan. 2010, doi: 10.1093/carcin/bgp263.
2. B. K. K. K. Jinadasa, C. Elliott, and G. D. T. M. Jayasinghe, "A review of the presence of formaldehyde in fish and seafood," *Food Control*, vol. 136, p. 108882, Jun. 2022, doi: 10.1016/j.foodcont.2022.108882.
3. S. Roy *et al.*, "Adulteration in spices—a threat to human health and well being," *American Journal of Applied Biotechnology Research*, vol. 1, no. 3, pp. 25–28, Jul. 2020, doi: 10.15864/ajabtr.1303.
4. K. Gairola, S. Gururani, A. Bahuguna, V. Garia, R. Pujari, and S. K. Dubey, "Natural products targeting cancer stem cells: Implications for cancer chemoprevention and therapeutics," *Journal of Food Biochemistry*, vol. 45, no. 7, p. e13772, 2021, doi: 10.1111/jfbc.13772.
5. L. Zhu and L. Chen, "Progress in research on paclitaxel and tumor immunotherapy," *Cell Molecular Biology Letters*, vol. 24, no. 1, p. 40, Jun. 2019, doi: 10.1186/s11658-019-0164-y.
6. C. Fuchs, E. P. Mitchell, and P. M. Hoff, "Irinotecan in the treatment of colorectal cancer," *Cancer Treatment Reviews*, vol. 32, no. 7, pp. 491–503, Nov. 2006, doi: 10.1016/j.ctrv.2006.07.001.
7. S. Barnett *et al.*, "Vincristine dosing, drug exposure and therapeutic drug monitoring in neonate and infant cancer patients," *European Journal of Cancer*, vol. 164, pp. 127–136, Mar. 2022, doi: 10.1016/j.ejca.2021.09.014.
8. J. Wang *et al.*, "Adebrelimab or placebo plus carboplatin and etoposide as first-line treatment for extensive-stage small-cell lung cancer (CAPSTONE-1): A multicentre, randomised, double-blind, placebo-controlled, phase 3 trial," *The Lancet Oncology*, May 2022, doi: 10.1016/S1470-2045(22)00224-8.
9. N. H. Oberlies and D. J. Kroll, "Camptothecin and Taxol: Historic Achievements in Natural Products Research," *Journal Natural Products*, vol. 67, no. 2, pp. 129–135, Feb. 2004, doi: 10.1021/np030498t.
10. K. Kandhari, H. Agraval, A. Sharma, U. C. S. Yadav, and R. P. Singh, "Flavonoids and Cancer Stem Cells Maintenance and Growth," in *Functional Food and Human Health*, V. Rani and U. C. S. Yadav, Eds. Singapore: Springer, 2018, pp. 587–622, doi: 10.1007/978-981-13-1123-9_26.

11. S. Timbrell *et al.*, "FAK inhibition alone or in combination with adjuvant therapies reduces cancer stem cell activity," *NPJ Breast Cancer*, vol. 7, no. 1, Art. no. 1, May 2021, doi: 10.1038/s41523-021-00263-3.

12. D. W. Clark and K. Palle, "Aldehyde dehydrogenases in cancer stem cells: Potential as therapeutic targets," *Annals of Translational Medicine*, vol. 4, no. 24, p. 518, Dec. 2016, doi: 10.21037/atm.2016.11.82.

13. "Targeting cancer stem cell pathways for cancer therapy | Signal Transduction and Targeted Therapy." www.nature.com/articles/s41392-020-0110-5 (accessed Jun. 25, 2022).

14. M. Katoh, "Canonical and non-canonical WNT signaling in cancer stem cells and their niches: Cellular heterogeneity, omics reprogramming, targeted therapy and tumor plasticity (Review)," *International Journal of Oncology*, vol. 51, no. 5, pp. 1357–1369, Nov. 2017, doi: 10.3892/ijo.2017.4129.

15. J. Burnett, B. Newman, and D. Sun, "Targeting Cancer Stem Cells with Natural Products," *Current Drug Targets*, vol. 13, no. 8, pp. 1054–1064, Jul. 2012, doi: 10.2174/138945012802009062.

16. N. Takebe *et al.*, "Targeting Notch, Hedgehog, and Wnt pathways in cancer stem cells: Clinical update," *Nature Review Clinical Oncology*, vol. 12, no. 8, Art. no. 8, Aug. 2015, doi: 10.1038/nrclinonc.2015.61.

17. M. Yuan, G. Zhang, W. Bai, X. Han, C. Li, and S. Bian, "The role of bioactive compounds in natural products extracted from plants in cancer treatment and their mechanisms related to anticancer effects," *Oxidative Medicine and Cellular Longevity*, vol. 2022, p. e1429869, Feb. 2022, doi: 10.1155/2022/1429869.

18. S. Jae Young and S. Nanjoo, "Targeting cancer stem cells in solid tumors by vitamin D," *Journal of Steroid Biochemistry and Molecular Biology*, vol. 148, pp. 79–85, Apr. 2015, doi: 10.1016/j.jsbmb.2014.10.007.

19. Y. Cho and Y. K. Kim, "Cancer stem cells as a potential target to overcome multidrug resistance," *Frontiers in Oncology*, vol. 10, p. 764, Jun. 2020, doi: 10.3389/fonc.2020.00764.

20. A. S. Choudhari, P. C. Mandave, M. Deshpande, P. Ranjekar, and O. Prakash, "Phytochemicals in cancer treatment: From preclinical studies to clinical practice," *Frontiers in Pharmacology,* vol. 10, 2020, doi: 10.3389/fphar.2019.01614.

21. J. Fu, Z. Wu, J. Liu, and T. Wu, "Vitamin C: A stem cell promoter in cancer metastasis and immunotherapy," *Biomedicine & Pharmacotherapy*, vol. 131, p. 110588, Nov. 2020, doi: 10.1016/j.biopha.2020.110588.

22. B. Abiri and M. Vafa, "Vitamin C and cancer: The role of vitamin C in disease progression and quality of life in cancer patients," *Nutrition and Cancer*, vol. 73, no. 8, pp. 1282–1292, Sep. 2021, doi: 10.1080/01635581.2020.1795692.

23. B. L. Sailo, K. Banik, G. Padmavathi, M. Javadi, D. Bordoloi, and A. B. Kunnumakkara, "Tocotrienols: The promising analogues of vitamin E for cancer therapeutics," *Pharmacological Research*, vol. 130, pp. 259–272, Apr. 2018, doi: 10.1016/j.phrs.2018.02.017.

24. K.-L. Pang and K.-Y. Chin, "Emerging anticancer potentials of selenium on osteosarcoma," *International Journal of Molecular Sciences*, vol. 20, no. 21, Art. no. 21, Jan. 2019, doi: 10.3390/ijms20215318.

25. G. Murdolo, D. Bartolini, C. Tortoioli, M. Piroddi, P. Torquato, and F. Galli, "Chapter nine—selenium and cancer stem cells," in *Advances in Cancer Research*, vol. 136, K. D. Tew and F. Galli, Eds. Academic Press, 2017, pp. 235–257. doi: 10.1016/bs.acr.2017.07.006.

26. G. Ramadan, M. A. Al-Kahtani, and W. M. El-Sayed, "Anti-inflammatory and anti-oxidant properties of curcuma longa (turmeric) versus zingiber officinale (ginger) rhizomes in rat adjuvant-induced arthritis," *Inflammation*, vol. 34, no. 4, pp. 291–301, Aug. 2011, doi: 10.1007/s10753-010-9278-0.

27. M. Ghandadi and A. Sahebkar, "Curcumin: An effective inhibitor of interleukin-6," *Current Pharmaceutical Design*, vol. 23, no. 6, pp. 921–931, Feb. 2017.

28. H. Hidaka *et al.*, "Curcumin inhibits interleukin 8 production and enhances interleukin 8 receptor expression on the cell surface," *Cancer*, vol. 95, no. 6, pp. 1206–1214, 2002, doi: 10.1002/cncr.10812.

29. Y. Panahi *et al.*, "Evidence of curcumin and curcumin analogue effects in skin diseases: A narrative review," *Journal of Cellular Physiology*, vol. 234, no. 2, pp. 1165–1178, 2019, doi: 10.1002/jcp.27096.

30. M. Ashrafizadeh *et al.*, "Curcumin therapeutic modulation of the Wnt signaling pathway," *Current Pharmaceutical Biotechnology*, vol. 21, no. 11, pp. 1006–1015, Sep. 2020, doi: 10.2174/1389201021666200305115101.

31. H. Tabassum, A. Ahmad, and I. Z. Ahmad, "Nigella sativa L. and its bioactive constituents as hepatoprotectant: A review," *Current Pharmaceutical Biotechnology*, vol. 19, no. 1, pp. 43–67, Jan. 2018, doi: 10.2174/1389201019666180427110007.

32. Z. Fatfat, M. Fatfat, and H. Gali-Muhtasib, "Therapeutic potential of thymoquinone in combination therapy against cancer and cancer stem cells," *World Journal of Clinical Oncology*, vol. 12, no. 7, pp. 522–543, Jul. 2021, doi: 10.5306/wjco.v12.i7.522.

33. A. H. Alkhathlan *et al.*, "Evaluation of the anticancer activity of phytomolecules conjugated gold nanoparticles synthesized by aqueous extracts of zingiber officinale (ginger) and nigella sativa L. seeds (black cumin)," *Materials*, vol. 14, no. 12, Art. no. 12, Jan. 2021, doi: 10.3390/ma14123368.

34. S. Prasad and A. K. Tyagi, "Ginger and its constituents: Role in prevention and treatment of gastrointesti-nal Cancer," *Gastroenterol Research and Practices*, vol. 2015, p. 142979, 2015, doi: 10.1155/2015/142979.

35. C. Imo and J. S. Za'aku, "Medicinal properties of ginger and garlic: A review," *Current Trends in Biochemical Engineering and Biotechnology*, vol. 18, no. 2, pp. 1–7, Feb. 2019, doi: 10.19080/CTBEB.2019.18.555985.

36. A. Shang *et al.*, "Bioactive compounds and biological functions of garlic (Allium sativum L.)," *Foods*, vol. 8, no. 7, p. 246, Jul. 2019, doi: 10.3390/foods8070246.

37. E. Sarkhani, N. Najafzadeh, N. Tata, M. Dastan, M. Mazani, and M. Arzanlou, "Molecular mechanisms of methylsulfonylmethane and allicin in the inhibition of CD44± breast cancer cells growth," *Journal of Functional Foods*, vol. 39, pp. 50–57, Dec. 2017, doi: 10.1016/j.jff.2017.10.007.

38. S.-H. Kim, C. H. Kaschula, N. Priedigkeit, A. V. Lee, and S. V. Singh, "Forkhead box Q1 is a novel target of breast cancer stem cell inhibition by diallyl trisulfide *," *Journal of Biological Chemistry*, vol. 291, no. 26, pp. 13495–13508, Jun. 2016, doi: 10.1074/jbc.M116.715219.

39. A. Tiwari, S. J. Modi, S. Y. Gabhe, and V. M. Kulkarni, "Evaluation of piperine against cancer stem cells (CSCs) of hepatocellular carcinoma: Insights into epithelial-mesenchymal transition (EMT)," *Bioorganic Chemistry*, vol. 110, p. 104776, May 2021, doi: 10.1016/j.bioorg.2021.104776.

40. A. Amin *et al.*, "Saffron and its major ingredients' effect on colon cancer cells with mismatch repair deficiency and microsatellite instability," *Molecules*, vol. 26, no. 13, Art. no. 13, Jan. 2021, doi: 10.3390/molecules26133855.

41. W. G. Gutheil, G. Reed, A. Ray, S. Anant, and A. Dhar, "Crocetin: An agent derived from saffron for prevention and therapy for cancer," *Current Pharmaceutical Biotechnology*, vol. 13, no. 1, pp. 173–179, Jan. 2012, doi: 10.2174/138920112798868566.

42. W. Zhou *et al.*, "Dietary polyphenol quercetin targets pancreatic cancer stem cells," *International Journal of Oncology*, vol. 37, no. 3, pp. 551–561, Sep. 2010, doi: 10.3892/ijo_00000704.

43. K. K. Khoja *et al.*, "Fenugreek, a naturally occurring edible spice, kills MCF-7 human breast can-cer cells via an apoptotic pathway," *Asian Pacific Journal of Cancer Prevention*, vol. 12, no. 12, pp. 3299–3304, 2011.

44. G. Bhuvanalakshmi *et al.*, "Breast cancer stem-like cells are inhibited by diosgenin, a steroidal sapo-nin, by the attenuation of the Wnt β-catenin signaling via the wnt antagonist secreted frizzled related protein-4," *Frontiers in Pharmacology*, vol. 8, p. 124, 2017, doi: 10.3389/fphar.2017.00124.

45. M. A. Raza *et al.*, "The medicinal and aromatic activities of cinchona: A review," *Asian Journal of Advances in Research*, pp. 42–45, Jun. 2021.

46. W. Liu, Y. Qi, L. Liu, Y. Tang, J. Wei, and L. Zhou, "Suppression of tumor cell proliferation by quinine via the inhibition of the tumor necrosis factor receptor-associated factor 6-AKT interaction," *Molecular Medicine Reports*, vol. 14, no. 3, pp. 2171–2179, Sep. 2016, doi: 10.3892/mmr.2016.5492.

47. F. Yan *et al.*, "Pristimerin-induced uveal melanoma cell death via inhibiting PI3K/Akt/FoxO3a signal-ling pathway," *Journal of Cellular and Molecular Medicine*, vol. 24, no. 11, pp. 6208–6219, 2020, doi: 10.1111/jcmm.15249.

48. M. J. Nirmala, L. Durai, V. Gopakumar, and R. Nagarajan, "Anticancer and antibacterial effects of a clove bud essential oil-based nanoscale emulsion system," *International Journal of Nanomedicine*, vol. 14, pp. 6439–6450, Aug. 2019, doi: 10.2147/IJN.S211047.

49. M. Liu *et al.*, "Active fraction of clove induces apoptosis via PI3K/Akt/mTOR-mediated autophagy in human colorectal cancer HCT-116 cells," *International Journal of Oncology*, vol. 53, no. 3, pp. 1363–1373, Sep. 2018, doi: 10.3892/ijo.2018.4465.

50. S. Batool, R. A. Khera, M. A. Hanif, and M. A. Ayub, "Bay leaf," *Medicinal Plants of South Asia*, pp. 63–74, 2020, doi: 10.1016/B978-0-08-102659-5.00005-7.

51. P. Choudhury, A. Barua, A. Roy, R. Pattanayak, M. Bhattacharyya, and P. Saha, "Eugenol emerges as an elixir by targeting β-catenin, the central cancer stem cell regulator in lung carcinogenesis: An in vivo and in vitro rationale," *Food Function*, vol. 12, no. 3, pp. 1063–1078, Feb. 2021, doi: 10.1039/D0FO02105A.

52. S. Jamieson, C. E. Wallace, N. Das, P. Bhattacharyya, and A. Bishayee, "Guava (Psidium guajava L.): A glorious plant with cancer preventive and therapeutic potential," *Critical Reviews in Food Science and Nutrition*, vol. 0, no. 0, pp. 1–32, Jul. 2021, doi: 10.1080/10408398.2021.1945531.

53. N. Irrera *et al.*, "β-caryophyllene inhibits cell proliferation through a direct modulation of CB2 recep-tors in glioblastoma cells," *Cancers*, vol. 12, no. 4, Art. no. 4, Apr. 2020, doi: 10.3390/cancers12041038.

54. K. Fidyt, A. Fiedorowicz, L. Strządała, and A. Szumny, "β-caryophyllene and β-caryophyllene oxide—natural compounds of anticancer and analgesic properties," *Cancer Medicine*, vol. 5, no. 10, pp. 3007–3017, 2016, doi: 10.1002/cam4.816.

55. B. Kamalidehghan, S. Ghafouri-Fard, E. Motevaseli, and F. Ahmadipour, "Inhibition of human prostate cancer (PC-3) cells and targeting of PC-3-derived prostate cancer stem cells with koenimbin, a natural dietary compound from Murraya koenigii (L) Spreng," *Drug Design Development and Theory*, vol. 12, pp. 1119–1133, May 2018, doi: 10.2147/DDDT.S156826.

56. F. Ahmadipour *et al.*, "Koenimbin, a natural dietary compound of Murraya koenigii (L) Spreng: Inhibition of MCF7 breast cancer cells and targeting of derived MCF7 breast cancer stem cells (CD44+/CD24–/low): An in vitro study," *Drug Design Development and Theory*, vol. 9, pp. 1193–1208, Feb. 2015, doi: 10.2147/DDDT.S72127.

57. L. Zhang, X. Wen, M. Li, S. Li, and H. Zhao, "Targeting cancer stem cells and signaling pathways by resveratrol and pterostilbene," *BioFactors*, vol. 44, no. 1, pp. 61–68, 2018, doi: 10.1002/biof.1398.

58. Y. Li, M. S. Wicha, S. J. Schwartz, and D. Sun, "Implications of cancer stem cell theory for cancer chemoprevention by natural dietary compounds," *The Journal of Nutritional Biochemistry*, vol. 22, no. 9, pp. 799–806, Sep. 2011, doi: 10.1016/j.jnutbio.2010.11.001.

59. G. Bonuccelli, F. Sotgia, and M. P. Lisanti, "Matcha green tea (MGT) inhibits the propagation of cancer stem cells (CSCs), by targeting mitochondrial metabolism, glycolysis and multiple cell signalling pathways," *Aging (Albany NY)*, vol. 10, no. 8, pp. 1867–1883, Aug. 2018, doi: 10.18632/aging.101483.

60. J. Hou *et al.*, "α-Pinene induces apoptotic cell death via caspase activation in human ovarian cancer cells," *Medical Science Monitor*, vol. 25, pp. 6631–6638, Sep. 2019, doi: 10.12659/MSM.916419.

61. H. Jo *et al.*, "α-Pinene enhances the anticancer activity of natural killer cells via ERK/AKT pathway," *International Journal of Molecular Sciences*, vol. 22, no. 2, Art. no. 2, Jan. 2021, doi: 10.3390/ijms22020656.

62. A. K. Singh, N. Sharma, M. Ghosh, Y. H. Park, and D. K. Jeong, "Emerging importance of dietary phytochemicals in fight against cancer: Role in targeting cancer stem cells," *Critical Reviews in Food Science and Nutrition*, vol. 57, no. 16, pp. 3449–3463, Nov. 2017, doi: 10.1080/10408398.2015.1129310.

63. P. Fan *et al.*, "Genistein decreases the breast cancer stem-like cell population through Hedgehog pathway," *Stem Cell Research Theory*, vol. 4, no. 6, p. 146, Dec. 2013, doi: 10.1186/scrt357.

64. S. M. Kerwin, "Soy saponins and the anticancer effects of soybeans and soy-based foods," *Current Medicinal Chemistry—Anti-Cancer Agents*, vol. 4, no. 3, pp. 263–272.

65. A. Górska, D. Przystupski, M. J. Niemczura, and J. Kulbacka, "Probiotic bacteria: A promising tool in cancer prevention and therapy," *Current Microbiology*, vol. 76, no. 8, pp. 939–949, 2019, doi: 10.1007/s00284-019-01679-8.

66. M. Sharifi, A. Moridnia, D. Mortazavi, M. Salehi, M. Bagheri, and A. Sheikhi, "Kefir: A powerful probiotics with anticancer properties," *Medical Oncology*, vol. 34, no. 11, p. 183, Sep. 2017, doi: 10.1007/s12032-017-1044-9.

67. D. Drider, F. Bendali, K. Naghmouchi, and M. L. Chikindas, "Bacteriocins: Not only antibacterial agents," *Probiotics & Antimicrobial Proteins*, vol. 8, no. 4, pp. 177–182, Dec. 2016, doi: 10.1007/s12602-016-9223-0.

68. P. Baindara, S. Korpole, and V. Grover, "Bacteriocins: Perspective for the development of novel anticancer drugs," *Applied Microbiology Biotechnology*, vol. 102, no. 24, pp. 10393–10408, Dec. 2018, doi: 10.1007/s00253-018-9420-8.

5 Functional Foods in Cancer Chemoprevention and Therapeutics

Shreya Joshi, Pranshu Sharma, and Atul Kumar Joshi

CONTENTS

5.1 Introduction 76
5.2 Functional Food..77
    5.2.1 Regulations and Legislation 77
    5.2.2 Technological Challenges...78
    5.2.3 Health Benefits and Market Trend 79
5.3 Cancer and Its Pathophysiology...79
5.4 Classification of Cancer Chemopreventive and Therapeutically Used Functional Foods 82
    5.4.1 Classification Based on Origin 83
        5.4.1.1 Plant Origin...83
        5.4.1.2 Animal Origin...86
        5.4.1.3 Microbial Origin 86
    5.4.2 Classification Based on Chemical Class of Phytoconstituents.....................................86
        5.4.2.1 Phenolics 87
        5.4.2.2 Alkaloids...88
        5.4.2.3 Terpenes and Triterpenoids..88
        5.4.2.4 Carotenoids 88
        5.4.2.5 Organosulfur Compounds...88
        5.4.2.6 Quinones 88
        5.4.2.7 Other Miscellaneous Compounds...88
    5.4.3 Classification Based on Mechanism of Action 89
        5.4.3.1 Angiogenesis Inhibitors 89
        5.4.3.2 Metastasis Inhibitors..89
        5.4.3.3 Free Radicals and ROS Inhibitors 89
        5.4.3.4 Antiproliferative Compounds 90
        5.4.3.5 Apoptosis Inducers...90
        5.4.3.6 DNA Methylation Inducers...90
        5.4.3.7 Matrix Metalloproteinase Inhibitors..90
    5.4.4 Classification Based on Processing Methods 90
        5.4.4.1 Basic Food from a Natural Source..90
        5.4.4.2 Processed Food 90
        5.4.4.3 Fortified Food 91
        5.4.4.4 Enhanced Food 91
        5.4.4.5 Isolated and Purified Preparation of Active Phytochemical from Food Constituents (Herbal/Product/Dosage Form)....................................91
5.5 Summary 92
5.6 Acknowledgments...92
5.7 References...92

DOI: 10.1201/b23311-5

5.1 INTRODUCTION

The ancient science Ayurveda clearly mentions the health benefits obtained through a balanced diet and lifestyle regime. It was not only prevalent in India during Vedic times but was also followed in other ancient civilizations; for example, the use of honey for its beneficial effect has been mentioned in almost all ancient civilizations [1]. This concept of a healthy diet in the form of "functional food" was again put forward in the past few decades and was first introduced in Japan for those food items that provide nutrition to the body and also possess certain specific disease-reducing potential, from cancer to obesity, because of specific ingredients in them. This special food may be the same as the conventional food taken in meals or may be processed or fortified food. However, its inception was slow till the mid-1990s but gained popularity in the 21st century due to the emerging techno-logical innovations for the development of new products keeping in view the consumer's demand for healthy foods. The development of any desired functional food is restricted through different regulations and legislations that have been formulated all across the globe, depending on the legal aspects available to each country; thus, as such, no particular definition of functional food can be given [2–4].

Even though a tremendous amount of scientific research and investigations are carried out each day globally to diagnose and cure cancer, still, the morbidity rate due to cancer is not declining, nor is it controlled to large extent. Cancer is an abnormal cell growth in the body resulting due to internal factors or external factors commonly called carcinogens, as shown in Figure 5.1.

Cancer is not a single-step disease; it progresses slowly starting with a mutation in a single cell and then its proliferation, invasion, and metastasis, and its treatment depends on the severity of the disease, although it generally includes surgery, radiotherapy, or systemic therapy [5]. The use of spe-cific chemicals or nutrients for the prevention of cancer is called chemoprevention. Tamoxifen was the first chemopreventive drug approved by FDA (Food and Drug Administration) for breast cancer, although some serious side effects were also noticed with its consumption [6]. The daily dietary intake of fruits and vegetables that might protect from cancer was first proposed by Wattenberg as they are enriched with specific phytoconstituents [7]. The inverse relationship between dietary

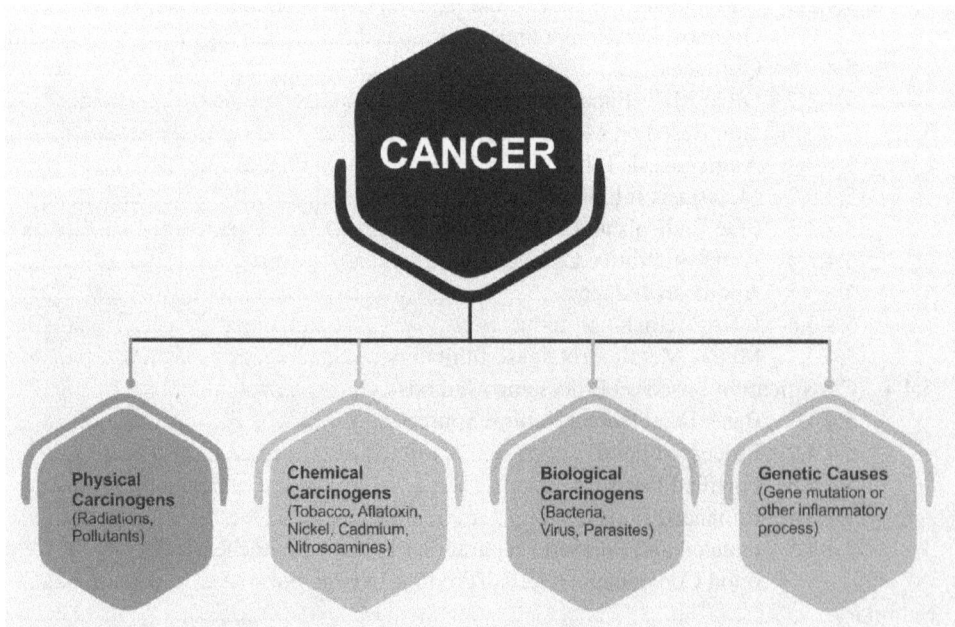

FIGURE 5.1 Causative agents of carcinogenesis.

factors and cancer rates have been shown in different studies; up to 90% of cancer cases are due to lifestyle and environmental issues, and the remaining cases are caused due to genetic defects. Therefore, approximately 30–40% of cancer cases can be prevented and cured to some extent by dietary means [8]. More than 250,00 different phytoconstituents are known to have the anticancer potential, either from animal sources or even processed or fortified food. These can interfere with the pathophysiological cycle of cancer probably starting from the damage of DNA till the formation of a malignant tumor. However, still, there is a limitation of in-depth knowledge and is a subject of research what dose at which step is required [9].

5.2 FUNCTIONAL FOOD

The plants have been used as dietary food and for the treatment of diseases not only today but even 6,000 years ago in India, China, and other civilizations. Even today almost all tribal people in India use local plants for the treatment of common diseases. Ethnobiologically, food can be classified as main food, used to satisfy hunger; supplementary food, used with main food as additional food, such as juices and salads; food ingredients, such as spices and condiments; and also medicinal food, which is specifically used for the treatment of the disease, like the fruits of bel (*Agel marmelos*) and leaves of tulsi (*Ocimum tenuiflorum*) [10]. These plants are not only used as a single component but also used in combinations, such as *Emblica officinalis*, *Terminalia belerica*, and *Terminalia chebula*, commonly called as *triphala* in India and consumed as a digestive aid [11]. The dietary alterations lead to adulteration in the microbial environment within intestines, which ultimately increases the risk of many chronic diseases [12]. Revealing the traditional concept of food in this modern era, the emphasis is laid on the scientific findings of the different food sources to improve the health of an individual, bringing a transition from adequate to optimal nutrition. In 1991, Japanese Ministry of Health and Welfare described functional foods as those that present specific health effects (1) whose food constituents are added or removed and (2) whose allergens have been removed. The concept of combining other dietary supplements and fortified food with traditional food also attained popularity all across the globe and can be classified accordingly, as shown in Figure 5.2 [13, 14].

The widely used concept today is not only fortified foods but also enriched foods, such as folate-enriched bread. Cruciferous vegetables, like cabbage, enhanced with selenium are used in the chemoprevention of cancer. Besides these, in some foods, harmful components are reduced or replaced with beneficial components [15]. The overuse of pharmaceuticals in daily life is mainly responsible for the increase in the demand for functional foods [16]. The clinical trial data and epidemiological studies present relevant information about formulated diet to serve as a functional food for preventing and minimizing the effects of chronic disease [17]; however, according to the American Dietetic Association (ADA), these should not be considered superior to natural foods [18].

5.2.1 REGULATIONS AND LEGISLATION

There are different regulations and legislations in various countries meeting the demand and need for functional foods. However, in general, the legislation states that the food products and supplements will be recognized as functional foods if there is scientific evidence to prove the health claim, and the developed functional foods should be safe per all the criteria defined in the food regulations of a country. The FDA suggests that functional foods can be grouped into different categories as required under the jurisdiction of Dietary Supplement and Health and Education Act (DSHEA) of 1994. It allows using these special foods without permission/authorization, although the marketing companies should clearly mention the ingredients and describe their mechanism of action affecting cellular functions. Per ADA recommendations, the designed functional food should be safe in terms of scientific research, and the claim made by the manufacturer must be true, implying all good manufacturing practices (GMP). The labels should publish the ingredients, nutrient content, and information about chemical function and structure [19]. Though regulations in Europe, the United

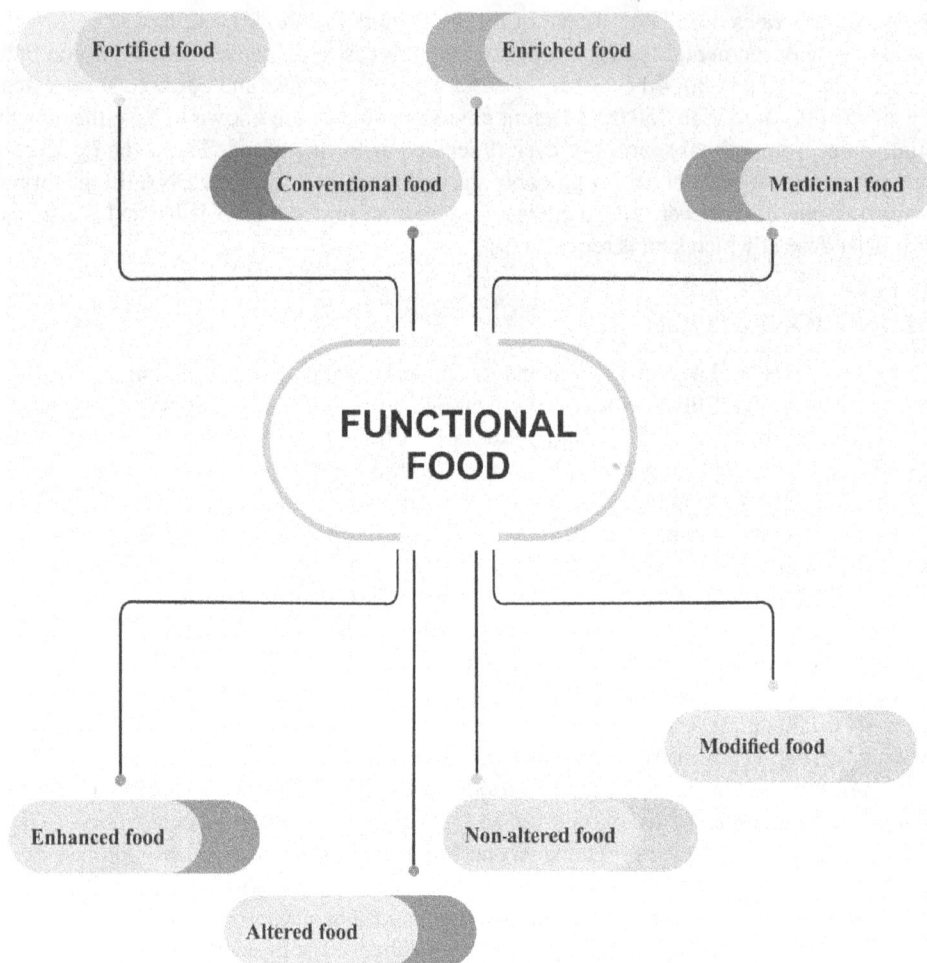

FIGURE 5.2   Different types of functional foods.

States, and the United Kingdom are streamlined and have become conducive to the development of these products, in India, the situation is different as both the regulatory system and the food sector are gradually coming up as international players in the market. In India, regulations are operationalized by the industry, and the FSSAI (Food Safety and Standards Authority of India) provides license and registration to those food companies which successfully meet all testing and quality standards per the guidelines. The FSSAI has notified the Food Safety and Standards Regulations (Food for Special Dietary, Functional foods, Health Supplements, Nutraceuticals, Food for Special Medical Purpose, and Novel Food) in the Official Gazette 2016. The prime focus of all the companies remains the same: to meet the quality compliances per the national and international standards of high purity, bioavailability, and ensuring toxicological safety. The manufacturers must be accredited by GMP certification and ISO 22000 FSMS for food safety, as these authorities confirm the quality of products through valid documentation and audit process [20].

5.2.2   TECHNOLOGICAL CHALLENGES

In this era of technological advancement, any food product can be made functional by the use of certain basic scientific and technical interventions, which depend on the need and requirement of the desired product. Basically, to design and develop a functional food, the conceptual knowledge of its mechanism of action should be known, which can be confirmed if any epidemiological data validated by statically proven results describing the relation between specific food constituent and its health benefit is available [21–23]. The scientific research carried out in different institutions involving molecular biology and genetic engineering, biodiversity, bioanalysis, and bioinformatics has resulted in the development of many biofortified crops [24]. Cruciferous crops (cauliflower and cabbage) have been developed with selectively bio-protecting selenium in it, which mainly acts as a chemopreventive agent. This technology presents a dual practical relevance covering healthcare and having an agronomic advantage as it provides food-chain-safe levels of selenium and also protects crops against any kind of environmental stress including adverse effects of drought [15, 25]. These modified cabbage and cauliflower, due to the presence of high levels of sulforaphane and isothiocyanates, prevent the growth of cancerous cells, thereby reducing the risk of tumor formation and also helping to prevent cancer of the prostrate, breast, lung, colon, and so on [26]. The development of functional foods not only requires scientific research and complex technology but safety parameters, and its integration in the food chain and the preservation/enhancement of all its parameters are also considered. There are numerous technological challenges while developing functional foods—that is, the functionality of a raw material or traditional food by adding new food constituents in it and the maximal preservation of constituents to be done in order to optimize functional food components in raw materials and also to increase their bioavailability. Besides these, quality assessments through efficient monitoring of all functional components through newly developed sensitive markers, the development of bioreactors for different immobilized enzymes or live microorganisms, and effective monitoring of them throughout the entire process of manufacturing and processing are the needs of the hour for the development of probiotics especially [27].

5.2.3 Health Benefits and Market Trend

The functional food market, from a relatively slow start, is continuously reaching new heights each day. In the past few years, it was worth over $21.3 billion in the United States and over $8 billion in Europe, attracting multinational companies to invest in this sector. Per Assocham's report, India holds only 2% of world nutraceutical products per capita spent—$2.5 compared to the global average of $21. This nutraceutical sector integrates with the beverage and functional food market as is consumed in day-to-day life. In today's scenario, the specific data on the availability and consumption of functional foods is not available for all countries, so the complete assessment of health promotion and disease prevention activities by the year 2050 is tough to predict [14, 28, 29].

5.3 CANCER AND ITS PATHOPHYSIOLOGY

Cancer is considered as abnormal cell growth occurring at any stage in life and any part of the body. The different carcinogens, commonly known as initiating agents, can bring about mutations in the cells, which ultimately result in the development of cancer. Approximately 80–90% of lung cancer is caused by a common carcinogen, tobacco, and is responsible for approximately one-third of all cancer deaths [30]. The carcinogens not only cause mutations in cells but can also stimulate cell proliferation and are known as tumor promoters. The compounds like phorbol esters activate protein kinase and stimulate cell proliferation, leading to the development of cancer. Hormones are also considered as tumor promoters in some cases; for example, the proliferation of uterine endometrium cells is stimulated by estrogen, increasing the risk of endometrial cancer in women [31]. Approximately 10–20% of cancer incidences are caused by viruses commonly causing cervical and liver cancer. It was predicted by global demographic characteristics that approximately 420 million new cases will be reported by 2025, increasing gradually, and about 18 million cases of cancer were recorded in 2018 worldwide [32]. Generally, to distinguish between cancer pathology, mainly two types of tumors are categorized: (1) A benign tumor retains its original location and does not spread/invade the neighboring cells, like a common skin wart. (2) Malignant tumor cells invade neighboring normal cells and spread via lymphatic or circulatory system to other parts of body (metastasis). These malignant tumors are commonly called cancers, and their invading and metastasizing properties make them too dangerous. Both these tumor types are thus classified according to the cell type from which they arise. A brief classification of cancers based on their places of origin is depicted in Figure 5.3 [5].

The first step in carcinogenesis is initiation, which results mainly from genetic alteration (mutation) or can also be by the carcinogen interaction with the genetic constitution of a cell. In this mutation, the gene is copied twice, lost, or damaged, and as a result, cell division occurs rapidly and the cell stops producing proteins required for cell division or produces them excessively and forms an abnormal growth, which leads to abnormal proliferation of a single cell. At the molecular level, the mutation of the DNA strand or the inappropriate expression of a gene leads to the decrease the methylation of DNA, which makes it a loosely bound structure around a histone protein, affecting the chromatin structure, the expression of the gene, and ultimately, the stability of genome. The progression is the second step as there can be additional mutation within these abnormal cells, and these cells divide continuously, and as they are immature cells, they do not develop into major cells capable of vital functions. They can easily spread via the lymphatic or circulatory system and grow and damage different tissues and organs. In this stage, these abnormal cells are actually cloned tumor cells that have evolved based on increased growth rate or survival, and metastasis spreads cancerous cells to other body parts and, as a result, slowly forms new tumors. The lymph nodes located in the underarms, neck, groin, and so on are easily attacked by cancerous cells via lymphatic circulation, whereas the liver, bones, and brain develop cancer through bloodstream, and the cancer is thus considered metastatic. For example, if lung

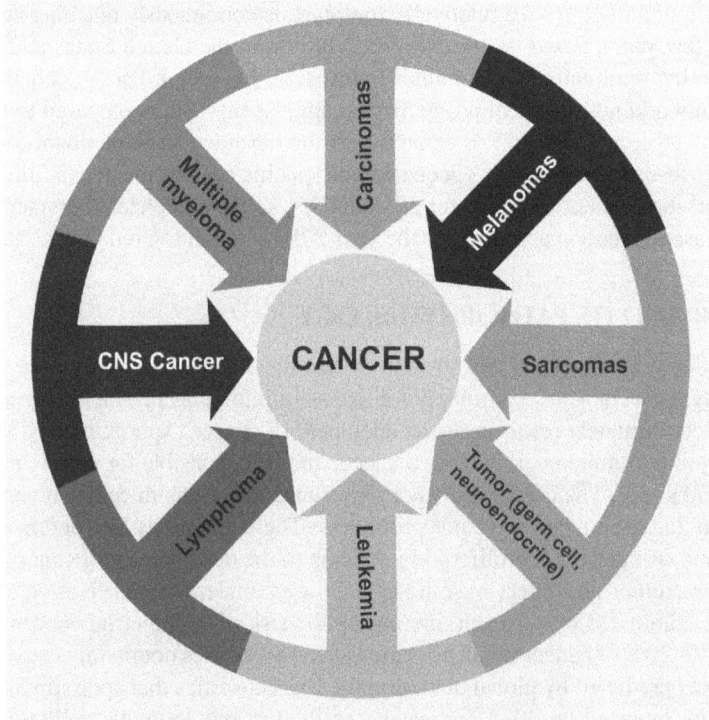

FIGURE 5.3 Different types of cancer.

cancer spreads to any other part of body, like the breast, then it is not called breast cancer but metastatic lung cancer. Besides this, in some cases, it has been observed that certain growth factors are produced from the cancer cells that stimulate their own proliferation, leading to continuous auto-stimulation of cell division, commonly known as autocrine stimulation, enabling cancer cells to be somehow less dependent on other growth factors [5, 9, 33]. The spreading of cancer cells is a complex process involving many enzymes and other substances. The cells secrete protease, which helps in invading the other cells, digesting the extracellular matrix surrounding normal cells. The collagenase helps to penetrate basal laminae and digest the underlying connective tissues. Besides enzymes, cancer cells secrete certain growth factors, helping in angiogenesis, the formation of new blood vessels, which are required to supply nutrients and oxygen to the proliferating cells. The mechanism is bit complex as these new vessels also stimulate the proliferation of endothelial cells present in the walls of capillaries of nearby surrounding tissues, and thus, the outgrowth of new capillaries occurs into the tumor cells, which finally enable cancerous cells to enter circulatory system, and metastasis begins. The pathophysiology of cancer also involves free radicals, matrix metalloproteinases, and so on, which are responsible for expression of different pathological conditions of it. The oxidants and free radicals are generated during physiological and pathological conditions and are in balanced condition in normal physiological conditions, but during pathological conditions, either due to overproduction or depletion of antioxidant substances, there exists an imbalance between free radical accumulation and neutralization [32, 34]. The different biomolecules, like lipids, DNA, proteins, are damaged by the reactive species by either imitating or promoting oxidative process. These biomolecules are essential in different biochemical pathways operating in the body, essentially required for growth and development. Any damage/alteration to them results in different pathophysiological process, including cardiovascular problems, cataract, rheumatoid arthritis, and even some neurodegenerative conditions; however, they also play an important role in cell signaling and gene expression regulation. The different transformations in the body as a result of mutations leading to cancerous growth give no signs or symptoms of diseases. However, few symptoms prevailing in the body, like change in bowel movements, persistent cough, blood in stool, change in urination, blood-tinged saliva, unexplained anemia, weight loss, breast lump or discharge, lumps in testicles, and so on, are probable indications for it. There are other symptoms that reflect the probable location of cancer, such as change in color, shape, size, and thickness of wart or mole; excessively sore throat; and thickening of the lump in the testicles or breasts. The initial diagnosis of cancer is done by certain screening tests, as shown in Figure 5.4. However, the tests vary, depending on the basis and advanced symptoms of the patient [5, 33].

The diagnosis of a patient determines the stage of cancer and the strategy required for either prevention or treatment of cancer (Figure 5.5).

The treatment also can require a combination of two or more strategies, depending on the severity of it. Surgery with radiation therapy is generally suitable as a treatment initiative. Surgeries are used for solid tumors as they are contained in a particular area and are not applicable for blood cancer or metastatic cancer. In radiotherapy, high doses of radiation are used either to kill a cell or shrink it, thereby minimizing the growth of cancerous cells, which results in DNA damage, thus ensuring cell death. However, this therapy has many side effects in normal cells, which can be damaged by radiation [5].

Chemoprevention is an essential part during cancer treatment procedure as it requires the understanding of the molecular mechanism behind the disease and the knowledge to stop the process at any level—initiation, promotion, or progression—with the help of specific molecules (nutrients or chemicals) [35]. The synthetic chemopreventive agents present more or less side effects, and in certain cases, tumor drug resistance has also been formed. Thus, to minimize these problems, there is a need to develop a complete range of functional foods with lesser adverse effects. It has been found that bioactive phytoconstituents present in the dietary food, even in low amounts, are able to have a great impact on cancer treatment [12, 36].

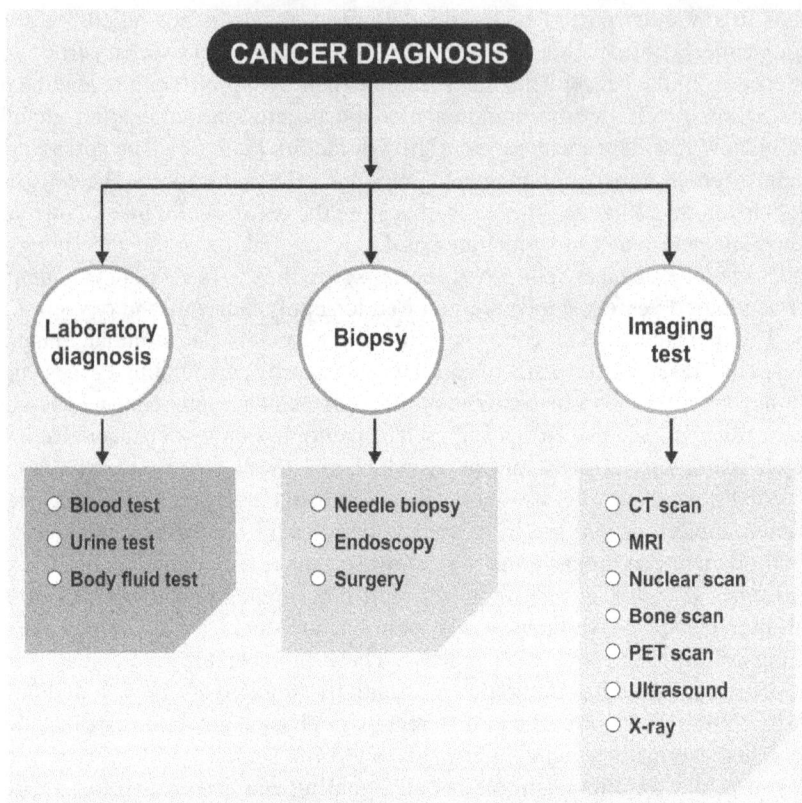

FIGURE 5.4   Different methods for cancer diagnosis.

5.4   CLASSIFICATION OF CANCER CHEMOPREVENTIVE AND THERAPEUTICALLY USED FUNCTIONAL FOODS

Chemoprevention is an effective means for the treatment of cancer and can be categorized as primary, secondary, and tertiary. The use of specific dietary phytoconstituents and non-steroidal anti-inflammatory drugs (NSAIDs) can be suitable for those who have a high risk of developing cancer or with no cancer and are regarded as primary chemoprevention. However, secondary chemopreventive measures are given to patients gradually progressing toward invasive cancer or with premalignant lesions. Usually, both these measures can be considered as one (i.e., primary only). To prevent the recurrence, compounds like tamoxifen are administered in breast cancer, known as tertiary chemoprevention. There are certain phytochemicals which either inhibit carcinogens or avoid the development of carcinogens from precursor substances and are termed as blocking agents. There are also chemicals that regularly suppress the neoplasia in cells at certain doses of carcinogens termed as "suppressing agents." The consumption of either pure extracts from plants or consumption of whole plant foods to prevent cancer is also known as "green chemoprevention" [37, 38]. The chemopreventive functional foods can be commonly classified as shown in Figure 5.6.

To know the molecular mechanism behind the neoplastic process of cancer is the main aim of cancer chemoprevention. The epidemiologic studies conducted during last three decades clearly indicate a direct relation between the dietary intake with bioactive compounds and a reduce risk of cancer [39] as they exhibit close association with either cancer onset, prognosis after its onset, and quality of life led by the patients after the treatment. Even though some individual constituents did not show activity when used, in combination they exhibited some biological activity [40].

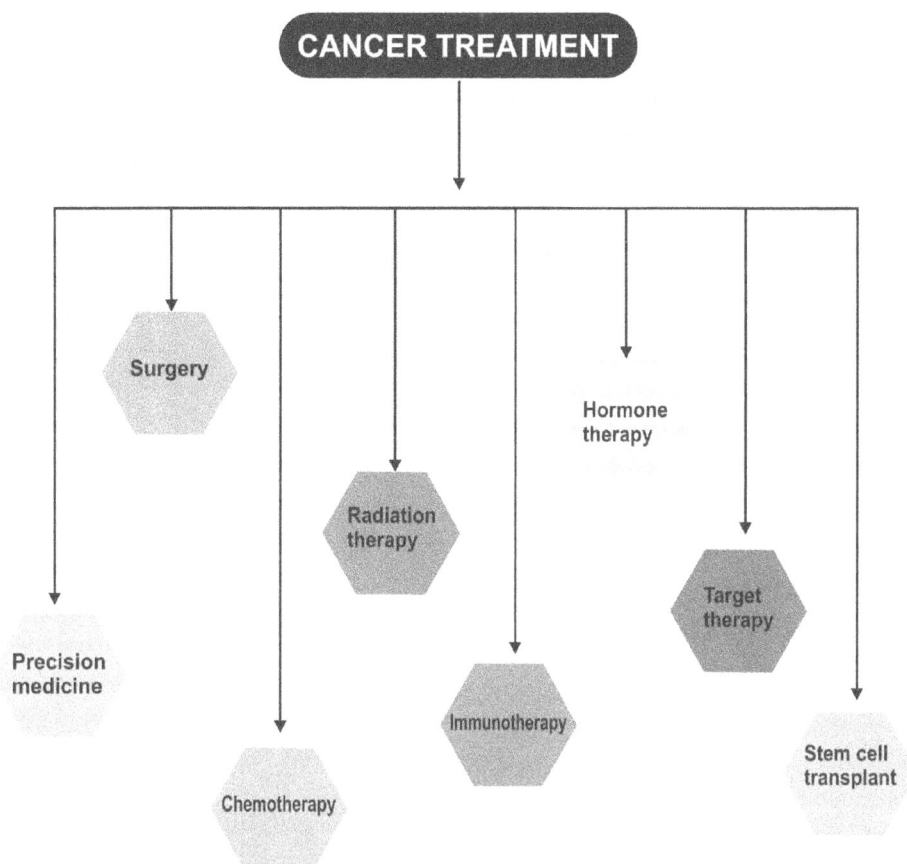

FIGURE 5.5 Commonly used treatment strategy for cancer.

5.4.1 CLASSIFICATION BASED ON ORIGIN

The natural compounds derived from either source (plants, animals, and microbes) or consumed in any form (conventional food, functional food) for prevention and therapy of cancer is due to presence of enormous number of active phytochemicals in them. These can interfere within the pathophysiological cycle of cancer, as shown in Figure 5.7, probably starting from DNA damage till the formation of malignant neoplasm.

These compounds act as antiproliferative agents, apoptosis inhibiting factors, free radicals, antiangiogenic agents, and so on, thereby exhibiting their anticancer potential. They also show anti-inflammatory properties due to their free radical and antioxidant capacity, which in turn also affect various signaling pathways and exhibit numerous biological activities. The active ingredients belonging to different chemical classes are capable of bringing some detrimental influences and other beneficial synergic activities, and thus, it becomes very difficult to identify the contribution of these phytochemicals in plant-based diet acting as cancer chemopreventive agents. However, recent surveys show that physicians and patients are very much inclined to use natural food items for complementary and alternative medicine [32, 41, 42].

5.4.1.1 Plant Origin

In some studies, it has been found that in Asia, the incidences of colon, breast, prostate, and lung cancers are lower as compared with those in America and Europe. This low incidence can be

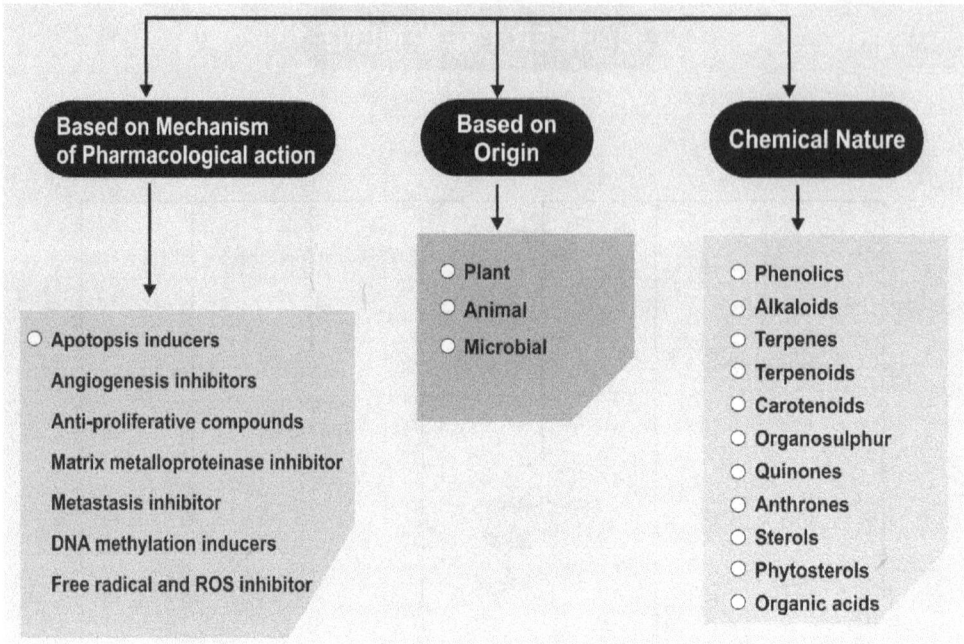

FIGURE 5.6   Classification of chemopreventive functional foods.

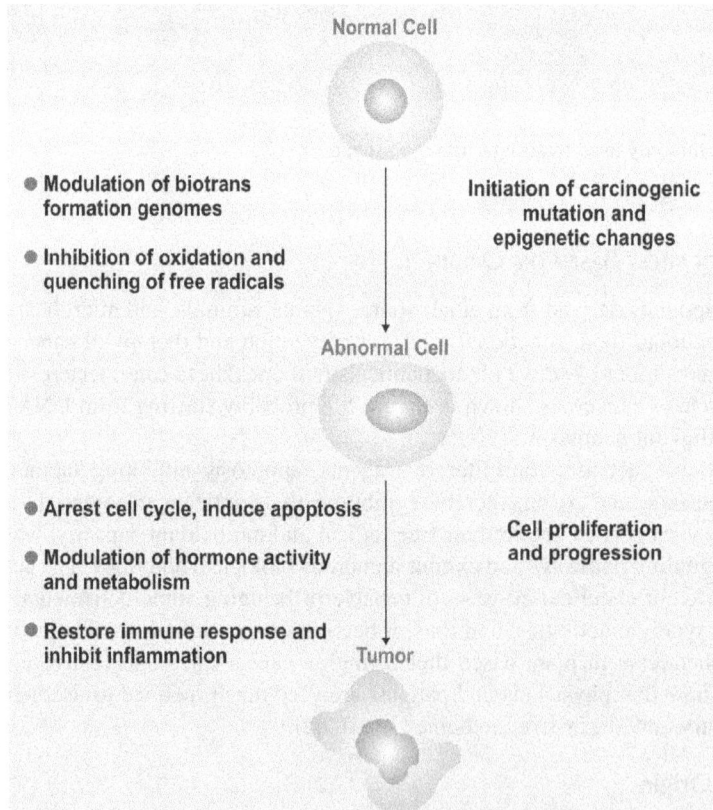

FIGURE 5.7   Probable mechanisms shown by phytochemicals in cancer chemoprevention.

probably due to the beneficial effects of plant-based diet in their daily food. The commonly used fruits, vegetables, and cereals having different phytochemicals exhibiting anticancer activity are shown in Figure 5.8 [32, 43]; however, it is incomplete as each day new research highlights the role of another isolated natural phytochemical from plant sources.

The US FDA has approximately approved 70% of plant-based (natural source) anticancer drugs [44]. Commonly used are paclitaxel and docetaxel derived from natural phytochemicals as vinca alkaloids (vinblastine, vincristine, and taxanes). Podophyllotoxin derivatives, such as etoposide, teniposide, and camptothecin, are also plant-derived anticancer agents. The ongoing research highlights the use of dietary sources as effective cancer chemopreventive agents validated by authentic results. The scientific investigations reveal that cruciferous vegetables, rice bran, barley grain and grass powder, rye, wheat, and oats hold promising roles as functional foods during cancer treatment [41, 45–47]. A novel peptide, lunasin, isolated initially from soy but also found in other cereals [48], also exhibits chemopreventive activities. Since ancient times, in Ayurveda and Latin American folk medicine, long pepper has been used as medicine and spice, and the current research confirms its significant chemopreventive and chemotherapeutic potential. The citrus fruits are widely used for their vitamin C content, which exhibits pro-oxidative behavior via different chemical reactions

FIGURE 5.8   Some commonly used cancer chemopreventive dietary food.

leading to formation of hydrogen peroxide exerting cytotoxic effect on cancerous cells. Similarly, the leaves of tea and seeds of coffee are rich sources of phytochemicals and have been reported for their anticancer potential [12, 49]. With the ongoing research conducted each new day, the list of plants used as cancer chemopreventives is long due to a wide range of phytochemicals. However, in a nutshell, it can be inferred that the biodiversity has provided us to explore as much as we can.

5.4.1.2 Animal Origin

The food from animal sources are also good chemopreventive agents as in numerous activities the anticancer potential of fish oil and dairy products have been reported. The fish oils from tuna, salmon, and sardines are mainly rich in omega-3, polyunsaturated fats, and the like, interfering in the process of carcinogenesis via different mechanisms [50]. Dairy products are a rich source of calcium, casein, conjugated linoleic acids, butyric acid, and different whey proteins. It has been shown that in the presence of calcium, the proliferating cells forcefully either differentiate or die, thereby reducing proliferation. Casein and whey proteins also exhibit beneficial role in treatment of various types of cancer [44, 51, 52]. In the case of cooked meat products, there have been reports of many mutagens and carcinogens. However, the compound CLA, first reported in 1987 from lamb and beef, has been found effective in suppressing stomach tumors and mammary carcinogenesis in mice and rats, respectively [53]. In Europe, the alkaloid tetrahydroisoquinoline and trabectedin from sea squid (*Ecteinascidia turbinata*) have been approved in treating patients with soft carcinoma [54].

5.4.1.3 Microbial Origin

In past few decades, mushrooms and probiotics are highly recommended as chemopreventive functional foods. The recommendable antiproliferation effects have been shown by biopolymer BE3 and lectin from *Boletus edulis*, *Tricholoma matsutake* inhibited tumor growth by inducing apoptosis [55], and *Phellinus linteus* exhibited antitumor activity. Rapamycin obtained from *Streptomyces hygroscopicus* exhibited immunosuppressive and antiproliferative activities in mammalian cells [56], and anthracyclines from *Streptomyces peucetius* var. *caesisus* are used to treat leukemia and solid tumors in lung, breast, ovarian cancers and sarcomas. The compounds mitomycin and actinomycin D obtained from *Streptomyces* and bleomycin, first marine compound is also]used for cancer treatment [57]. A new and emerging field within functional food development is the concept of pre-, pro-, and postbiotics, and these are regarded as the fastest growing group of dietary functional food supplements worldwide [58]. A prebiotic is a nonviable component of the diet in an intact form, which, when it reaches the intestine of the host, is selectively fermented by the beneficial bacteria such as *Lactobacillus* and *Bifidobacterium*. A probiotic, on the other hand, is a live microbial feed supplement that beneficially affects the host animal by improving its intestinal microbial balance. These are generally fermented milk products containing some selected strains of *Lactobacillus*, *Bifidobacterium*, *Saccharomyces*, *Lactococcus*, and *Streptococcus* and also have been proven efficient against cancer [59]. The probiotic supplements have been developed with a range of fortification approaches to further improve health benefits, such as the green tea fortified with soy [60], addition of linoleic acid, supplementation with vitamin D as an aid to weight loss, addition to fruit juices, and also utilization as an adjuvant or a prophylactic therapy in cancer treatment [61]. The postbiotics are referred as nonviable bacterial cells and can comprise whole-cell extracts or cell-free extracts or culture supernatant or purified cell wall components and are administered in the same way as probiotics and also have shown significant immunomodulatory effects [58].

5.4.2 CLASSIFICATION BASED ON CHEMICAL CLASS OF PHYTOCONSTITUENTS

Different chemical compounds are produced as a result of different biochemical reactions in the body of all living organisms. These compounds are classified mainly based on the parent compound from which other secondary metabolites are further produced, depending on numerous factors responsible for them. However, the genetic constitution and geographical conditions play an important role. Thus, the classification of these phytochemicals can be variable per the researcher and one specific format is difficult to maintain. Figure 5.9 depicts some important class of chemopreventive phytochemicals [9, 32, 42].

5.4.2.1 Phenolics

Phenolics are secondary metabolites mainly found in almost all plant-based food items. These phenolic compounds are further classified based on their structural pattern containing the basic moiety of phenolic group. There are more than 4,000 diverse flavonoids known so far and can be further classified. Almost all phenolic flavonoids promisingly show high antioxidant activity and possess potential to minimize the risk of almost all chronic diseases [62]. Phenolics are nucleophilic moieties and play an important role in inhibiting promotion process, metabolic activation, cell proliferation and differentiation, signal transduction pathway, oncogene expression involving cyclooxygenase-2, phase I enzyme, xanthine oxidase enzymes, and other cell adhesion and invasion activities. These are also inducers of cell cycle arrest and tumor suppressor gene expression and also enhance the detoxification of catalase, superoxide dismutase, glutathione peroxidase, and phase II enzyme. These compounds also inhibit kinases by reducing the hyperproliferation of epithelial cells and also prevent the formation of carcinogenic nitrosamines and nitroamides in food [9, 32, 42].

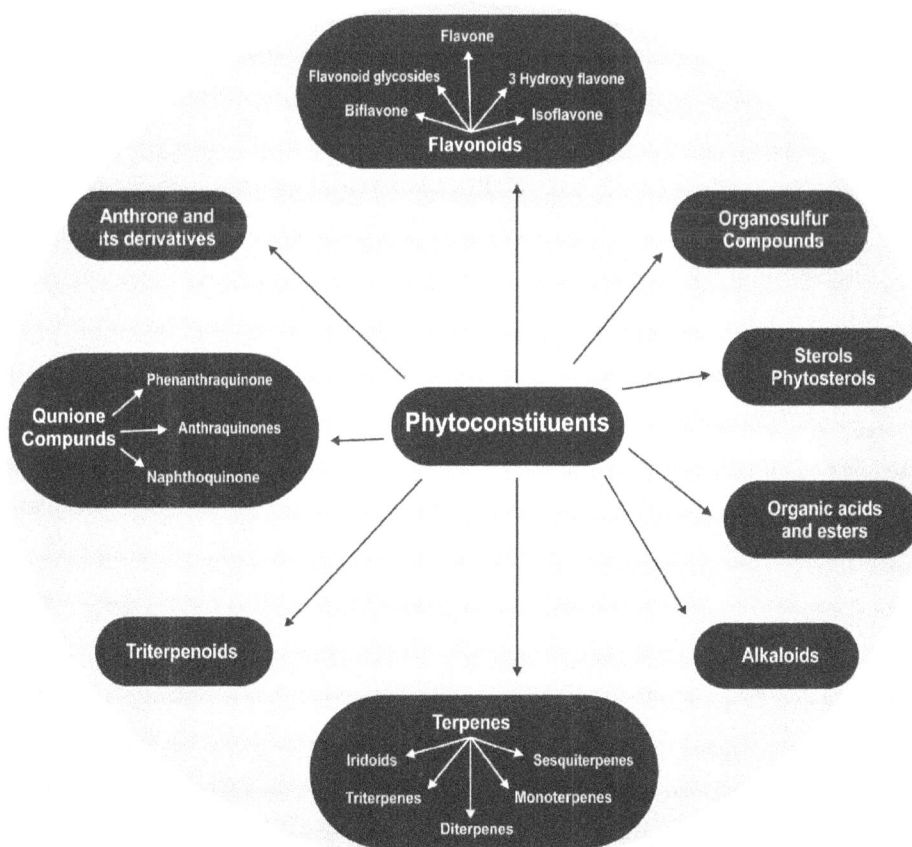

FIGURE 5.9 Different classes of phytoconstituents.

5.4.2.2 Alkaloids

Alkaloids are a class of organic compounds containing the nitrogen ring structure and are further classified based on their structural characteristics and are mainly of the pyrrolizidine and dendrobine type. These compounds exhibit anticancer activity as they prevent enzymatic activity of topoisomerase, mainly responsible for imitating DNA, inducing apoptosis and p53 gene expression. These are also responsible for inhibiting the growth of tumor cells and modifying their metabolism [42, 63, 64].

5.4.2.3 Terpenes and Triterpenoids

Terpenes are volatile unsaturated hydrocarbons found in essential oils of mainly citrus and coniferous trees. The parent moiety of all terpenes is 10 carbons and are further sub-classified. However, triterpenoids are composed of 30 carbons and are found in different fungi, ferns, animals, higher plants, and marine organisms. Monoterpenes are the resultant product formed as a result of polymerization of two isoprene units and their oxygen containing saturated derivatives as limonene from citrus fruits. The difference between almost all terpenes lies in the arrangement of their isoprene units as in sesquiterpene presence of three isoprene units, along with various chains and rings exists. These compounds are mainly present in the volatile plant root oil as curcumol, present in curcuma rhizomes [63]. Lucidenic acid and ganoderic acid from the members of Ganodermataceae possess cell toxicity and cell apoptosis and cell cycle arrest properties [65].

5.4.2.4 Carotenoids

Carotenoids are organic compounds having 40-carbon skeleton in the form of isoprene moieties cyclized at one or both ends, with oxygen containing groups as functional moiety, present usually as trans conjugated double bonds at the center of the structure, thereby presenting variable hydrogenation degree. These unique structural arrangements lead to specific chemical functions and optical and rotational properties, which enable this molecule to react with free radicals. Carotenoids are thus regarded as one of the most important natural pigments for their provitamin and antioxidant properties and prominently act as inducers of differentiation. These are widely present in pepper, coffee, tea leaves, tomatoes, and so on [66, 67].

5.4.2.5 Organosulfur Compounds

These natural organic phytochemicals are rich in sulfur present as a main functional group, responsible for particular aroma of these natural compounds, and are also known for their anticancer activity. The organosulfur compounds exhibit anticancer activity as prevent carcinogenesis stimulated by N-nitrosodiethylamine, and among all compounds, the strongest activity was shown by diallyl disulfide [68].

These compounds are widely present in members of *Allium* and are responsible for the induction of carcinogen detoxification, apoptosis, and the arrest of the proliferation of tumor cells and cell cycle. These compounds are also good free radical scavengers and inhibits DNA adduct formation [42].

5.4.2.6 Quinones

Quinones commonly occur as biological pigments in lower and higher plants, fungi, certain bacteria, and certain animals. The structural moiety of these quinones in general is cyclized; however, various variations exist, depending on the other functional groups attached to them, and they are further classified accordingly. The quinones exert anticancer activity as they are capable of forming active oxygen species, and also, some get converted to DNA binding semiquinone free radicals [69].

5.4.2.7 Other Miscellaneous Compounds

The ongoing research conducted each day to search for new anticancer compounds makes the classification of chemopreventive phytochemicals not final. Based on some previous research on phytochemicals present in any plant species, a new study can be conducted, and thus, the new chemicals become countless. Besides the compounds mentioned in detail, to some extent in the text, there are many other phytochemicals used as different dietary sources interfering in different steps of carcinogenesis and thus act as important chemopreventive agent. L-asparagine is catalyzed by L-asparaginase, which then activates apoptosis and autophagy in acute myeloid leukemia cells. Taxanes, including docetaxel and paclitaxel, exhibit anticancer effects as they are spindle poisons and thus inhibit cell division. Phytosterols are plant steroids mainly found in grains, beans, and rapeseed oil. Dioscin from the roots of the ginger plant and solasonine from black nightshade are effective in cancer chemoprevention. The fruits, leaves, and roots of many plants, like raspberries, plums, grapes, pomegranate, eucalyptus, and rhubarb, are rich in organic acids and esters that play an important role in adding medicinal attributes to these [42, 70].

5.4.3 CLASSIFICATION BASED ON MECHANISM OF ACTION

The phytochemicals from different sources play an important role in cancer chemoprevention as they are responsible for different inhibition mechanisms and certain other mechanisms both at the molecular and biochemical levels in such a manner that they ultimately inhibit/stop cancerous growth. These can be the activation of apoptosis in cancer cells, inhibition of tumor angiogenesis or activating the suppressive effect on cell migration, invasion, procarcinogen bioactivity repression prevention of DNA damage, or any detoxification activity.

5.4.3.1 Angiogenesis Inhibitors

Angiogenesis is simply the formation of new blood vessels from existing ones. It is basically dysfunctional imbalanced blood vessel growth resulting in the pathophysiological condition of cancer. It has been observed that certain natural plant- and animal-based food products are actively involved in the modulation of angiogenesis and are considered as angiogenesis modulators. Foods like garlic, walnuts, turmeric, green tea, and grapes are rich source of phytochemicals belonging to different class of chemical compounds and act as antiangiogenic agents. The compounds—like acacetin, aloin, arenobufagin, aspfalcholide, bigelovin, boswellicacid, beta-eudesmol, caffeicacid, ellagicacid, emodin, farnesiferol C, glyceollins, plumbagin, pterogynidne, punarnavine, quercetin, raddeanin A, santalol, vincristine, withaferin A, xanthohumol, and zerumbaneetc—are potent inhibitors of angiogenesis and are responsible for inhibition of VEGF-induced endothelial proliferation, PDGF receptors or TRAFG, HIF-1 α- MMP9 signaling pathways, and so on. Besides these, phytoconstituents like artemisnin, artesunate, barbatolic acid, β-escin sodium, celastrol, herboxidiene, honokiol, indole-3-carbinol, leucosesterterpenone, platycodin D, rhamnazin, rhein, rottlenin, salvicine, secalonic acid D, stibinin, sprengerinin C, streptochlorin, and tabectedin are considered as antiangiogenic compounds acting via different mechanism [32, 42]. The compounds like epigallocatechin-3- gallate inhibits VEGF signaling and thereby prevents the formation of new blood vessels. Similarly, isoliquintgenin is responsible for the inhibition of neovascularization and tube formation. Compounds like paclitaxel (Taxol) and docetaxel (Taxotere) have been approved as antiangiogenic drug by FDA (US Food and Drug Administration) [42, 71, 72].

5.4.3.2 Metastasis Inhibitors

Metastasis is the development of secondary malignant growth and its inhibition is one of the prominent anticancer therapies. The dietary food like fish oil rich in omega-3 polyunsaturated fatty acids exhibits anticancer activity as inhibits metastasis [42].

5.4.3.3 Free Radicals and ROS Inhibitors

The phytoconstituents present in natural food mainly act to enhance the level of antioxidation within the body and clear these reactive oxygen species in cancerous cells. These may also inhibit the growth of cells playing a vital role in cancer and can be considered as chemopreventive agents.

A dietary intake of citrus fruits, cruciferous vegetables, onions, tomatoes, green tea, blueberries, and blackberries are good source of antioxidants and are helpful to rebalance oxidative processes [42, 73].

5.4.3.4   Antiproliferative Compounds

There are several functional foods exhibiting antiproliferative mechanisms, thereby inhibiting the factors responsible for cell proliferation. Bitter melon, *Boletus edulis*, and dairy products, such as yogurt, cheese, and milk, which contain calcium, have antiproliferative proliferative [51, 55].

5.4.3.5   Apoptosis Inducers

Apoptosis is an energy-dependent, highly regulated mechanism commonly known as programmed cell death. The dietary intake of certain bioactive foods like *Tricholoma matsutake* proved beneficial for treatment of oral cancer [55].

5.4.3.6   DNA Methylation Inducers

The initiation of carcinogenesis leads to decreased methylation of DNA strands, ultimately affecting chromatin structure, leading to unstable genome stability and gene expression. These all are responsible for formation of inappropriate proteins, thus probably leading to the development of cancer. Green leafy vegetables, mushrooms, cheese, certain cereals, and strawberries are rich sources of folic acid, methionine, and choline responsible for methylation of DNA [42].

5.4.3.7   Matrix Metalloproteinase Inhibitors

Matrix metalloproteinases represent a family of zinc endopeptidases secreted by the cells and are involved in different biological processes like apoptosis and angiogenesis, responsible for the expression of pathological conditions of cancer. Phytochemicals have been scientifically proven to be inhibitors of different class of metalloproteinases. The active ingredient of turmeric, curcumin, showed inhibition of MMP-14 and MMP-2. Similarly, phenolic compounds from Amla inhibited MMP-1, and polyphenols from tea inhibited MMP-1, MMP-2, MMP-3, and MMP-9, reflecting their potential as chemopreventive agents [42, 73].

5.4.4   CLASSIFICATION BASED ON PROCESSING METHODS

The food processing technologies involve fermentation, drying, and canning, depending on the need and requirement of the desired products. Some processing methods involve sophisticated technologies to maintain the quality of the desired product. However, some are simple in process and can require even simple physical methods, like physical pressing to obtain oil from seeds.

5.4.4.1   Basic Food from a Natural Source

Basic natural food can be in raw form (salads, dressings, spices) or can be consumed as different spices or oils of different seeds. This allows the intake of natural phytochemicals without any alteration done in their chemical structure or physical attributes. The oils obtained from coconut, avocado, olives, and mustard are rich in different fatty acid composition. Whole grains rich in phyto-estrogens, such as lignin, plant stanols and sterols, minerals and vitamins, and apple cider vinegar, as well as honey are just few examples of natural functional foods consumed in our daily diet, enabling us to remain healthy [1, 69].

5.4.4.2   Processed Food

The natural foods, irrespective of their sources, revered for their convenience, novelty, palatability, and healthfulness, are termed as altered food items/processed foods. Processed food is an outcome of different technological process, such as pasteurizing, freezing, fermentation, and cooking. There are both risks and benefits, depending on the context, as there can be chances of formation of toxic compounds in the processing of meat due to certain biochemical reactions. Heat treatment reduces the microbial activity in food, thereby increasing the shelf life of processed food. Overall, processed food cannot be classified by using a single approach, and the classification basis can be discussed in lieu of the degree of processing, extent of change from its natural state, or even methods used by the food industries [74]. Usually, processed foods are not considered and developed as cancer chemopreventive agents. However, baked goods, such as biscuits, breads, and cakes, are considered as vehicles for the development of functional food products. The recent trend of research is now based on substitution strategies—for example, the replacement of wheat flour with a range of vegetable/fruit byproducts rich in nutrients by the use of suitable technology like encapsulation for sensitive bioactive compounds. Nanoencapsulation technology involves the incorporation of desired natural bioactive substances, such as vitamins, polyphenols, volatile additives, enzymes, and oils, into nanosized capsules to mask flavors and aromas, aiming to deliver these compounds easily to target sites within the body [75, 76].

5.4.4.3 Fortified Food

The development of fortified food was to benefit the population by minimizing nutritional deficiencies without any unintended harm to their health. The fortification of food is based on scientific evidence, taking into account of human rights for intervention and implementation as it can be a subject to ethical challenges. The deficiencies of vitamin A, folic acid, iron, iodine, and zinc in human diet are more or less minimized by fortifying salt, vegetable oil, rice, and wheat as they are absolutely important in diet. Recent research has also helped in the development of fortified food as chemopreventive and therapeutic agents in cancer. For example, lunasin was developed and commercialized as a dietary supplement for various health benefits, including cancer treatment [77].

5.4.4.4 Enhanced Food

Enhanced foods are considered the most direct supplementation method in the form of preparations, with any specific phytoconstituent required for the treatment of disease. These are generally produced through genetic modification done for a purposeful change of substances to obtain a desired result, usually in crops or animal feed, resulting in the development and production of new varieties with high yield and nutrition-enhanced traits. The development of golden rice to provide more vitamin A and feeding hens a special diet to improve the omega-3 fatty acids in eggs are some examples of producing enhanced foods [78]. The cancer patient undergoing treatment, be it radiotherapy or chemical treatment, is at risk of selenium deficiency. This deficiency can be minimized by the use of selenium-rich dietary supplements or some specific medicines that can also help to reduce the risk of developing certain types of cancer. Per EU regulations and guidelines, the selenium-enriched yeast can be developed as dietary supplement to overcome the deficiency and meet the demand [79].

5.4.4.5 Isolated and Purified Preparation of Active Phytochemical from Food Constituents (Herbal/Product/Dosage Form)

The ongoing researches for the development of cancer chemopreventive functional foods also involve isolated pure compounds to be therapeutically used. The different natural phytoconstituents, like catechins, capsaicin, cucurbitacin B, isoflavones, lycopenes, piperlongumine, benzyl isothiocyanate, and phenethylisothiocyanate have shown inhibitory effects on cancer cells, thereby indicating their potential role as chemopreventive agents. Curcumin can induce apoptosis in different cell lines and inhibit tumor formation, as shown in animal models. There are several studies that indicate that curcumin might be employed as a therapeutic agent against several molecular targets. The compounds isoflavone and genistein have been proven to be effective against prostate and breast cancer, and their in vitro anticancer effects have also been elucidated and proposed. The peptide lunasin, naturally present in soybean, has shown promising chemopreventive properties. The treatment with it does not affect normal cell proliferation or morphology but is capable to prevent mammalian cell transformation induced either by chemical carcinogens, ras-oncogene, or viral oncogene E1A. This compound also exerts cytotoxic effect against leukemia cells (L12120) by arresting the cell cycle at G2/M phase and inducing apoptosis [80–83]. However, not only isolated compounds present in different forms of plants like cruciferous vegetables, but also the organic extracts of leaves etc., also exhibits cytotoxic potential and helps in reducing cancer risk at several sites [84, 85].

5.5 SUMMARY

Since time immemorial, men had been utilizing food not only as a source of nutrition but also as treatments against disease. This concept has again gained popularity and now is considered as alternative, complementary, or integrative medicine and is commonly called as functional foods. Different foods are classified in a number of ways, and one exact definition cannot be given due to different legislations and regulations all across the globe, although the basic aim to develop and utilize them remains the same. In the current scenario, great scientific efforts and research have to be investigated in this field to discover new sources and design novel functional foods. The restricted literature about the biochemical factors, mechanisms, and biochemical processes involved right from onset of cancer till its progression are still unclear. The molecular complexity of them also play a unique role; at what step and at what dose the chemopreventive agent blocks the pathways remains still unclear. Bioactive phytoconstituents are mainly from plants, although those from animal and microbial sources play an important role in cancer chemoprevention; however, the complete mechanism of action for all these have not been completely elucidated. In a nutshell, to sum up all the chemopreventive functional foods becomes extremely difficult; however, a concise description of different types of functional foods and the pathophysiology of cancer have been discussed. It becomes very necessary to develop a research methodology to shed light on the molecular mechanism involved in the pathophysiology of cancer, along with targeted chemopreventive functional foods.

5.6 ACKNOWLEDGMENTS

The authors express their sincere gratitude to Mr. Ajeet Chauhan, Senior Graphics Designer, University of Patanjali, Haridwar, Uttarakhand, India, for designing such phenomenal graphics.

5.7 REFERENCES

1. Liyanage, D., and BauddhalokaMawatha. 2017. "Health benefits and traditional uses of honey: A review." *Journal of Apitherapy* 2(1) 9–14.doi: 10.5455/ja.20170208043727
2. Arai, Soichi, Toshihiko Osawa, Hajime Ohigashi, Masaaki Yoshikawa, Shuichi Kaminogawa, Michiko Watanabe, Tadashi Ogawa et al. 2001. "A mainstay of functional food science in Japan—history, present status, and future outlook." *Bioscience, Biotechnology, and Biochemistry* 65(1): 1–13. https://doi.org/10.1271/bbb.65.1
3. Grochowicz, Józef, Anna Fabisiak, and Dorota Nowak. 2018. "Market of functional food—legal regulations and development perspectives." *ZeszytyProblemowePostępówNaukRolniczych* 595: 51–67. https://doi.org/10.22630/ZPPNR.2018.595.35
4. Adefegha, Stephen Adeniyi. 2018. "Functional foods and nutraceuticals as dietary intervention in chronic diseases; novel perspectives for health promotion and disease prevention." *Journal of Dietary Supplements* 15(6): 977–1009. https://doi.org/10.1080/19390211.2017.1401573
5. Saini, Anupam, Manish Kumar, Shailendra Bhatt, Vipin Saini, and Anuj Malik. 2020. "Cancer causes and treatments." *International Journal of Pharmaceutical Sciences and Research* 11(7): 3121–3134. https://doi.org//10.13040/IJPSR.0975-8232.11(7).3121-34
6. Waters, Erika A., Timothy S. McNeel, WortaMcCaskill Stevens, and Andrew N. Freedman. 2012. "Use of tamoxifen and raloxifene for breast cancer chemoprevention in." *Breast Cancer Research and Treatment* 134(2): 875–880. https://doi.org/10.1007/s10549-012-2089-2
7. Wattenberg, Lee W. 1966. "Chemoprophylaxis of carcinogenesis: A review." *Cancer Research* 26(7): 1520–1526.

8. Koller, P. 1968. "Heredity and cancer." *Nature* 217: 199–200. https://doi.org/10.1038/217199b0

9. Anand, Preetha, Ajaikumar B. Kunnumakara, Chitra Sundaram, Kuzhuvelil B. Harikumar, Sheeja T. Tharakan, Oiki S. Lai, Bokyung Sung, and Bharat B. Aggarwal. 2008. "Cancer is a preventable disease that requires major lifestyle changes." *Pharmaceutical Research* 25(9): 2097–2116. https://doi.org/10.1007/s11095-008-9661-9.

10. Goswami, Hit Kishore, and Hitendra Kumar Ram. 2017. "Ancient food habits dictate that food can be medicine but medicine cannot be 'food'!!" *Medicines* 4(4): 82. https://doi.org/10.3390/medicines4040082

11. Baliga, Manjeshwar Shrinath, Sharake Meera, Benson Mathai, Manoj Ponadka Rai, Vikas Pawar, and Princy Louis Palatty. 2012. "Scientific validation of the ethnomedicinal properties of the Ayurvedic drug Triphala: A review." *Chinese Journal of Integrative Medicine* 18(12): 946–954. https://doi.org/10.1007/s11655-012-1299-x

12. Singh, Rasnik K., Hsin-Wen Chang, D. I. Yan, Kristina M. Lee, DeryaUcmak, Kirsten Wong, Michael Abrouk et al. 2017. "Influence of diet on the gut microbiome and implications for human health." *Journal of Translational Medicine* 15(1): 1–17. https://doi.org/10.1186/s12967-017-1175-y; https://doi.org/10.1146/annrev-food-062520-093642

13. Granato, Daniel, Francisco J. Barba, Danijela Bursać Kovačević, José M. Lorenzo, Adriano G. Cruz, and Predrag Putnik. 2020. "Functional foods: Product development, technological trends, efficacy testing, and safety." *Annual Review of Food Science and Technology* 11: 93–118. https://doi.org/10.1146/annrev-food-062520-093642

14. Birch, Catherine S., and Graham A. Bonwick. 2019. "Ensuring the future of functional foods." *International Journal of Food Science & Technology* 54(5): 1467–1485. https://doi.org/10.1111/ijfs.14060

15. Revelou, Panagiota-Kyriaki, Marinos Xagoraris, Maroula G. Kokotou, and Violetta Constantinou-Kokotou. 2021. "Cruciferous vegetables as functional foods: Effects of selenium biotransformation." *International Journal of Vegetable Science*: 1–2. https://doi.org/10.1080/19315260.2021.1957052

16. Sotos-Prieto, Mercedes, Daniele Del Rio, Greg Drescher, Ramon Estruch Chavanne Hanson Timothy Harlan, Frank B. Hu et al. 2022. "Mediterranean diet –promotion and dissemination of healthy eating: Proceedings of an exploratory seminar at the Radcliffe institute for advanced study." *International Journal of Food Sciences and Nutrition* 73(2): 158–171. https://doi.org/10.1080/09637486.2021.1941804

17. Konstantinidi, Melina, and Antonios E. Koutelidakis. 2019. "Functional foods and bioactive compounds: A review of its possible role on weight management and obesity's metabolic consequences." *Medicines* 6(3): 94. https://doi.org/10.3390/medicines6030094

18. Garcia-Alvarez, Alicia, Bernadette Egan, Simone De Klein, Lorena Dima, Franco M. Maggi, Merja Isoniemi, Lourdes Ribas-Barba et al. 2014. "Usage of plant food supplements across six European countries: Findings from the PlantLIBRA consumer survey." *PloS One* 9(3): e92265. https://doi.org/10.1371/journal.pone.0092265

19. Dickinson, Annette. 2011. "History and overview of DSHEA." *Fitoterapia* 82(1): 5–10. https://doi.org/10.1016/j.fitote.2010.09.001

20. Banerjee, Swapan. 2018. "Dietary supplements market in India is rapidly growing-An overview." *IMS Management Journal* 10(1): 1–6.

21. Niva, Mari. 2007. "'All foods affect health': Understandings of functional foods and healthy eating among health-oriented Finns." *Appetite* 48(3): 384–393. https://doi.org/10.1016/j.appet.2006.10.006

22. Diplock, A.T., P.J. Aggett, M. Ashwell, F. Bornet, E.B. Fern, and M.B. Roberfroid. 1999. "Scientific concepts of functional foods in Europe: Concensus document." *British Journal of Nutrition* 81(1): S1–S27.

23. Plasek, Brigitta, Zoltán Lakner, Gyula Kasza, and Ágoston Temesi. 2019. "Consumer evaluation of the role of functional food products in disease prevention and the characteristics of target groups." *Nutrients* 12(1): 69. https://doi.org/10.3390/nu12010069

24. Butnariu, Monica, and Ioan Sarac. 2019. "Functional food." *International Journal of Nutrition* 3(3): 7–16. https://doi.org/10.14302/issn.2379-7835.ijn-19-2615

25. Bañuelos, Gary S., John Freeman, and Irvin Arroyo. 2020. "Accumulation and speciation of selenium in biofortified vegetables grown under high boron and saline field conditions." *Food Chemistry* X(5): 100073. https://doi.org/10.1016/j.fochx.2019.100073.

26. Nandini, D. B., Roopa S. Rao, B. S. Deepak, and Praveen B. Reddy. 2020. "Sulforaphane in broccoli: The green chemoprevention!! Role in cancer prevention and therapy." *Journal of Oral and Maxillofacial Pathology* 24(2): 405. https://doi.org/10.4103/jomfp.JOMFP_126_19

27. Putnik, Predrag, Jose M. Lorenzo, Francisco J. Barba, Shahin Roohinejad, Anet Režek Jambrak, Daniel Granato, Domenico Montesano, and Danijela Bursać Kovačević. 2018. "Novel food processing and extraction technologies of high-added value compounds from plant materials." *Foods* 7(7): 106. https://doi.org/10.3390/foods7070106

28. Farid, Mohamed, Kota Kodama, TeruyoArato, Takashi Okazaki, Tetsuaki Oda, Hideko Ikeda, and Shintaro Sengoku. 2019. "Comparative study of functional food regulations in Japan and globally." *Global Journal of Health Science* 11: 132–145. doi:10.5539/gjhs.v11n6p132

29. Deborah, Bentivoglio, Margherita Rotordam, StaffolaniGiacomo, Chiaraluce Giulia, and Finco Adele. 2021. "Understanding consumption choices of innovative products: An outlook on the Italian functional food market." *AIMS Agriculture and Food* 6(3): 818–837. https://doi.org/10.3934/agrfood.2021050

30. Mucci, L.A., S. Wedren, R.M. Tamimi, D. Trichopoulos, and H-O. Adami. 2001. "The role of gene–environment interaction in the aetiology of human cancer: Examples from cancers of the large bowel, lung and breast." *Journal of Internal Medicine* 249(6): 477–493. https://doi.org/10.1046/j.1365-2796.2001.00839.x

31. Rodriguez, Adriana C., Zannel Blanchard, Kathryn A. Maurer, and Jason Gertz. 2019. "Estrogen signaling in endometrial cancer: A key oncogenic pathway with several open questions." *Hormones and Cancer* 10(2): 51–63. https://doi.org/10.2147/IJWH.S40942

32. Varol, M. 2017. "Angiogenesis as an important target in cancer therapies." In *Researches on Science and Art in 21st Century Turkey, 1971–1981*. Turkey: Grece Publishing. www.researchgate.net/publication/321679954_

33. Cooper, Geoffrey M., Robert E. Hausman, and Robert E. Hausman. 2007. *The Cell: A Molecular Approach*. Vol. 4. Washington, DC, USA: ASM Press. https://doi.org/10.1086/513338

34. Felmeden,D.C., A.D. Blann, andG.Y.H. Lip. 2003. "Angiogenesis: Basic pathophysiology and implications for disease." *European Heart Journal* 24(7): 586–603. https://doi.org/10.1016/s0195-668x(02)00635-8

35. Vogel, Victor G., Joseph P. Costantino, D. Lawrence Wickerham, Walter M. Cronin, Reena S. Cecchini, James N. Atkins, Therese B. Bevers et al. 2010. "Update of the national surgical adjuvant breast and bowel project study of tamoxifen and raloxifene (STAR) P-2 trial: Preventing breast cancer." *Cancer Prevention Research* 3(6): 696–706. https://doi.org/10.1158/1940-6207.CAPR-10-0076.

36. Glade, Michael J. 1997. "Food, nutrition, and the prevention of cancer: A global perspective. American institute for cancer research/world cancer research fund, American institute for cancer research, 1997." *Nutrition* 15: 523–526. https://doi.org/10.3390/metabo10040123

37. Penny, Lewis K., and Heather M. Wallace. 2015. "The challenges for cancer chemoprevention." *Chemical Society Reviews* 44(24): 8836–8847. https://doi.org/10.1039/C5CS00705D

38. Fahey, Jed W., Paul Talalay, and Thomas W. Kensler. 2012. "Notes from the field: 'green' chemoprevention as frugal medicine." *Cancer Prevention Research* 5(2): 179–188. https://doi.org/10.1158/1940-6207.CAPR-11-0572

39. Islam, SM Rafiqul, and TowfidalJahan Siddiqua. 2020. "Functional foods in cancer chemoprevention and therapy: Recent epidemiological findings." *Functional Foods in Cancer Prevention and Therapy*: 405–433. https://doi.org/10.1016/C2017-0-04743-2

40. Lewandowska, Urszula, Sylwia Gorlach, Katarzyna Owczarek, Elżbieta Hrabec, and Karolina Szewczyk. 2014. "Synergistic interactions between anticancer chemotherapeutics and phenolic compounds and anticancer synergy between polyphenols." *Advances in Hygiene & Experimental Medicine/PostepyHigieny i MedycynyDoswiadczalnej* 68: 528–540. https://doi.org/10.5604/01.3001.0003.1229

41. Sak, Katrin, and Hele Everaus. 2017. "Established human cell lines as models to study anti-leukemic effects of flavonoids." *Current Genomics* 18(1): 3–26. https://doi.org/10.2174/1389202917666160803165447

42. Gupta, P. 2020. "Targeted cancer therapy with bioactive foods and their products." In *Functional Foods in Cancer Prevention and Therapy* (pp. 33–46). London: Academic Press. https://doi.org/10.1016/B978-0-12-816151-7.00002-8

43. Giovannucci, Edward, and Walter C. Willett. 1994. "Dietary factors and risk of colon cancer." *Annals of Medicine* 26(6): 443–452. https://doi.org/10.1093/jn/131.11.3027S

44. Amin, ARM Ruhul, Omer Kucuk, Fadlo R. Khuri, and Dong M. Shin. 2009. "Perspectives for cancer prevention with natural compounds." *Journal of ClinicalOncology* 27(16): 2712. https://doi.org/10.1200/JCO.2008.20.6235

45. Jeffery, Elizabeth H., and Anna-Sigrid Keck. 2008. "Translating knowledge generated by epidemiological and in vitro studies into dietary cancer prevention." *Molecular Nutrition & Food Research* 52(S1): S7–S17. https://doi.org/10.1021/jf9034817

46. Henderson, Angela J., Cadie A. Ollila, Ajay Kumar, Erica C. Borresen, KomalRaina, Rajesh Agarwal, and Elizabeth P. Ryan. 2012. "Chemopreventive properties of dietary rice bran: Current status and future prospects." *Advances in Nutrition* 3(5): 643–653. https://doi.org/10.3945/an.112.002303

47. Venugopal, Shonima, and Uma M. Iyer. 2010. "Management of diabetic dyslipidemia with subatmospheric dehydrated barley grass powder." *International Journal of Green Pharmacy* 4(4). https://doi.org/10.4103/0973-8258.7413

48. Nakurte, Ilva, Inga Kirhnere, Jana Namniece, Kristine Saleniece, LigaKrigere, PeterisMekss, ZaigaVicupe, Mara Bleidere, Linda Legzdina, and Ruta Muceniece. 2013. "Detection of the lunasin peptide in oats (Avenasativa L)." *Journal of Cereal Science* 57(3): 319–324. https://doi.org/10.1016/j.jcs.2012.12.008

49. W. Watson, Gregory, Laura M. Beaver, David E. Williams, Roderick H. Dashwood, and Emily Ho. 2013. "Phytochemicals from cruciferous vegetables, epigenetics, and prostate cancer prevention." *The AAPS Journal* 15(4): 951–961. https://doi.org/10.1208/s12248-013-9504-4

50. Lovegrove, C., K. Ahmed, B. Challacombe, M.S. Khan, R. Popert, and P. Dasgupta. 2015. "Systematic review of prostate cancer risk and association with consumption of fish and fish-oils: Analysis of 495,321 participants." *International Journal of Clinical Practice* 69(1): 87–105.https://doi.org/10.1111/ijcp.12514

51. Lamprecht, Sergio A., and Martin Lipkin. 2001. "Cellular mechanisms of calcium and vitamin D in the inhibition of colorectal carcinogenesis." *Annals of the New York Academy of Sciences* 952(1): 73–87. doi/10.1111/j.1749-6632.2001.tb02729.x

52. deKok, Theo M., Simone G. van Breda, and Margaret M. Manson. 2008. "Mechanisms of combined action of different chemopreventive dietary compounds." *European Journal of Nutrition* 47(2): 51–59. https://doi.org/10.1007/s00394-008-2006-y

53. Ip, Clement, and Joseph A. Scimeca. 1997. "Conjugated linoleic acid and linoleic acid are distinctive modulators of mammary carcinogenesis." *Nutrition and Cancer* 27: 131–135. https://doi.org/10.1080/01635589709514514

54. Mans, Dennis RA. 2016. "Exploring the global animal biodiversity in the search for new drugs—marine invertebrates." *Journal of Translational Science* 2(3): 170–179. https://doi.org/10.15761/JTS.100013

55. Shin, Ji, Jun Sung Kim, In-Sun Hong, and Sung-Dae Cho. 2012. "Bak is a key molecule in apoptosis induced by methanol extracts of Codonopsislanceolata and Tricholomamatsutake in HSC-2 human oral cancer cells." *Oncology Letters* 4(6): 1379–1383. https://doi.org/10.3892/ol.2012.898

56. Kuščer, Enej, Nigel Coates, Iain Challis, Matt Gregory, Barrie Wilkinson, Rose Sheridan, and Hrvoje Petkovic. 2007. "Roles of rapH and rapG in positive regulation of rapamycin biosynthesis in Streptomyces hygroscopicus." *Journal of Bacteriology* 189(13): 4756–4763. https://doi.org/10.1128/JB.00129-07

57. Tan, LohTeng-Hern, Chim-Kei Chan, Kok-Gan Chan, Priyia Pusparajah, Tahir Mehmood Khan, Hooi-Leng Ser, Learn-Han Lee, and Bey-Hing Goh. 2019. "Streptomyces sp. MUM256: A source for apoptosis inducing and cell cycle-arresting bioactive compounds against colon cancer cells." *Cancers* 11(11): 1742. https://doi.org/10.3390/cancers11111742

58. Scarpellini, Emidio, EmanueleRinninella, Martina Basilico, Esther Colomier, Carlo Rasetti, TizianaLarussa, PierangeloSantori, and Ludovico Abenavoli. 2021. "From pre-and probiotics to postbiotics: A narrative review." *International Journal of Environmental Research and Public Health* 19(1): 37. https://doi.org/10.3390/ijerph19010037

59. Gao, Jie, Xiyu Li, Guohua Zhang, Faizan Ahmed Sadiq, Jesus Simal-Gandara, Jianbo Xiao, and Yaxin Sang. 2021. "Probiotics in the dairy industry—Advances and opportunities." *Comprehensive Reviews in Food Science and Food Safety* 20(4): 3937–3982. https://doi.org/10.1111/1541-4337.12755

60. Moumita, Sahoo, Bhaskar Das, ArchanaSundaray, SanghamitraSatpathi, P. Thangaraj, S. Marimuthu, and R. Jayabalan. 2018. "Study of soy-fortified green tea curd formulated using potential hypocholesterolemic and hypotensive probiotics isolated from locally made curd." *Food Chemistry* 268: 558–566. https://doi.org/10.1016/j.foodchem.2018.06.114

61. Chadare, Flora Josiane, RodrigueIdohou, Eunice Nago, Marius Affonfere, Julienne Agossadou, ToyiKévinFassinou, ChristelKénou et al. 2019. "Conventional and food-to-food fortification: An appraisal of past practices and lessons learned." *Food Science Nutrition* 7(9): 2781–2795. https://doi.org/10.1002/fsn3.1133

62. Serra, Aida, Alba Macià, Maria-Paz Romero, JordiReguant, Nadia Ortega, and Maria-José Motilva. 2012. "Metabolic pathways of the colonic metabolism of flavonoids (flavonols, flavones and flavanones) and phenolic acids." *Food Chemistry* 130(2): 383–393. https://doi.org/10.1016/j.foodchem.2011.07.055

63. Pagare, Saurabh, Manila Bhatia, Niraj Tripathi, Sonal Pagare, and Y.K. Bansal. 2015. "Secondary metabolites of plants and their role: Overview." *Current Trends in Biotechnology and Pharmacy* 9(3): 293–304. www.researchgate.net/publication/283132113

64 Brown, Christopher J., Sonia Lain, Chandra S. Verma, Alan R. Fersht, and David P. Lane. 2009. "Awakening guardian angels: Drugging the p53 pathway." *Nature Reviews Cancer* 9(12): 862–873. https://doi.org/10.1038/nrc2763

65 Smith, M.T. 1985. "Quinones as mutagens, carcinogens and anticancer agents: Introduction and overview." 665–672. https://doi.org/10.1080/15287398509530776

66 Stahl, Wilhelm, and Helmut Sies. "Antioxidant activity of carotenoids." *Molecular Aspects of Medicine* 24, no. 6 (2003): 345–351. https://doi.org/10.1016/S0098-2997(03)00030-X

67. Valko, Marian, Dieter Leibfritz, Jan Moncol, Mark T.D. Cronin, Milan Mazur, and Joshua Telser. 2007. "Free radicals and antioxidants in normal physiological functions and human disease." *The International Journal of Biochemistry & Cell Biology* 39(1): 44–84. https://doi.org/10.1016/j.biocel.2006.07.001

68. Rajkapoor, Balasubramanian, Narayanan Murugesh, Dechen Chodon, and Dhanapal Sakthisekaran. 2005. "Chemoprevention of N-nitrosodiethylamine induced phenobarbitol promoted liver tumors in rat by extract of Indigoferaaspalathoides." *Biological and Pharmaceutical Bulletin* 28(2): 364–366. https://doi.org/10.1248/bpb.28.364

69. CFR Ferreira, I., J.A. Vaz, M.H. Vasconcelos, and A. Martins. 2010. "Compounds from wild mushrooms with antitumor potential." *Anti-cancer Agents in Medicinal Chemistry (Formerly Current Medicinal Chemistry –Anti-Cancer-Agents)* 10(5): 424–436. www.ingentaconnect.com/content/ben/acamc/2010/00000010/00000005/art00006

70. Zhou, Qiang, Wei Song, and Wei Xiao. 2017. "Dioscin induces demethylation of DAPK-1 and RASSF-1alpha genes via the antioxidant capacity, resulting in apoptosis of bladder cancer T24 cells." *EXCLI Journal* 16: 101. https://doi.org/10.17179/excli2016-571

71. Wu, Shanshan, JihuaXue, Ying Yang, Haihong Zhu, Feng Chen, Jing Wang, Guohua Lou et al. "Isoliquiritigenin inhibits interferon-γ-inducible genes expression in hepatocytes through down-regulating activation of JAK1/STAT1, IRF3/MyD88, ERK/MAPK, JNK/MAPK and PI3K/Aktsignaling pathways." *Cellular Physiology and Biochemistry* 37(2): 501–514. https://doi.org/10.1159/000430372

72. Moyle, Christina W.A., Ana B. Cerezo, Mark S. Winterbone, Wendy J. Hollands, Yuri Alexeev, Paul W. Needs, and Paul A. Kroon. 2015. "Potent inhibition of VEGFR-2 activation by tight binding of green tea epigallocatechingallate and apple procyanidins to VEGF: Relevance to angiogenesis." *Molecular Nutrition &Food Research* 59(3): 401–412. https://doi.org/10.1002/mnfr.201400478

73. Mannello, Ferdinando, Francesca Luchetti, ElisabettaFalcieri, and Stefano Papa. 2005. "Multiple roles of matrix metalloproteinases during apoptosis." *Apoptosis* 10(1): 19–24. https://doi.org/10.1007/s10495-005-6058-7

74. Sadler, Christina R., Terri Grassby, Kathryn Hart, Monique Raats, MilkaSokolović, and Lada Timotijevic. 2021. "Processed food classification: Conceptualisation and challenges." *Trends in Food Science & Technology* 112: 149–162. https://doi.org/10.1016/j.tifs.2021.02.059

75. Akhavan, Sahar, ElhamAssadpour, ImanKatouzian, and Seid Mahdi Jafari. 2018. "Lipid nano scale cargos for the protection and delivery of food bioactive ingredients and nutraceuticals." *Trends in Food Science & Technology* 74: 132–146. https://doi.org/10.1016/j.tifs.2018.02.001

76. Assadpour, Elham, and Seid Mahdi Jafari. 2019. "A systematic review on nanoencapsulation of food bioactive ingredients and nutraceuticals by various nanocarriers." *Critical Reviews in Food Science and Nutrition* 59(19): 3129–3151. https://doi.org/10.1080/10408398.2018.1484687

77. Hurrell, R. 2002. "How to ensure adequate iron absorption from iron-fortified food." *Nutrition Reviews* 60: S7–S15. https://doi.org/10.3929/ethz-b-000422926

78. Hefferon, KathleenL. 2015. "Nutritionally enhanced food crops; progress and perspectives." *International Journal of Molecular Sciences* 16(2): 3895–3914. https://doi.org/10.3390/ijms16023895

79. Kieliszek, Marek, and Stanisław Błażejak. 2016. "Current knowledge on the importance of selenium in food for living organisms: A review." *Molecules* 21(5): 609. https://doi.org/10.3390/molecules21050609

80. Xia, Yong, Ruijiao Chen, Guangzhen Lu, Changlin Li, SenLian, Taek-Won Kang, and Young Do Jung. 2021. "Natural phytochemicals in bladder cancer prevention and therapy." *Frontiers in Oncology* 11. https://doi.org/10.3389/fonc.2021.652033

81. Kwon, Ki Han, AvantikaBarve, Siwang Yu, Mou-Tuan HUANG, and Ah-Ng Tony Kong. 2007. "Cancer chemoprevention by phytochemicals: Potential molecular targets, biomarkers and animal models 1." *ActaPharmacologicaSinica* 28(9): 1409–1421.

82. Hsieh, Chia-Chien, Cristina Martínez-Villaluenga, Ben O. de Lumen, and Blanca Hernández-Ledesma. 2018. "Updating the research on the chemopreventive and therapeutic role of the peptide lunasin." *Journal of the Science of Food and Agriculture* 98(6): 2070–2079. https://doi.org/10.1002/jsfa.8719

83. Ding, Jing, Jing-Jing Wang, Chen Huang, Li Wang, Shining Deng, Tian-Le Xu, Wei-Hong Ge, Wei-Guang Li, and Fei Li. 2014. "Curcumol from RhizomaCurcumae suppresses epileptic seizure by facilitation of GABA(A)receptors." *Neuropharmacology* 81: 244–255. https://doi.org/10.1016/j.neuropharm.2014.02.009.

84. Sengottuvelu, S., and M. Pharm. 2022. "Cytotoxic consequences of Andrographisechiodies as a novel anticancer entity against cancer cell line." *Turkish Journal of Physiotherapy and Rehabilitation* 32: 3.36. https://doi.org/10.1093/ecam/nep135

85. Talalay, Paul, and Jed W. Fahey. 2001. "Phytochemicals from cruciferous plants protect against cancer by modulating carcinogen metabolism." *The Journal of Nutrition* 131(11): 3027S–3033S. https://doi.org/10.1093/jn/131.11.3027S

6 Phytochemicals
Potential Source of Interventions in Cancer Prevention and Treatment

*Sonu Kumar Mahawer, Himani, Sushila Arya,
Tanuja Kabdal, Ravendra Kumar, and Om Prakash*

CONTENTS

6.1 Introduction 97
6.2 Cancer...98
6.3 Causes of Cancer 98
6.4 Types and Symptoms of Cancer 99
6.5 Phytochemicals in Cancer Prevention 100
    6.5.1 Curcumin..100
    6.5.2 Lycopene..100
    6.5.3 Isothiocyanates (ITCs)..100
    6.5.4 Genistein 101
    6.5.5 Resveratrol 101
    6.5.6 Other Phytochemicals in Cancer Prevention..102
6.6 Phytochemicals in Cancer Treatment...102
    6.6.1 Preclinically Studied Anticancer Phytochemicals......................................102
    6.6.2 Clinically Studied Anticancer Phytochemicals...102
6.7 Conclusion and Future Prospects 105
6.8 References..106

6.1 INTRODUCTION

Cancer is an illness involving abnormal cell growth and their sufficient potential to attack and metastasize in other body parts. The primary cause of cancer is thought to be changes in the genes regulating normal bodily function [1]. According to studies, cancer causes 13% of all deaths, a number that is rising daily as the population grows and more people engage in cancer-causing habits, including smoking, drinking alcohol, and eating unhealthily [2, 3]. Most cancers are those of the lung, stomach, colon, breast, liver, and blood [4].

For thousands of years, people have used plants to either prevent or treat a variety of diseases. Nowadays, thousands of scientific reports are available regarding the non-nutritional components found in plants and plant-based products that have the potential to prevent and/or treatment of degenerative illnesses like cancer. The diverse class of such compounds is known as phytochemicals, which include carotenoids, flavonoids, phenolic acid indoles, sulfur compounds, and many more. More than 10,000 phytochemicals have been identified and defined, and about 6,000 are contained within the chemical class of flavonoids [5]. These compounds are widely found in plant-based food products, several dietary supplements, or plant-based medicines.

Phytochemicals have been utilized for cancer therapy from ancient times as these are safe, less toxic, and easily available. It was demonstrated in earlier studies that fruit and vegetable consumption was associated with abating the risk of cancer. Effective compounds not only interrupt cancer-related abnormal signaling pathways but also synergize with cancer therapies [6]. Therefore, the capability of phytochemicals to prevent and treat cancer gains the attention of the researcher.

In this chapter, we are focusing on promising phytochemicals and their role in cancer prevention and/or cancer treatment based on the literature available on preclinically and clinically proven aspects.

6.2 CANCER

The term *cancer* is used to describe different types of anomalous, unhealthy cells growing in an uncontrolled manner in the body tissues. Malevolent tumors are formed in cancer, leading to the abnormal functioning of the body organ. Cancer cells can spread to the neighboring tissues or through the bloodstream or lymph system to other body parts.

6.3 CAUSES OF CANCER

There are several types of factors that are believed to be a cause of cancer. Cancer-causing factors are broadly divided into two categories: interior factors, such as inherited mutations, hormonal, and immune conditions, and exterior/environmental/acquired factors, such as tobacco and alcohol consumption, unhealthy diet, sun and UV radiation, infectious microbes, and workplace factors [7, 8]. Factor-wise percent contribution is represented in Figure 6.1. It shows that about only 5–10%

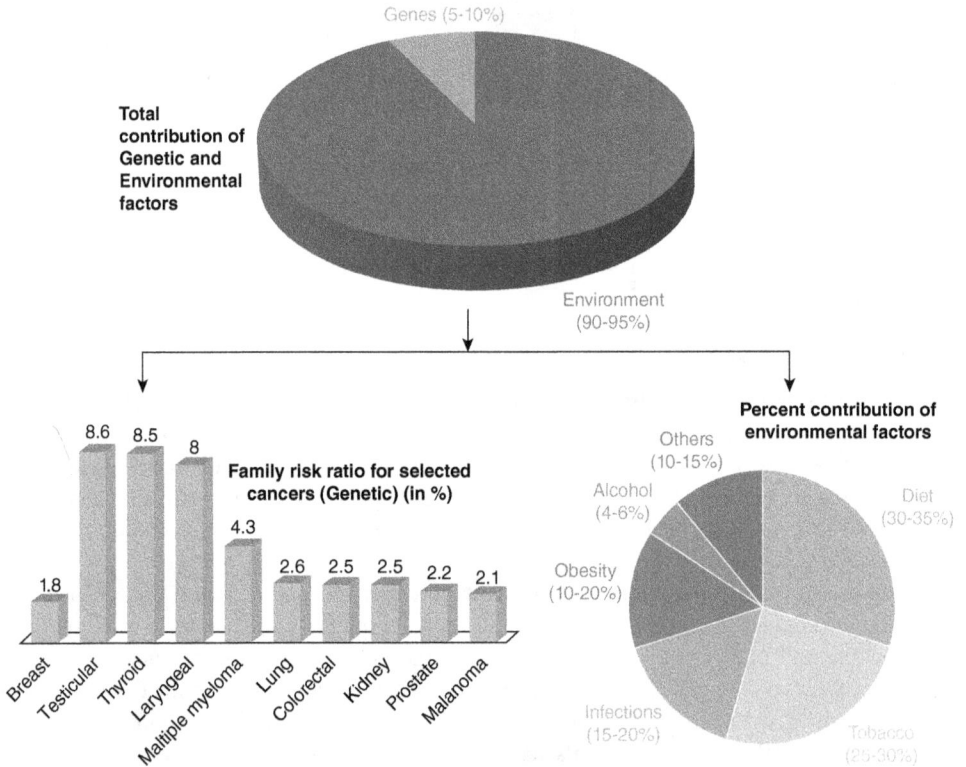

FIGURE 6.1 Percent contribution of genetic and environmental factors in causing cancer (adopted and modified from Anand et al., 2008; Katyal and Sharma, 2019) [7, 8].

of causes of cancer are contributed by genetic factors, and the rest of 90–95% are contributed by external or environmental factors. The contribution of individual factors among genetic and environmental are also depicted in Figure 6.1.

6.4 TYPES AND SYMPTOMS OF CANCER

There are numerous types of cancers classified based on different factors—for instance, based on sites of origin, such as lung, breast, and liver cancer; based on tissue types, such as leukemia, carcinoma, sarcoma, and myeloma; based on the intensity of abnormality in cells, such as grade 1, grade 2, and grade 3; and based on the stage (tumor size, T), such as T1 to T4 [9]. Different types of cancers described by Cancer Research UK (n.d.) [10] and National Cancer Institute US (n.d.) [11] are presented in Figure 6.2. Different types of cancer have different types of symptoms. However, apart from the developing changes in the affected area where it began, some common signs and symptoms are pain in the affected area, weight loss, weakness, fatigue, fever, unusual change in the

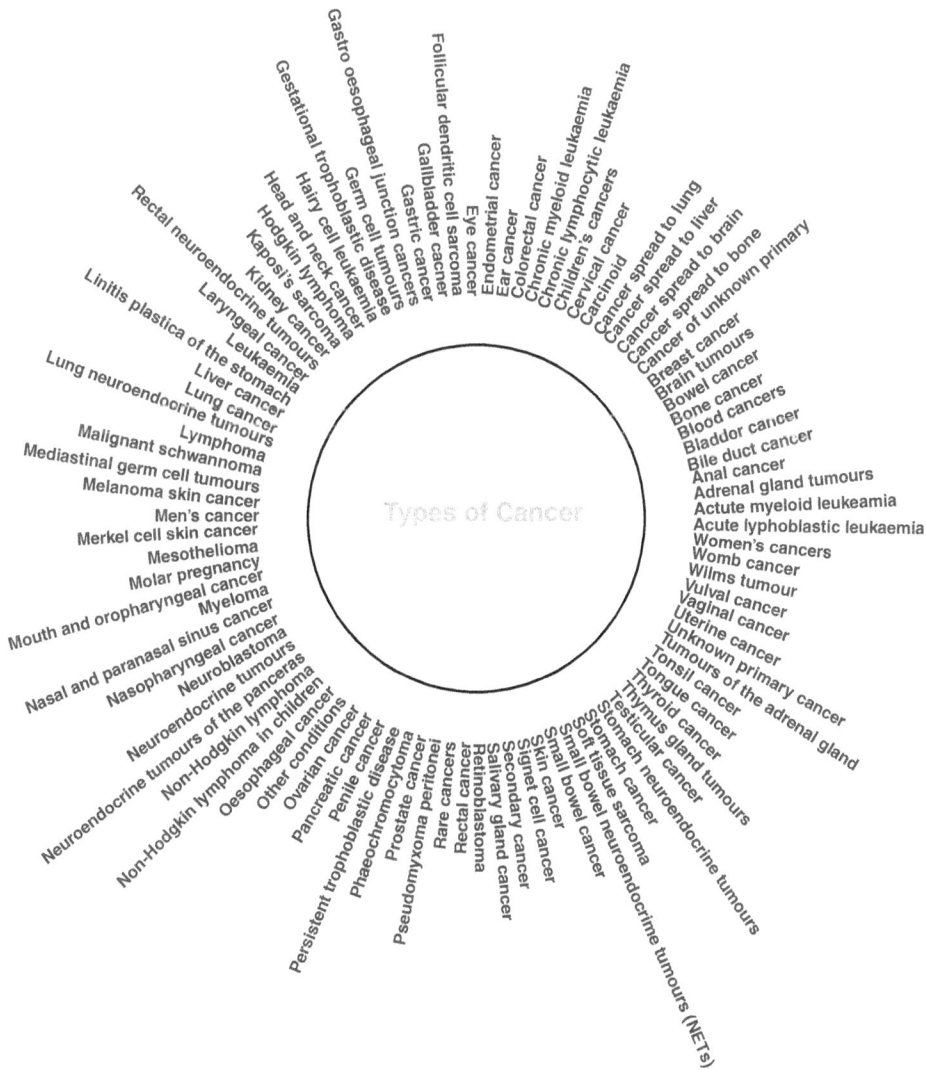

FIGURE 6.2 Types of cancers.

skin, unhealing sores, coughing or hoarseness that persists, unusual bleeding, anemia, shortness of breath, and depression.

6.5 PHYTOCHEMICALS IN CANCER PREVENTION

There are several ways to avoid the occurrence of cancer. Food and nutrition have been reported to contain key factors to decrease the cancer risks where 30–40% risk of cancer can be evaded through preventive measures. Epidemiological, preclinical, clinical, and investigational studies reveal the beneficial effects of dietary factors on gastrointestinal cancers. The main preventative components are discovered to be fiber, omega-3 fatty acids, and polyphenols (together with resveratrol) (Dominique and Latruffe). Some important phytochemicals studied and proven for cancer prevention are briefly described in the following sections.

6.5.1 CURCUMIN

Curcumin is a yellow crystalline powder derived from turmeric that has anticancer effects, inhibiting cell development in a variety of cancers, such as cholangiocarcinoma and uterine leiomyosarcoma. It has anti-inflammatory, antioxidant, and antitumor properties. It has a positive effect because it inhibits NF-kB, which in turn inhibits pro-inflammatory pathways [12]. Curcumin inhibits the appearance of TGF- and IL-10 in T regulatory cells, lowering their activity while increasing the capability of T effector cells to destroy cancer cells. It suppresses the oncoproteins of the human papillomavirus, which causes cervical cancer [13]. Curcumin's many molecular targets make it capable of suppressing most cancer stages and symptoms.

6.5.2 LYCOPENE

Lycopene is a natural pigment that gives tomatoes and other foods, including red carrots, pink grapefruit, watermelons, papaya, and pink guava, their red color [14]. Lycopene is a powerful dietary antioxidant that has been shown to protect against various diseases, including tumors, diabetes, osteoporosis, high blood pressure, heart disease, and neurological diseases [15]. Lycopene is the most capable carotenoid in terms of preventing and treating cancer. Patients with hormone-refractory metastatic prostate cancer received daily dosages of 10 mg lycopene for three months; this resulted in lowered PSA, tumor, bone pain, and urinary tract symptoms. Lycopene inhibits tumor growth by lowering tumor neo-angiogenesis, as evidenced by the reduced countenance of tumor tissue biomarkers linked to cell toxicity, growth, and distinction in patient samples with increasing lycopene consumption [16]. It has been discovered to accumulate more heavily in prostate tissue than in other tissues, which may explain how it prevents prostate cancer cells from dividing and undergoing apoptosis [17]. Lycopene, in conjunction with melatonin, has been found in several trials to have a substantial chemopreventive effect via antioxidant and anti-inflammatory activities [18]. Lycopene alters intracellular signaling pathways, prevents DNA and lipids membrane from oxidizing, modifies intercellular communication, and enhances immune function. Lycopene increases the action of quinacrine on breast cancer cells by blocking Wnt-TCF signaling [19].

6.5.3 ISOTHIOCYANATES (ITCs)

Isothiocyanates are natural constituents of great therapeutic significance that occur as conjugates in the cruciferous vegetable genus *Brassica*. Isothiocyanates are known for their chemopreventive properties, and they facilitate anticarcinogenic activity by inhibiting carcinogen stimulation and boosting detoxification. The high quantity of glucosinolates in cruciferous vegetables, which store ITCs, has anticancer properties and lowers the risks of colorectal, colon, and stomach cancers. ITCs slow the growth of tumor by inducing oxidative-stress-mediated mortality, cell cycle disruption, angiogenesis, and metastasis inhibition [20]. They can block phase I enzymes that stimulate possible carcinogens, induce phase II enzymes that purify potential carcinogens, and regulate cell proliferation and apoptosis. The isothiocyanate benzyl isothiocyanate (BITC) is one of the primary groups of isothiocyanates that affects multiple important signaling ways that are thought to be cancer promises. Furthermore, BITC makes tumors more susceptible to chemotherapy and has anticancer properties in several type of human cancers [21]. BITC inhibits the movement and invasion of human colon cancer cells. MMP-2/9 and urokinase-type plasminogen activator (uPA) are downregulated by BITC, which is connected to the protein kinase C (PKC) and MAPK signaling methods [22]. It was previously reported that BITC inhibited the PI3K/AKT/FOXO pathway, which inhibited pancreatic tumor growth. BITC also inhibits STAT3-mediated HIF-1/VEGF/Rho-GTPases, which inhibits angiogenesis and invasion in pancreatic cancers [23]. BITC has anticancer effects by potentiating p53 signaling and suppressing the capability of breast cancer cells to form mammospheres. Phenethyl isothiocyanate (PEITC) is another isothiocyanate mainly found in cruciferous family plants. PEITC is an anticancer compound found in cruciferous vegetables that have been widely examined in prostate and breast cancer, glioblastoma, and leukemia. Consumption of cruciferous vegetables like broccoli and garden cress has been linked to chemoprevention in a variety of animal models [24]. PEITC, which is found to have tumor-exploitative ability by encouraging G2/M cell cycle arrest and death in prostate cancer cells, was found to reactivate RASSF1A [25].

6.5.4 GENISTEIN

Soybean is a great supplier of genistein, having concentrations ranging from 560 to 3,810 mg per kilogram of soy [26]. It has been demonstrated that genistein significantly inhibits the growth of leukemia, lymphoma, and gastric and breast cancer, among other cancers.

Asia has a lower relative incidence of prostate and breast cancer than Western countries since soy is the main source of isoflavones in the diet [17]. When compared to estradiol, it has a 4% binding affinity for ERα and an 87% binding affinity for ERβ. Other health benefits of genistein include lowering the risk of cardiovascular disease, avoiding osteoporosis, alleviating postmenopausal symptoms, and reducing body mass and fat tissue. Genistein's anticancer properties include effects on tumor prevention and progression. It reduces the genotoxicity caused by 7,12-dimethylbenz-[α] anthracene (DMBA) [27]. In vitro, genistein, combined with chemotherapy medicines like cisplatin, erlotinib, doxorubicin, bleomycin, docetaxel, and gemcitabine, as well as phytochemicals such as indole-3-carbinol, has a synergistic impact. Docetaxel and gemcitabine also have a synergistic effect with genistein in tumor models. By interacting with ER+ cells, genistein was discovered to be an effective differentiation inducer in breast cancer stem cells. The PI3K/Akt pathway mediates genistein's effect on differentiation. By suppressing c-erbB-2, MMP-2, and MMP-9 in breast cancer, another study [18] established the antiangiogenic and antimetastatic properties of genistein.

6.5.5 RESVERATROL

Resveratrol (RSV), also known as trans-3,4′,5-trihydroxystilbene, is a naturally occurring nonflavonoid polyphenol that is abundant in the root of *Polygonum cuspidatum*, the Japanese knotweed [28]. Besides, vine plants produce high amounts of RSV in response to biotic (*Botrytis cinerea*) and abiotic stresses and also by other edible plants, like hops, peanuts, and various berries (Dominique and Latruffe).

Its effects toward precancerous or cancer cells appear to be aided by a variety of biochemical and molecular processes. By controlling the signal transduction pathways that regulate cell division and development, angiogenesis, inflammation, apoptosis, and metastasis, resveratrol affects all three stages of carcinogenesis—initiation, promotion, and progression—making anticancer therapy possible [29]. Resveratrol triggers the convergence of several intracellular pathways, including the Fas receptor or CD95 signaling system, the apoptotic pathway, the nuclear factor b pathway, the phosphoinositol-3-kinase/Akt pathway, and the SIRT1-regulated pathway [29, 30]. Jang et al. showed that resveratrol's first chemopreventive action was to reduce multistage carcinogenesis. Numerous preclinical animal investigations were then conducted as a result of the component's subsequent in vitro demonstration of its capability to prevent the multiplying of many types of human tumor cells [31].

6.5.6   OTHER PHYTOCHEMICALS IN CANCER PREVENTION

Apart from the discussed phytochemical compounds, there are many more have been identified and reported to be effective in cancer prevention. Table 6.1 lists a few of these substances along with their origins and cancer preventive roles.

6.6   PHYTOCHEMICALS IN CANCER TREATMENT

Keeping in mind the adverse effects of the existing treatment methods for cancer, including surgical removal, radiation treatment, and systemic chemotherapy, phytochemicals and their derivatives have been found to be a promising option to rally treatment efficacy in cancer patients with a reduction in adverse reactions. Several naturally occurring phytochemicals are biologically active with remarkable antitumor potential.

From 1940 to 2014, around half of all approved anticancer medicines originate directly or indirectly from natural sources [35]. Both in vitro and in vivo tests have been done to see how effective the phytochemicals are at fighting cancer. Phytochemicals contain overlapping and complementary mechanisms that slow down the carcinogenic process by reducing oxidative stress, reducing cancerous cell proliferation and survival, reducing tumor invasion and metastasis and angiogenesis, and lowering cancerous cell survival and proliferation. Examples of the molecular targets and signal transduction pathways that are impacted by phytochemicals include membrane receptors, kinases, downstream tumor-activator or tumor-suppressor proteins, transcriptional factors, microRNAs (miRNAs), cyclins, and caspases [36].

6.6.1   PRECLINICALLY STUDIED ANTICANCER PHYTOCHEMICALS

Preclinical study or nonclinical study is a stage of drug development that is done prior to the clinical studies for screening the activity, mode of action, pharmacological and pharmacokinetic properties, molecular targets and pathways, efficacy, safety, and toxicity of the isolated phytochemicals so that they can be explored for identifying possible lead compounds for developing novel agents for cancer treatment. A number of evidence show the preclinical effectiveness of a variety of phytochemicals, and some are illustrated in Table 6.2.

6.6.2   CLINICALLY STUDIED ANTICANCER PHYTOCHEMICALS

Clinical studies employing phytochemicals to treat cancer are still in the early phases, despite the enormous number of anticancer molecules now being discovered. The clinical study of phytochemicals against cancer treatment mainly focuses on three major aspects:

- Improving how cancerous cells respond to standard chemo and radiation therapies
- Reducing the negative effects associated with standard cancer treatments
- Looking for undesirable interactions with conventional therapy

Currently, clinically under trial phytochemicals against different types of cancers are summarized in Table 6.3. Because of their ability to spontaneously cross bio-membrane barriers due to their nano size and surface reactivity, nanoparticle-based plant extract anticancer remedies have been revealed to

TABLE 6.1

Role of Other Phytochemicals in Cancer Prevention

Compound	Source	Cancer	Proposed Anticancer Mechanism	References
Anethole	Fennel	Oral cancer	Promoting cell apoptosis, reducing the formation of reactive oxygen species (ROS), inhibiting Cdc2 dephosphorylation and that of MAPKs, triggering autophagy, and boosting intracellular glutathione (GSH) activity	[32]
Capsaicin	*Capsicum*	Pancreatic cancer	Blocking AP1, NF-KB, and STAT3 signaling; cell cycle arrest; β-catenin signaling inhibition	[33]
Catechins	Green tea and other beverages	Neuroblastoma, breast and prostate cancer	Cell cycle at G2 phase, guard toward oxidative stress, affecting STAT3-NFκB and PI3K/AKT/mTOR pathways	[33]
CucurbitacinB	Cucurbitaceous plants	Colorectal cancer, lung cancer, neuroblastoma, breast cancer, pancreatic cancer	Inhibiting JAK-STAT3, HER2-integrin and MAPK signaling pathways	[33]
Benzyl isothiocyanate (BITC)	Garlic mustard, pilu oil, papaya seeds	Leukemia, breast cancer, prostate cancer, lung cancer, pancreatic cancer, colon cancer, hepatocellular carcinoma	G2/M cell cycle arrest and apoptosis, downregulating MMP-2/9 through PKC and MAPK signaling pathway, inhibiting PI3K/AKT/FOXO pathway, inhibiting STAT3-mediated HIF-1α/VEGF/Rho-GTPases	[33]
PEITC	Cruciferous vegetables	Glioblastoma, prostate cancer, breast cancer, cervical cancer, leukemia	ROS activation, G2/M cell cycle arrest, and apoptosis; downregulating HER2 and STAT3 signaling	[33]
Isoflavone	Chickpeas, beans, lentils, and soy	Leukemia; lymphoma; gastric, breast, prostate, head, and neck carcinoma; non-small-cell lung cancer	Inhibiting c-erB-2, MMP-2, and MMP-9 signaling pathways; affecting IGF-1R/p-Akt signaling transduction	[33]
Diosgenin	Fenugreek, roots of wild yam	Colon carcinoma, osteosarcoma, leukemia, human rheumatoid arthritis	Inhibiting RANKL-induced osteoclastogenesis, suppressing TNF-induced invasion, blocking the proliferation of tumor cells	[34]
Piperlongumine	Roots of long pepper	Multiple myeloma; melanoma; pancreatic, colon, breast, and prostate cancer; oral squamous cell carcinoma	Autophagy-mediated apoptosis by inhibition of PIK3/Akt/mTOR	[33]
Zerumbone	Asian ginger	Colon cancer, breast cancer	Hindering the initiation of NF-κB- and NF-κB-regulated gene expression	[34]
Ursolic acid	Rosemary	Breast carcinoma, melanoma, hepatoma, prostate carcinoma, acute myelogenous leukemia	Inducing apoptosis; inhibiting the activation of upstream kinases c-Src, JAK1, JAK2, and ERK1/2; building up of cells in G1/G0 phase; blocking STAT3 activation	[34]

TABLE 6.2

Phytochemicals in Preclinical Trials for Cancer Treatment

Phytochemical	Type	Source (Family)	Molecular Target	Mode of Action	References
6-Shogaol	Phenylpropanoid	Ginger	Akt and STAT signaling pathway	Reducing the growth of NCI-H1650 lung cancer cells by enhancing apoptosis and reducing cell proliferation	[37]
Allicin	Organosulfur	*Allium sativum* (Amaryllidaceae)	STAT3 signaling pathway	Lowering HuCCT-1 cell proliferation, penetration, and epithelial-mesenchymal transition (EMT) by diminishing matrix metalloproteinase (MMP)-2 and MMP-9 levels and the activation of the STAT3 signaling pathway	[38]
Artemisinin	Sesquiterpene lactone	*Artemisia annua* (Asteraceae)	SMMC-7721	Ceasing proliferation by inhibiting the PI3K/AKT and mTOR signaling pathways; promoting apoptosis by upregulating pro-apoptotic proteins cleaved caspase-3 and PARP while downregulating anti-apoptotic proteins XIAP and survivin; preventing migratory and invasive behavior and increasing cell-cell adhesion, which prevents metastasis	[39]
Gingerol	Polyphenol	*Zingiber officinale* (Roscoe)	Intrinsic apoptosis pathway	Inhibiting mouse brain metastatic 4T1Br4 mammary carcinoma cells' ability to spread to other organs, including the lung, brain, and bone	[40]
Curcumin	Phytopolyphenol	*Curcuma longa* (Zingiberaceae)	Modulating cell signaling and gene expression regulatory pathways	Inhibiting tumor growth in mice when subcutaneously injected with human A375 melanoma cells	[41, 42]
Nimbolide	Triterpene	*Azadirachta indica* (Meliaceae)	PI3K/AKT/mTOR and ERK signaling	Increasing ROS production, inhibiting proliferation and metastasis through mitochondrial-mediated apoptotic cell death, which was used to induce apoptosis in an athymic nu/nu mouse model of pancreatic cancer	[43]
Resveratrol	Phenol	*Polygonum cuspidatum* (Polygonaceae)	Regulating cell cycle and apoptosis pathways	Suppressing the growth of a range of tumor cells, including those in the breast, liver, lung, colon, and prostate	[44]
Ursolic acid	Triterpenoids	*Oldenlandia diffusa* (Rubiacea)	Ki-67, CD31, and miR-29a	Dramatically lowering the expression of drug-resistant gene, increasing apoptosis and RO production, significantly reducing cell proliferation	[45]

TABLE 6.3

Phytochemicals Evaluated in Clinical Trials on Various Cancers [46]

Phytochemicals	Class	Source (Family)	Cancer Type
Alvocidib	Flavone	*Dysoxylum binectariferum* (Meliaceae)	Treatment for kidney cancer, prostate cancer, and acute myeloid leukemia
Berberine	Alkaloid	Barberry (Berberidaceae)	Colorectal cancer
Curcumin	Polyphenol	Turmeric (Zingiberaceae)	Advanced and metastatic breast cancer
Epigallocatechin	Flavonoid	*Camellia sinensis* (Theaceae)	Colorectal cancer
R-(−)-Gossypol acetic acid	Polyphenol	*Gossypium herbaceum* (Malvaceae)	Multiple myeloma, lung cancer, prostate cancer
Paclitaxel (Taxol)	Diterpenoid	*Taxus brevifolia* (Taxaceae)	Refractory breast cancer
Lycopene	Carotenoid	Tomato (Solanaceae)	Metastatic colorectal cancer
Resveratrol	Stilbenoid	Japanese knotweed (Polygonaceae)	Low-grade GI neuroendocrine tumors
Vincristine sulfate	Alkaloid	*Vinca rosea* Linn. (Apocynaceae)	Lymphoblastic lymphoma
Genistein	Isoflavone	*Glycine max* L. (Leguminaceae)	Fourth-stage breast cancer metastatic colorectal cancer, initial prostate cancer

TABLE 6.4

Phytochemical-Based Nano-Formulations Under Clinical Trials for Cancer Treatment [46]

Phytochemicals	Trade Name	Nano-Formulation	Cancer Type
Paclitaxel	Abraxane	Albumin-stabilized nanoparticle formulation	Pancreatic cancer, breast cancer, and non-small-cell lung cancer
	Lipusu	Paclitaxel liposome injection of cholesterol and lecithin	Ovarian cancer, breast cancer, non-small-cell lung cancer
	NK 105	Paclitaxel incorporating micellar nanoparticle	Breast cancer, solid tumors
Vinorelbine tartrate	Navelbine/NanoVNB	Liposomal vinorelbine tartrate	Non-small-cell lung cancer, breast cancer, ovarian cancer
Curcumin	Lipocurc	Liposomal curcumin	Solid tumors
Docetaxel	LE-DT/ATI-1123	Liposomal docetaxel	Solid tumors
Docetaxel	SYP-0709	Docetaxel- polymeric nanoparticles (DOCPNP)	Advanced solid cancer
Docetaxel	CRLX301	Docetaxel-loaded nanopharmaceuticals	Solid tumor
Camptothecin	L9-NC	Liposomal 9- nitrocamptothecin	Endometrial cancer, Ewing's sarcoma

be more effective compared to conventional therapy in treating cancer. This is due to the fact that their delayed elimination allows them to accumulate to the required levels in tumors without endangering surrounding normal tissues from drug toxicity. Numerous nano-formulations based on phytochemicals are undergoing clinical studies for their potential use in cancer therapy, as shown in Table 6.4.

6.7 CONCLUSION AND FUTURE PROSPECTS

For the preventive and therapeutic management of a variety of diseases like cancer, many herbs have been well established and demonstrated to be effective. Some unique chemical elements found in various plant sections are responsible for these traits of plants. Many of such substances have been discovered and are still being used for both cancer treatment and prevention. Researchers are working continuously to identify such potential phytochemicals and their mechanisms of action. However, further studies are required in the direction of identification of the most potent phytochemicals and their targeted mechanisms of action for the prevention and cure for different types of cancer. Preclinical and postclinical studies have to be framed for plant-based treatment options for cancer as phytochemicals can be a promising substitute to the harmful chemotherapies and radiotherapies.

6.8 REFERENCES

1. Ranjan, A., S. Ramachandran, N. Gupta, I. Kaushik, S. Wright, S. Srivastava, H. Das, S. Srivastava, S. Prasad and S.K. Srivastava. 2019. Role of phytochemicals in cancer prevention. *Int. J. Mol. Sci.* 20(20): 4981.
2. Zhang, J. and S.F. Zhou. 2012. Can we discover "really safe and effective" anticancer drugs. *Adv Pharmacoepidemiol. Drug Saf.* 1(05): 2167–1052.
3. Schottenfeld D., J.L. Beebe-Dimmer, P.A. Buffler and G.S. Omenn. 2013. Current perspective on the global and United States cancer burden attributable to lifestyle and environmental risk factors. *Annu. Rev. Public Health.* 34: 97–117.
4. Chen W., R. Zheng, H. Zeng, S. Zhang and J. He. 2015. Annual report on status of cancer in China, 2011. *Chin. J. Cancer Res.* 27(1): 2.
5. Russo M., C. Spagnuolo, I. Tedesco and G.L. Russo. 2010. Phytochemicals in cancer prevention and therapy: Truth or dare? *Toxins.* 2(4): 517–551.
6. Pratheeshkumar P., Y.O. Son, P. Korangath, K.A. Manu and K.S. Siveen. 2015. Phytochemicals in cancer prevention and therapy. *Biomed Res. Int.* 2015: 1–2.
7. Anand, P., A.B. Kunnumakara, C. Sundaram, K.B. Harikumar, S.T. Tharakan, O.S. Lai, B. Sung and B.B. Aggarwal. 2008. Cancer is a preventable disease that requires major lifestyle changes. *Pharm. Res.* 25(9): 2097–116.
8. Katyal, P. and S. Sharma. 2019. Emerging alkaloids against cancer: A peep into factors, regulation, and molecular mechanisms. In *Bioactive Natural Products for the Management of Cancer: From Bench to Bedside.* Springer, Singapore. 37–60
9. Mandal, A. 2019. Cancer classification. *News-Medical,* viewed 26 April 2022, www.news-medical.net/health/Cancer-Classification.aspx.
10. Cancer Research UK (n.d.), retrieve from www.cancerresearchuk.org/about-cancer/type on 26th April, 2022.
11. National Cancer Institute, US (n.d.), retrieved from www.cancer.gov/types on 26th April, 2022.
12. Shehzad A., F. Wahid and Y.S. Lee. 2010. Curcumin in cancer chemoprevention: Molecular targets, pharmacokinetics, bioavailability, and clinical trials. *Arch. Pharm.* (Weinheim). 343: 489–499.
13. Bhattacharyya, S., D. Md Sakib Hossain, S. Mohanty, G. Sankar Sen, S. Chattopadhyay, S. Banerjee, J. Chakraborty, K. Das, D. Sarkar, T. Das and G. Sa. 2010. Curcumin reverses T cell-mediated adaptive immune dysfunctions in tumor-bearing hosts. *Cell. Mol Immunol.* 7: 306–315.
14. Chen, J., Y. Duan, X. Zhang, Y. Ye, B. Ge and J. Chen. 2015. Genistein induces apoptosis by the inactivation of the IGF-1R/p-Akt signaling pathway in MCF-7 human breast cancer cells. *Food Funct.* 6: 995–1000.
15. Milligan, S.A., P. Burke, D.T. Coleman, R.L. Bigelow, J.J. Steffan, J.L. Carroll, B.J. Williams and J.A. Cardelli. 2009. The green tea polyphenol EGCG potentiates the antiproliferative activity of c-Met and epidermal growth factor receptor inhibitors in non-small cell lung cancer cells. *Clin. Cancer Res.* 15: 4885–4894.
16. Tang, S.N., C. Singh, D. Nall, D. Meeker, S. Shankar and R.K. Srivastava. 2010. The dietary bioflavonoid quercetin synergizes with epigallocathechin gallate (EGCG) to inhibit prostate cancer stem cell characteristics, invasion, migration and epithelial-mesenchymal transition. *J. Mol Signal.* 5: 14. Banerjee S., Y. Li, Z. Wang and F.H. Sarkar. 2008. Multi-targeted therapy of cancer by genistein. *Cancer Lett.* 269(2): 226–242.
17. Banerjee, S., Y. Li, Z. Wang and F.H. Sarkar. 2008. Multi-targeted therapy of cancer by genistein. *Cancer Lett.* 269(2): 226–242.
18. Duffy, C., K. Perez and A. Partridge. 2007. Implications of phytoestrogen intake for breast cancer. *CA Cancer J. Clin.* 57: 260–277.

19. Milner, J.A., McDonald, S.S., Anderson, D.E. and P. Greenwald. 2001. Molecular targets for nutrients involved with cancer prevention. *Nutr. Cancer. 41*: 1–16.
20. Michaud, D.S., P. Pietinen, P.R. Taylor, M. Virtanen, J. Virtamo and D. Albanes. 2002. Intakes of fruits and vegetables, carotenoids and vitamins A, E, C in relation to the risk of bladder cancer in the ATBC cohort study. *Br. J. Cancer. 87*: 960–965.
21. Wu, K., J.W. Erdman, S.J. Schwartz, E.A. Platz, M. Leitzmann, S.K. Clinton, V. DeGroff, W.C. Willett and E. Giovannucci. 2004. Plasma and dietary carotenoids, and the risk of prostate cancer: A nested case-control study. *Cancer Epidemiol. Biomark. Prev. 13*: 260–269.
22. Moy, K.A., J.M. Yuan, F.L. Chung, X.L. Wang, D. Van Den Berg, R. Wang, Y.T. Gao and M.C. Yu. 2009. Isothiocyanates, glutathione S-transferase M1 and T1 polymorphisms and gastric cancer risk: A prospective study of men in Shanghai, China. *Int. J. Cancer. 125*: 2652–2659.
23. Mense, S.M., T.K. Hei, R.K. Ganju and H.K. 2008. Bhat. Phytoestrogens and breast cancer prevention: Possible mechanisms of action. *Environ. Health Perspect. 116*: 426–433.
24. Jenab, M., E. Riboli, P. Ferrari, M. Friesen, J. Sabate, T. Norat, N. Slimani, A. Tjønneland, A. Olsen, K. Overvad and M.C. Boutron-Ruault. 2006. Plasma and dietary carotenoid, retinol and tocopherol levels and the risk of gastric adenocarcinomas in the European prospective investigation into cancer and nutrition. *Br. J. Cancer. 95*: 406–415.
25. Hirvonen, T., J. Virtamo, P. Korhonen, D. Albanes and P. Pietinen. 2001. Flavonol and flavone intake and the risk of cancer in male smokers (Finland). *Cancer Causes Control. 12*: 789–796.
26. Fletcher, R.J. 2003. Food sources of phyto-oestrogens and their precursors in Europe. *Br. J. Nutr. 89*(S1): S39–S43.
27. Pugalendhi, P., S. Manoharan, K. Panjamurthy, S. Balakrishnan and M.R. Nirmal. 2009. Antigenotoxic effect of genistein against 7,12-dimethylbenz [a]anthracene induced genotoxicity in bone marrow cells of female Wistar rats. *Pharmacol. Rep. 61*: 296–303.
28. Nonomura, S., H. Kanagawa, and A. Makimoto. 1963. Chemical constituents of polygonaceous plants. I. studies on the components of ko-j o-kon. (polygonum cuspidatum sieb. et zucc.). *Yakugaku zasshi: Journal of the Pharmaceutical Society of Japan. 83*: 988–990.
29. Berman, A.Y., R.A. Motechin, M.Y. Wiesenfeld, and M.K. Holz. 2017. The therapeutic potential of resveratrol: A review of clinical trials. *NPJ Precis. Oncol. 1*(1): 1–9.
30. Shukla, Y. and R. Singh. 2011. Resveratrol and cellular mechanisms of cancer prevention. *Ann. N. Y. Acad. Sci. 1215*(1): 1–8.
31. Bishayee, A. 2009. Cancer prevention and treatment with resveratrol: From rodent studies to clinical trials. *Cancer Prev. Res. 2*(5): 409–418.
32. Contant, C., M. Rouabhia, L. Loubaki, F. Chandad and A. Semlali. 2021. Anethole induces anti-oral cancer activity by triggering apoptosis, autophagy and oxidative stress and by modulation of multiple signaling pathways. *Sci. Rep. 11*(1): 1–14.
33. Ranjan, A., S. Ramachandran, N. Gupta, I. Kaushik, S. Wright, S. Srivastava, H. Das, S. Srivastava, S. Prasad and S.K. Srivastava. 2019. Role of phytochemicals in cancer prevention. *Int. J. Mol. Sci. 20*(20): 4981.
34. Aggarwal, B.B., A.B. Kunnumakkara, K.B. Harikumar, S.T. Tharakan, B. Sung and P. Anand. 2008. Potential of spice-derived phytochemicals for cancer prevention. *Planta Med. 74*(13): 1560–1569.
35. Newman, D.J. and G.M. Cragg. 2016. Natural products as sources of new drugs from 1981 to 2014. *J. Nat. Prod. 79*(3): 629–661.
36. Choudhari, A.S., P.C. Mandave, M. Deshpande, P. Ranjekar and O. Prakash. 2020. Phytochemicals in cancer treatment: From preclinical studies to clinical practice. *Front. Pharmacol.* 1614.
37. Kim, M.O., M.H. Lee, N. Oi, S.H. Kim, K.B. Bae and Z. Huang. 2014. [6]-shogaol inhibits growth and induces apoptosis of non-small cell lung cancer cells by directly regulating Akt1/2. *Carcinogenesis. 35*(3): 683–691.
38. Chen, H., B. Zhu, L. Zhao, Y. Liu, F. Zhao and J. Feng. 2018. Allicin inhibits proliferation and invasion in vitro and in vivo via SHP-1-mediated STAT3 signaling in cholangiocarcinoma. *Cell Physiol. Biochem. 47*(2): 641–653.
39. Jung, K.H., M. Rumman, H. Yan, M.J. Cheon, J.G. Choi, X. Jin, S. Park, M.S. Oh and S.S. Hong. 2018. An ethyl acetate fraction of Artemisia capillaris (ACE-63) induced apoptosis and anti-angiogenesis via inhibition of PI3K/AKT signaling in hepatocellular carcinoma. *Phytother. Res. 32*: 2034–2046.
40. Martin, A., A.M. Fuzer, A.B. Becceneri, J.A. da Silva, R. Tomasin and D. Denoyer. 2017. [10]-gingerol induces apoptosis and inhibits metastatic dissemination of triple negative breast cancer in vivo. *Oncotarget. 8*(42): 72260–72271.

41. Kunnumakkara, A.B., D. Bordoloi, C. Harsha, K. Banik, S.C. Gupta and B.B. Aggarwal. 2017. Curcumin mediates anticancer effects by modulating multiple cell signaling pathways. *Clin. Sci.* (Lond) *131*(15): 1781–1799. doi: 10.1042/CS20160935.

42. Zhao, G., X. Han, S. Zheng, Z. Li, Y. Sha, J. Ni, Z. Sun, S. Qiao and Z. Song. 2016. Curcumin induces autophagy, inhibits proliferation and invasion by downregulating AKT/mTOR signaling pathway in human melanoma cells. *Oncol. Rep. 35*(2): 1065–1074. doi: 10.3892/or.2015.4413.

43. Subramani, R., E. Gonzalez, A. Arumugam, S. Nandy, V. Gonzalez and J. Medel. 2016. Nimbolide inhibits pancreatic cancer growth and metastasis through ROS-mediated apoptosis and inhibition of epithelial-to-mesenchymal transition. *Sci. Rep.* 6: 19819. doi: 10.1038/srep19819.

44. Banerjee, S., C. Bueso-Ramos and B.B. Aggarwal. 2002. Suppression of 7,12- dimethylbenz(a)anthracene-induced mammary carcinogenesis in rats by resveratrol: Role of nuclear factor-kB, cyclooxygenase 2, and matrix metalloprotease 9. *Cancer Res. 62*: 4945–4954.

45. Zhang, Y., L. Huang, H. Shi, H. Chen, J. Tao and R. Shen. 2018. Ursolic acid enhances the therapeutic effects of oxaliplatin in colorectal cancer by inhibition of drug resistance. *Cancer Sci. 109*(1): 94–102. doi: 10.1111/cas.13425.

46. Dhupal, M. and D. Chowdhury. 2020. Phytochemical-based nanomedicine for advanced cancer theranostics: Perspectives on clinical trials to clinical use. *Int. J. Nanomedicine. 15*: 9125.

7 Nano-Formulation-Based Approaches for Chemoprevention

Dipanjana Ghosh, Wean Sin Cheow, and Rohit Saluja

CONTENTS

7.1 Introduction 109
7.2 Nano-Formulations: Design, Physicochemical Properties, and Advantages
over Conventional Therapeutic Formulations.. 110
    7.2.1 Design 110
        7.2.1.1 Polymer-Based Particles 110
        7.2.1.2 Non-polymeric Particles 110
        7.2.1.3 Lipid-Based Particles 112
        7.2.1.4 Nanocrystalline Particles 113
    7.2.2 Physicochemical Properties 113
    7.2.3 Advantages over Conventional Therapeutic Formulations............................... 113
7.3 Targeting Cancer Tissues by Nano-Formulation.. 114
    7.3.1 Active Targeting... 114
    7.3.2 Passive Targeting 115
7.4 Nano-Formulation-Based Chemoprevention in Various Cancer Types 115
    7.4.1 Pancreatic Cancer 115
    7.4.2 Breast Cancer.. 116
    7.4.3 Colorectal Cancer 116
    7.4.4 Lung and Bronchial Cancer... 117
    7.4.5 Miscellaneous Cancers 118
7.5 Clinical Insights.. 118
7.6 Conclusion and Future Perspective... 120
7.7 References.. 120

7.1 INTRODUCTION

Cancer, being one of the topmost deadly diseases throughout the world, is increasingly becoming a considerable threat to the human civilization [1]. Per the latest cancer statistics by the National Cancer Institute, NIH, in 2020, the new cancer cases diagnosed in United States were 1,806,590 and deaths were 6,06,520 [2]. Treatment of cancer is mainly limited to surgery, radiotherapy, chemotherapy, and immunotherapy that often remain ineffective due to drug resistance, cytotoxicity, and lack of specificity for the site of actions [3].

Nano-formulations of an anticancer drug can be advantageous to minimize the side effects and enhance the drug efficacy by achieving improved pharmacokinetic features, such as controlled release, targeted delivery, enhanced permeability and retention effect, high drug sensitivity, and low drug resistance [3]. Although nano-formulation-based chemoprevention is presently in the stage of infancy, the increasing number of scientific literature reported each day reflects the growing potential of the field. In brief, nanotechnology-based anticancer formulations can offer improved survival opportunities in cancer patients.

This chapter focuses on the nano-formulations for cancer treatment and describes the background, structural, and functional details of the nanoparticles and the nanocarriers being used for the nano-formulation development.

7.2 NANO-FORMULATIONS: DESIGN, PHYSICOCHEMICAL PROPERTIES, AND ADVANTAGES OVER CONVENTIONAL THERAPEUTIC FORMULATIONS

Nano-formulations comprise particles in the range of ≤100 nm to <1,000 nm and are thus termed as nanoparticles (NPs) [4]. NPs began its journey in the 1950s, with the design of a polymer-drug conjugate designed by Jatzkewitz [5], followed by the design of liposomes in mid-1960s by Bangham [6]. Albumin-bound paclitaxel NPs were approved by US Food and Drug Administration (FDA) in 2005, under the marketed name of Abraxane, for the treating breast cancer [7] and also for the treatment of lung cancer in subsequent times [8].

7.2.1 Design

Based on the excipient composition or the nanocarriers used, nano-formulations are broadly categorized into two major classes: nanostructured and nanocrystalline (Figure 7.1). Nanostructured materials can further be subclassified into polymeric, non-polymeric, and lipid-based NPs (Table 7.1). These four categories of nanocarriers are described in detail in the following sections.

7.2.1.1 Polymer-Based Particles

Polymer-based NPs vary within the size range of 1–1,000 nm where the active drug components can either be entrapped within the polymeric core or be adsorbed onto the surface (Figure 7.1). Nanocapsules are examples of the former type where the drug is dissolved in the oily core of the polymer-based NPs, whereas nanospheres are examples of the latter type, where the drug is adsorbed onto the polymer surface. Polymer-based NPs can either be produced by polymerizing the monomers or dispersing the preformed polymers into the solvent base. For both preparation methods, aqueous colloidal suspensions are obtained as the end product. For preparing polymer-based NPs, the method of choices includes emulsification/solvent diffusion, solvent evaporation, reverse salting-out, and nanoprecipitation. The physical properties of the polymer-based NPs, such as the morphology, particle size distribution, chemical composition and crystallinity, and zeta potential, vary widely and can be assessed by electrophoresis, chromatography, dynamic light scattering (DLS), electron microscopy, near-infrared spectroscopy, or photon correlation spectroscopy (PCS) [9].

Some examples of polymer-based NPs used for anticancer treatment include rapamycin-loaded polysorbate 80-coated poly(lactide-co-glycolide) (PLGA) nanospheres used for antiglioma activity [10], curcumin (Cur) loaded PLGA nanocapsules used for the treatment of pancreatic cancer [11], Cur-loaded colloidal nanocapsules made with polyethylene glycol (PEG) and polypropylene glycol (PPG) used for anticancer activity [12], paclitaxel (PTX) loaded nanocapsules made with PCL (poly[ε-caprolactone])-PEG-PCL used for the treatment of lung cancers and PTX-loaded PLGA-PEG nanocapsules (NCs) for the treatment of pancreatic, breast, ovarian, and brain cancers [13].

7.2.1.2 Non-polymeric Particles

Non-polymeric NPs have a widespread surface conjugation chemistry; however, they come with compromised biocompatibility and biodegradability (Figure 7.1). The broadly studied non-polymeric NPs include carbon nanotubes (CNTs), nanodiamonds (NDs), metallic NPs, quantum dots (QDs), and silica-based NPs [14].

A

Nanocarrier	
Drug	
Gene	
Targeting ligand	
Antibody	
Chemical moiety	
Photodynamic/ Photothermal agent	
Diagnostic/ Imaging agent	
Coating (Lipid/Polymeric	

B

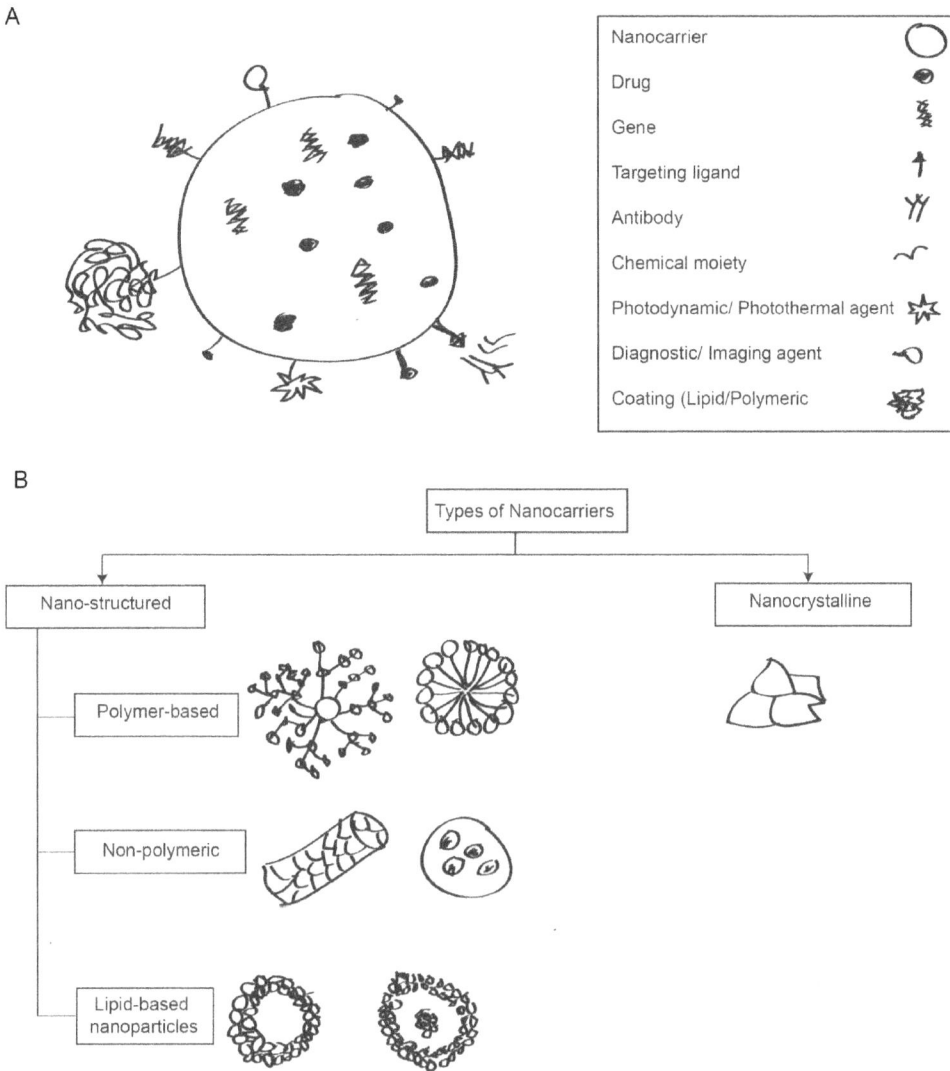

FIGURE 7.1 Schematic representation of a therapeutic nanoparticle and various nanocarriers: (A) Therapeutic nanoparticles; (B) Various types of nanocarriers.

Carbon-based tubular structures of 1–100 nm length and 1 nm diameter are termed as CNTs and can be internalized by endocytosis or by intracellular penetration through plasma membrane. CNTs hence can be used for effective delivery of therapeutics. NDs are also carbon-based nanomaterials with <100 nm diameter and unique surface electrostatic properties that can be utilized for various biomedical applications like magnetic resonance imaging (MRI) and delivery of anticancer drugs via surface immobilization of several biomolecules.

Metallic NPs belong to 1–100 nm size ranges and are composed of various metals like iron, cobalt, nickel, gold, and their respective oxides. While serving as a carrier for therapeutic agents, metallic NPs can incorporate unique features, such as the magnetic properties, which, via regulation of an external magnetic field, helps these NPs to specifically deliver the drug to a target site. Also, the magnetic hyperthermia produced due to the effect of an alternating magnetic field is used for tumor ablation for cancer treatment [15]. The metallic NPs that are extensively used for the treatment and diagnosis of cancer, includes the gold NPs (AuNPs) that possess moderately low

TABLE 7.1

Types of Nanoparticles and Their Physicochemical Properties

Nanoparticles		Size	Charge	Shape	Surface Properties	Solubility	Reference
Nanostructured							
	Polymer-based	1–1,000 nm	Depends on the polymeric particles used	• Spherical (nanospheres) • Thin polymeric envelop (nanocapsules)	Surface modified	Water-soluble	[9]
	Non-polymeric	1–100 nm	Negative (gold NPs)	• Cylindrical-spherical (wide range)	Surface	Water-soluble	[19]
	Lipid-based	0.5–1,000 nm	Negative	Hemispherical to spherical	Surface modified	Improved water solubility	[20]
Nanocrystalline		1–10 μm	Neutral	Desired shape	Crystalline or amorphous structure	Increase in saturation solubility	[21]

cytotoxicity owing to the inert characteristics of gold. Apart from that, titanium dioxide (TiO_2) and zinc NPs are also used for phototherapy for malignant cells [16, 17].

Silica-based NPs have a porous nature and thereby provide a large surface area that favors water absorption and improve stability and controlled release of the therapeutic agents. β-cyclodextrin-loaded silica NPs are used for releasing the encapsulated drug at the tumor tissue [18]. QDs are composed of semiconducting materials with a 2–10 nm diameter and are known to produce characteristic fluorescence colors. However, the therapeutic use of QDs is still disputed.

7.2.1.3 Lipid-Based Particles

Lipid-based NPs (LBNPs) have widely been used for therapeutic applications since they can transport both hydrophobic and hydrophilic molecules with prolonged half-life, very low cytotoxicity, and controlled drug release. Widely used LBNPs for cancer treatment include liposomes, solid lipid NPs (SLNs), and nanostructured lipid carriers (NLCs) (Figure 7.1).

Liposomes are the mostly studied LBNPs composed of phospholipid bilayers that exhibit excellent biocompatibility and biodegradability. In contact with water, they form vesicles encapsulating the anticancer drug loaded onto it [22]. Other than phospholipids, cholesterol is another component of liposomes that present single to multiple bilayers with varying size ranges and nomenclatures. Examples are multilaminar vesicles (MLVs) with multiple bilayers of sizes 0.5–10 nm, large unilamellar vesicles LUVs with single bilayers of sizes >100 nm, and small unilamellar vesicles (SUVs) of intermediate size range 10–100 nm [22]. Doxorubicin (DOX) and curcumin (CUR) loaded liposomes were studied for anticancer treatment [23, 24].

SLNs are made up of physiological lipids that persist to stay at a solid state at room temperature and body temperature and give rise to a novel colloidal drug delivery system of 50–1,000 nm size range. SLNs provide prolonged stability, specific targeting, controlled release, and reduced toxicity to the drugs irrespective of their nature of hydrophilicity [25]. However, they have limited capacity for drug loading and expulsion [26]. Few examples of SLNs for delivering anticancer drugs include, niclosamide [27] and talazoparib [28] loaded SLNs for treating breast cancer cells particularly the triple negative ones (TNBC).

Nanostructured lipid carriers (NLCs) are developed from SLNs and are composed of lipids, such as glyceryl dioleate, glyceryl tricaprylate, ethyl oleate, and isopropyl myristate. The particle size is similar to SLNs. NLCs can be surface-modified, can target specifically to the site, offer controlled drug release, and have low toxicity.

7.2.1.4 Nanocrystalline Particles

Nanocrystalline particles are crystals with a nanometer size range with 100% drug component without any carrier material and can be dispersed in liquid media as nanosuspensions. This aqueous or non-aqueous suspension can be stabilized by the use of surfactants or polymeric stabilizers. Nanocrystalline particles are produced by three variable methods—milling, precipitation, and homogenization—or a combination of these methods. Examples of some nanocrystalline formulations for anticancer treatment include megestrol for breast cancer and thymectacin (Theralux) in non-Hodgkin's lymphoma (NHL) [21].

7.2.2 Physicochemical Properties

NPs hold unique physicochemical properties, including charge, size, shape, and surface properties, that largely impact the bioavailability and intracellular access of these drug carriers. Since NPs have particle size in the nanometer range, they can be rapidly eliminated from the circulation through reticuloendothelial system (RES) [29] and thus contribute to improved biodistribution. However, particles up to 15 μm can accumulate in the bone marrow, liver, and spleen [30]. Furthermore, the smaller particle size of NPs helps in rapid and smooth cellular internalization through phagocytosis, micropinocytosis, and caveolar and clathrin-mediated endocytosis [31], which are also facilitated by the unique geometry of the particles. In addition, studies on mammalian cell line model suggested that NPs can be internalized rapidly and particles with similar sizes, but different shapes differ in their rate of internalization [32]. Hence, along with size, shape is also a vital parameter for NPs.

Particle shapes were shown to be critical for controlling the degradation properties of the NPs [33] and their cellular uptake [34]. Hemispherical shapes help to formulate particles that accomplish sustained release properties. Spherical particles, however, show susceptibility upon degradation [35]. Deformation of the spherical NPs help to play a key role to avoid spleen filtration [36]. Another study showed that the size and shape of NPs are vital for in vivo biodistribution [37].

Surface properties of the NPs contribute efficiently to the longevity of the particles in blood circulation time. This property contributes to the opsonization of the NPs where NPs are associated with immunoglobulin and complement proteins that helps to identify the particles by macrophages. Positively charged surfaces induce cell membrane penetrability and enhance intracellular uptake, whereas negatively charged NPs result in faster RES clearance [38]. Neutrally charged particles or particles with polyethylene glycol (PEG) coat induce reduction in particle uptake by RES [39] and thus exhibit prolonged blood circulation in rats. Another method of surface modification is with specific receptor recognizing ligands or with monoclonal antibodies. Some of the examples of surface modified NPs include 2C5-modified doxorubicin-loaded liposomes that were shown to improve the therapeutic efficacy in the xenograft models for brain tumor [40] and modified HER2-specific antibody NPs that deliver drugs to HER2-expressing cells [41]. These studies further support that the surface characteristics of NPs are vital for avoiding their swift clearance from circulation before reaching target tumor site.

7.2.3 Advantages over Conventional Therapeutic Formulations

NPs with diameter range 10–100 nm are usually used for cancer treatment as these particles can effectively deliver drugs compared to the conventional therapeutic formulations. NPs can achieve enhanced permeability and retention and controlled therapeutic efficacy [42]. Particles <10 nm in diameter can easily leak out from normal vasculature, damage the normal cells, and be filtered by the RES [43]. In contrary, the particles that are >100 nm in diameter, are usually cleared from circulation by delayed phagocytosis [37]. In addition, the shape and surface properties of NPs, as discussed in previous section, influence the therapeutic efficacy, prolonged release, and bioavailability of the therapeutic component. Taken together, the NPs can provide better therapeutic efficacy of the anticancer formulation.

7.3 TARGETING CANCER TISSUES BY NANO-FORMULATION

Site-specific drug targeting into cancer cells is a distinguished characteristics of NPs for effective delivery of the drug since this helps to improve the therapeutic efficacy and protect the normal cells from side effects. Several studies were conducted to investigate the design of NP-based drugs for site-specific targeting. For further clarity in understanding the tumor tissue targeting, it is essential to further understand the tumor biology and the crosstalk between the nanocarriers and the tumor cells. The mechanism of drug targeting is broadly classified into active and passive targeting.

7.3.1 ACTIVE TARGETING

In the surface of the cancer cells, there are specific target molecules that are overexpressed compared to the normal cells. Active targeting utilizes this particular characteristic of the cancer cells where the surface ligands on NPs are typically chosen based on their interactions with these overexpressed molecules [44]. These target molecules mostly include receptors that interact with the NP ligands, leading to its internalization followed by successful release of the therapeutic components [45]. This kind of targeting is suitable for delivery of macromolecules, such as proteins, siRNAs, and miRNAs (Figure 7.2).

Targeting moieties on the surface of the cancer cells can be peptides, amino acids, carbohydrates, vitamins, and monoclonal antibodies [46]. Widely studied receptor moieties on the cancer cell surface include folate receptors, transferrin receptors, glycoproteins, and epidermal growth factor receptors (EGFRs). One widely studied example of active targeting method for anticancer drug delivery includes transferrin-conjugated NPs [47], which have shown higher cellular uptake than unmodified NPs [48] and proven to be crucial for overcoming the drug-resistant chemotherapy [49]. Another example is of folate receptors (FR-a and FR-b), overexpressed in most of the human cancers [50], which are targeted by folate conjugated anticancer nanomaterials for the treatment of cancer [51]. Additionally, cancer cells express non-immunological proteins that can selectively bind to specific carbohydrates [52] that can be targeted by lectins conjugated to NPs. Targeting epidermal growth factor receptors (EGFRs) for anticancer treatment [53] includes the inhibition of human epidermal receptor-2 (HER-2) for the treatment of breast and gastric cancer that are positive for HER-2 [54]. Moreover, one can achieve improved target specificity by conjugating two different cancer-specific ligands into a single NP [55].

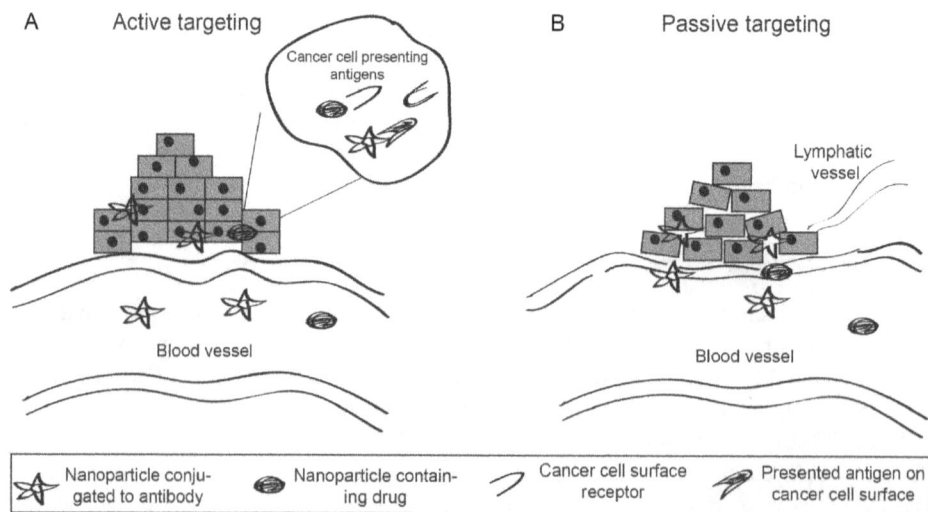

FIGURE 7.2 Targeting cancer cells by nano-formulations: (A) Active targeting; (B) Passive targeting.

7.3.2 PASSIVE TARGETING

In passive targeting, different characteristics of tumor and normal tissues are utilized to success-fully deliver the drug to the target site (Figure 7.2). In cancer tissues, high proliferation rate induces rapid vascularization in the vessel walls that lead to enhanced permeability of tumor-associated blood vessels compared to normal tissue microenvironment [56]. This property enables the NPs to accumulate within the tumor cells, as they leak out from the blood vessels. Moreover, cancer tissues are associated with poor lymphatic drainage and thus increase the retention effect of the supplied macromolecules. These enhanced permeability and retention effect (EPR) of the tumor microenvironment [57] allows the passive targeting of the NPs to the cancer tissues. This EPR effect is reliant on the size of the NPs, where smaller NPs have better permeability towards the cancer tissues but not towards the normal tissues [58]. Larger NPs are cleared by the immune system faster [59]. Simultaneously, cancer cell metabolism also plays vital role in effective delivery of NPs to the cancer tissues [60]. For example, glycolysis creates an acidic environment by reducing the pH of the tumor microenvironment, which helps to trigger the drug release from some pH-sensitive NPs at the close proximity of the tumor tissue [61]. However, nonspecific drug distribution, the EPR effect, and variable permeability of the vasculature across various tumors are some of the limitations of the passive targeting [62].

7.4 NANO-FORMULATION-BASED CHEMOPREVENTION IN VARIOUS CANCER TYPES

7.4.1 PANCREATIC CANCER

Pancreatic cancer (PaCa) relates to the malignancies originating from the exocrine as well as endo-crine tissues of pancreas. Per the latest statistical report of American Cancer Society [63], PaCa has evolved as one of the major leading causes of cancer associated deaths in United States since this remains asymptomatic till the end stage. Thus, the treatment and management of PaCa remains challenging, although a combination of Gemcitabine and Folfirinox (fluorouracil, leucovorin, iri-notecan, and oxaliplatin combination therapy) is used as the first line of treatment [64]. Nano-formulations, particularly the lipid nanocarriers, were widely studied with an aim to better manage the disease. Examples include the combination of aspirin, curcumin, and free sulforaphane for-mulated with solid lipid NPs, which showed significant reduction in the proliferation in cell line models of PaCa (Panc-1 and Mia PaCa-2) [65]. There are reports on effective reduction of tumor incidence in transgenic mice with induced pancreatic cancer when treated with nano-formulations [66]. Examples include self micro-emulsifying drug delivery system (SMEDDS) that was reported to enhance chemoprevention in Panc-1 and Mia PaCa-2 PC cell lines [67]. There are several studies that report about enhanced chemoprevention of potent phytochemicals like curcumin and ellagic acid when formulated as liposomes, dendrimers, or micelles and tested on PaCa cell lines [68]. These observations can further be extended for translational use. Another nanoviral vector, deliver-ing melanoma differentiation associated gene-7/Interleukin-24 (mda-7/IL-24), was found to exhibit suppression of tumor growth at primary and distant sites in pancreatic cancer cells, in the presence of perillyl alcohol [69].

For PaCa, many of the nano-formulations have also crossed the path of clinical evaluation and being approved by the FDA for being in the market. One example is Abraxane, an albumin-bound nano-formulation of paclitaxel that has been approved by the FDA in 2013 as a first line treatment for metastatic PaCa in combination with gemcitabine [70]. A liposomal preparation of irinotecan, marketed as Onivyde, has also received FDA approval in 2015 as a second line treatment for meta-static PaCa patients [71].

These strategies will show future direction for clinical solution for chemoprevention and treatment and play a vital role in arresting PaCa relapse or improved survival of metastatic PaCa patients [72].

7.4.2 Breast Cancer

Breast cancer is reported as the cancer of highest incidence and among the cancer associated deaths in women, breast cancer stands fourth in the United States [63]. At the time of diagnosis, in about 35–40% cases, the tumor has already metastasized to the bone and lymph nodes. Hence, treatment and management of the disease at an advanced stage remains challenging. The use of nano-formulations for the management of breast cancer has been explored and reported in several studies, some of which are described below.

Hormonal drugs incorporated in NPs are used for drug delivery in breast cancer tissues in a site-specific manner [73]. Polyethylene glycol conjugated NPs containing folic acid were reported for their faster entry to the breast cells compared to their conventional formulations [74]. Tamoxifen-loaded SLNs showed higher effectiveness in an in vivo study on rat mammary tumor, induced by LA7 cell [75]. Radiolabeled gold NPs are used to treat and diagnose solid breast tumors [76]. They have also been developed as the carriers of immunotherapeutic drugs and macromolecules like oligonucleotide and peptides that facilitate targeted drug delivery with reduced toxicity [77]. Modified polymersomes containing doxorubicin have been synthesized and tested to have better half-life and reduced cytotoxicity in breast cancer cells compared to conventional delivery systems [78]. pH-sensitive biodegradable polymers were reported to be used for targeted paclitaxel delivery to the solid breast tumors having acidic microenvironment [79]. Magnetic NPs are used for diagnosis purpose using noninvasive techniques, such as MRI and magnetic relaxometry [80]. Polymers such as dextran, chitosan, polyvinyl alcohol, and caprolactone are used to target delivery at the breast cancer tissues [81]. Nanographene oxide–methylene blue formulations incorporated with photodynamic and photothermal treatment were reported to prevent metastasis of breast cancer to the distant organs, such as the liver, lungs, and spleen [82].

Several NPs for breast cancer treatment are in clinical use as well. One example is the use of QDs for diagnosis and active targeting of multiple biomarkers for breast cancer. Other than that, trastuzumab, an FDA-approved humanized monoclonal antibody that actively targets HER-2, is widely used for the treating metastatic breast cancers [83]. Streptavidin-conjugated superparamagnetic NPs are commercially available for nuclear imaging in breast cancer [84]. Tamoxifen, doxorubicin, and paclitaxel NPs are also in clinical practice and helps to limit adverse effects of conventional formulations [85]. To support further research on nanotechnology for breast cancer prevention, the US government has taken an initiative on funding mechanisms and new administrative policies.

7.4.3 Colorectal Cancer

Colorectal cancer (CRC) initiates as benign adenomatous polyps at the wall of the intestine, where prolonged non-attendance of the polyp might give rise to malignancy [86]. Per the latest American statistics on cancer, CRC contributes to the second highest number of cancer-related deaths in the United States [63]. Surgical resection of the tumor is recommended as the first line of treatment [87], whereas among the medicinal care, NSAIDs, particularly aspirin, alone and in combination, has been approved as the drug of maximum chemopreventive activity for CRC [88] with no or minimum side effects. Nano-formulation of aspirin can further enhance the therapeutic efficacy of the drug at lower doses. Among other NSAIDs, celecoxib, when prepared as polymeric NPs using ethyl cellulose with sodium caseinate/bile salt, lipid hybrid NPs and microemulsions, were found to improve bioavailability and reduce the therapeutic dose [89]. Naturally derived phytochemicals, such as curcumin, when formulated in polymeric NP nano-formulations, can help to improve bioavailability of the drug [90].

Several other nano-formulations have been developed and studied for targeting CRC in multiple cell line models [91]. For example, nanosized maghemite particle conjugated with antibodies against carcinoembryonic antigen (CEA) were tested on CRC cell lines expressing high (LS174T) or low (HCT116) levels of CEA [92]. Dextran and PEG-coated superparamagnetic iron oxide NPs (abf-SPION) conjugated with single-chain Fv antibody fragments (scFv) was studied for its efficacy to target CEA in multiple CRC cell lines [93]. In another study, the humanized anti-CEA monoclonal antibody A5B7 was formulated as dye-doped silica NPs conjugated with polyamidoamine dendrimers and tested in CRC cell lines and in murine xenografts model [94]. Similarly, nano-formulations targeting death receptor 5 (DR5) in CRC were developed as the DR5 antibody conjugated, conatumumab (AMG 655) coated NPs [95] and photosensitizer meso-tetra(N-methyl-4-pyridyl) porphine tetra tosylate chitosan/alginate NPs [96]. Both the formulations were tested on HCT116 cell lines. Polymer capsules conjugated with humanized A33 monoclonal antibody (huA33 mAb) [97] and scFv-conjugated gold and iron oxide hybrid NPs [98] were targeted against A33 antigen and tested on multiple CRC cell lines. Poly(lactide-coglycolide) NP loaded with camptothecin was studied for targeting Fas receptor (CD95/Apo-1) using respective antibody conjugate and studied in HCT116 cell line [99]. Nano-formulations targeting folate receptors (FRs) include chitosan NPs loaded with 5-ALA (5-aminolaevulinic acid) [100] and FA-CS (folate-chitosan) conjugated NPs [101], which were studied on HT29 and Caco-2 CRC cell lines, that overexpresses folate receptors. HPMA-copolymer-doxorubicin conjugated peptide GE11 targets EGFR, and this nano-formulation is studied in multiple CRC cell lines [102]. T22-empowered protein-only NPs conjugated with 18-mer peptide T22 (T22-GFP-H6) can target against CXC chemokine receptor 4 (CXCR4) and were tested on HeLa cells [103]. Chitosan NPs encapsulating oxaliplatin (L-OHP) conjugated to hyaluronic acid (HA) can target HA receptor and this was tested on HT-29 cell line and in C57BL mice [104]. Mesoporous silica NPs (MSNs) coated with poly-(L-lysine) and HA can target CD44 receptor in HCT116 cell line [105]. In CRC metastasis mice models, reconstituted HDL (rHDL) NPs loaded with siRNA, conjugated with apolipoprotein A-I (Apo A-1), were found to target scavenger receptor type B1 (SR-B1) [106].

However, although several nano-formulations have been developed and studied for their therapeutic effect in vitro for CRC, none of them have reached the clinical stages. Therefore, extensive clinical studies will be required on the developed nano-formulations to confirm their clinical efficacy to progress in the field of translational nanomedicine.

7.4.4 LUNG AND BRONCHIAL CANCER

The highest incidence of cancer-related deaths in United States are reported to be from lung and bronchial cancers [63]. Like in all other cancers, nano-formulations contribute to the enhancement of therapeutic efficacy and targeted drug delivery in lung and bronchial cancers too [107].

The major drug of choice for treatment of lung cancer is cisplatin, which demonstrates nephrotoxicity when incorporated in high doses [108] as in conventional formulations. The liposomal formulation of cisplatin, called lipoplatin, was developed in 2004 and was found to reduce this systemic nephrotoxicity [109]. Liposomal paclitaxel formulation was found to target lung cancer cells and help to overcome drug resistance [110]. Liposomal formulations were also reported to successfully deliver therapeutic vaccines such as Biomira Liposomal Protein 25 (BLP25) [111] or effective gene delivery such as 1,2-dioleoyl-3-trimethylammonium propane (DOTAP): cholesterol (Chol.) carrying p53, TUSC2/FUS1, or mda-7/IL-24, the tumor suppressor genes to the metastatic tumor sites [112]. Polymeric NPs, such as PEG-modified polylactic acid NPs loaded with taxanes, were shown to improve the chemoradiation therapy in lung cancer cells and in xenograft models [113]. PLGA hybrid polymers were reported to co-deliver paclitaxel and STAT3 siRNA successfully to the drug-resistant A549 cell line [114]. Metal-based NPs, such as gold NPs conjugated with methotrexate (MTX), contribute to better efficacy and prolonged tumor retention in mouse models of Lewis lung carcinoma [115]. Gold NPs can also be used in photodynamic therapy PDT for delivering water-soluble PDT agent purpurin-18-N-methyl-D-glucamine (Pu-18-NMGA) [116]. Magnetic NPs were developed to detect micrometastasis in lung cancer in a distinct study by Wang et al. [117]. MSNs were reported to deliver radionuclide isotope holmium-165 (Ho^{165}) in a xenograft tumor model [118].

Several of these nano-formulations have entered or crossed the path of clinical trials as well. BLP25 liposome vaccine for lung neoplasms or non-small-cell lung carcinoma has crossed phase II clinical trial and was under phase III trial in 2011 [119]. The other nano-formulations, although successfully studied for their efficacy in laboratory scale, are still facing challenges for translating their use into clinical practice.

7.4.5 Miscellaneous Cancers

Nano-formulation-based chemoprevention has also been studied in other rare cancer types such as head, neck, and oral squamous cell carcinoma and skin, prostate, and liver cancer [120].

In recent years, extracellular vesicles, such as exosomes, microvescicles, and apoptotic bodies, as originated from mammalian tumor cells, have gained wide popularity as chemopreventive tools or carriers for chemotherapeutic drugs, RNAs, and peptides [121]. Phytoconstituents having anticancer potential, such as polyphenol, catechin, and so on, have also been formulated with nanocarriers to improve the solubility, stability, and bioavailability of the active constituents [122].

For oral squamous cell carcinoma, local treatment is required at the desired site, and various nano-polymeric drug delivery systems, such as naringenin NPs, ellagic acid, and chitosan NPs, have shown enhanced efficacy for this purpose [123].

For head and neck carcinoma, PEGylated nanoliposomes of paclitaxel, 5-fluorouracil, and resveratrol have shown to have improved therapeutic efficacy of their controlled release properties [123]. In another study, salvianolic acid B, a natural chemopreventive agent as encapsulated in phospholipid complex loaded NPs, showed improved cellular uptake compared to conventional dosage forms for head and neck cell carcinoma [124].

For skin cancer management, nanoemulsion of 5-fluorouracil, PLGA NPs, bromelain polymeric NPs, solid lipid NPs of doxorubicin, and 5-flurouracil have been reported by different groups [125].

Prostate cancer prevention and management can be achieved using targeted polymeric NPs using biocompatible polymer polylactic coglycolic acid–polyethylene glycol-A (PLGA-PEG-A) that binds specifically to the prostate cancer cells via membrane antigen binding. This results in improved bioavailability and reduced toxicity, leading to enhanced therapeutic efficacy [126].

Gold-conjugated green tea NPs showed hepatoprotective behavior against tumor-induced cellular damage [127]. A bioflavonoid, hesperetin, is a drug of choice for liver cancer treatment. However, its poor solubility, bioavailability, and biocompatibility issues restrain it from giving exact efficacy that could be improved by designing a nano-formulation of the same, using gold NPs [128].

Other than these, NP-based formulations have shown its promising role for protection against UV radiation. Examples include silver NPs and ultra-flexible NPs of an antioxidant diindolylmethane derivative [129].

7.5 CLINICAL INSIGHTS

As described in the earlier sections, nano-formulations are widely being developed and tested in laboratory scale for improved cancer therapy by delivering the drug in a site-specific manner as well as prolonged release of the drug from the site of delivery. Many of these NPs are being approved by the United States Food and Drug Administration (US FDA) and are currently at various phases of clinical trials. The first approved nano-formulation for cancer treatment was of sterically stabilized liposomal formulation of doxorubicin (Doxil and Caelyx) composed of phospholipids, cholesterol, and a lipopolymer (PEG) [85]. The said formulation, due to its reduced nanometer size range, shows extended retention and permeability [130]. Marqibo is another FDA-approved liposomal formulation used for the treatment of acute lymphatic leukemia [131]. The active drug component of the formulation is vincristine sulfate, and the liposome is composed of sphingomyelin and cholesterol that improves therapeutic effect compared to conventional preparation of vincristine [132]. Another FDA-approved antibody-drug conjugate used for the treating HER^{2+} breast cancer is Kadcyla, which is a covalent conjugate of maytansine derivative DM1 with transtuzumab (known as Herceptin) [133]. Transtuzumab is recognized by the HER receptors present at the surface of the breast cancer cells to which DM1 is delivered and the internalized DM1 triggers apoptosis. Albumin-conjugated nano-formulation of paclitaxel (Abraxane) [134] and micellar nano-formulation of rapamycin (Rapamune) are other examples of FDA-approved NPs for anticancer treatment [135].

Simultaneously, there are several lipid-based nano-formulations that are presently under clinical trials [136]. Examples include Lipoplatin (by Regulon Inc.; a cisplatin formulated in cholesterol, dipalmitoyl phosphatidyl glycerol (DPPG), soy phosphatidylcholine (SPC-3), and methoxy-PEG-distearoyl phosphatidyl ethanolamine (mPEG2000-DSPE). The said formulation, during its preclinical trials, revealed less nephrotoxicity compared to its conventional formulation and thus can be used for the treatment of PaCa, breast cancer, and advanced gastric cancer as announced by the European Medicines Agency (EMA) [137–139]. Table 7.2 summarizes the therapeutic nano-formulations currently in the preclinical or clinical trials or in market for the anticancer treatment.

Nano-formulation-based chemoprevention is advancing toward development of nanovaccines and nanoplatforms for early detection of premalignant markers [140]. Nanovaccines and artificial antigen-presenting cells (aAPCs) [141] target the immunosuppressed tumor microenvironment (TME). Alternatively, NPs can also be used as adjuvants to enhance antigen presentation that ultimately helps in activation of cytotoxic T cell antitumor function [142]. Liposomes, gold NPs, PLGA NPs, micelles, and dendrimers all can deliver the tumor-associated antigens (TAAs) at the cytoplasm and enhances the immune response against tumor cells [143]. However, despite all these advancement in nano-formulations, the translational role of them in cancer prevention still awaits a prolonged route depending on the clinical regulatory considerations.

TABLE 7.2
Nano-Formulations in Commercial Use or in Clinical Trials for Chemoprevention

Nano-Formulation/Marketed Name	Type of Nanoparticle	Conjugated Drug	Cancer Type	Evaluation Status	Reference
Abraxane	Albumin bound	Paclitaxel	Pancreatic cancer, breast cancer	FDA approved	[70]
Onivyde	Liposome	Irinotecan	Pancreatic cancer	FDA approved	[71]
Kadcyla	Humanized monoclonal antibody trastuzumab-drug conjugate	Emtansine (DM1)	Breast cancer	FDA approved	[83, 138]
Superparamagnetic NPs	Streptavidin-conjugated	HER2/neu antibody conjugated	Breast cancer	FDA approved	[84]
Doxil, Caelyx	Liposome, PEGylated liposome	Doxorubicin	Breast cancer	FDA approved	[85]
Marqibo	Liposomal formulation	Vincristine sulfate	Acute lymphatic leukemia	FDA approved	[131]
Rapamune	Micellar	Rapamycin	Renal cancer, lung cancer	FDA approved	[135]
L-BLP25	Liposomal vaccine	Lipopeptides	Non-small-cell lung cancer (NSCLC)	Phase III clinical trial	[119]
Lipoplatin	Lipid NPs	Cisplatin	Pancreatic adenocarcinoma, breast cancer, and advanced gastric cancer	Phase IV clinical trial	[139]

7.6   CONCLUSION AND FUTURE PERSPECTIVE

The nanoparticle-based delivery of therapeutic agents was extensively studied during the last decade for the treatment of various diseases, including cancer. Nano-formulations showed its potential in chemoprevention, by delivering the drug in a targeted manner to the tumor site, resulting in enhanced therapeutic efficacy and reduced side effects. As discussed in the chapter, several nano-formulations have already received FDA approval and many are in clinical trials for the treatment of various cancer types, such as pancreatic cancer, breast cancer, lung cancer, and lymphoid leukemia. Also, there are few cancer types, such as colorectal cancer, that still await a commercial nano-formulation for the treatment, although extensive research have already been carried out.

Although majority of the nano-formulations were initially designed to treat or prevent single disease type, a combination of various types of NPs for the development of multi-therapeutic NPs are currently in trend. Despite their targeted delivery and reduced side effects, very little is known about the metabolism, clearance, and toxicity of NPs. Similarly, NP encapsulation of clinical drugs are vastly being studied; however, encapsulating other biomolecules, such as genes, enzymes, and DNA or RNA molecules, into NPs is still less explored and thereby leaves a significant scope of research in this field. Additionally, understanding the mechanism of action of the therapeutic NPs at the cellular and molecular level will help to establish novel treatment, prevention, and diagnosis strategies.

Moreover, the manufacturing cost of the nano-formulations is another vital topic that needs to be addressed. Shortage of sufficient financial investment due to compromised profitable output delays the bench-to-bedside journey of the nano-formulations.

Simultaneously, nanoplatform-based early diagnosis approaches are under exploration and so as the nanovaccines for cancer prevention [144]. Nonetheless, the bench-to-bedside translation of these advanced nanotechnology-based research will still take some time, considering the clinical and regulatory approval required for the same. In view of this, the clinical translation of nanotherapeutics for chemoprevention can be expected in near future.

7.7   REFERENCES

1. Bray F, Jemal A, Grey N, Ferlay J, Forman D. Global cancer transitions according to the Human Development Index (2008–2030): A population-based study. *Lancet Oncology* 2012;13:790–801.
2. NIH National Cancer Institute. Cancer Statistics [Internet] 2020. Available from: www.cancer.gov/about-cancer/understanding/statistics
3. Desai P, Thumma NJ, Wagh PR, Zhan S, Ann D, Wang J, et al. Cancer chemoprevention using nanotechnology-based approaches. *Frontiers in Pharmacology* [internet] 2020;11:1–9. Available from: www.frontiersin.org/article/10.3389/fphar.2020.00323/full
4. Jeevanandam J, Barhoum A, Chan YS, Dufresne A, Danquah MK. Review on nanoparticles and nano-structured materials: History, sources, toxicity and regulations. *Beilstein Journal of Nanotechnology* 2018;9:1050–1074.
5. Jatzkewitz H. [Incorporation of physiologically-active substances into a colloidal blood plasma substitute. I. Incorporation of mescaline peptide into polyvinylpyrrolidone]. *Hoppe-Seyler's Zeitschrift fur physiologische Chemie* [Internet] 1954;297:149–156. Available from: www.ncbi.nlm.nih.gov/pubmed/13221275
6. Bangham AD, Horne RW. Negative staining of phospholipids and their structural modification by surface-active agents as observed in the electron microscope. *Journal of Molecular Biology* [Internet] 1964;8:660-IN10. Available from: https://linkinghub.elsevier.com/retrieve/pii/S0022283664801157
7. Gradishar WJ, Tjulandin S, Davidson N, Shaw H, Desai N, Bhar P, et al. Phase III trial of nanoparticle albumin-bound paclitaxel compared with polyethylated castor oil–based paclitaxel in women with breast cancer. *Journal of Clinical Oncology* [Internet] 2005;23:7794–803. Available from: http://ascopubs.org/doi/10.1200/JCO.2005.04.937
8. Casaluce F, Sgambato A, Rossi A, Mulshine JL. The US FDA has approved Abraxane® for the treatment of non-small-cell lung cancer Aurora: A new light for targeted therapy in small-cell lung cancer. *Lung Cancer* 2012;1:251–254.

9. Zielińska A, Carreiró F, Oliveira AM, Neves A, Pires B, Venkatesh DN, et al. Polymeric nanoparticles: Production, characterization, toxicology and ecotoxicology. *Molecules* 2020;25.

10. Escalona-Rayo O, Fuentes-Vázquez P, Jardon-Xicotencatl S, García-Tovar CG, Mendoza-Elvira S, Quintanar-Guerrero D. Rapamycin-loaded polysorbate 80-coated PLGA nanoparticles: Optimization of formulation variables and in vitro anti-glioma assessment. *Journal of Drug Delivery Science and Technology* 2019;52:488–499.

11. Gao M, Long X, Du J, Teng M, Zhang W, Wang Y, et al. Enhanced curcumin solubility and antibacterial activity by encapsulation in PLGA oily core nanocapsules. *Food & Function* 2020;11:448–455.

12. Bechnak L, Khalil C, Kurdi R el, Khnayzer RS, Patra D. Curcumin encapsulated colloidal amphiphilic block co-polymeric nanocapsules: Colloidal nanocapsules enhance photodynamic and anticancer activities of curcumin. *Photochemical & Photobiological Sciences: Official Journal of the European Photochemistry Association and the European Society for Photobiology* 2020;19:1088–1098.

13. Avramović N, Mandić B, Savić-Radojević A, Simić T. Polymeric nanocarriers of drug delivery systems in cancer therapy. *Pharmaceutics* 2020;12.

14. Jiang Y, Huo S, Hardie J, Liang XJ, Rotello VM. Progress and perspective of inorganic nanoparticle-based siRNA delivery systems. *Expert Opinion on Drug Delivery* 2016;13:547–559.

15. Shetake N, Balla M, Kumar A, Pandey B. Magnetic hyperthermia therapy: An emerging modality of cancer treatment in combination with radiotherapy. *Journal of Radiation and Cancer Research* [Internet] 2016;7:13–17. Available from: www.journalrcr.org/article.asp?issn=2588-9273

16. Liu Z, Davis C, Cai W, He L, Chen X, Dai H. Circulation and long-term fate of functionalized, biocompatible single-walled carbon nanotubes in mice probed by Raman spectroscopy. *Proceedings of the National Academy of Sciences* [Internet] 2008;105:1410–1415. Available from: https://doi.org/10.1073/pnas.0707654105

17. Rasmussen JW, Martinez E, Louka P, Wingett DG. Zinc oxide nanoparticles for selective destruction of tumor cells and potential for drug delivery applications. *Expert Opinion on Drug Delivery* 2010;7:1063–1077.

18. Mura S, Nicolas J, Couvreur P. Stimuli-responsive nanocarriers for drug delivery. *Nature Materials* 2013;12:991–1003.

19. Chouikrat R, Seve A, Vanderesse R, Benachour H, Barberi-Heyob M, Richeter S, et al. Non polymeric nanoparticles for photodynamic therapy applications: Recent developments. *Current Medicinal Chemistry* 2012;19:781–792.

20. García-Pinel B, Porras-Alcalá C, Ortega-Rodríguez A, Sarabia F, Prados J, Melguizo C, et al. Lipid-based nanoparticles: Application and recent advances in cancer treatment. *Nanomaterials* [Internet] 2019;9:638. Available from: www.mdpi.com/2079-4991/9/4/638

21. Junghanns JUAH, Müller RH. Nanocrystal technology, drug delivery and clinical applications. *International Journal of Nanomedicine* [Internet] 2008;3:295–309. Available from: www.ncbi.nlm.nih.gov/pubmed/18990939

22. Yingchoncharoen P, Kalinowski DS, Richardson DR. Lipid-based drug delivery systems in cancer therapy: What is available and what is yet to come. *Pharmacological Reviews* [Internet] 2016;68:701 LP—787. Available from: http://pharmrev.aspetjournals.org/content/68/3/701.abstract

23. Deshpande P, Jhaveri A, Pattni B, Biswas S, Torchilin V. Transferrin and octaarginine modified dual-functional liposomes with improved cancer cell targeting and enhanced intracellular delivery for the treatment of ovarian cancer. *Drug Delivery* [Internet] 2018;25:517–532. Available from: https://doi.org/10.1080/10717544.2018.1435747

24. Sesarman A, Tefas L, Sylvester B, Licarete E, Rauca V, Luput L, et al. Co-delivery of curcumin and doxorubicin in PEGylated liposomes favored the antineoplastic C26 murine colon carcinoma microenvironment. *Drug Delivery and Translational Research* [Internet] 2019;9:260–272. Available from: https://doi.org/10.1007/s13346-018-00598-8

25. Mydin RBSMN, Moshawih S. Nanoparticles in nanomedicine application: Lipid-based nanoparticles and their safety concerns [Internet]. In: Siddiquee S, Melvin GJH, Rahman MdM, editors. *Nanotechnology: Applications in Energy, Drug and Food.* Cham: Springer International Publishing; 2019. page 227–232. Available from: https://doi.org/10.1007/978-3-319-99602-8_10

26. Rajabi M, Mousa SA. Lipid nanoparticles and their application in nanomedicine. *Current Pharmaceutical Biotechnology* 2016;17:662–672.

27. Pindiprolu SKSS, Chintamaneni PK, Krishnamurthy PT, Ratna Sree Ganapathineedi K. Formulation-optimization of solid lipid nanocarrier system of STAT3 inhibitor to improve its activity in triple negative breast cancer cells. *Drug Development and Industrial Pharmacy* [Internet] 2019;45:304–313. Available from: https://doi.org/10.1080/03639045.2018.1539496

28. Guney Eskiler G, Cecener G, Egeli U, Tunca B. Synthetically Lethal BMN 673 (Talazoparib) Loaded solid lipid nanoparticles for BRCA1 mutant triple negative breast cancer. *Pharmaceutical Research* 2018;35:218.

29. Alexis F, Pridgen E, Molnar LK, Farokhzad OC. Factors affecting the clearance and biodistribution of polymeric nanoparticles. *Molecular Pharmaceutics* 2008;5:505–515.

30. Petros RA, DeSimone JM. Strategies in the design of nanoparticles for therapeutic applications. *Nature Reviews Drug Discovery* 2010;9:615–627.

31. Decuzzi P, Godin B, Tanaka T, Lee SY, Chiappini C, Liu X, et al. Size and shape effects in the biodistribution of intravascularly injected particles. *Journal of Controlled Release: Official Journal of the Controlled Release Society* 2010;141:320–327.

32. Gratton SEA, Ropp PA, Pohlhaus PD, Luft JC, Madden VJ, Napier ME, et al. The effect of particle design on cellular internalization pathways. *Proceedings of the National Academy of Sciences* 2008;105:11613–11618.

33. Bawa R, Siegel RA, Marasca B, Karel M, Langer R. An explanation for the controlled release of macromolecules from polymers. *Journal of Controlled Release* 1985;1:259–267.

34. Panyam J, Dali MM, Sahoo SK, Ma W, Chakravarthi SS, Amidon GL, et al. Polymer degradation and in vitro release of a model protein from poly(D,L-lactide-co-glycolide) nano- and microparticles. *Journal of Controlled Release: Official Journal of the Controlled Release Society* 2003;92:173–187.

35. Champion JA, Katare YK, Mitragotri S. Particle shape: A new design parameter for micro- and nanoscale drug delivery carriers. *Journal of Controlled Release: Official Journal of the Controlled Release Society* 2007;121:3–9.

36. Moghimi SM, Hunter AC, Murray JC. Long-circulating and target-specific nanoparticles: Theory to practice. *Pharmacological Reviews* 2001;53:283–318.

37. Decuzzi P, Pasqualini R, Arap W, Ferrari M. Intravascular delivery of particulate systems: Does geometry really matter? *Pharmaceutical Research* 2009;26:235–243.

38. Zahr AS, Davis CA, Pishko M v. Macrophage uptake of core-shell nanoparticles surface modified with poly(ethylene glycol). *Langmuir* 2006;22:8178–8185.

39. Otsuka H, Nagasaki Y, Kataoka K. PEGylated nanoparticles for biological and pharmaceutical applications. *Advanced Drug Delivery Reviews* 2003;55:403–419.

40. Gupta B, Torchilin VP. Monoclonal antibody 2C5-modified doxorubicin-loaded liposomes with significantly enhanced therapeutic activity against intracranial human brain U-87 MG tumor xenografts in nude mice. *Cancer, Immunology, Immunotherapy* 2007;56:1215–1223.

41. Kirpotin DB, Drummond DC, Shao Y, Shalaby MR, Hong K, Nielsen UB, et al. Antibody targeting of long-circulating lipidic nanoparticles does not increase tumor localization but does increase internalization in animal models. *Cancer Research* 2006;66:6732–6740.

42. Bahrami B, Hojjat-Farsangi M, Mohammadi H, Anvari E, Ghalamfarsa G, Yousefi M, et al. Nanoparticles and targeted drug delivery in cancer therapy. *Immunology Letters* [Internet] 2017;190:64–83. Available from: www.sciencedirect.com/science/article/pii/S0165247817301761

43. Venturoli D, Rippe B. Ficoll and dextran vs. globular proteins as probes for testing glomerular permselectivity: Effects of molecular size, shape, charge, and deformability. *American Journal of Physiology-Renal Physiology* [Internet] 2005;288:F605–613. Available from: https://doi.org/10.1152/ajprenal.00171.2004

44. Kamaly N, Xiao Z, Valencia PM, Radovic-Moreno AF, Farokhzad OC. Targeted polymeric therapeutic nanoparticles: Design, development and clinical translation. *Chemical Society Reviews* [Internet] 2012;41:2971–3010. Available from: http://dx.doi.org/10.1039/C2CS15344K

45. Farokhzad OC, Langer R. Impact of nanotechnology on drug delivery. *ACS Nano* [Internet] 2009;3:16–20. Available from: https://doi.org/10.1021/nn900002m

46. Danhier F, Feron O, Préat V. To exploit the tumor microenvironment: Passive and active tumor targeting of nanocarriers for anti-cancer drug delivery. *Journal of Controlled Release* [Internet] 2010;148:135–146. Available from: www.sciencedirect.com/science/article/pii/S0168365910007108

47. Santi M, Maccari G, Mereghetti P, Voliani V, Rocchiccioli S, Ucciferri N, et al. Rational design of a transferrin-binding peptide sequence tailored to targeted nanoparticle internalization. *Bioconjugate Chemistry* [Internet] 2017;28:471–480. Available from: https://doi.org/10.1021/acs.bioconjchem.6b00611

48. Cui Y na, Xu Q xing, Davoodi P, Wang D ping, Wang CH. Enhanced intracellular delivery and controlled drug release of magnetic PLGA nanoparticles modified with transferrin. *Acta Pharmacologica Sinica* [Internet] 2017;38:943–953. Available from: https://doi.org/10.1038/aps.2017.45

49. Soe ZC, Kwon JB, Thapa RK, Ou W, Nguyen HT, Gautam M, et al. Transferrin-conjugated polymeric nanoparticle for receptor-mediated delivery of doxorubicin in doxorubicin-resistant breast cancer cells. *Pharmaceutics* 2019;11.
50. Low PS, Kularatne SA. Folate-targeted therapeutic and imaging agents for cancer. *Current Opinion in Chemical Biology* [Internet] 2009;13:256–262. Available from: www.sciencedirect.com/science/article/pii/S1367593109000465
51. Muralidharan R, Babu A, Amreddy N, Basalingappa K, Mehta M, Chen A, et al. Folate receptor-targeted nanoparticle delivery of HuR-RNAi suppresses lung cancer cell proliferation and migration. *Journal of Nanobiotechnology* [Internet] 2016;14:47. Available from: https://doi.org/10.1186/s12951-016-0201-1
52. Minko T. Drug targeting to the colon with lectins and neoglycoconjugates. *Advanced Drug Delivery Reviews* [Internet] 2004;56:491–509. Available from: www.sciencedirect.com/science/article/pii/S0169409X03002321
53. Sigismund S, Avanzato D, Lanzetti L. Emerging functions of the EGFR in cancer. *Molecular Oncology* [Internet] 2018;12:3–20. Available from: https://doi.org/10.1002/1878-0261.12155
54. Alexis F, Basto P, Levy-Nissenbaum E, Radovic-Moreno AF, Zhang L, Pridgen E, et al. HER-2-Targeted nanoparticle–affibody bioconjugates for cancer therapy. *ChemMedChem* [Internet] 2008;3:1839–1843. Available from: https://doi.org/10.1002/cmdc.200800122
55. Balasubramanian S, Ravindran Girija A, Nagaoka Y, Iwai S, Suzuki M, Kizhikkilot V, et al. Curcumin and 5-Fluorouracil-loaded, folate- and transferrin-decorated polymeric magnetic nanoformulation: A synergistic cancer therapeutic approach, accelerated by magnetic hyperthermia. *International Journal of Nanomedicine* 2014;9:437–459.
56. Carmeliet P, Jain RK. Angiogenesis in cancer and other diseases. *Nature* [Internet] 2000;407:249–257. Available from: https://doi.org/10.1038/35025220
57. Maeda H. The enhanced permeability and retention (EPR) effect in tumor vasculature: The key role of tumor-selective macromolecular drug targeting. *Advances in Enzyme Regulation* [Internet] 2001;41:189–207. Available from: www.sciencedirect.com/science/article/pii/S0065257100000133
58. Carita CA, Eloy OJ, Chorilli M, Lee JR, Leonardi RG. Recent advances and perspectives in liposomes for cutaneous drug delivery. *Current Medicinal Chemistry* [Internet] 2018;25:606–635. Available from: www.eurekaselect.com/article/86266
59. Sykes EA, Chen J, Zheng G, Chan WCW. Investigating the impact of nanoparticle size on active and passive tumor targeting efficiency. *ACS Nano* [Internet] 2014;8:5696–5706. Available from: https://doi.org/10.1021/nn500299p
60. Pelicano H, Martin DS, Xu RH, Huang P. Glycolysis inhibition for anticancer treatment. *Oncogene* [Internet] 2006;25:4633–4646. Available from: https://doi.org/10.1038/sj.onc.1209597
61. Lim EK, Chung HB, Chung JS. Recent advances in pH-sensitive polymeric nanoparticles for smart drug delivery in cancer therapy. *Current Drug Targets* [Internet] 2018;19:300–317. Available from: www.eurekaselect.com/article/76176
62. Jain RK. Barriers to drug delivery in solid tumors. *Scientific American* 1994;271:58–65.
63. Cancer Statistics Center, American Cancer Society [Internet]. American Cancer Society 2019 cited 2019 Dec 12]. Available from: https://cancerstatisticscenter.cancer.org
64. Desai P, Ann D, Wang J, Prabhu S. Pancreatic cancer: Recent advances in nanoformulation-based therapies. *Critical Reviews™ in Therapeutic Drug Carrier Systems* 2019;36.
65. Prabhu S, Kanthamneni N, Wang J. Synergistic chemoprevention of colorectal cancer using colon-targeted polymer nanoparticles. *Cancer Research* 2007;67:1.
66. Thakkar A, Sutaria D, Grandhi BK, Wang J, Prabhu S. The molecular mechanism of action of aspirin, curcumin and sulforaphane combinations in the chemoprevention of pancreatic cancer. *Oncology Reports* 2013;29:1671–1677.
67. Desai P, Thakkar A, Ann D, Wang J, Prabhu S. Loratadine self-microemulsifying drug delivery systems (SMEDDS) in combination with sulforaphane for the synergistic chemoprevention of pancreatic cancer. *Drug Delivery and Translational Research* 2019;9:641–651.
68. Wei Y, Wang Y, Xia D, Guo S, Wang F, Zhang X, et al. Thermosensitive liposomal codelivery of HSA–paclitaxel and HSA–ellagic acid complexes for enhanced drug perfusion and efficacy against pancreatic cancer. *ACS Applied Materials & Interfaces* 2017;9:25138–25151.
69. Bhutia SK, Das SK, Azab B, Menezes ME, Dent P, Wang XY, et al. Targeting breast cancer-initiating/stem cells with melanoma differentiation-associated gene-7/interleukin-24. *International Journal of Cancer* 2013;133:2726–2736.

70. von Hoff DD, Ervin T, Arena FP, Chiorean EG, Infante J, Moore M, et al. Increased survival in pancreatic cancer with nab-Paclitaxel plus gemcitabine. *New England Journal of Medicine* [Internet] 2013;369:1691–1703. Available from: https://doi.org/10.1056/NEJMoa1304369

71. Wang-Gillam A, Li CP, Bodoky G, Dean A, Shan YS, Jameson G, et al. Nanoliposomal irinotecan with fluorouracil and folinic acid in metastatic pancreatic cancer after previous gemcitabine-based therapy (NAPOLI-1): A global, randomised, open-label, phase 3 trial. *Lancet* 2016;387:545–557.

72. Sarkar S, Azab B, Quinn BA, Shen X, Dent P, Klibanov AL, et al. Chemoprevention gene therapy (CGT) of pancreatic cancer using perillyl alcohol and a novel chimeric serotype cancer terminator virus. *Current Molecular Medicine* 2014;14:125–140.

73. Yezhelyev M, Yacoub R, O'Regan R. Inorganic nanoparticles for predictive oncology of breast cancer. *Nanomedicine (Lond)* 2009;4:83–103.

74. Leuschner C, Kumar CSSR, Hansel W, Soboyejo W, Zhou J, Hormes J. LHRH-conjugated magnetic iron oxide nanoparticles for detection of breast cancer metastases. *Breast Cancer Research and Treatment* 2006;99:163–176.

75. Abbasalipourkabir R, Salehzadeh A, Abdullah R. Antitumor activity of tamoxifen loaded solid lipid nanoparticles on induced mammary tumor gland in sprague-dawley rats. *African Journal of Biotechnology* 2010;9:7337–7345.

76. Luna-Gutiérrez M, Ferro-Flores G, Ocampo-García BE, Santos- Cuevas CL, Jiménez-Mancilla N, de León-Rodríguez LM, et al. A therapeutic system of 177Lu-labeled gold nanoparticles-RGD internalized in breast cancer cells. *Journal of the Mexican Chemical Society* [Internet] 2013;57:421–429. Available from: https://linkinghub.elsevier.com/retrieve/pii/S0002937837902084

77. Almeida JPM, Figueroa ER, Drezek RA. Gold nanoparticle mediated cancer immunotherapy. *Nanomedicine* 2014;10:503–514.

78. Cho HJ, Yoon IS, Yoon HY, Koo H, Jin YJ, Ko SH, et al. Polyethylene glycol-conjugated hyaluronic acid-ceramide self-assembled nanoparticles for targeted delivery of doxorubicin. *Biomaterials* [Internet] 2012;33:1190–1200. Available from: www.sciencedirect.com/science/article/pii/S0142961211012804

79. Potineni A, Lynn DM, Langer R, Amiji MM. Poly(ethylene oxide)-modified poly(β-amino ester) nanoparticles as a pH-sensitive biodegradable system for paclitaxel delivery. *Journal of Controlled Release* [Internet] 2003;86:223–234. Available from: www.sciencedirect.com/science/article/pii/S0168365902003747

80. Adolphi NL, Butler KS, Lovato DM, Tessier TE, Trujillo JE, Hathaway HJ, et al. Imaging of Her2-targeted magnetic nanoparticles for breast cancer detection: Comparison of SQUID-detected magnetic relaxometry and MRI. Contrast Media. *Molecular Imaging* 2012;7:308–319.

81. Wagner V, Dullaart A, Bock AK, Zweck A. The emerging nanomedicine landscape. *Nature Biotechnology* 2006;24:1211–1217.

82. dos Santos MSC, Gouvêa AL, de Moura LD, Paterno LG, de Souza PEN, Bastos AP, et al. Nanographene oxide-methylene blue as phototherapies platform for breast tumor ablation and metastasis prevention in a syngeneic orthotopic murine model. *Journal of Nanobiotechnology* 2018;16:1–17.

83. Peddi PF, Hurvitz SA. Trastuzumab emtansine: The first targeted chemotherapy for treatment of breast cancer. *Future Oncology* 2013;9:319–326.

84. Artemov D, Mori N, Okollie B, Bhujwalla ZM. MR molecular imaging of the Her-2/neu receptor in breast cancer cells using targeted iron oxide nanoparticles. *Magnetic Resonance in Medicine* 2003;49:403–408.

85. Blanco E, Ferrari M. Emerging nanotherapeutic strategies in breast cancer. *Breast* 2014;23:10–18.

86. Testa U, Pelosi E, Castelli G. Colorectal cancer: Genetic abnormalities, tumor progression, tumor heterogeneity, clonal evolution and tumor-initiating cells. *Medical Sciences* 2018;6:31.

87. Vernon SW. Participation in colorectal cancer screening: A review. *Journal of the National Cancer Institute* 1997;89:1406–1422.

88. Umezawa S, Higurashi T, Komiya Y, Arimoto J, Horita N, Kaneko T, et al. Chemoprevention of colorectal cancer: Past, present, and future. *Cancer Science* 2019;110:3018–3026.

89. Margulis-Goshen K, Weitman M, Major DT, Magdassi S. Inhibition of crystallization and growth of celecoxib nanoparticles formed from volatile microemulsions. *Journal of Pharmaceutical Sciences* 2011;100:4390–4400.

90. Jayaprakasha GK, Murthy KNC, Patil BS. Enhanced colon cancer chemoprevention of curcumin by nanoencapsulation with whey protein. *European Journal of Pharmacology* 2016;789:291–300.

91. Cisterna BA, Kamaly N, Choi W il, Tavakkoli A, Farokhzad OC, Vilos C. Targeted nanoparticles for colorectal cancer. *Nanomedicine* [Internet] 2016;11:2443–2456. Available from: www.futuremedicine.com/doi/10.2217/nnm-2016-0194

92. Lacava Z, Campos da Paz, Almeida Santos, Santos, Silva, Souza, et al. Anti-CEA loaded maghemite nanoparticles as a theragnostic device for colorectal cancer. *International Journal of Nanomedicine* [Internet] 2012;7:5271. Available from: https://pubmed.ncbi.nlm.nih.gov/23055733

93. Vigor KL, Kyrtatos PG, Minogue S, Al-Jamal KT, Kogelberg H, Tolner B, et al. Nanoparticles functionalized with recombinant single chain Fv antibody fragments (scFv) for the magnetic resonance imaging of cancer cells. *Biomaterials* 2010;31:1307–1315.

94. Tiernan JP, Ingram N, Marston G, Perry SL, Rushworth J v, Coletta PL, et al. CEA-targeted nanoparticles allow specific in vivo fluorescent imaging of colorectal cancer models. *Nanomedicine (Lond)* 2015;10:1223–1231.

95. Fay F, McLaughlin KM, Small DM, Fennell DA, Johnston PG, Longley DB, et al. Conatumumab (AMG 655) coated nanoparticles for targeted pro-apoptotic drug delivery. *Biomaterials* 2011;32:8645–8653.

96. Abdelghany SM, Schmid D, Deacon J, Jaworski J, Fay F, McLaughlin KM, et al. Enhanced antitumor activity of the photosensitizer meso-Tetra(N-methyl-4-pyridyl) porphine tetra tosylate through encapsulation in antibody-targeted chitosan/alginate nanoparticles. *Biomacromolecules* 2013;14:302–310.

97. Cortez C, Tomaskovic-Crook E, Johnston APR, Scott AM, Nice EC, Heath JK, et al. Influence of size, surface, cell line, and kinetic properties on the specific binding of A33 antigen-targeted multilayered particles and capsules to colorectal cancer cells. *ACS Nano* 2007;1:93–102.

98. Kirui DK, Rey DA, Batt CA. Gold hybrid nanoparticles for targeted phototherapy and cancer imaging. *Nanotechnology* 2010;21:105105.

99. McCarron PA, Marouf WM, Quinn DJ, Fay F, Burden RE, Olwill SA, et al. Antibody targeting of camptothecin-loaded PLGA nanoparticles to tumor cells. *Bioconjugate Chemistry* 2008;19:1561–1569.

100. Yang SJ, Lin FH, Tsai KC, Wei MF, Tsai HM, Wong JM, et al. Folic acid-conjugated chitosan nanoparticles enhanced protoporphyrin IX accumulation in colorectal cancer cells. *Bioconjugate Chemistry* 2010;21:679–689.

101. Li P, Wang Y, Zeng F, Chen L, Peng Z, Kong LX. Synthesis and characterization of folate conjugated chitosan and cellular uptake of its nanoparticles in HT-29 cells. *Carbohydrate Research* 2011;346:801–806.

102. Kopansky E, Shamay Y, David A. Peptide-directed HPMA copolymer-doxorubicin conjugates as targeted therapeutics for colorectal cancer. *Journal of Drug Target* 2011;19:933–943.

103. Unzueta U, Céspedes MV, Ferrer-Miralles N, Casanova I, Cedano J, Corchero JL, et al. Intracellular CXCR4+ cell targeting with T22-empowered protein-only nanoparticles. *International Journal of Nanomedicine* 2012;7:4533–4544.

104. Jain A, Jain SK, Ganesh N, Barve J, Beg AM. Design and development of ligand-appended polysaccharidic nanoparticles for the delivery of oxaliplatin in colorectal cancer. *Nanomedicine* 2010;6:179–190.

105. Gary-Bobo M, Brevet D, Benkirane-Jessel N, Raehm L, Maillard P, Garcia M, et al. Hyaluronic acid-functionalized mesoporous silica nanoparticles for efficient photodynamic therapy of cancer cells. *Photodiagnosis and Photodynamic Therapy* 2012;9:256–260.

106. Shahzad MMK, Mangala LS, Han HD, Lu C, Bottsford-Miller J, Nishimura M, et al. Targeted delivery of small interfering RNA using reconstituted high-density lipoprotein nanoparticles. *Neoplasia* 2011;13:309–319.

107. Babu A, Templeton AK, Munshi A, Ramesh R. Nanoparticle-based drug delivery for therapy of lung cancer: Progress and challenges. *Journal of Nanomaterials* [Internet] 2013;2013:863951. Available from: https://doi.org/10.1155/2013/863951

108. Yao X, Panichpisal K, Kurtzman N, Nugent K. Cisplatin nephrotoxicity: A review. *The American Journal of the Medical Sciences* 2007;334:115–124.

109. Boulikas T. Low toxicity and anticancer activity of a novel liposomal cisplatin (Lipoplatin) in mouse xenografts. *Oncology Reports* [Internet] 2004;12:3–12. Available from: https://doi.org/10.3892/or.12.1.3

110. Zhou J, Zhao WY, Ma X, Ju RJ, Li XY, Li N, et al. The anticancer efficacy of paclitaxel liposomes modified with mitochondrial targeting conjugate in resistant lung cancer. *Biomaterials* 2013;34:3626–3638.

111. North S, Butts C. Vaccination with BLP25 liposome vaccine to treat non-small cell lung and prostate cancers. *Expert Review of Vaccines* [Internet] 2005;4:249–257. Available from: https://doi.org/10.1586/14760584.4.3.249

112. Lu C, Stewart DJ, Lee JJ, Ji L, Ramesh R, Jayachandran G, et al. Phase I clinical trial of systemically administered TUSC2 (FUS1)-nanoparticles mediating functional gene transfer in humans. *PLoS One* 2012;7:e34833.

113. Jung J, Park SJ, Chung HK, Kang HW, Lee SW, Seo MH, et al. Polymeric nanoparticles containing taxanes enhance chemoradiotherapeutic efficacy in non-small cell lung cancer. *International Journal of Radiation Oncology* Biology* Physics* 2012;84:e77–83.

114. Su WP, Cheng FY, Shieh DB, Yeh CS, Su WC. PLGA nanoparticles codeliver paclitaxel and Stat3 siRNA to overcome cellular resistance in lung cancer cells. *International Journal of Nanomedicine* 2012;7:4269.

115. Chen YH, Tsai CY, Huang PY, Chang MY, Cheng PC, Chou CH, et al. Methotrexate conjugated to gold nanoparticles inhibits tumor growth in a syngeneic lung tumor model. *Molecular Pharmaceutics* 2007;4:713–722.

116. Lkhagvadulam B, Kim JH, Yoon I, Shim YK. Size-dependent photodynamic activity of gold nanoparticles conjugate of water soluble purpurin-18-N-methyl-D-glucamine. *BioMed Research International* 2013;2013.

117. Wang Y, Zhang Y, Du Z, Wu M, Zhang G. Detection of micrometastases in lung cancer with magnetic nanoparticles and quantum dots. *International Journal of Nanomedicine* 2012;7:2315.

118. di Pasqua AJ, Miller ML, Lu X, Peng L, Jay M. Tumor accumulation of neutron-activatable holmium-containing mesoporous silica nanoparticles in an orthotopic non-small cell lung cancer mouse model. *Inorganica Chimica Acta* 2012;393:334–336.

119. Wu YL, Park K, Soo RA, Sun Y, Tyroller K, Wages D, et al. INSPIRE: A phase III study of the BLP25 liposome vaccine (L-BLP25) in Asian patients with unresectable stage III non-small cell lung cancer. *BMC Cancer* 2011;11:430.

120. Crooker K, Aliani R, Ananth M, Arnold L, Anant S, Thomas SM. A review of promising natural chemopreventive agents for head and neck cancer. *Cancer Prevention Research* 2018;11:441–450.

121. Wu M, Wang G, Hu W, Yao Y, Yu XF. Emerging roles and therapeutic value of exosomes in cancer metastasis. *Molecular Cancer* 2019;18:1–11.

122. Tyagi N, De R, Begun J, Popat A. Cancer therapeutics with epigallocatechin-3-gallate encapsulated in biopolymeric nanoparticles. *International Journal of Pharmaceutics* 2017;518:220–227.

123. Desai KGH. Polymeric drug delivery systems for intraoral site-specific chemoprevention of oral cancer. *Journal of Biomedical Materials Research Part B: Applied Biomaterials* 2018;106:1383–413.

124. Li H, Shi L, Wei J, Zhang C, Zhou Z, Wu L, et al. Cellular uptake and anticancer activity of salvianolic acid B phospholipid complex loaded nanoparticles in head and neck cancer and precancer cells. *Colloids and Surfaces B: Biointerfaces* 2016;147:65–72.

125. Ravikumar P, Tatke P. Advances in encapsulated dermal formulations in chemoprevention of melanoma: An overview. *Journal of Cosmetic Dermatology* 2019;18:1606–1612.

126. Sanna V, Singh CK, Jashari R, Adhami VM, Chamcheu JC, Rady I, et al. Targeted nanoparticles encapsulating (−)-epigallocatechin-3-gallate for prostate cancer prevention and therapy. *Scientific Reports* 2017;7:1–15.

127. Mukherjee S, Ghosh S, Das DK, Chakraborty P, Choudhury S, Gupta P, et al. Gold-conjugated green tea nanoparticles for enhanced anti-tumor activities and hepatoprotection—Synthesis, characterization and in vitro evaluation. *The Journal of Nutritional Biochemistry* 2015;26:1283–1297.

128. Gokuladhas K, Jayakumar S, Rajan B, Elamaran R, Pramila CS, Gopikrishnan M, et al. Exploring the potential role of chemopreventive agent, hesperetin conjugated pegylated gold nanoparticles in diethylnitrosamine-induced hepatocellular carcinoma in male wistar albino rats. *Indian Journal of Clinical Biochemistry* 2016;31:171–184.

129. Bagde A, Mondal A, Singh M. Drug delivery strategies for chemoprevention of UVB-induced skin cancer: A review. *Photodermatol Photoimmunol Photomed* 2018;34:60–68.

130. Barenholz Y. Doxil®-the first FDA-approved nano-drug: Lessons learned. *Journal of Controlled Release: Official Journal of the Controlled Release Society* 2012;160:117–134.

131. Silverman JA, Deitcher SR. Marqibo® (vincristine sulfate liposome injection) improves the pharmacokinetics and pharmacodynamics of vincristine. *Cancer Chemotherapy and Pharmacology* [Internet] 2013;71:555–564. Available from: https://doi.org/10.1007/s00280-012-2042-4

132. O'Brien S, Schiller G, Lister J, Damon L, Goldberg S, Aulitzky W, et al. High-dose vincristine sulfate liposome injection for advanced, relapsed, and refractory adult Philadelphia chromosome-negative acute lymphoblastic leukemia. *Journal of Clinical Oncology: Official Journal of the American Society of Clinical Oncology* 2013;31:676–683.

133. Xu Z, Guo D, Jiang Z, Tong R, Jiang P, Bai L, et al. Novel HER2-targeting antibody-drug conjugates of trastuzumab beyond T-DM1 in breast cancer: Trastuzumab deruxtecan(DS-8201a) and (vic-)trastuzumab duocarmazine (SYD985). *European Journal of Medicinal Chemistry* 2019;183:111682.

134. Gradishar WJ. Albumin-bound paclitaxel: A next-generation taxane. *Expert Opinion on Pharmacotherapy* 2006;7:1041–1053.

135. Lopez-Soler RI, Chen P, Nair L, Ata A, Patel S, Conti DJ. Sirolimus use improves cancer-free survival following transplantation: A single center 12-year analysis. *Transplantation Reports* [Internet] 2020;5:100040. Available from: www.sciencedirect.com/science/article/pii/S2451959620300020

136. Misra R, Acharya S, Sahoo SK. Cancer nanotechnology: Application of nanotechnology in cancer therapy. *Drug Discovery Today* [Internet] 2010;15:842–850. Available from: http://dx.doi.org/10.1016/j.drudis.2010.08.006

137. Stathopoulos GP, Boulikas T. Lipoplatin formulation review article. *Journal of Drug Delivery* 2012;2012:581363.

138. Byrne JD, Betancourt T, Brannon-Peppas L. Active targeting schemes for nanoparticle systems in cancer therapeutics. *Advanced Drug Delivery Reviews* 2008;60:1615–1626.

139. Boulikas T. Clinical overview on Lipoplatin: A successful liposomal formulation of cisplatin. *Expert Opinion on Investigational Drugs* 2009;18:1197–218.

140. Bentolila LA, Ebenstein Y, Weiss S. Quantum dots for in vivo small-animal imaging. *Journal of Nuclear Medicine* [Internet] 2009;50:493–496. Available from: https://pubmed.ncbi.nlm.nih.gov/19289434

141. Zang X, Zhao X, Hu H, Qiao M, Deng Y, Chen D. Nanoparticles for tumor immunotherapy. *European Journal of Pharmaceutics and Biopharmaceutics* [Internet] 2017;115:243–256. Available from: www.sciencedirect.com/science/article/pii/S0939641116308086

142. Yang R, Xu J, Xu L, Sun X, Chen Q, Zhao Y, et al. Cancer cell membrane-coated adjuvant nanoparticles with mannose modification for effective anticancer vaccination. *ACS Nano* 2018;12:5121–5129.

143. Guo Y, Wang D, Song Q, Wu T, Zhuang X, Bao Y, et al. Erythrocyte membrane-enveloped polymeric nanoparticles as nanovaccine for induction of antitumor immunity against melanoma. *ACS Nano* 2015;9:6918–6933.

144. Kheirollahpour M, Mehrabi M, Dounighi NM, Mohammadi M, Masoudi A. Nanoparticles and vaccine development. *Pharmaceutical Nanotechnology* 2020;8:6–21.

8 Potential of Natural-Product-Based Nano-Formulations as Chemopreventive Agents

Sneh Gautam, Pushpa Lohani, Shiv Dutt Purohit, Sonu Ambwani, and Poonam Maan

CONTENTS

8.1 Introduction 129
8.2 Natural Plant Products with Anticancer Properties 130
8.3 Challenges Associated with Natural Products to Use as Chemopreventive Agents 131
8.4 Various Nano-Formulations for Chemoprevention 131
    8.4.1 Liposomes.. 131
    8.4.2 Nanoparticles 133
    8.4.3 Dendrimers 134
    8.4.4 Carbon Nanotubes 135
    8.4.5 Nanomicelles 135
8.5 Conclusions and Future Prospects 136
8.6 References.. 136

8.1 INTRODUCTION

Cancer is a major public health problem that affects a wide range of diseases in all age populations. Cancer eradication is a big challenge in low- and middle-income nations due to the limitations of resources [1]. Worldwide, 19.3 million new cancer cases and nearly 10 million cancer deaths have been estimated in 2020, and it is expected to be 28.4 million cases in 2040 that would be 47% rise from 2020 [2]. Various therapeutic approaches like chemotherapy, surgery, and radiation have been evolved to treat different types of cancer. Different cytotoxic agents with chemotherapy are the most commonly adopted therapeutic approach to control various types of cancer [3]. However, all conventional therapeutic approaches are associated with several severe side effects, such as high systemic toxicity, multidrug resistance (MDR), and limited tolerability [4, 5]. Due to the severe side effects of conventional therapies, there is an intense need to explore a novel, more efficient, and selective approach to the treatment of various kinds of cancer. A large group of natural compounds derived from plants exhibited the significant antitumor, chemopreventive, and chemotherapeutic properties, estimated on basis of various cell culture and animal model studies [6, 7]. These compounds belong to the various class of organic compounds, including terpenes, alkaloids, polyphenols, and organosulfur [6]. Phenolic acid and flavonoids are commonly found polyphenols in plants. These natural compounds are called phytochemicals that are abundantly present in various plant-based food products, like vegetables and fruits, and are frequently taken up by humans in their daily diet. For a long time, the phytochemicals are being used as anticancer agents, and presently more than half of drugs used in clinical trials for chemotherapy are either natural in origin or their semisynthetic derivates [6, 8, 9]. However, the phytochemicals have various health benefits along with tremendous anticancer properties but are also associated with many disadvantages like slow

DOI: 10.1201/b23311-8

absorption, low solubility in an aqueous environment, low bioavailability, and short retention period in an animal body [10]. Besides this, after administration in the body, the intrinsic structure of the phytochemicals changes due to various physical and chemical barriers of a living system that ultimately affect the anticancer properties of these compounds [6, 10]. Thus, there is an utmost need to investigate some novel strategies that can protect the basic structure of these compounds and maximize their antiproliferative, chemopreventive, and chemotherapeutic properties. Recently, various nano-formulations, such as nanoparticles, nanosphere, dendrimers, liposomes, carbon nanotubes, and nanomicelles, are being explored for encapsulating the phytochemicals and their target delivery so that the inherent structure of these compounds can be protected. Nanospecies that have a size range of 10–100 nm possesses high surface-area-to-volume ratio, which facilitates their penetration through the plasma membrane and efficient absorption into a cell [11, 12]. These characteristic properties of nanospecies play a vital role to conquer the limitation associated with phytochemicals. Thus, several nano-formulations having various nanospecies with phytochemicals have been used for cancer therapy, which was found superior in bioavailability, drug retention time, and solubility in an aqueous environment along with minimum systemic toxicity [11–13]. Phytochemicals encapsulated in nanospecies defend against the negative impact of physiological conditions like pH fluctuations, biochemical breakdown, and enzymatic attack [6]. Nanoencapsulation of phytochemicals allows their target drug delivery at tumor site as well as protects their functionality by preserving their original structure [14–16]. The majority of phytochemicals encapsulated in nano-formulations for cancer treatment are still either in their early phases of research or preclinical/clinical pages.

In this chapter, various phytochemicals, nanospecies, and phytochemical-encapsulated nano-formulations and their recent development in cancer treatment have been discussed in brief.

8.2 NATURAL PLANT PRODUCTS WITH ANTICANCER PROPERTIES

A plant produces a wide range of organic chemicals, which include primary and secondary metabolites. Primary metabolites, such as protein, sugar, lipids, and nucleotides, perform essential functions for cell survival. However, secondary metabolites are the metabolic intermediates that are not necessary for the development and survival of the plants but generate the response against the stress condition and required for plant-environment interaction. Secondary metabolites perform various functions in plants and are evaluated as a source of insecticides, pigments, fragrances, flavoring agents, herbicides, UV protectants, antibiotics, and signaling molecules [17]. These secondary metabolites, which are also called as phytochemicals, are abundantly present in fruits and vegetables and taken up by humans in their diet, and they exhibit excellent antitumor properties, along with antioxidant, antimetastatic, antiproliferative, anti-inflammatory, and antiangiogenic activities [18]. Secondary metabolites, or phytochemicals, are placed into different classes based on their structure, including terpenes, phenolics, alkaloids, and glucosinolates [19, 20].

Terpenes are a wide group of natural products that are made up of isoprene units. However, terpenes are simply hydrocarbons, but terpenoids are oxygenated hydrocarbons. Terpenes have a general molecular formula of $(C_5H_8)_n$, where "n" is the number of connected isoprene units. Terpenes are hence also known as isoprenoid chemicals. The number of isoprene units in its structure determines their classification. Some terpenoids, such as many sterols, are primary metabolites. Some terpenoids, like gibberellins, brassinosteroids, and strigolactones, originated as secondary metabolites and work as plant hormones [21]. Artemisinin and paclitaxel are the most commonly known terpenoids that are well recognized for their antimalarial and antitumor properties, respectively. Paclitaxel is used as an active compound in chemotherapy for the treatment of various kinds of cancer (lung cancer, cervical cancer, and breast cancer) [22].

Further, chemical compounds that have an aromatic ring structure with one or more hydroxyl groups are called phenolics and polyphenolics. Based on their carbon skeleton, these compounds are classified into flavones, isoflavones, flavan-3-ols, anthocyanidins, flavonols, hydroxycinnamic acids, hydroxybenzoic acids, and stilbenes [6, 23]. Resveratrol is a phenolic compound that belongs to the class of flavonoids. It is extensively present in various fruits like grapes, peanuts, raspberries, and blueberries that decrease the risk of many diseases such as cancer and heart disease [21]. Furthermore, alkaloids, based on their structure and origin, are classified into three categories: true alkaloids, proto-alkaloids, and pseudo-alkaloids [24]. True alkaloids are produced from an amino acid and contain nitrogen atoms in their heterocyclic ring, whereas proto-alkaloids are deficient in nitrogen-containing heterocyclic rings. Pseudo-alkaloids are not produced from amino acids, but they contain terpene, purine, and steroid-like alkaloids. Based on the skeleton, alkaloids are distinguished into various classes, such as indole, tropane, pyrrole, piperidine, aporphine, indolizidine, pyrrolizidine, purine, norlupinane, pyrrolidine, imidazole, and steroids [24]. Vincristine and vinblastine are commonly used alkaloids for the treatment of various kinds of cancer [25]. Glucosinolates are the sulfur- and nitrogen-containing secondary metabolites generated from glucose, amino acids, and sulfate. Glucoraphanin is an example of a glucosinolate found in broccoli that reduces the risk of cancer [21]. Various phytochemicals and their anticancer effect are shown in Table 8.1.

8.3 CHALLENGES ASSOCIATED WITH NATURAL PRODUCTS TO USE AS CHEMOPREVENTIVE AGENTS

Although there are found numerous anticancer properties in phytochemicals, they also have many obstacles when used in the raw form [39, 40]. Natural polyphenolic compounds are associated with low solubility in an aqueous environment, reduced stability, and deprived oral bioavailability in the human body [41, 42]. Besides this, natural phytochemicals are excessively metabolized in the small intestine and liver and converted into their various conjugates/derivatives that deliver to the target site through blood [43]. As a result, the concentrations of these parent phytochemicals are substantially lower (in the nanomolar range) in human plasma than their pharmacologically active dosages that are utilized in preclinical trials. Thus, the amount of these phytochemicals is not sufficient to generate an active response against disease in the living human body [6, 44, 45]. Hence, for effective clinical application of these phytochemicals, there is no doubt that some novel strategies are required, which can protect these compounds from enzymatic reactions and efficiently deliver them to the target site.

Nowadays, nanotechnology is emerging as a new and novel approach to target these problems that are associated with the use of phytochemicals in their native form. In this approach, various nano-formulations are being prepared by encapsulating natural phytochemicals with various nanospecies, like nanoparticles, liposomes, dendrimers, nanomicelles, and carbon nanotubes, to protect the phytochemical from the enzymatic reaction. Further, due to their very small size (in the range 10–100 nm), the nanospecies (nanoparticles, liposomes, dendrimers, nanomicelles, and carbon nanotubes) have a high surface-area-to-volume ratio that enhances their reactivity at very high extent. Therefore, a low concentration of nano-formulations (having nanospecies with phytochemicals/semisynthetic compounds) is sufficient to treat the diseased tissue that causes the reduction in systemic toxicity inside the human body. Furthermore, a target delivery can be achieved very easily by attaching the nano-formulations with the ligand that successfully binds to the cell receptor of diseased tissue. Thus, by using ligand-receptor interaction, the target delivery of phytochemicals encapsulated in nano-formulations is possible. The properties of nano-formulations are shown in Figure 8.1.

8.4 VARIOUS NANO-FORMULATIONS FOR CHEMOPREVENTION

8.4.1 LIPOSOMES

Liposomes are lipid-based vesicles of size 10–1,000 nm that have an inner aqueous core and outer phospholipid bilayer. Due to their inner aqueous environment, hydrophilic drugs can be easily

TABLE 8.1

Showing Effect of Various Phytochemicals Along with the Therapeutic Agents on a Variety of Cancers (Data from [1] Under CC-BY License)

Plant Name	Phytochemical Class	Phytochemical Name	Therapeutic Agent	Cancer Type	Remarks	References
Curcuma longa	Polyphenol	Curcumin	Cisplatin	Bladder and ovarian cancer	Higher expression of P53 and P21 gene, trigger proapoptic p38 MAPK, caspase 3 activation	[26]
Curcuma longa	Polyphenol	Curcumin	Irinotecan	Colon cancer	Reactive oxygen species, activation of stress pathways	[27]
Onion	Flavonoid	Quercetin	Paclitaxel	Lung cancer	Cytotoxicity effect	[28]
Onion	Flavonoid	Quercetin	Paclitaxel	Breast cancer	Microtubule destruction, P-gp inhibition	[29]
Mulberries, grapes, peanuts	Polyphenol	Resveratrol	Sorafenib/ Cisplatin	Breast cancer	ROS-induced cell cycle inhibition, multikinase inhibition, activation caspase 3 and 9	[30]
Paris polyphylla	Saponin	Polyphyllin	Cisplatin	Stomach cancer	Suppression Akt, PP2A, and CIP2A pathways	[31]
Holy basil	Triterpene	Ursolic acid	Cisplatin	Liver cancer	Suppression ARE and Nrf2 pathways	[32]
Brucea javanica	Quassinoids	Dehydrobruceine B	Cisplatin	Lung cancer	Suppression of Nrf2, mitochondrial apoptosis	[27]
Green tea	Polyphenol	Epigallocatechin gallate	Cisplatin	Oral cancer	Suppression of STAT3 and Akt signaling pathways	[33]
Watercress	Sulfur-containing compound	Phenylethyl isothiocyanate	Cisplatin	Ovarian and biliary tract cancer	Reactive oxygen species, High H_2O_2 intracellular levels	[34]
Soybean	Isoflavone	Genistein (GEN) isolated isoflavone	–	Breast and skin cancer	A modulator of breast cancer, target p53, p21 checkpoint kinase 1 and 2, antioxidant activity	[35]
Salvia miltiorrhiza	Polyphenol	Salvianolic acid A	Cisplatin	Lung cancer	Suppression of protein kinase B	[36]
Alpinia galangal	Flavonoid	Galangin	Cisplatin	Lung cancer	Reduced level of NF-kB, regulation of STAT3 and	[37]
Mulberries, grapes, peanuts	Polyphenol	Resveratrol	Temozolomide	Skin cancer	ROS scavenger, NF-kB, p38 MAPK and ERK, COX-1 and COX-2 inhibitor	[38]
Catharanthus roseus	Alkaloid	Vincristine, vinblastine	–	Various types of cancer	Reduction in lymphocytic leukemia	[25]

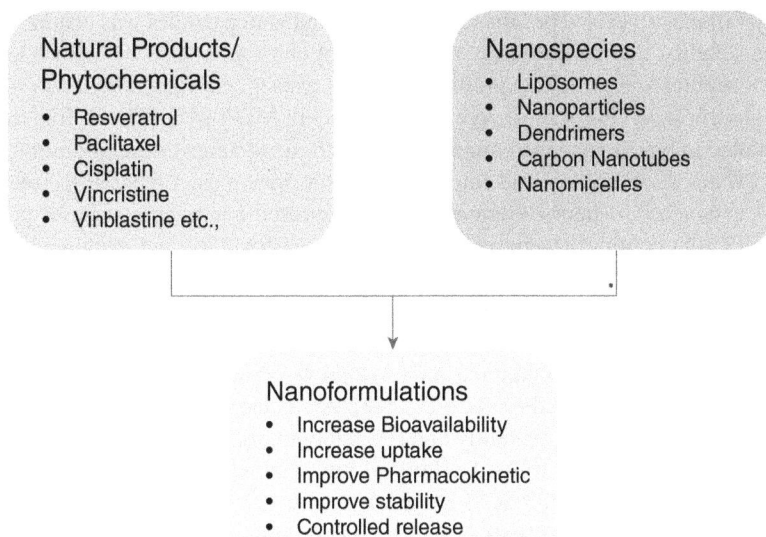

FIGURE 8.1 Properties of nano-formulations having phytochemicals and nanospecies.

incorporated in the inner core and hydrophobic drugs are safely integrated between the phospholipid bilayer [46, 47]. Liposomes are biocompatible, biodegradable, and non-immunogenic in nature. The pharmacokinetics of drugs is improved with liposomes owing to their biocompatibility in vivo [48]. However, the properties of liposomes that influence the fate of drugs inside the body can be optimized by changing the number and composition of lipid layers [48]. Besides this, the surface properties of synthesized liposomes for target drug delivery have been changed by ligating them with various types of compounds like polymers, lipids, and functional groups [1]. Shu et al. [49] synthesized liposomes laden with betulinic acid to cure myelogenous leukemia by targeting HepG2 cells. In order to improve transfection efficiency, the surface of synthesized liposomes with betulinic acid was modified with mannosylerythritol lipid-A (MEL-A), which improves the penetration of modified liposomes across the cell membrane and increases the cell apoptosis significantly [49]. Marqibo is an FDA-approved liposomal nano-formulation of vincristine [48] that entered the clinical phage. In preclinical studies, Marqibo showed an improved pharmacokinetics profile and was also found more active in cancer treatment than single vincristine in rats, mice, and dogs [50–52]. Similar to the preclinical studies, in clinical trials, Marqibo was also found more effective to improve the dysfunction of muscles and nerves than vincristine alone [53].

8.4.2 NANOPARTICLES

Nanoparticles are ultra-fine particles that have the size in the range of 10–100 nm. Nanoparticles can be synthesized in nanocapsules and nanospheres form [54]. Nanocapsule is a capsular structure of polymer with the hollow inner cavity in which phytochemicals/drugs can be encapsulated. However, the nanosphere is a polymer matrix structure in which drugs are distributed in the polymers [55, 56]. Polymers used for nanoparticle synthesis might be natural in origin, such as chitosan, gelatin, collagen, alginates, and cellulose, or synthetic, such as polyethylene glycol (PEG), polyethylene oxide (PEO), polycaprolactone (PCL), and polylactic acid (PLA) [1]. The surface of polymeric nanoparticles can be modified with PEG to improve the cellular uptake and escape the immune response against these nanoparticles [1]. In a previous study, PLA nanoparticles loaded with BioPerine were synthesized by Pillai et al. [57]. The surface of synthesized nanoparticles was coated with chitosan to enhance the stability, and furthermore, the surface of chitosan-coated nanoparticles was modified by PEG for suppressing protein expression in breast cancer cell lines. There was observed suppression in P-glycoprotein (P-gp) in the presence of PEG in MDA-MB 453 cell line that overcome the drug resistance in breast cancer treatment. The synthesized nano-formulation of CS-PEG-Bio-PLA exhibited water dispersibility and improved cellular uptake as compared to free BioPrine in MDA-MB 231 cells after 24 hours. Gera et al. [58] fabricated a nanocomposite of poly(lactic-co-glycolic acid) (PLGA) containing phytochemical extract (BRM270) and evaluated its anticancer properties against HepG2 human hepatoma cancer cells. Nanocomposites show higher toxic potential against HepG2 cells as compared to free BRM270. Nanocomposites with BRM270 extract suppress the expression of several proteins that are overexpressed through chemotherapy in HepG2 cells. Abraxane is a nano-formulation of nanoparticles loaded with paclitaxel and the only nanoparticle-based drug that have been approved by the FDA [48]. Abraxane has been synthesized with human serum albumin that is bound noncovalently and reversibly to paclitaxel [59]. Homogenization technology at high pressure was used to load paclitaxel into nanoparticles that synthesized Abraxane particles in a size range around 130 nm with approximately 10% paclitaxel loading [60]. Abraxane binds with gp60, a glycoprotein expresses over continuous endothelial cells that show a high affinity for albumin protein [61–63]. Further, gp60 exhibits a high affinity to SPARC (secreted protein, acidic and rich in cysteine), an extracellular glycoprotein that is highly expressed on the surface of cancer cells. This strategy enhanced Abraxane deposition in tumor [61–63]. In phase I of a clinical study, Abraxane was used in 19 metastatic breast cancer patients to study toxicity profile, pharmacokinetics properties, and maximum tolerated dose (MDR) that were found improved [48].

8.4.3 DENDRIMERS

A dendrimer is a nanosized polymeric spherical molecule that has a highly repetitive branched structure. It has an interior and exterior core; the exterior core has various functional surface groups [64]. Numerous types of polymers, such as poly(propylene imine) (PPI), PEG, melamine, polyamidoamine (PAMAM), poly(glycerol), poly(glycerol-co-succinic acid), poly(2,2-bis[hydroxymethyl] propionic acid), and triazine, are commonly used for dendrimers synthesis [65–69]. Due to their 3-architecture, low polydispersity, and excellent functionality, dendrimers have been used as drug carriers exceptionally [70]. During drug delivery, dendrimers shield the drugs from the biological environment and enhance the stability and durability of drugs inside the body [71, 72]. Multiple branches in dendrimers act as carriers and allow the accumulation of various biomolecules like proteins, amino acids, and numerous types of drugs [73, 74]. Further, the functional groups present on the surface of dendrimers are easily tunable, enabling the synthesis of dendrimers with amine, carboxyl, or hydroxyl groups. Thus, due to their three-dimensional architecture, monodisperse nature, and high functionality, dendrimers show reproducible, reliable pharmacokinetics, and superior uptake efficiency [75]. Therefore, dendrimers have been successfully explored in cancer therapy and delivery of anticancer drugs [76]. Conjugates of dendrimers and anticancer drugs have exhibited the capability to bypass efflux transporter, enhanced bioavailability of loaded drug molecules, and intracellular delivery of drugs [77]. The antiproliferative properties have been observed in cisplatin conjugates with dendrimers with low cytotoxicity [78]. For EpCAM-rich tumors, celastrol bioconjugates containing PEG, EpCAM aptamer, and polyamidoamine dendrimers have been synthesized by Ge et al. [79]. The cancerous cell line SW620 (colorectal cancer cells) was used in this study, and increased apoptosis was found on the SW620 cell line in vitro and in vivo (nude mice) conditions with high safety and great specificity [79]. Gao et al. [79] conjugated ursolic acid and folic acid inside the core of PAMAM for targeted drug delivery [79]. Folic acid binds to the folate receptor of HepG2 cells and improves cellular uptake. Due to electrostatic properties, PAMAM dendrimer pulls towards the HepG2 cells and enhances the cell toxicity effects of ursolic acid [80]. Thus, these studies indicated the target delivery of phytochemicals by dendrimer for cancer treatment.

8.4.4 CARBON NANOTUBES

Carbon nanotubes (CNTs) are carbon allotropes. They are cylindrical molecules and made of rolls of graphene sheet that have a high aspect ratio due to nano-range diameter (1–10 nm) but several micrometers in length. CNTs can be synthesized in single-walled carbon nanotubes (SWNTs) from a single graphene sheet and multiwalled carbon nanotubes (MWNTs) from multiple graphene sheets [81]. They have unique structural and functional properties, like high aspect ratio, high surface-area-to-volume ratio, size stability, and more significant surface functionality for a wide range of functional groups [82]. Owing to these reasons, various therapeutic agents can be easily accommodated on the surface of CNTs, which also help in target delivery [1]. Thus, various phytochemicals have been easily loaded in CNTs for several types of cancer treatment, which indicate it as a promising nanocarrier for target drug delivery [83]. Li et al. [83] synthesized SWNTs that were loaded with curcumin. The surface of CNTs was modified by phosphatidylcholine and polyvinylpyrrolidone, and the stability of CNTs was enhanced by lipid and polymeric coating. The synthesized nano-formulation showed an 18-fold increase in blood levels with enhanced antitumor properties in mice. Moreover, these CNTs also showed higher uptake by human prostate cancer cells. However, the lipid coating and polymer conjugation in the curcumin-loaded CNTs improved the drug delivery. Zhang et al. [84] studied the synergetic effect of CNTs and paclitaxel in human ovarian cancer and observed that cell death increased in human ovarian cancer OVCAR3 cells [84]. In another study, paclitaxel-loaded CNTs exhibited high activity against lung cancer [85]. Yang et al. [86] administrated magnetic MWNTs loaded with anticancer drug gemcitabine in mice subcutaneously and found high activity against the metastases of the lymph nodes. Sahoo et al. [87] functionalized MWNTs with polyvinyl alcohol and loaded with the anticancer camptothecin, which is weakly water-soluble and reported as a potential candidate for breast and skin cancer treatment [87].

8.4.5 NANOMICELLES

Nanomicelles are the self-assembled spherical structure of size 5–100 nm. They are commonly made of lipid and amphiphilic polymers [88, 89]. However, lipid-based nanomicelles are less stable than polymer-based nanomicelles. Therefore, a polymer-based nanomicelle system has been explored extensively for chemopreventive drug delivery application [90]. Polymer-based nanomicelles are composed of an interior hydrophobic core and exterior hydrophilic shell [43, 91]. They are biocompatible and easy to prepare and show high drug encapsulation efficiency and drug delivery to the target tissues [92, 93]. In polymer-based nanomicelles, water-insoluble drugs can be encapsulated in hydrophobic region, but water-soluble drugs are effectively loaded in the hydrophilic region [94, 95]. Thus, in these nanomicelles, both hydrophobic and hydrophilic drugs can be loaded. To treat oral cancer, Wang et al. [96] synthesized self-assembled nanomicelles of rebaudioside A loaded with honokiol and investigated its anticancer potential against HuH-7 cells. The synthesized nanomicelles inhibited the production of active ROS by cell cycle arrest and cell death. Additionally, rebaudioside A and honokiol also activated the extracellular-signal-regulated kinase signaling pathway by suppressing the DNA damage mechanism [96]. The most widely used hydrophilic element of polymeric micelles is PEG, which has been authorized for humans use by the US Food and Drug Administration (FDA) [97–99]. As it is biocompatible and biodegradable, does not produce any harmful metabolites, and is commercially available in molecular weights between 500 and 20,000 Da, it has been extensively used for modification of nano-formulations. PEG-based nanomicelles through surface modification can extend the duration of systemic circulation by reducing association with plasma proteins and nonspecific uptake by the reticuloendothelial system [100]. Due to PEG versatility, different drugs and targeting compounds can be attached to the same polymer chain. Presently, various PEG-derived polymeric micelles containing anticancer drugs like SP1049C, NK012, NK911, NK105, and Genexol-PM are under clinical trials [101, 102]. However, Genexol-PM has (paclitaxel based nano-formulation) received regulatory approval from South Korean authorities to treat breast cancer. Genexol-PM has been made by encapsulating paclitaxel in a diblock mono-methoxy poly(ethylene glycol)-block-poly(D, L -lactide) (mPEG-PDLLA) polymer by utilizing the solid dispersion process. The diameter of micelles was around 24 nm and with 16% w/w paclitaxel loading capability [103, 104]. Genexol-PM showed higher antitumor properties and enhanced maximum tolerated dose (MTD) than Taxol (commercial product of paclitaxel) in preclinical investigations [103]. In phase I clinical trials, 2.3-fold greater MTD was recorded in Genexol-PM than Taxol in humans [105]. Additionally, in further clinical investigations, Genexol-PM demonstrated higher efficacy, improved accessibility, and faster clearance than Taxol [106–108].

Thus, the use of phytochemicals with nano-formulations were found highly effective for cancer treatment and targeted drug delivery.

8.5 CONCLUSIONS AND FUTURE PROSPECTS

Phytochemicals present in fruits and vegetables have fabulous health benefits. They help in growth and development, maintain metabolic activity, and avoid the various diseased conditions. They have various properties, like antitumor, antiproliferative, antioxidant, anti-inflammatory, and anti-metastatic activity. In spite of their tremendous health benefits, phytochemicals cannot be used as a chemotherapeutic agent in raw form due to low solubility in an aqueous environment, low specificity, deprived bioavailability, and multidrug resistance. Therefore, to overcome the limitation of phytochemicals and take the advantages of their anticancer properties, various nano-formulations having nanospecies like nanoparticles, nanomicelles, dendrimers, liposomes, and phytochemicals have been synthesized, which showed the promising anticancer activity and target drug delivery. However, these nano-formulations showed the promising results, but their commercial prospects or clinical use are still limited due to some short of toxicity in human subjects; besides this, they are not fully validated till now. Further, owing to their small size, nano-formulations can easily cross the cell barrier, but there might a chance of their accumulation inside the cells and consequent their clearance from the body is also limited. Moreover, the oral route of administration is most accepted and popular but associated with enzymatic degradation, acidic stomach environment, and first-pass effect, which also need to be address in further research. Hence, by resolving these problems associated with the use of nano-formulations as an anticancer agent, it might be a wonderful approach in near future for cancer treatment.

8.6 REFERENCES

1. More, Mahesh P., Sagar R. Pardeshi, Chandrakantsing V. Pardeshi, Gaurav A. Sonawane, Mahesh N. Shinde, Prashant K. Deshmukh, Jitendra B. Naik, and Abhijeet D. Kulkarni. 2021. "Recent Advances in Phytochemical-Based Nano-Formulation for Drug-Resistant Cancer." *Medicine in Drug Discovery* 10: 100082. https://doi.org/10.1016/j.medidd.2021.100082.
2. Sung, Hyuna, Jacques Ferlay, Rebecca L. Siegel, Mathieu Laversanne, Isabelle Soerjomataram, Ahmedin Jemal, and Freddie Bray. 2021. "Global cancer statistics 2020: GLOBOCAN estimates of incidence and mortality worldwide for 36 cancers in 185 countries." *CA: A Cancer Journal for Clinicians* 71 (3): 209–249. https://doi.org/10.3322/caac.21660.
3. Wong, Ho Lun, Reina Bendayan, Andrew M. Rauth, Hui Yi Xue, Karlo Babakhanian, and Xiao Yu Wu. 2006. "A mechanistic study of enhanced doxorubicin uptake and retention in multidrug resistant breast cancer cells using a polymer-lipid hybrid nanoparticle system." *Journal of Pharmacology and Experimental Therapeutics* 317 (3): 1372–1381. https://doi.org/10.1124/jpet.106.101154.
4. Gottesman, Michael M., Tito Fojo, and Susan E. Bates. 2002. "Multidrug Resistance in Cancer: Role of ATP–Dependent Transporters." *Nature Reviews Cancer* 2 (1): 48–58. https://doi.org/10.1038/nrc706.
5. Szakács, Gergely, Jill K. Paterson, Joseph A. Ludwig, Catherine Booth-Genthe, and Michael M. Gottesman. 2006. "Targeting multidrug resistance in cancer." *Nature Reviews Drug Discovery* 5 (3): 219–234. https://doi.org/10.1038/nrd1984.
6. Kashyap, Dharambir, Hardeep Singh Tuli, Mukerrem Betul Yerer, Ajay Sharma, Katrin Sak, Saumya Srivastava, Anjana Pandey, Vivek Kumar Garg, Gautam Sethi, and Anupam Bishayee. 2021. "Natural

product-based nanoformulations for cancer therapy: Opportunities and challenges." *Seminars in Cancer Biology* 69: 5–23. https://doi.org/10.1016/j.semcancer.2019.08.014

7. Cragg, Gordon M., and John M. Pezzuto. 2016. "Natural products as a vital source for the discovery of cancer chemotherapeutic and chemopreventive agents." *Medical Principles and Practice* 25 (Suppl. 2): 41–59. https://doi.org/10.1159/000443404.

8. Amawi, Haneen, Charles R. Ashby, and Amit K. Tiwari. 2017. "Cancer chemoprevention through dietary flavonoids: What's limiting?" *Chinese Journal of Cancer* 36 (1). https://doi.org/10.1186/s40880-017-0217-4.

9. Kumar, Vikas, Prakash Bhatt, Mahfoozur Rahman, Gaurav Kaithwas, Hani Choudhry, Fahad Al-Abbasi, Firoz Anwar, and Amita Verma. 2017. "Fabrication, optimization, and characterization of umbelliferone β-D-galactopyranoside-loaded PLGA nanoparticles in treatment of hepatocellular carcinoma: In vitro and in vivo studies." *International Journal of Nanomedicine* 12: 6747–6758. https://doi.org/10.2147/ijn.s136629

10. Seca, Ana, and Diana Pinto. 2018. "Plant secondary metabolites as anticancer agents: Successes in clinical trials and therapeutic application." *International Journal of Molecular Sciences* 19 (1): 263. https://doi.org/10.3390/ijms19010263.

11. Ahmed, Ejaz, Muhammad Arshad, Muhammad Zakriyya Khan, Muhammad Shoaib Amjad, Huma Mehreen Sadaf, Iqra Riaz, Sidra Sabir, Nabila Ahmad and Sabaoon. 2017. "Secondary metabolites and their multidimensional prospective in plant life." *Journal of Pharmacognosy and Phytochemistry* 6: 205–214.

12. Rahman, Mahfoozur, Mohammad Zaki Ahmad, Imran Kazmi, Sohail Akhter, Yogesh Kumar, Farhan Jalees Ahmad, and Firoz Anwar. 2012. "Novel approach for the treatment of cancer: Theranostic nanomedicine." *Pharmacologia* 3 (9): 371–376. https://doi.org/10.5567/pharmacologia.2012.371.376.

13. Pandey, Preeti, Mahfoozur Rahman, Prakash Chandra Bhatt, Sarwar Beg, Basudev Paul, Abdul Hafeez, Fahad A Al-Abbasi, et al. 2018. "Implication of nano-antioxidant therapy for treatment of hepatocellular carcinoma using PLGA nanoparticles of rutin." *Nanomedicine* 13 (8): 849–870. https://doi.org/10.2217/nnm-2017-0306.

14. Shin, Seong-Ah, Sun Young Moon, Woe-Yeon Kim, Seung-Mann Paek, Hyun Ho Park, and Chang Sup Lee. 2018. "Structure-based classification and anti-cancer effects of plant metabolites." *International Journal of Molecular Sciences* 19 (9): 2651. https://doi.org/10.3390/ijms19092651.

15. Sharma, Ajay, Pooja Sharma, Hardeep Singh Tuli, and Anil K Sharma. 2018. "Phytochemical and pharmacological properties of flavonols." *eLS* 2018, 1–12. https://doi.org/10.1002/9780470015902.a0027666.

16. Aneja, Preeti, Mahfoozur Rahman, Sarwar Beg, ShivaliAneja, Vishal Dhingra, and Rupali Chugh. 2014. "Cancer targeted magic bullets for effective treatment of cancer." *Recent Patents on Anti-Infective Drug Discovery* 9 (2): 121–135. https://doi.org/10.2174/1574891x10666150415120506.

17. Sharma, Ajay, Dharambir Kashyap, Katrin Sak, Hardeep Singh Tuli, and Anil K. Sharma. 2018. "Therapeutic charm of quercetin and its derivatives: A review of research and patents." *Pharmaceutical Patent Analyst* 7 (1): 15–32. https://doi.org/10.4155/ppa-2017-0030.

18. Sak, Katrin. 2014. "Cytotoxicity of dietary flavonoids on different human cancer types." *Pharmacognosy Reviews* 8 (16): 122. https://doi.org/10.4103/0973-7847.134247.

19. Kashyap, Dharambir, Ajay Sharma, Hardeep S. Tuli, Sandeep Punia, and Anil K. Sharma. 2016. "Ursolic acid and oleanolic acid: Pentacyclic terpenoids with promising anti-inflammatory activities." *Recent Patents on Inflammation & Allergy Drug Discovery* 10 (1): 21–33. https://doi.org/10.2174/1872213x10666160711143904.

20. Kashyap, Dharambir, Ajay Sharma, Hardeep Singh Tuli, Katrin Sak, Sandeep Punia, and Tapan K. Mukherjee. 2017. "Kaempferol—a dietary anticancer molecule with multiple mechanisms of action: Recent trends and advancements." *Journal of Functional Foods* 30: 203–219. https://doi.org/10.1016/j.jff.2017.01.022.

21. https://en.wikipedia.org/wiki/Secondary_metabolite

22. https://dtp.cancer.gov/timeline/flash/success_stories/s2_taxol.htm

23. Yadav, Priya, Vishal Jaswal, Ajay Sharma, Dharambir Kashyap, Hardeep S. Tuli, Vivek K. Garg, Shonkor K. Das, and R. Srinivas. 2018. "Celastrol as a pentacyclic triterpenoid with chemopreventive properties." *Pharmaceutical Patent Analyst* 7 (4): 155–167. https://doi.org/10.4155/ppa-2017-0035.

24. Gupta, Ajai Prakash, Pankaj Pandotra, Manoj Kushwaha, Saima Khan, Rajni Sharma, and Suphla Gupta. 2015. "Alkaloids: A source of anticancer agents from nature." *Studies in Natural Products Chemistry* 46: 341–445. https://doi.org/10.1016/b978-0-444-63462-7.00009-9.

25. Cragg, Gordon M., and David J. Newman. 2005. "Plants as a source of anti-cancer agents." *Journal of Ethnopharmacology* 100 (1–2): 72–79. https://doi.org/10.1016/j.jep.2005.05.011.

26. Park, Bong Hee, Joung Eun Lim, Hwang Gyun Jeon, Seong Il Seo, Hyun Moo Lee, Han Yong Choi, Seong Soo Jeon, and Byong Chang Jeong. 2016. "Curcumin potentiates antitumor activity of cisplatin in bladder cancer cell lines via ROS-mediated activation of ERK1/2." *Oncotarget* 7 (39): 63870–63886. https://doi.org/10.18632/oncotarget.11563.

27. Huang, Yan-Feng, Da-Jian Zhu, Xiao-Wu Chen, Qi-Kang Chen, Zhen-Tao Luo, Chang-Chun Liu, Guo-Xin Wang, Wei-Jie Zhang, and Nv-Zhu Liao. 2017. "Curcumin enhances the effects of irinotecan on colorectal cancer cells through the generation of reactive oxygen species and activation of the endoplasmic reticulum stress pathway." *Oncotarget* 8 (25): 40264–40275. https://doi.org/10.18632/oncotarget.16828

28. Wang, Yonghong, Hongli Yu, Saisai Wang, Chengcheng Gai, Xiaoming Cui, Zhilu Xu, Wentong Li, and Weifen Zhang. 2021. "Targeted delivery of quercetin by nanoparticles based on chitosan sensitizing paclitaxel-resistant lung cancer cells to paclitaxel." *Materials Science and Engineering: C* 119: 111442. https://doi.org/10.1016/j.msec.2020.111442.

29. Liu, Mengyao, Manfei Fu, Xiaoye Yang, Guoyong Jia, Xiaoqun Shi, Jianbo Ji, Xianghong Liu, and Guangxi Zhai. 2020. "Paclitaxel and quercetin co-loaded functional mesoporous silica nanoparticles overcoming multidrug resistance in breast cancer." *Colloids and Surfaces B: Biointerfaces* 196: 111284. https://doi.org/10.1016/j.colsurfb.2020.111284.

30. Lee, Yoon-Jin, Gina J. Lee, Sun Shin Yi, Su-HakHeo, Cho-Rong Park, Hae-Seon Nam, Moon-Kyun Cho, and Sang-Han Lee. 2016. "Cisplatin and resveratrol induce apoptosis and autophagy following oxidative stress in malignant mesothelioma cells." *Food and Chemical Toxicology: An International Journal Published for the British Industrial Biological Research Association* 97: 96–107. https://doi.org/10.1016/j.fct.2016.08.033.

31. Zhang, Yunfei, Ping Huang, Xuewen Liu, Yuchen Xiang, Te Zhang, Yezi Wu, Jiaxin Xu, et al. 2018. "Polyphyllin I inhibits growth and invasion of cisplatin-resistant gastric cancer cells by partially inhibiting CIP2A/PP2A/Akt signaling axis." *Journal of Pharmacological Sciences* 137 (3): 305–312. https://doi.org/10.1016/j.jphs.2018.07.008.

32. Wu, Shouhai, Tianpeng Zhang, and Jingsheng Du. 2016 "Ursolic acid sensitizes cisplatin-resistant HepG2/DDP cells to cisplatin via inhibiting Nrf2/ARE pathway." *Drug Design, Development and Therapy* 10: 3471.

33. Yuan, Chien-Han, Chi-Ting Horng, Chiu-Fang Lee, Ni-Na Chiang, Fuu-Jen Tsai, Chi-Cheng Lu, Jo-Hua Chiang, Yuan-Man Hsu, Jai-Sing Yang, and Fu-An Chen. 2016. "Epigallocatechin gallate sensitizes cisplatin-resistant oral cancer CAR cell apoptosis and autophagy through stimulating AKT/STAT3 pathway and suppressing multidrug resistance 1 signaling." *Environmental Toxicology* 32 (3): 845–855. https://doi.org/10.1002/tox.22284.

34. Li, Qiwei, Ming Zhan, Wei Chen, Benpeng Zhao, Kai Yang, Jie Yang, Jing Yi, et al. 2016. "Phenylethyl isothiocyanate reverses cisplatin resistance in biliary tract cancer cells via glutathionylation-dependent degradation of Mcl-1." *Oncotarget* 7 (9): 10271–10282. https://doi.org/10.18632/oncotarget.7171.

35. Iqbal, Javed, BanzeerAhsan Abbasi, AliTalha Khalil, Barkat Ali, Tariq Mahmood, Sobia Kanwal, SayedAfzal Shah, and Wajid Ali. 2018. "Dietary isoflavones, the modulator of breast carcinogenesis: Current landscape and future perspectives." *Asian Pacific Journal of Tropical Medicine* 11 (3): 186. https://doi.org/10.4103/1995-7645.228432.

36. Tang, Xia-li, Li Yan, Ling Zhu, De-min Jiao, Jun Chen, and Qing-yong Chen. 2017. "Salvianolic acid a reverses cisplatin resistance in lung cancer A549 cells by targeting C-met and attenuating Akt/MTOR pathway." *Journal of Pharmacological Sciences* 135 (1): 1–7. https://doi.org/10.1016/j.jphs.2017.06.006.

37. Yu, Shuo, Lian-sheng Gong, Nian-feng Li, Yi-feng Pan, and Lun Zhang. 2018. "Galangin (GG) combined with cisplatin (DDP) to suppress human lung cancer by inhibition of STAT3-regulated NF-KB and Bcl-2/BaxSignaling pathways." *Biomedicine & Pharmacotherapy* 97: 213–224. https://doi.org/10.1016/j.biopha.2017.10.059.

38. Guthrie, Ariane R., H-H. Sherry Chow, and Jessica A. Martinez. 2017. "Effects of resveratrol on drug- and carcinogen-metabolizing enzymes, implications for cancer prevention." *Pharmacology Research & Perspectives* 5 (1): e00294. https://doi.org/10.1002/prp2.294

39. Leonarduzzi, G., G. Testa, B. Sottero, P. Gamba, and G. Poli. 2010. "Design and development of nanovehicle-based delivery systems for preventive or therapeutic supplementation with flavonoids." *Current Medicinal Chemistry* 17 (1): 74–95. https://doi.org/10.2174/092986710789957760.

40. Jeetah, Roubeena, Archana Bhaw-Luximon, and DhanjayJhurry. 2014. "Nanopharmaceutics: Phytochemical-based controlled or sustained drug-delivery systems for cancer treatment." *Journal of Biomedical Nanotechnology* 10 (9): 1810–1840. https://doi.org/10.1166/jbn.2014.1884.

41. Gohulkumar, M., K. Gurushankar, N. Rajendra Prasad, and N. Krishnakumar. 2014. "Enhanced cytotoxicity and apoptosis-induced anticancer effect of silibinin-loaded nanoparticles in oral carcinoma (KB) cells." *Materials Science and Engineering: C* 41: 274–282. https://doi.org/10.1016/j.msec.2014.04.056.

42. Khan, N., D. J. Bharali, V. M. Adhami, I. A. Siddiqui, H. Cui, S. M. Shabana, S. A. Mousa, and H. Mukhtar. 2013. "Oral administration of naturally occurring chitosan-based nanoformulated green tea polyphenol EGCG effectively inhibits prostate cancer cell growth in a xenograft model." *Carcinogenesis* 35 (2): 415–423. https://doi.org/10.1093/carcin/bgt321.

43. Zhang, Xiaolan, Yixian Huang, and Song Li. 2014. "Nanomicellar carriers for targeted delivery of anticancer agents." *Therapeutic Delivery* 5 (1): 53–68. https://doi.org/10.4155/tde.13.135

44. Park, Eun-Jung, and John M. Pezzuto. 2012. "Flavonoids in cancer prevention." *Anti-Cancer Agents in Medicinal Chemistry* 12 (8): 836–851. https://doi.org/10.2174/187152012802650075.

45. Rodriguez-Mateos, Ana, David Vauzour, Christian G. Krueger, DhanansayanShanmuganayagam, Jess Reed, Luca Calani, Pedro Mena, Daniele Del Rio, and Alan Crozier. 2014. "Bioavailability, bioactivity and impact on health of dietary flavonoids and related compounds: An update." *Archives of Toxicology* 88 (10): 1803–1853. https://doi.org/10.1007/s00204-014-1330-7.

46. Hofheinz, Ralf-Dieter, Senta Ulrike Gnad-Vogt, Ulrich Beyer, and Andreas Hochhaus. 2005. "Liposomal encapsulated anti-cancer drugs." *Anti-Cancer Drugs* 16 (7): 691–707. https://doi.org/10.1097/01.cad.0000167902.53039.5a.

47. Tattersall, Martin, and Stephen Clarke. 2003. "Developments in drug delivery." *Current Opinion in Oncology* 15 (4): 293–299. https://doi.org/10.1097/00001622-200307000-00003.

48. Xie, Jing, Zhaogang Yang, Chenguang Zhou, Jing Zhu, Robert J. Lee, and Lesheng Teng. 2016. "Nanotechnology for the delivery of phytochemicals in cancer therapy." *Biotechnology Advances* 34 (4): 343–353. https://doi.org/10.1016/j.biotechadv.2016.04.002.

49. Shu, Qin, Jianan Wu, and Qihe Chen. 2019. "Synthesis, characterization of liposomes modified with biosurfactant MEL-A loading betulinic acid and its anticancer effect in HepG2 cell." *Molecules* 24 (21): 3939. https://doi.org/10.3390/molecules24213939.

50. Castle, M.C., and J.A.R. Mead. 1978. "Investigations of the metabolic fate of tritiated vincristine in the rat by high-pressure liquid chromatography." *Biochemical Pharmacology* 27 (1): 37–44. https://doi.org/10.1016/0006-2952(78)90254-x.

51. Krishna, Rajesh, Murray S. Webb, Ginette St. Onge, and Lawrence D. Mayer. 2001 "Liposomal and nonliposomal drug pharmacokinetics after administration of liposome-encapsulated vincristine and their contribution to drug tissue distribution properties." *Journal of Pharmacology and Experimental Therapeutics* 298(3): 1206–1212.

52. Webb, Murray S., Patricia Logan, Peter M. Kanter, Ginette St.-Onge, Karen Gelmon, Troy Harasym, Lawrence D. Mayer, and Marcel B. Bally. 1998. "Preclinical pharmacology, toxicology and efficacy of sphingomyelin/cholesterol liposomal vincristine for therapeutic treatment of cancer." *Cancer Chemotherapy and Pharmacology* 42 (6): 461–470. https://doi.org/10.1007/s002800050846.

53. Hagemeister, Fredrick, Maria Alma Rodriguez, Steven R. Deitcher, Anas Younes, Luis Fayad, Andre Goy, Nam H. Dang, et al. 2013. "Long term results of a phase 2 study of vincristine sulfate liposome injection (Marqibo®) substituted for non-liposomal vincristine in cyclophosphamide, doxorubicin, vincristine, prednisone with or without rituximab for patients with untreated aggressive." *British Journal of Haematology* 162 (5): 631–638. https://doi.org/10.1111/bjh.12446.

54. Ige, Pradum, Sagar Pardeshi, and Raju Sonawane. 2018. "Development of PH-dependent nanospheres for nebulisation- in vitro diffusion, aerodynamic and cytotoxicity studies." *Drug Research* 68 (12): 680–686. https://doi.org/10.1055/a-0595-7678.

55. Navya, P.N., Anubhav Kaphle, S.P. Srinivas, Suresh Kumar Bhargava, Vincent M. Rotello, and Hemant Kumar Daima. 2019. "Current Trends and Challenges in Cancer Management and Therapy Using Designer Nanomaterials." *Nano Convergence* 6 (1): 23. https://doi.org/10.1186/s40580-019-0193-2.

56. Naik, Jitendra B., Sagar R. Pardeshi, Rahul P. Patil, Pritam B. Patil, and Arun Mujumdar. 2020. "Mucoadhesive micro-/nano carriers in ophthalmic drug delivery: An overview." *BioNanoScience* 10 (3): 564–582. https://doi.org/10.1007/s12668-020-00752-y.

57. Pillai, Sindhu C, Ankita Borah, Amandeep Jindal, Eden Mariam Jacob, Yohei Yamamoto, and D. Sakthi Kumar. 2020. "BioPerine encapsulated nanoformulation for overcoming drug-resistant breast cancers." *Asian Journal of Pharmaceutical Sciences* 15(6): 701–712. https://doi.org/10.1016/j.ajps.2020.04.001.

58. Gera, Meeta, Nameun Kim, Mrinmoy Ghosh, Neelesh Sharma, Do Luong Huynh, NisansalaChandimali, Hyebin Koh, et al. 2019. "Synthesis and evaluation of the antiproliferative efficacy of BRM270

phytocomposite nanoparticles against human hepatoma cancer cell lines." *Materials Science and Engineering: C* 97: 166–176. https://doi.org/10.1016/j.msec.2018.11.055.

59. Yardley, Denise A. 2013. "Nab-paclitaxel mechanisms of action and delivery." *Journal of Controlled Release* 170 (3): 365–372. https://doi.org/10.1016/j.jconrel.2013.05.041.
60. Stinchcombe, Thomas E. 2007. "Nanoparticle albumin-bound paclitaxel: A novel Cremphor-EL®-free formulation of paclitaxel." *Nanomedicine* 2 (4): 415–423. https://doi.org/10.2217/17435889.2.4.415.
61. Elsadek, Bakheet, and Felix Kratz. 2012. "Impact of albumin on drug delivery—new applications on the horizon." *Journal of Controlled Release* 157 (1): 4–28. https://doi.org/10.1016/j.jconrel.2011.09.069.
62. Petrelli, Fausto, Karen Borgonovo, and Sandro Barni. 2010. "Targeted delivery for breast cancer therapy: The history of nanoparticle-albumin-bound paclitaxel." *Expert Opinion on Pharmacotherapy* 11 (8): 1413–1432. https://doi.org/10.1517/14656561003796562.
63. Reddy, L. Harivardhan, and Didier Bazile. 2014. "Drug delivery design for intravenous route with integrated physicochemistry, pharmacokinetics and pharmacodynamics: Illustration with the case of taxane therapeutics." *Advanced Drug Delivery Reviews* 71: 34–57. https://doi.org/10.1016/j.addr.2013.10.007.
64. Jain, Keerti, Ashwni Kumar Verma, Prabhat Ranjan Mishra, and Narendra Kumar Jain. 2015. "Characterization and evaluation of amphotericin B loaded MDP conjugated poly(propylene imine) dendrimers." *Nanomedicine: Nanotechnology, Biology and Medicine* 11 (3): 705–713. https://doi.org/10.1016/j.nano.2014.11.008.
65. Franiak-Pietryga, Ida, Barbara Ziemba, Bradley Messmer, and Dorota Skowronska-Krawczyk. 2018. "Dendrimers as drug nanocarriers: The future of gene therapy and targeted therapies in cancer." *Dendrimers: Fundamentals and Applications* (25): 7–27. https://doi.org/10.5772/intechopen.75774
66. Bhadra, D., S. Bhadra, S. Jain, and N.K. Jain. 2003. "A PEGylated dendritic nanoparticulate carrier of fluorouracil." *International Journal of Pharmaceutics* 257 (1–2): 111–124. https://doi.org/10.1016/s0378-5173(03)00132-7.
67. He, Hai, Yan Li, Xin-Ru Jia, Ju Du, Xue Ying, Wan-Liang Lu, Jin-Ning Lou, and Yan Wei. 2011. "PEGylated poly(amidoamine) dendrimer-based dual-targeting carrier for treating brain tumors." *Biomaterials* 32 (2): 478–487. https://doi.org/10.1016/j.biomaterials.2010.09.002.
68. Crampton, Hannah L, and Eric E Simanek. 2007. "Dendrimers as drug delivery vehicles: Non-covalent interactions of bioactive compounds with dendrimers." *Polymer International* 56 (4): 489–496. https://doi.org/10.1002/pi.2230.
69. Tomalia, D.A., L.A. Reyna, and S. Svenson. 2007. "Dendrimers as multi-purpose nanodevices for oncology drug delivery and diagnostic imaging." *Biochemical Society Transactions* 35 (1): 61–67. https://doi.org/10.1042/bst0350061.
70. Jain, Keerti, and Narendra Kumar Jain. 2014. "Surface engineered dendrimers as antiangiogenic agent and carrier for anticancer drug: Dual attack on cancer." *Journal of Nanoscience and Nanotechnology* 14 (7): 5075–5087. https://doi.org/10.1166/jnn.2014.8677.
71. Jain, Anupriya, Keerti Jain, Prashant Kesharwani, and Narendra K. Jain. 2013a. "Low density lipoproteins mediated nanoplatforms for cancer targeting." *Journal of Nanoparticle Research* 15 (1888): 1–38. https://doi.org/10.1007/s11051-013-1888-7
72. Jain, Anupriya, Keerti Jain, Neelesh Kumar Mehra, and N. K. Jain. 2013b. "Lipoproteins tethered dendrimeric nanoconstructs for effective targeting to cancer cells." *Journal of Nanoparticle Research* 15 (10): 1–18. https://doi.org/10.1007/s11051-013-2003-9
73. Sharma, Ashok Kumar, AvinashGothwal, Prashant Kesharwani, Hashem Alsaab, Arun K. Iyer, and Umesh Gupta. 2017. "Dendrimer nanoarchitectures for cancer diagnosis and anticancer drug delivery." *Drug Discovery Today* 22 (2): 314–326. https://doi.org/10.1016/j.drudis.2016.09.013.
74. Xiong, Zhijuan, Mingwu Shen, and Xiangyang Shi. 2018. "Dendrimer-based strategies for cancer therapy: Recent advances and future perspectives." *Science China Materials* 61 (11): 1387–1403. https://doi.org/10.1007/s40843-018-9271-4.
75. Singh, Jaspreet, Keerti Jain, Neelesh Kumar Mehra, and N.K. Jain. 2016. "Dendrimers in Anticancer drug delivery: Mechanism of interaction of drug and dendrimers." *Artificial Cells, Nanomedicine, and Biotechnology* 44 (7): 1626–1634. https://doi.org/10.3109/21691401.2015.1129625
76. Li, Jing, Fei Yu, Yi Chen, and David Oupický. 2015. "Polymeric drugs: Advances in the development of pharmacologically active polymers." *Journal of Controlled Release* 219: 369–382. https://doi.org/10.1016/j.jconrel.2015.09.043
77. Uram, Łukasz, Magdalena Szuster, Aleksandra Filipowicz, Krzysztof Gargasz, StanisławWołowiec, and ElżbietaWałajtys-Rode. 2015. "Different patterns of nuclear and mitochondrial penetration by the G3 PAMAM dendrimer and its biotin–pyridoxal bioconjugate BC-PAMAM in normal and cancer cells in vitro." *International Journal of Nanomedicine* 10: 5647–5661. https://doi.org/10.2147/ijn.s87307

78. Nguyen, Hoang, Ngoc Hoa Nguyen, Ngoc Quyen Tran, and Cuu Khoa Nguyen. 2015. "Improved method for preparing cisplatin-dendrimer nanocomplex and its behavior against NCI-H460 lung cancer cell." *Journal of Nanoscience and Nanotechnology* 15 (6): 4106–4110. https://doi.org/10.1166/jnn.2015.9808.

79. Ge, Pengjin, Boning Niu, Yuehuang Wu, Weixia Xu, Mingyu Li, Huisong Sun, Hu Zhou, Xiaokun Zhang, and Jingjing Xie. 2020. "Enhanced cancer therapy of celastrol in vitro and in vivo by smart dendrimers delivery with specificity and biosafety." *Chemical Engineering Journal* 383: 123228. https://doi.org/10.1016/j.cej.2019.123228

80. Gao, Yu, Zhihong Li, XiaodongXie, Chaoqun Wang, Jiali You, Fan Mo, BiyuJin, et al. 2015. "Dendrimeric anticancer prodrugs for targeted delivery of ursolic acid to folate receptor-expressing cancer cells: Synthesis and biological evaluation." *European Journal of Pharmaceutical Sciences* 70: 55–63. https://doi.org/10.1016/j.ejps.2015.01.007.

81. Dinarvand, Rassoul, Paulo Cesar de Morais, and Antony D'Emanuele. 2012. "Nanoparticles for targeted delivery of active agents against tumor cells." *Journal of Drug Delivery* Editorial: 1–2. https://doi.org/10.1155/2012/528123.

82. Son, Kuk Hui, JeongHee Hong, and Jin Woo Lee. 2016. "Carbon nanotubes as cancer therapeutic carriers and mediators." *International Journal of Nanomedicine* 11: 5163–5185. https://doi.org/10.2147/IJN.S112660.

83. Li, Haixia, Nan Zhang, Yongwei Hao, Yali Wang, Shasha Jia, and Hongling Zhang. 2019. "Enhancement of curcumin antitumor efficacy and further photothermal ablation of tumor growth by single-walled carbon nanotubes delivery system in vivo." *Drug Delivery* 26 (1): 1017–1026. https://doi.org/10.1080/10717544.2019.1672829.

84. Zhang, Wenjing, Daoqiang Zhang, Jianhua Tan, and Haibo Cong. 2012. "Carbon nanotube exposure sensitize human ovarian cancer cells to paclitaxel." *Journal of Nanoscience and Nanotechnology* 12 (9): 7211–7214. https://doi.org/10.1166/jnn.2012.6506.

85. Arya, Neha, Aditya Arora, K.S. Vasu, A.K. Sood, and Dhirendra S. Katti. 2013. "Combination of single walled carbon nanotubes/graphene oxide with paclitaxel: A reactive oxygen species mediated synergism for treatment of lung cancer." *Nanoscale* 5 (7): 2818. https://doi.org/10.1039/c3nr33190c.

86. Yang, Feng, Chen Jin, Dong Yang, Yongjian Jiang, Ji Li, Yang Di, Jianhua Hu, Changchun Wang, Quanxing Ni, and Deliang Fu. 2011. "Magnetic functionalised carbon nanotubes as drug vehicles for cancer lymph node metastasis treatment." *European Journal of Cancer* 47 (12): 1873–1882. https://doi.org/10.1016/j.ejca.2011.03.018.

87. Sahoo, Nanda Gopal, Hongqian Bao, Yongzheng Pan, Mintu Pal, MitaliKakran, Henry Kuo Feng Cheng, Lin Li, and Lay Poh Tan. 2011. "Functionalized carbon nanomaterials as nanocarriers for loading and delivery of a poorly water-soluble anticancer drug: A comparative study." *Chemical Communications* 47 (18): 5235. https://doi.org/10.1039/c1cc00075f.

88. Schütz, Catherine A, Lucienne Juillerat-Jeanneret, Heinz Mueller, Iseult Lynch, and Michael Riediker. 2013. "Therapeutic nanoparticles in clinics and under clinical evaluation." *Nanomedicine* 8 (3): 449–467. https://doi.org/10.2217/nnm.13.8.

89. Svenson, Sonke. 2012. "Clinical translation of nanomedicines." *Current Opinion in Solid State and Materials Science* 16 (6): 287–294. https://doi.org/10.1016/j.cossms.2012.10.001.

90. Wei, Tuo, Chao Chen, Juan Liua, Cheng Liub, Paola Posocco, Xiaoxuan Liu, Qiang Cheng, ShuaidongHuo, Zicai Liang, Maurizio Fermeglia, Sabrina Pricl, Xing-JieLianga, Palma Rocchic, and Ling Peng. 2015. "Anticancer drug nanomicelles formed by self-assembling amphiphilic dendrimer to combat cancer drug resistance." *Proceedings of the National Academy of Sciences* 112 (10): 2978–2983. www.pnas.org/cgi/doi/10.1073/pnas.1418494112

91. Mohamed, Salma, Neha N Parayath, Sebastien Taurin, and Khaled Greish. 2014. "Polymeric nanomicelles: Versatile platform for targeted delivery in cancer." *Therapeutic Delivery* 5 (10): 1101–1121. https://doi.org/10.4155/tde.14.69.

92. Gong, Jian, Meiwan Chen, Ying Zheng, Shengpeng Wang, Yitao Wang. 2012. Polymeric micelles drug delivery system in oncology. *Journal of Controlled Release* 159: 312–323. https://doi.org/10.1016/j.jconrel.2011.12.012

93. Li, Yanping, Ting Zhang, Qinhui Liu, and Jinhan He. 2019. "PEG-derivatized dual-functional nanomicelles for improved cancer therapy." *Frontiers in Pharmacology* 10: 808. https://doi.org/10.3389/fphar.2019.00808

94. Kedar, Uttam, Prasanna Phutane, SupriyaShidhaye, and Vilasrao Kadam. 2010. "Advances in polymeric micelles for drug delivery and tumor targeting." *Nanomedicine: Nanotechnology, Biology and Medicine* 6 (6): 714–729. https://doi.org/10.1016/j.nano.2010.05.005.

95. Tanbour, Rafeeq, Ana M. Martins, William G. Pitt, and Ghaleb A. Husseini. 2016. "Drug delivery systems based on polymeric micelles and ultrasound: A review." *Current Pharmaceutical Design* 22 (19): 2796–2807. https://doi.org/10.2174/1381612822666160217125215

96. Wang, Jun, Hui Yang, Qiqi Li, Xianggen Wu, Guohu Di, Junting Fan, Dongxu Wei, and Chuanlong Guo. 2020. "Novel nanomicelles based on rebaudioside A: A potential nanoplatform for oral delivery of honokiol with enhanced oral bioavailability and antitumor activity." *International Journal of Pharmaceutics* 590: 119899. https://doi.org/10.1016/j.ijpharm.2020.119899

97. Qu, Jin, Xin Zhao, Yongping Liang, Tianlong Zhang, Peter X. Ma, and Baolin Guo. 2018. "Antibacterial adhesive injectable hydrogels with rapid self-healing, extensibility and compressibility as wound dressing for joints skin wound healing." *Biomaterials* 183: 185–199. https://doi.org/10.1016/j.biomaterials.2018.08.044

98. Molineux, G. 2002. "Pegylation: Engineering improved pharmaceuticals for enhanced therapy." *Cancer Treatment Reviews* 28: 13–16. https://doi.org/10.1016/s0305-7372(02)80004-4.

99. Duncan, Ruth. 2014. "Polymer therapeutics: Top 10 selling pharmaceuticals—what next?" *Journal of Controlled Release* 190: 371–380. https://doi.org/10.1016/j.jconrel.2014.05.001.

100. Bae, Younsoo, and Kazunori Kataoka. 2009. "Intelligent polymeric micelles from functional poly(ethylene glycol)-poly(amino acid) block copolymers." *Advanced Drug Delivery Reviews* 61 (10): 768–784. https://doi.org/10.1016/j.addr.2009.04.016.

101. Cabral, Horacio, and Kazunori Kataoka. 2014. "Progress of drug-loaded polymeric micelles into clinical studies." *Journal of Controlled Release* 190: 465–476. https://doi.org/10.1016/j.jconrel.2014.06.042.

102. Deshmukh, Anand S., Pratik N. Chauhan, Malleshappa N. Noolvi, Kiran Chaturvedi, KuntalGanguly, Shyam S. Shukla, Mallikarjuna N. Nadagouda, and Tejraj M. Aminabhavi. 2017. "Polymeric micelles: Basic research to clinical practice." *International Journal of Pharmaceutics* 532 (1): 249–268. https://doi.org/10.1016/j.ijpharm.2017.09.005.

103. Kim, Sung Chul, Dong Wook Kim, Yong Ho Shim, Joon Seok Bang, Hun Seung Oh, Sung Wan Kim, and Min Hyo Seo. 2001. "In vivo evaluation of polymeric micellar paclitaxel formulation: Toxicity and efficacy." *Journal of Controlled Release* 72 (1–3): 191–202. https://doi.org/10.1016/s0168-3659(01)00275-9.

104. Yasugi, Kenji, Yukio Nagasaki, Masao Kato, and Kazunori Kataoka. 1999. "Preparation and characterization of polymer micelles from poly(ethylene glycol)-poly(D,L-lactide) block copolymers as potential drug carrier." *Journal of Controlled Release* 62 (1–2): 89–100. https://doi.org/10.1016/s0168-3659(99)00028-0.

105. Kim, T.-Y. 2004. "Phase I and pharmacokinetic study of Genexol-PM, a cremophor-free, polymeric micelle-formulated paclitaxel, in patients with advanced malignancies." *Clinical Cancer Research* 10 (11): 3708–3716. https://doi.org/10.1158/1078-0432.ccr-03-0655.

106. Gatzemeier, Ulrich, Joachim von Pawel, Maya Gottfried, G.P. M. ten Velde, Karin Mattson, FilipoDeMarinis, Peter Harper, et al. 2000. "Phase III comparative study of high-dose cisplatin versus a combination of paclitaxel and cisplatin in patients with advanced non–small-cell lung cancer." *Journal of Clinical Oncology* 18 (19): 3390–3399. https://doi.org/10.1200/jco.2000.18.19.3390.

107. Lee, Keun Seok, Hyun Cheol Chung, Seock Ah Im, Yeon Hee Park, Chul Soo Kim, Sung-Bae Kim, Sun Young Rha, Min Young Lee, and Jungsil Ro. 2007. "Multicenter phase II trial of genexol-PM, a cremophor-free, polymeric micelle formulation of paclitaxel, in patients with metastatic breast cancer." *Breast Cancer Research and Treatment* 108 (2): 241–250. https://doi.org/10.1007/s10549-007-9591-y.

108. Lim, W.T., E.H. Tan, C.K. Toh, S.W. Hee, S.S. Leong, P.C.S. Ang, N.S. Wong, and B. Chowbay. 2010. "Phase I pharmacokinetic study of a weekly liposomal paclitaxel formulation (Genexol®-PM) in patients with solid tumors." *Annals of Oncology* 21 (2): 382–388. https://doi.org/10.1093/annonc/mdp315.

9 Nano-Formulations of Chemotherapeutic Drugs
Recent Outlook, Advances, and Challenges in Cancer Management

Deepa Bisht, Shilpa Maddheshiya, Seema Nara, and Manisha Sachan

CONTENTS

9.1 Introduction 144
9.2 Drawbacks of Conventional Chemotherapy 145
9.3 Nano-Formulations as Delivery Systems for Chemotherapeutic Drugs 146
9.4 Characteristics and Synthesis Routes of Different Nano-Formulations.............. 147
    9.4.1 Liposomes.. 147
        9.4.1.1 Classification of Liposomes 147
        9.4.1.2 Liposome Synthesis Methods 148
        9.4.1.3 Thin-Film Method 148
        9.4.1.4 Reverse-Phase Evaporation Method 148
        9.4.1.5 Freeze-Thaw Method 148
        9.4.1.6 Sonication Method... 148
        9.4.1.7 Calcium-Induced Fusion Method 149
    9.4.2 Polymeric Nanoparticles.. 149
        9.4.2.1 Methods of Polymeric Nanoparticles Synthesis 149
        9.4.2.2 Solvent Evaporation 149
        9.4.2.3 Emulsion Polymerization 150
    9.4.3 Inorganic Nanoparticles... 150
        9.4.3.1 Carbon Nanoparticles 150
        9.4.3.2 Mesoporous Silica Nanoparticles (MSNs).................................... 150
    9.4.4 Dendrimer.. 151
        9.4.4.1 Dendrimer Synthesis Methods... 151
        9.4.4.2 Convergent Growth Method 152
        9.4.4.3 Divergent Growth Method 152
9.5 Clinical Applications of Different Nano-Formulated Chemotherapeutic
    Drugs 152
    9.5.1 Advancement in the Clinical Trials of Different Nanodrugs 152
9.6 Nano-Formulated Drugs: Exploration of Cellular Pathway 153
    9.6.1 Modes of Nanoparticles Entry into the Cell.. 153
        9.6.1.1 Phagocytosis 154
        9.6.1.2 Micropinocytosis 154
        9.6.1.3 Clathrin-Mediated Endocytosis (CME).. 155
        9.6.1.4 Caveolae-Mediated Endocytosis.. 155

DOI: 10.1201/b23311-9

      9.6.1.5   Clathrin- and Caveolae-Independent Endocytosis 155

      9.6.1.6   Additional Entryway Processes 155

   9.6.2   Techniques for Studying Cellular Interactions of NPs 155

9.7   Recent Advances in Nano-Co-delivery for Cancer Treatment............................. 155

9.8   Toxicity Issues, Challenges, and Regulations Associated with
Nano-Formulated Drugs.. 156

9.9   Conclusion 157

9.10 References.. 158

9.1 INTRODUCTION

The worldwide incidence of cancer increases dramatically with stringent treatment constraints. The projected number of new cancer cases in 2020, including all types is 1,92,92,789, among both sexes and all age groups. According to the Global Cancer Cbservatory, the estimated number of new cancer cases in the year 2040 will be around 2,88,87,940 [1]. WHO data shows a high cancer mortality rate, as exemplified by ten million deaths in 2020 [2]. Different cell types like myofibroblasts, neuroendocrine cells, cancer-linked fibroblasts, adipose tissues, immunological cells, inflammation with blood, extracellular matrix (ECM), and lymphatic vascular networks are part of the dynamic and intricate process of tumorigenesis. Within the niche of the tumor microenvironment, these mechanisms create a complicated crosstalk network. As a result, cancer is defined as a class of disorder caused by cells' unregulated growth and migration [3].

It is envisioned that pharmacology, nanoscience, and the medical system will be heavily reliant on therapeutic nanoparticle-based formulation technology [4]. The increasing prevalence and popularization of the discipline of nanotechnology is due to the development and progress of different nanostructures and their properties. The domain of nanomedicine includes all the nanosized materials or instruments for theranostic applications [5].

Chemotherapy, resection, radiation therapy, antiangiogenic therapy, immunotherapy, endocrine therapy, and most newly, gene therapy is altogether treatment options for cancer, yet so many patients still succumb to death [6]. As the most effective cancer treatment modality, chemotherapy is one of the initial options for various cancers. Although the site-oriented distribution of chemotherapeutic drugs usually results in substantial tumor compression with repercussions like complete remission, unfavorable harm to healthy cells, and multidrug resistance (MDR), which continue to be a significantly challenging task in the curative treatment of cancer. Consequently, resulting in comprehensive research on targeted therapy. This domain has significant research with numerous inevitable side effects [7].

In nano-formulation of chemotherapeutic drugs, the solubility, entrapment, and encapsulation of drugs are linked to the drug delivery nanocarriers to enhance bioavailability, absorption, and minimization of toxicological issues [8]. In the creation of nanodrugs, a few key aspects must be considered. The formulations must make it feasible for the drug to finally reach its target from the point of administration and shield it from harmful environmental conditions, like acidity, enzymatic damage, and biological breakdown. In addition, the formulation must discharge the cargo in its active mode within the target site, enabling even smaller dosages to obtain the optimum pharmacological impact [9]. Liposome, polymeric, inorganic, and dendrimer nano-formulations have been examined extensively. Because of their effective pharmacokinetic profile, prolonged circulation period, and targeted deposition into tumor tissue via the EPR effect (enhanced permeability and retention), they have excellent efficacy and increased tumor selectivity [10, 11]. This chapter highlights current nano-formulations for chemotherapeutic drug delivery and their features, cellular uptake methods, synthesis routes, and clinical applications, together with the issues, challenges, and regulations of nanodrugs in healthcare.

9.2 DRAWBACKS OF CONVENTIONAL CHEMOTHERAPY

Conventional chemotherapy is the leading primary cancer therapy and is very imprecise in targeting medications or drugs to cancerous cells, resulting in unnecessary side effects in normal cells [12]. While cytotoxic agents are widely used for systemic treatment of recurring cancers, they have numerous drawbacks, such as poor aqueous solubility, imprecise bioavailability, severe toxic effects on healthy cells, and inadequate drug levels at tumors or malignant cells, as well as the development of multidrug resistance [13, 14]. As a result, the need for alternative treatment modalities goes unfulfilled. The following points highlight the major drawbacks associated with chemotherapy:

- Selectivity: The absence of selectivity for tumor tissue causes massive harm to non-cancerous cells, leading to serious adverse effects like immunosuppression, mucositis, fertility problems, nausea, and additional malignancies. Furthermore, because anticancer drugs have a massive distribution capacity, the dosage form is nonspecific to tumors, culminating in an aberrant accumulation of antitumor medicines in healthy cells [15].
- Mode of action: Another major drawback is specificity for the mode of action of the drug. Most chemotherapy medications work on general pathways represented by both cancerous and healthy cells instead of internal routes exclusive to tumor cells. Consequently, these medications' toxic and cytostatic mechanisms also affect healthy non-cancerous cells. For example, epirubicin (EPI), a hepatocellular carcinoma (HCC) drug, damages DNA by altering the cleavage-rejoining balance and raising the quantity of DNA topoisomerase II covalent groups. As a result, the p53 DNA-damage detector along with active caspases trigger apoptotic cell death [16, 17].
- Cytotoxicity: Chemotherapeutic medicines cause cytotoxicity because low-molecular-weight medications have a significant pharmacokinetics volume of spread. Reduced-weight chemicals are readily eliminated from the body. As a result, a larger dose is needed to provide therapeutic benefit. Because chemotherapeutic medicines have low therapeutic efficacy, the required dosage for successful therapy is frequently oversized, resulting in widespread dose-dependent adverse effects [18].
- Solubility: Chemotherapeutic medications are difficult to formulate and produce due to their low solubility in water. Due to its sizable hydrophobic nature and poor solubility (50.5 mg/L), paclitaxel's chemotherapeutic utility has been restricted. Significantly, with the addition of lipid-soluble moieties that have selectivity for the specific receptor, other chemotherapeutic drugs have poor solubility [19].
- Multidrug resistance: Chemotherapy has reduced anticancer medication potency due to inbuilt or adaptive chemoresistance mechanisms [20]. A tumor's interstitium is the space between its cells characterized by elevated hydrostatic pressure, increasing outward convective interstitial flow, which ultimately leads to the removal of the drug in contrast to normal tissues. Furthermore, if the medicine is delivered to the patient correctly and if the tumoral interstitium is not removed, the efficacy of the treatment may be reduced. Multidrug resistance has developed in all types of cancerous cells. Upregulation of the cellular membranes P-glycoprotein (P-gp), which is potent to keep medicines out of the cell, is a salient feature of MDR. To circumvent P-gp-facilitated MDR, several techniques have been planned, including encapsulating anticancer medicines in nanomaterials and co-administration of P-gp antagonists [21].
- Transport: The delivery of traditional chemotherapeutic medications to the tumor site is a hurdle. The medication's biophysical features, such as mass, surface characteristics, and charges, significantly impact its transit [22]. Another stumbling block is pathophysiological tumor diversity, which prevents homogeneous medication distribution throughout the tumor site. Furthermore, the acidic tumor microenvironment leads acid-sensitive medicines to degrade.

9.3 NANO-FORMULATIONS AS DELIVERY SYSTEMS FOR CHEMOTHERAPEUTIC DRUGS

Nano-formulation is a novel and an upcoming domain of nanotechnology that has a promising role in the biomedical sector for the prognosis and diagnosis of tumor, leading toward its safe and effective treatment by specific targeting [23]. Chemotherapeutic drugs are conventionally administered in free solution form as an infusion or intravenous bolus for tumor treatment. Despite their prolonged usage and the introduction of various novel strategies for increased clinical effectiveness, the outcome of conventional chemotherapy response is disappointing because administrated drugs frequently bind to cells or tissues unpredictably and in an indiscriminate way as a result only a small percentage of drugs reach to the cancerous or tumorous tissue [24]. This may increase systemic toxicity and reduce the therapeutic efficacy. The lack of specificity in the case of drug biodistribution and pharmacokinetics in cells or tissue imposes many limitations for successful cancer chemotherapy. For instance, doxorubicin (DOX), a broad-spectrum anticancer drug, kills the cancerous cells by inhibiting the topoisomerase II function. Administration of DOX as a free drug by the conventional approach leads to many adverse and undesirable effects on patients like cardiotoxicity, alopecia, myelosuppression, typhlitis, nausea, and vomiting [25]. The nano-formulation-based platform successfully delivers the chemotherapeutics drugs into the cancerous tissue because it can exploit the pathophysiology of the tumor microenvironment, overcome biological barriers, and then target tumorous tissue, which ultimately enhances the therapeutic efficacy [26]. Nano-formulation-based drug delivery strategies have various advantages over conventional delivery as represented in (Table 9.1).

Nano-formulation-based drug delivery offers (1) enhanced therapeutic index of the entrapped drug delivery, (2) improved efficacy of the drugs by maintaining steady-state therapeutic levels of loaded medicine for a prolonged time, (3) reduced toxicity of drugs owing to regulated drug release, and (4) improved drug's pharmacology by enhancing the solubility and stability of drugs [27]. For specificity, nanovesicles can be coated with targeting ligands or moiety, which target and bind to overexpressed receptors on cancerous cells or tissue. Additionally, the physicochemical nature of nano-formulations can be tailored by changing composition (organic, inorganic, or composites), size range, forms (spherical, dot, or spine shaped), and surface qualities (charges of the surface, functional group, biocompatible covering) [28]. The nano-based platform can also synergize with other therapies, such as photothermal and photodynamic therapy, and kill the cancerous cells with high efficiency. The goal of using nano-formulation in medicine administration is to target a site and successfully treats a disease with minimal adverse effects.

TABLE 9.1

Advantages of Nano-Formulation-Based Drug Delivery over Conventional Dug Delivery

Parameter	Conventional Delivery	Nano-Based Delivery
Dosage required	High	Low
Resistance	Rapid drug resistance	Low drug resistance
Toxicity	Relatively high	Relatively low
Side effects	Undesirable effects on normal tissue	Minimal side effects
Specificity	Less specific	Highly specific
Imaging	Not possible	Possible
Multimode therapy	No	Yes
Therapeutic efficacy	Low	High
Patient compliance	Poor	High
Therapeutic efficacy	Low	High

9.4 CHARACTERISTICS AND SYNTHESIS ROUTES OF DIFFERENT NANO-FORMULATIONS

Till now various nano-formulation as chemotherapeutic drug delivery platform has been reported using a variety of nanoparticles like inorganic nanoparticles, dendrimers, polymers, micelles, liposomes, protein-drug nanoconjugates, and other polymer-drug nanoconjugates [29].

9.4.1 LIPOSOMES

In 1968, for the first time, Bangham described liposomes, demonstrating that phospholipids in water solution spontaneously form multilamellar concentric bilayer vesicles [30]. These lipid-based formulations have a diameter of 40–500 nm and are the most studied carrier system derived from either plant or animal sources. These are well employed in wide applications, such as pharmaceuticals, biomedicine, and cosmetics, as they can entrap polar and nonpolar drugs due to their amphipathic nature in aqueous media. Polar drugs are entrapped within the aqueous core, while nonpolar compounds are loaded in the bilayer membrane. Various advantages offered by liposome-based nano-formulations for drug delivery include their low-cost production at a large scale, sustainable drug release, targeted drug administration, low toxicity due to their biodegradable and biocompatible nature, protection of the drug from harsh conditions, overcoming multidrug resistance, enhancing the therapeutic index of loaded drugs, and enhanced half-life circulation of chemotherapeutic drugs [31]. Chemotherapeutic drugs encapsulated in liposomal structures can reduce the drug uptake in normal tissue, improving its therapeutic index. For active targeting, a ligand (like antibody-based) method can be used, which is accomplished by coating the liposomal surface with antibodies that are specific to cancer tissue or cells (immunoliposomes, or ILP). For example, Thangapazham et al. prepared prostate-membrane-specific antigen (PMSA) curcumin-loaded liposomes by sonication size of 100–150 nm. Compared to typical liposomes, these antibody-attached nano-formulations had a higher capability to target the tumorous or cancerous tissue [32]. Recently pH-stimulated PEGylated liposomes have been reported to administrate bovine serum albumin (BSA) in bladder epithelium. A liposome-based nano-formulation for chemoimmunotherapy has been designed for co-loading IDO-1 and DOX drugs. DOX/IND-liposome was self-assembled by phospholipid-conjugated IND, followed by the DOX loading. The study demonstrates that DOX/IND liposomes increase the anti-breast-cancer immune response more efficiently than DOX-liposomes. In this way, ligand-decorated liposomes can offer selective drug accumulation leading to the development of immunoliposomes and stimuli-responsive liposomal systems for regulated drug delivery [33–35].

9.4.1.1 Classification of Liposomes

Liposomes can be classified based on their structure, lipid composition, number of the lipid bilayers, and their method of synthesis and application. Based on the layer (also called lamellae), they contain either one phospholipid bilayer membrane (unilamellar) or more than one phospholipid bilayer (multilamellar). Unilamellar is again subclassified into large unilamellar vesicles and small unilamellar vesicles. The size and membrane of a vesicle are significant factors in regulating the half-life of liposomes in circulation and the concentration of drug entrapment in the liposome [36]. Based on preparation approach, liposomes can be of various types, such as VET (vesicle developed by extrusion technique), DRV (dehydration rehydration approach), REV (single- or oligolamellar vesicle made by reverse-phase evaporation approach), MLV-REV (multilamellar vesicle synthesized by reverse-phase evaporation approach), and FATMLV (frozen and thawed multilamellar vesicles) [37]. Also, the liposome can be fusogenic liposome, long circulatory liposome, pH-sensitive liposome, conventional liposome, cationic liposome, magnetic liposome, temperature- and heat-sensitive liposome, and immunoliposome. In this way, the very diverse, unique, and flexible nature of the liposome distinguishes it as the preferred carrier for the delivery of chemotherapeutic drugs [38].

9.4.1.2 Liposome Synthesis Methods

Liposomes can be made in a variety of ways, all of which include combining lipids with aqueous media and affecting liposome properties, like size, lamellarity, and encapsulation efficiency (EE). The most used methods for preparing liposomes involve four basic steps: (1) drying down lipids from organic solvent, (2) dispersion of the lipid in aqueous media, (3) purification of the resultant liposome, and (4) analyzing the final product and quality control assay [39]. A simple flow diagram for the preparation of liposomes is depicted in (Figure 9.1).

9.4.1.3 Thin-Film Method

The thin-film method is one of the most well-known methods for liposome preparation. This method includes soaking a thin lipid film in an organic solvent and then sucking the organic solvent out, resulting in liposomes. The solid lipid mixture is hydrated with an aqueous buffer after the solvent is completely removed. Liposomes are formed when lipids swell and hydrate spontaneously. The encapsulating substance (drugs) can be added to the lipids before the thin film is prepared (for lipophilic compounds) or to the water/buffer solution before the thin film is formed (for hydrophilic compounds) [40]. Encapsulation efficiency is low with this approach.

9.4.1.4 Reverse-Phase Evaporation Method

The mixture of lipid is placed in a spherical backside flask, and the organic solvents are evaporated using a rotary evaporator at reduced pressure. When the residual solvent is evaporated by rotary evaporation at low pressure, liposomes are formed. A medium having less ionic strength, such as 0.01 M sodium chloride, can achieve excellent encapsulation efficiency (up to 65%). Large macro-molecules can be efficiently encapsulated using this approach [41].

9.4.1.5 Freeze-Thaw Method

This method involves liposomes prepared by the film method being vortexed with the solute to be trapped until the entire film is suspended. Liposomes are then frozen in dry ice–ethanol (−80°C) or liquid nitrogen, thawed, and vortexed again. The freeze-thaw cycle is repeated. Encapsulation efficiencies ranging from 20% to 30% were achieved [42].

9.4.1.6 Sonication Method

The approach is likely usually employed in SUV synthesis. Sonication of an aqueous dispersion of phospholipid is done by either using bath sonication or probe sonication. The drawback of this method is its low internal volume/encapsulation efficacy [36].

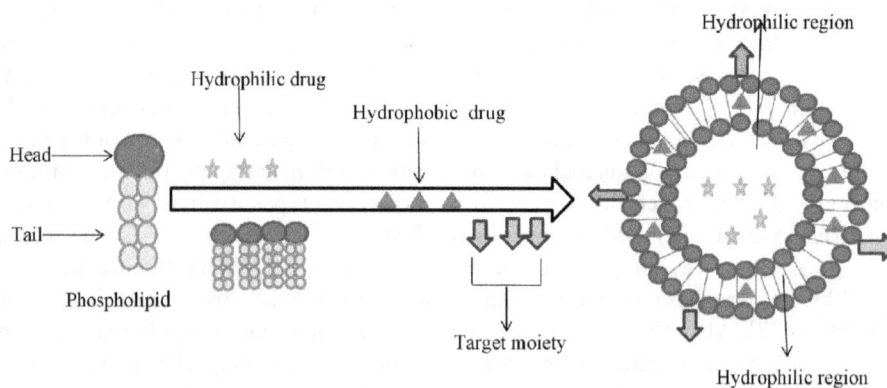

FIGURE 9.1 Schematic representation of liposome formation and entrapment of drug molecules.

9.4.1.7   Calcium-Induced Fusion Method

This approach is reported for the synthesis of LUV liposomes from acidic phospholipids, and calcium is supplied to SUV liposomes, which causes them to fuse and create a multilamellar vesicle. In the last step introduction of EDTA to the preparations causes LUV liposomes to form [40].

9.4.2   Polymeric Nanoparticles

Nano-formulations prepared from the polymers, size 10–1,000 nm are called polymeric nanoparticles (PNPs). Over the past few decades, synthetic and natural polymers, such as gelatin, chitosan, albumin, poly(amino acids), and poly(lactic acid) (PLA), have been used to synthesize PNPs for anticancer drug delivery [29]. Depending upon the synthesis approach, PNPs can be nanospheres or nanocapsules. In nanocapsular PNPs, chemotherapeutics drugs are confined to water or liquid core of oil encapsulated by a polymeric layer or membrane, whereas in nanospherical PNPs, drugs are confined in the matrix of the polymer; see Figure 9.3. Drug adsorption on the surface is also conceivable with both these types of PNPs [43]. PNPs offer many unique and attractive properties, such as their nano size, biocompatibility, biodegradability, enhanced drug payload, higher stability in vivo, long blood circulation time, controlled drug release, and easy surface functionalization using specific ligands that target desired cells or tissue. In 1979, PNPs were first attempted to be used as vehicles for effective delivery of doxorubicin to the site of action. For example, Abraxane, an FDA-approved paclitaxel-loaded albumin-based nano-formulation, was developed for the lung cancer treatment synergized with carboplatin in patients who are not candidates for curative surgery or radiation therapy. Paclitaxel (PTX) loaded PEGylated polylactic glycolic acid (PLGA) NPs with improved therapeutic index were found to be three times as toxic as Taxol as indicated in an in vivo study on transplantable liver tumor-bearing mice; PTX-entrapped nano-formulation inhibited tumor growth to a greater extent. This was due to the EPR phenomena of PTX-entrapped nano-formulation and their capacity to maintain the drug levels in the blood for a prolonged period [44].

9.4.2.1   Methods of Polymeric Nanoparticles Synthesis

Depending on the composition and required features of PNPs, a variety of synthesis methods have been reported. Polymeric nanoparticles are prepared either by direct monomeric polymerization or from pre-existed ones. The approach comprises direct polymerization of monomers, including emulsification polymerization, microemulsion polymerization, miniemulsion polymerization, interfacial polymerization, and controlled/living radical polymerization. The approach involving the dispersion of preformed polymers includes dialysis, solvent evaporation, and nanoprecipitation supercritical fluid technology [42]. The following are some frequently used methods for the preparation of polymeric nanoparticles.

9.4.2.2   Solvent Evaporation

This is the first and most widely used technique for the preparation of polymeric nanoparticles in which polymeric emulsions are synthesized by dissolving the polymer in a volatile solvent. The most used solvents were chloroform and dichloromethane, but for safety and toxicological reasons, ethyl acetate solvent is currently preferred. The emulsion changes into a nanoparticle suspension when the polymer's solvent evaporates, allowing the nanoparticle suspension to diffuse into the emulsion's continuous phase. In the traditional approach, single emulsions, such as oil-in-water (o/w), or double-emulsions, such as (water-in-oil)-in-water ([w/o]/w), were used to synthesize emulsions. High-speed homogenization or ultrasonication is used in these methods, followed by evaporation of the solvent at room temperature or reduced pressure with continuous magnetic stirring. The solidified nanoparticles can then be collected by ultracentrifugation and cleaned with distilled water to eliminate any additives, such as surfactants. Finally, the product is lyophilized [41].

9.4.2.3 Emulsion Polymerization

This is the fastest and most popular method to produce polymeric nanoparticles from monomers and is readily ascendable. Based on whether an organic or aqueous continuous phase is used, the procedure is divided into two groups. The monomer dispersion into insoluble substances is part of the continuous organic phase approach. To prevent clumping during the initial phases of polymerization, protective or surfactant-soluble polymers can be used. An anionic polymerization method is used to commence chain growth when initiated monomer ions or monomer radicals clash with additional monomer molecules. Phase separation and the creation of solid particles can occur before or after the polymerization reaction is completed [45].

9.4.3 INORGANIC NANOPARTICLES

Recently, inorganic nanoparticles achieved great attention due to their distinctive physicochemical properties, such as optical and magnetic properties, enhanced stability, easy surface modification, and inertness. Inorganic nanoparticles, such as gold, carbon-based iron oxide, and hafnium oxide NPs, are being explored in cancer treatment. In Europe, iron oxide NPs are already approved for the treatment of glioblastoma. In recent years, several inorganic materials, like nanodiamonds and graphene, are also used in cancer therapy [46].

9.4.3.1 Carbon Nanoparticles

Several carbon-based nano-formulations, such as carbon nanotubes, fullerene, graphene, and quantum dots, have been studied as potential drug delivery platforms [47]. The carbon nanotubes (CNTs) as novel drug delivery platforms were marked by immediate demonstration of the ability to penetrate the drug molecules into cells. CNTs were prepared by rolling graphene sheets into the tube-like structure. The name single-wall nanotubes and multiple-wall nanotubes are based on the number of graphene layers being rolled. But there is a limitation with CNTs as drug vehicles, which is their low solubility and cytotoxicity. CNTs can be surface functionalized, which renders them more soluble in water, and the toxicity exerted by CNTs can be overcome by using biocompatible agents, such as polyethylene glycol [48]. CNTs are one of the best anticancer drug vehicles due to their cell penetration ability, distinctive physicochemical properties, stability, optimum drug loading and surface functionalization, and structural flexibility. Anticancer medications can be enclosed in the inner chamber of CNTs linked to the surface of CNTs, either covalently or noncovalently. Some reported examples of CNTs as chemotherapeutics drugs carriers delivered methotrexate, paclitaxel, doxorubicin, and cisplatin with high efficacy [49, 50].

9.4.3.2 Mesoporous Silica Nanoparticles (MSNs)

In recent decades, silica (SiO_2) has seen extensive used in biomedical sector due to its ease of synthesis and large-scale synthesis. Numerous silica materials have been reported but mesoporous silica (MSNs) have great potential in drug delivery as they can hold large amounts of chemicals or drugs because they have a honeycomb-like structure with many pores. MSNs display numerous attractive characteristics, such as large specific surface area, controllable pore size, high loading capacity, chemical stability, good thermal capability, biocompatibility, and flexibility for drug entrapment, which makes them promising nano-vehicles for drug delivery [51]. The mesoporous structure allows MSNs to entrap many drugs and the nano-range size enables them to concentrate in tumor tissues via passive targeting and versatile surface modification of MSNs with ligands specific to the target site allowing them to bind and act on tumor tissues by active targeting mechanism [52]. Zhang et al. (2018) synthesized bienzyme responsive multifunctional MSNs as drug carriers loaded with DOX and modified with hyaluronic acid for targeting breast cancer. This nano-formulation was coated with a layer of gelatin and functionalized with biocompatible polyethylene glycol to increase circulation time as well as specific cellular uptake. This nanoplatform (MSN@HA-gelatin-PEG

[MHGP]) is triggered by bienzyme metalloprotease-2 (MMP-2) and hyaluronidases (HAase) and in vitro studies reported that PEG acts as a protector of healthy cells. The controlled release of DOX begins with the action of HAase, which leads to the elimination of HA and the development of promising chemotherapy [53].

9.4.4 DENDRIMER

Dendrimers are attractive macromolecules with a highly branched structure with a diameter in the range of 1.5–14.5 nm. These can be synthesized using either synthetic or natural compounds like amino acids, sugars, and nucleotides. Dendrimers possess well-recognized three structural regions: a central junction called core with various internal repeating units covalently connected to the nucleus (known as generations, G) and a distal chemical structure that offers the multifunctional surface on dendrimers [54]

The interesting design of the dendrimers acts as a promising novel drugs delivery platform owing to their uniqueness, like distinctive molecular weight, numerous branching, multivalent nature, round shapes (branched, linear, or cross-linked), monodisperse structures, multifunctional structure, internal chamber available to loaded molecules, control over macromolecular growth, and increased solubilization [55]. Drugs can be entrapped in either dendrimer's core or surface, and ligands can be attached to the multifunctional surface of dendrimers to achieve specific contact between drugs and target cells or tissue. Among various polymers possessing dendritic structures poly(propylene imine) (PPI) and poly(amidoamine) (PAMAM) based dendrimers are the most widely reported in the biomedical sector. However, there are also many other dendritic architectures, including polypeptide scaffolds and polyester, which have been developed with improved biodegradability [56].

9.4.4.1 Dendrimer Synthesis Methods

Dendrimers are prepared by two approaches called convergent and divergent methods. In the divergent method, the repeating layers are added around a central core, whereas in the convergent synthesis approach, the individual segments of dendrimers are prepared and attached at the end step. Figure 9.2 represents the flow diagram of the synthesis method of the dendrimer.

a) Convergent approach

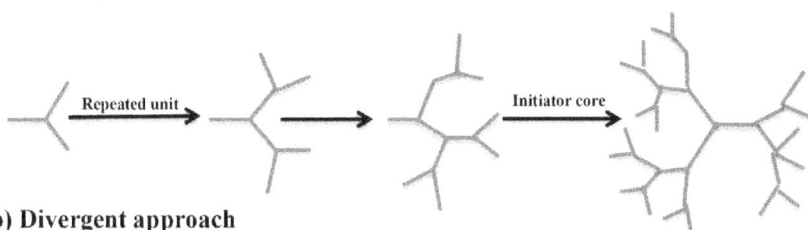

Repeated unit

Initiator core

b) Divergent approach

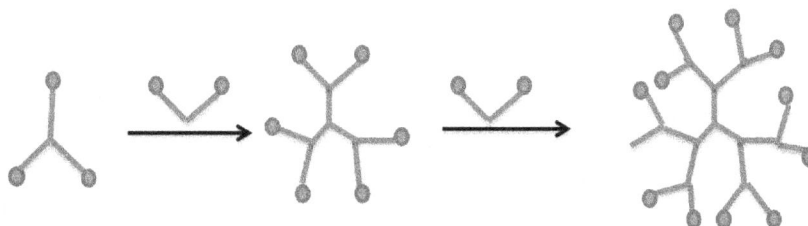

FIGURE 9.2   Synthesis method of dendrimers: (a) convergent approach; (b) divergent approach [57].

9.4.4.2 Convergent Growth Method

This approach begins from the periphery of the dendrimers and proceeds toward the central core by coupling monomers. Because of its sequential assembly of building pieces, convergent synthesis is sometimes referred to as the "organic chemist's approach." Frechet and his colleagues were the first to use this process to make poly(aryl ether) dendrimers. Poly(aryl alkyne) dendrimers, poly(phenylene) dendrimers, poly(alkyl ester) dendrimers, and poly(aryl alkene) dendrimers are the other types of dendrimers that are prepared by convergent growth approach. Because of the existence of numerous active functional groups on the surface, the dendrimers produced using this method is structurally pure and flexible. However, the time-taken stepwise synthesis has drawbacks in terms of scaling up dendrimer manufacturing [58, 59].

9.4.4.3 Divergent Growth Method

This approach involves the preparation of the dendrimers beginning from the core and proceeded molecule toward the periphery in a gradual manner. In each step, a new layer of the branching unit is joined, increasing the generation number in the dendrimers by one. It is ascending synthesis having two basic operations: first, monomers are linked, and then later, the end group of monomers is deprotected or converted to generate a new reactive surface and then again linked to a new monomer unit. Although this approach is simple, there is a limitation that it requires an excessive amount of reagents and a uniform structure is difficult to design. Several dendritic structures such, as poly(propylene imine) (PPI), poly(amidoamine) (PAMAM), and phosphorus-based dendrimers, can be prepared by this approach [60, 61].

9.5 CLINICAL APPLICATIONS OF DIFFERENT NANO-FORMULATED CHEMOTHERAPEUTIC DRUGS

Lipid- and polymer-based drugs, as represented in (Table 9.2), along with inorganic nanoparticles and dendrimers, all have been employed in the nano-formulation of different cancer types. Most nano-formulated medicines are tested in phase I trials in solid tumor patients. Advanced trials including phases II and III are being investigated for a few cancer types.

Even though nanostructures enhance the pharmacokinetic profile and bioactivity of a drug molecule and reduce generalized toxic effects via passive targeting when compared to standard compositions for poorly water-soluble anticancer drugs, nanostructures face issues in illustrating optimized treatment effectiveness due to reduced tumor preservation owing to excessive interstitial compression, slow buildup, and slower levels of discharge of entrapped drug, that also leads to reduced treatment effectiveness and development of multidrug resistance [72].

9.5.1 Advancement in the Clinical Trials of Different Nanodrugs

The most widely used lipid-based nanocarriers are liposomes and micelles. Various antibodies were distributed on the surface of liposomes in various preclinical tests. These nanostructures were shown to transport cytotoxic chemicals precisely to cells that express the target antigen in animal models of human cancer. The association of the specific antigen on the cell surface with the antibody on the lipid nanoparticles surface was required for cellular uptake. For example, GAH-coated (in which GAH is a human monoclonal antibody) doxorubicin-loaded immunoliposomes (ILs) (MCC-465) were used in the first and earliest clinical study utilizing tailored liposomes in metastatic stomach cancer patients. GAH reacts to approximately 90% of human gastric tumors. GAH-ILs were administered for three phases up to six times and main dosage hazard was myelosuppression. Almost 10 of the 18 patients monitored had the stable illness; moreover, no recurrent cases were found [73].

Lipid-based nanodrugs are also used for the targeted delivery of many nucleic acids and materials with a very short half-life. Although no such drug has passed the clinical trials so far, the potency and efficacy of siRNA delivery with lipid nanocarriers have been applied in human trials. Phase

TABLE 9.2

Different Liposomal and Polymeric NPs in Various Stages of Clinical Trials

	Product Name	Chemotherapeutic Drug	Cancer Type	Clinical Trial Stage	References
1. Liposomes					
	a. 2B3–101	Doxorubicin	Brain and breast cancer	Phase I/IIa	[62]
	b. CPX-1	Irinotecan HCl floxuridine	Colorectal cancer	Phase II	[63]
	c. EGFR antisense DNA liposomes	EGFR antisense DNA	Head and neck cancer	Phase I	[64]
	d. Doxil/Caelyx	Doxorubicin	Metastatic ovarian cancer and advanced Kaposi's sarcoma	Approved	[65]
	e. Lipusu	Paclitaxel	Solid tumors, gastric cancer, metastatic breast cancer	Phase II	[66]
	f. Lipoplatin	Cisplatin	Non-small-cell lung cancer	Phase III	[67]
2. Polymeric NPs					
	a. Kadcyla	Emtansine	Breast cancer	Approved	[65]
	b. CRLX101	Camptothecin	Solid tumor	Phase I/II	[68]
	c. Lupron Depot	Leuprolid	Prostate and breast cancer	Approved	[65]
	d. AB1–008	Docetaxel	Metastatic breast cancer, prostate cancer	Phase II	[69]
	e. Lipotecan	TLC388	Liver cancer, renal cancer	Phase I/II	[70]
	f. Nanotax	Paclitaxel	Peritoneal neoplasms	Phase I	[71]

(Nano-formulation type)

I clinical trials on solid tumors are ongoing as exemplified by NCT01158079 in which RNAi is targeted to counter vascular endothelial growth factor (VEGF) and kinesin spindle protein (KSP)—in another example, NCT01262235, in which RNAi is targeted against polo-like kinase 1 (PLK1), and NCT00938574, which counter a protein kinase C, PKN3. All the nucleic acids in these nanocarriers potentially target oncogenes, such as Myc and Bcl-2, or the activators of tumor suppressors, such as P53. The one most prominent protein-based NPs in the clinic is albumin-NP-bound paclitaxel by the name nab-paclitaxel; Abraxane is very efficient for non-small lung cancer, pancreatic cancer, and breast cancer [74–77].

9.6 NANO-FORMULATED DRUGS: EXPLORATION OF CELLULAR PATHWAY

We start by attempting to understand why NPs acquire various physical-chemical characteristics in biofluids, like cell culture medium and blood. The surface assimilation of biomolecules especially proteins on NPs surface is popularly known as protein corona. So the bio-identity of the NPs is in association with the quantity, orientation, and types of protein present. Overall bio-identity of NPs is governed by three core elements: (1) NPs linked parameters, (2) biological parameters, and (c) temperature of adsorption. Targeting the cell with NPs is also governed by the environment surrounding the tumor cell, like ECM, temperature, pH, and other factors, including surface functionalization of NPs affecting their final destiny. The aberrant proliferation of tumor cells leads to an increase in interstitial pressure via low lymphatic drainage, compact ECM, and widespread fibrosis. For example, in gastrointestinal tract the severe pH conditions highly affect the functionality of NPs [78, 79].

9.6.1 MODES OF NANOPARTICLES ENTRY INTO THE CELL

Endocytosis is the process by which the NPs enter the cell via ECM and plasma membrane. Figure 9.3 shows different mechanisms of NPs' entry. The process of endocytosis is demonstrated

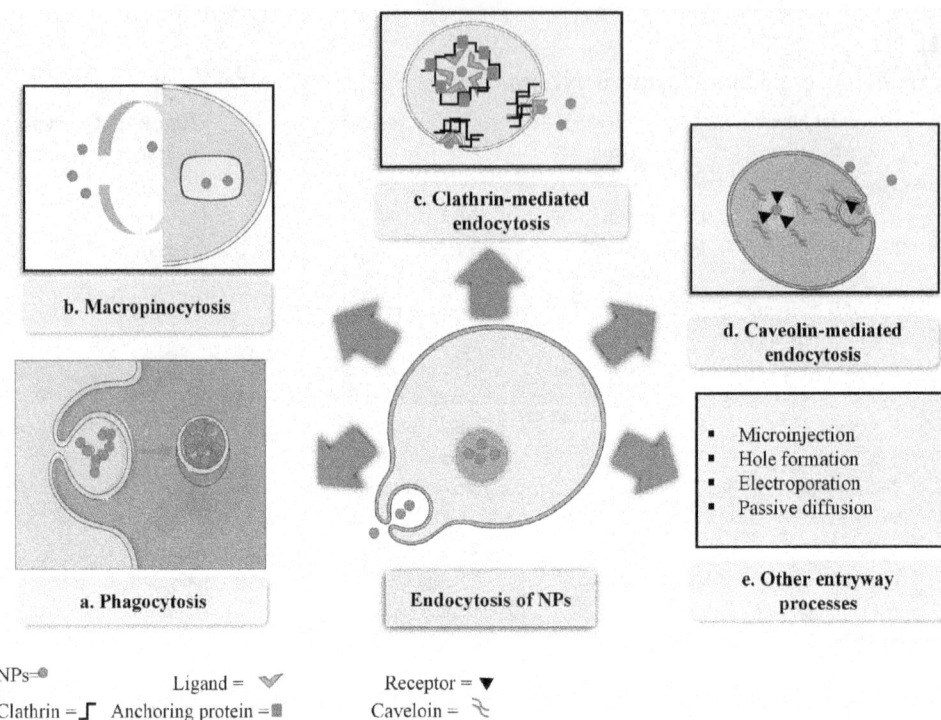

FIGURE 9.3   Different mechanisms of nanoparticles entry into the cell [78].

when NPs are engulfed in membrane folding, subsequently budding and pinching off to create endocytic vesicles, which are delivered to specialized internal trafficking divisions [79]. The five main types of endocytosis are following.

9.6.1.1 Phagocytosis

The various types of opsonins, like complement proteins, immunoglobins, and blood proteins, are adsorbed on the NPs' surface. Opsonized NPs get linked to phagocytes via a ligand receptor kind of interaction. This marks the beginning of downstream signaling for actin congregation, followed by extension of cell surface and finally the incorporation of substance for the formation of the phagosome. The duration of this process is mainly governed by parameters like the surface nature of the particle and the type of cells [80].

The pathway of phagocytosis is ruled by NPs' properties, like shape, size, and surface parameters. They are also influenced by the receptors; for instance, Fc dependent phagocytosis leads to the production of pro-inflammatory agents, which is not the case with complement-receptor-linked phagocytosis. So receptors are playing a crucial role in toxicity determination. In a well-cited example of Doxil, an FDA-approved anticancer formulation of liposome loaded with doxorubicin, the half-life was enhanced by PEGylation as it reduces the uptake by phagocytes. This overall improves the pharmacokinetic profile of nanodrug [81].

9.6.1.2 Micropinocytosis

Micropinocytosis is the expansion of the membrane formed by reorganization of the cytoskeleton, which fuses backward onto the plasma membrane for the creation of vesicles that enclosed a large amount of extracellular fluid. In micropinocytosis, all the substances and molecules in the fluid are processed into the endocytic vesicles without the specification of receptors, making it a form of major fluid absorption. It is significant for the uptake of large-sized NPs [82].

9.6.1.3 Clathrin-Mediated Endocytosis (CME)

Clathrin-linked endocytosis is the primary method for cellular nutrition and elements like choles-terol from LDLs and iron from transferrin. It can be possible via receptor-dependent and receptor-independent pathways. Different types of NPs are researched via this CME pathway. For example, NPs formed by poly(ethylene glycol co-lactide) (PEG-co-PLA) and D, L-polylactide (PLA) were enclosed via both routes—CME and caveolae-mediated endocytosis. The surface charge plays a crucial role in both the above mechanisms. In another research, silica nanotubes were studied via MDA-MB-231 cells for co-localization of clathrin, which plays a role as a marker for the CME pathway and lysotracker [83].

9.6.1.4 Caveolae-Mediated Endocytosis

Numerous biological activities rely on caveolae-mediated endocytosis, including transcytosis, cell sig-naling, and control of fatty acids, lipids, membrane proteins, and their tension. Caveolae-dependent endocytosis is thought to play vital role in diabetes, cancers, and viral infection. Caveolae are primar-ily flask-shaped invagination of membrane present widely in the epithelium and non-epithelial cells, like adipocytes and smooth muscle cells, its extension in cytoskeleton-linked masses. It dramatically enhances the surface area. This route was reported in nanodrug Abraxane, an albumin-bound type of paclitaxel. Albumin binds to gp60, and its receptor is present in the caveolae of endothelial cells, which makes its more accessible to interstitial space for effectively working against cancerous cells [84].

9.6.1.5 Clathrin- and Caveolae-Independent Endocytosis

Endocytosis, in the absence of both clathrin and caveolae, occurs in cells lacking them. Distinct cargos, like interleukin-2, cellular fluids, and human growth hormone, are taken up by such cells via other pathways, necessitating a specific type of lipid composition, mainly cholesterol. Popularly, folic acid is assimilated via this pathway. Folate-altered NPs are a very good example that uses this route. Folate binding takes place to its own receptor, which leads to its safe and successful delivery in the cytoplasm [85].

9.6.1.6 Additional Entryway Processes

It includes electroporation, passive diffusion, hole formation, and direct microinjection. For exam-ple, 4 nm of D-penicillamine-layered quantum dots (DPA-QDs) were penetrated through a passive process into the plasma membrane of red blood cells [86].

9.6.2 Techniques for Studying Cellular Interactions of NPs

In recent times, an in-depth study of NP interaction with the cell is mainly determined by the high-resolution imaging and chemical interaction study. Currently, different types of microscopy are employed, such as super-resolution fluorescence, atomic force, scanning electron microscopy (SEM), transmission electron microscopy (TEM), light-scattering, dark-field, photoacoustic, and correlative. Additionally, flow cytometry, surface-enhanced Raman scattering (SERS), and x-ray adsorption near-edge spectroscopy (XANES) are also popularly used to study the cellular mecha-nistic interaction of NPs and cellular interface. For example, confocal laser scanning microscopy detected the internal fluorescence from single-walled carbon nanotubes (SWNTs), along with mono-particle localizing software to study cellular uptake of SWNTs. In another study, TEM was used to study interactive nature of carbon nanotubes ingested by inhalation [87].

9.7 RECENT ADVANCES IN NANO-CO-DELIVERY FOR CANCER TREATMENT

Research has shown that developing a co-delivery mechanism inside a nanocarrier platform could permit concurrent discharge of various chemotherapeutic drugs at the targeted tumor site, improving combinatorial antitumor actions. Furthermore, the potent nanocarrier can be physicochemically changed to direct the encapsulated drugs to the desired location. As a result, nano co-delivery platforms support two significant benefits: first, greater control of medications bioavailability profiles, and second, selective and precise targeting of the desired cells. Co-delivery systems also offer a high level of safety. Co-delivery systems can help lessen the negative effects of high-dose specific treatment and conquer multidrug resistance (MDR), another major barrier to cancer care [88]. The important key points for physical-chemical features include the following:

- High solubility, excellent stability, tiny size, and good packing performance
- Capacity to incorporate different medications
- Protection power from bio-fluids (blood and cellular constituents)
- Excellent manufacturability with a prolonged storage life

James F. Holland, Emil Frei, and Emil J. Freireich first investigated the prospect of developing the first combo-chemotherapy for acute leukemia in 1965. In juvenile cases of acute lymphocytic leukemia, a combo of 6-mercaptopurine (6-MP), vincristine, methotrexate (MTX), and prednisone was found to be significantly effective in lowering tumor size and extending relapse [89]. A significant majority of medical investigations have demonstrated good synergistic results as compared to single-drug therapy.

Following are recent examples of co-delivery of chemotherapeutic drugs:

1. Elzoghby et al. studied the combination of letrozole and celecoxib in a phospholipid bilayer nanosytem and showed anticancer potential over individual drugs in breast cancer patients [90].
2. In another study, cisplatin with doxorubicin in the system of polymeric nanogels showed potential accumulation in targeted tumor tissues based on their biological distribution in breast cancer cell model MCF-7/ADR [91].
3. Yang et al. showed cisplatin and paclitaxel folate-linked lipid nanocarriers enhanced anticancer effects for chemotherapeutic therapy against head and neck cancer [92].
4. Mo et al. showed paclitaxel and carboplatin folic acid-PEG-conjugated phosphonated calix arene nano-formulation improved anticancer effects (in vitro and in vivo) in ovarian cancer by inactivating the epigenetic regulator JMJD3- HER2 axis [93].
5. In hyaluronan-based copolymer co-delivery of gefitinib with vorinostat showed antitumor potential and lesser side effects (in vivo) in lung cancer [94].
6. The combo of paclitaxel and doxorubicin fusogenic liposomal nano-formulation improved anticancer role (in vivo) with the better profile of cardiac toxicity in the animal model for breast cancer [95].

9.8　TOXICITY ISSUES, CHALLENGES, AND REGULATIONS ASSOCIATED WITH NANO-FORMULATED DRUGS

The ubiquitous usage of nanomaterials necessitates addressing safety concerns affecting both public health and the environment. Nanoparticle-based materials have so minute nano-scale size equivalent to cellular components associated with signaling. With advancements in the research in the nanotechnology domain, nanoparticles are known to have negative biologic interactions, which can have severe implications. The domain of nanotoxicology has proven as a distinct discipline of research due to all the potential side effects [96]

Therefore, evaluating the cytotoxicity of NPs with that of macroscopic materials is knotty. Cytotoxicity assays for nano-formulated drugs are fundamentally the same as that of conventional medications. As a result, current toxicity assessments of nanoparticles may be insufficient, and the creation of alternative toxicology tests for nanoparticle-based drugs should be the topic of prime importance. Diverse and distinct variables affect nanoparticle toxicities. So at the nano-biological junction, characteristic features like dimension, structure, surface characteristics, permeability, and dissolution power influence the behavior and efficacy of nanoparticle-based medications. The variable and dynamic factors considered, making it difficult to fully characterize and study the potential toxicity of nanoparticle-based technologies. Accessory stimulation of other processes, hemolytic anemia, swelling, oxidative damage, and mitochondrial dysfunction are common symptoms of nanoparticle-based toxicity [97].

To optimize nanodrug absorption and effectiveness, three critical criteria must be considered while manufacturing medications in nano-form. Such factors also include the designed medicine's durability, pharmaceutical transport, and disintegration mechanisms, plus FDA purity and standard restrictions. Such issues make employing nanostructures or nanodrugs difficult for pharmaceutical firms. The fundamental cause behind nano-poor formulation's therapeutic potential is its propensity to self-cumulate in lower drug concentration levels, compromising the formulation integrity and drug encapsulation heterogeneity owing to its miscibility. For instance, the nano-formulated drug doxorubicin has been shown to self-agglomerate, owing to excessive ionic strength in the medium, enhancing their size and affecting their efficiency and potential stability in biofluids [98].

An additional issue is still the inflating process of nano-formulations for the targeted administration of drugs, which has its own pros and cons. For example, a bulging process was exploited in nanocapsules to bulge and discharge medication to a target spot for cancer therapy. Inflammation, on the other hand, changes the volume of the nanodrug, which has an impact on its pharmacokinetic profile and dissolution. This drawback can be encountered by adding a topping compound to regulate the bulging process by wrapping the composition with a pH-sensitive layer that releases the desired drug at the target according to the pH. Last, the inadequacy of some nano-formulations to satisfy FDA performance rules and the difficulty in producing such in accordance with FDA's current good manufacturing standards make them unsuitable for bulk production at a higher scale. Moreover, the cost of manufacturing with standard requirements is also very high, which needs to be minimized for future studies and research [96].

Nano-formulations are prone to have a very significant influence on the healthcare industry; they are also anticipated to change current clinical norms, such as bioethics, environmental protection laws, and the healthcare management system. Both European Medicines Agency (EMEA) and the US Food and Drug Administration (US FDA) are indeed the two principal regulatory agencies regulating nano-formulations in healthcare. Some governing agencies and laws and regulations are often used in some countries to govern the use of nanotechnology formulation in the pharmacological industry, such as the US FDA, Nanotechnology Interest Group (NTIG), Centre for Drug Evaluation and Research (CDER), Centre for Devices and Radiological Health (CDRH), Nanotechnology Task Force, and Centre for Biologics Evaluation and Research (CBER). These all are from the USA region, and similarly, from the European Union, are as follows: New and Emerging Technologies (N&ET) Working Group, Innovation Task Force (ITF), and Committee for Medicinal Products for Human Use (CHMP). Governing agencies and regulations have been developed in the European Union and the United States to govern the usage of hazardous nanodrugs [99].

9.9 CONCLUSION

The once-fanciful ideals of what nanoscience may bring to biomedicine, particularly in the division of nanopharmaceuticals, have now become a realism. The present chapter highlights the exploration of nanodrugs' synthesis routes, uptake mechanisms, toxicity issues with challenges, and regulation policies.

With substantial expenditures made in the nanotechnology domain over the last two decades, revolutionary breakthroughs have opened the trail for the authorization of plentiful nanopharmaceutical products, with several now in the market. Immense sums have indeed been invested in academic centers devoted to the development of nanostructures in recent times, and researchers are migrating to the discipline, with a focus on nanomedicine. Nano-formulations of drugs would continue to rise in fame as they prove their worth and legality with several challenges. Oncology faces numerous drawbacks since present therapeutic choices were inadequate. It is crucial to realize, too, that while nanotechnology has a lot of promise, this might not be the answer to curing all cancer malignancies.

9.10   REFERENCES

1. Global Cancer Observatory. https://gco.iarc.fr/ (Accessed on 10 April 2022).
2. WHO Key Facts available online. www.who.int/news-room/fact-sheets/detail/cancer (Accessed on 12 April 2022)
3. Sonugür, F. Gizem, and Hakan Akbulut. 2019. "The role of tumour microenvironment in genomic instability of malignant tumours." *Frontiers in Genetics* 10 (2019): 1063. https://doi.org/10.3389/fgene.2019.01063
4. Kumar, Raj, Sameer V. Dalvi, and Prem Felix Siril. 2020. "Nanoparticle-based drugs and formulations: Current status and emerging applications." *ACS Applied Nano Materials* 3 (6): 4944–4961. https://doi.org/10.1021/acsanm.0c00606
5. Salama, Lavinia, Elizabeth R. Pastor, Tyler Stone, and Shaker A. Mousa. 2020. "Emerging nano-pharmaceuticals and nanonutraceuticals in cancer management." *Biomedicines* 8 (9): 347. https://doi.org/10.3390/biomedicines8090347
6. Lin, Ruo-Kai, Chiu-Yi Wu, Jer-Wei Chang, Li-Jung Juan, Han-Shui Hsu, Chih-Yi Chen, Yun-Yueh Lu et al. 2010. "Dysregulation of p53/Sp1 control leads to DNA methyltransferase-1 overexpression in lung cancer." *Cancer Research* 70 (14): 5807–5817. https://doi.org/10.1158/0008-5472.CAN-09-4161
7. Eftekhari, Reza B., Niloufar Maghsoudnia, Shabnam Samimi, Ali Zamzami, and Farid Abedin Dorkoosh. 2019. "Co-delivery nanosystems for cancer treatment: A review." *Pharmaceutical Nanotechnology* 7 (2): 90–112. https://doi.org/10.2174/2211738507666190321112237
8. Soppimath, Kumaresh S., Tejraj M. Aminabhavi, Anandrao R. Kulkarni, and Walter E. Rudzinski. 2001. "Biodegradable polymeric nanoparticles as drug delivery devices." *Journal of Controlled Release* 70 (1–2): 1–20. https://doi.org/10.1016/S0168-3659(00)00339-4
9. Ranjan, Shivendu, Nandita Dasgupta, Arkadyuti Roy Chakraborty, S. Melvin Samuel, Chidambaram Ramalingam, Rishi Shanker, and Ashutosh Kumar. 2014. "Nanoscience and nanotechnologies in food industries: Opportunities and research trends." *Journal of Nanoparticle Research* 16 (6): 1–23. http://dx.doi.org/10.1007/s11051-014-2464-5
10. Zaro, Jennica L. 2015. "Lipid-based drug carriers for prodrugs to enhance drug delivery." *The AAPS Journal* 17 (1): 83–92. https://dx.doi.org/10.1208%2Fs12248-014-9670-z
11. Maeda, Hiroshi. 2015. "Toward a full understanding of the EPR effect in primary and metastatic tumours as well as issues related to its heterogeneity." *Advanced Drug Delivery Reviews* 91: 3–6. https://doi.org/10.1016/j.addr.2015.01.002
12. Rajora, Amit Kumar, Divyashree Ravishankar, Hongbo Zhang, and Jessica M. Rosenholm. 2020. "Recent advances and impact of chemotherapeutic and antiangiogenic nanoformulations for combination cancer therapy." *Pharmaceutics* 12 (6): 592. https://doi.org/10.3390/pharmaceutics12060592
13. Peer, Dan, Jeffrey M. Karp, Seungpyo Hong, Omid C. Farokhzad, Rimona Margalit, and Robert Langer. 2007. "Nanocarriers as an emerging platform for cancer therapy." *Nature Nanotechnology* 2 (12): 751–760. http://dx.doi.org/10.1038/nnano.2007.387
14. Kwon, Glen S. 2003. "Polymeric micelles for delivery of poorly water-soluble compounds." *Critical Reviews™ in Therapeutic Drug Carrier Systems* 20 (5): 57–403. https://doi.org/10.1615/critrevtherdrugcarriersyst.v20.i5.20
15. Giordano, Karin F., and Aminah Jatoi. 2005. "The cancer anorexia/weight loss syndrome: Therapeutic challenges." *Current Oncology Reports* 7 (4): 271–276. https://doi.org/10.1007/s11912-005-0050-9
16. Mays, Ashley N., Neil Osheroff, Yuanyuan Xiao, Joseph L. Wiemels, Carolyn A. Felix, Jo Ann W. Byl, Kandeepan Saravanamuttu et al. 2010. "Evidence for direct involvement of epirubicin in the formation of chromosomal translocations in t (15; 17) therapy-related acute promyelocytic leukemia." *Blood, The Journal of the American Society of Hematology* 115 (2): 326–330. https://doi.org/10.1182/blood-2009-07-235051
17. Šimůnek, Tomáš, Martin Štěrba, Olga Popelová, Michaela Adamcová, Radomír Hrdina, and Vladimír Geršl. 2009. "Anthracycline-induced cardiotoxicity: Overview of studies examining the roles of oxidative stress and free cellular iron." *Pharmacological Reports* 61 (1): 154–171. https://doi.org/10.1016/S1734-1140(09)70018-0

18. Torchilin, Vladimir P. 2000. "Drug targeting." *European Journal of Pharmaceutical Sciences* 11: S81–S91. https://doi.org/10.1016/S0928-0987(00)00166-4

19. Gao, Yu, Zhihong Li, Xiaodong Xie, Chaoqun Wang, Jiali You, Fan Mo, Biyu Jin et al. 2015. "Dendrimeric anticancer prodrugs for targeted delivery of ursolic acid to folate receptor-expressing cancer cells: Synthesis and biological evaluation." *European Journal of Pharmaceutical Sciences* 70: 55–63. https://doi.org/10.1016/j.ejps.2015.01.007

20. Szakács, Gergely, Jill K. Paterson, Joseph A. Ludwig, Catherine Booth-Genthe, and Michael M. Gottesman. 2006. "Targeting multidrug resistance in cancer." *Nature Reviews Drug Discovery* 5 (3): 219–234. http://dx.doi.org/10.1038/nrd1984

21. Krishna, Rajesh, and Lawrence D. Mayer. 2000. "Multidrug resistance (MDR) in cancer: Mechanisms, reversal using modulators of MDR and the role of MDR modulators in influencing the pharmacokinetics of anticancer drugs." *European Journal of Pharmaceutical Sciences* 11 (4): 265–283. https://doi.org/10.1016/S0928-0987%2800%2900114-7

22. Park, Jae Hyung, Seulki Lee, Jong-Ho Kim, Kyeongsoon Park, Kwangmeyung Kim, and Ick Chan Kwon. 2008. "Polymeric nanomedicine for cancer therapy." *Progress in Polymer Science* 33 (1): 113–137. http://dx.doi.org/10.1016/j.progpolymsci.2007.09.003

23. Patra, Jayanta Kumar, Gitishree Das, Leonardo Fernandes Fraceto, Estefania Vangelie Ramos Campos, Maria del Pilar Rodriguez-Torres, Laura Susana Acosta-Torres, Luis Armando Diaz-Torres et al. 2018. "Nano based drug delivery systems: Recent developments and future prospects." *Journal of Nanobiotechnology* 16 (1): 1–33. https://doi.org/10.1186/s12951-018-0392-8

24. Wong, Ho Lun, Reina Bendayan, Andrew M. Rauth, Yongqiang Li, and Xiao Yu Wu. 2007. "Chemotherapy with anticancer drugs encapsulated in solid lipid nanoparticles." *Advanced Drug Delivery Reviews* 59 (6): 491–504. https://doi.org/10.1016/j.addr.2007.04.008

25. Haghiralsadat, Fateme, Ghasem Amoabediny, Mohammad Hasan Sheikhha, Tymour Forouzanfar, Marco N. Helder, and Behrouz Zandieh-Doulabi. 2017. "A novel approach on drug delivery: Investigation of a new nano-formulation of liposomal doxorubicin and biological evaluation of entrapped doxorubicin on various osteosarcoma cell lines." *Cell Journal (Yakhteh)* 19 (1): 55. https://doi.org/10.22074/cellj.2017.4502

26. Babu, Anish, Amanda K. Templeton, Anupama Munshi, and Rajagopal Ramesh. 2013. "Nanoparticle-based drug delivery for therapy of lung cancer: Progress and challenges." *Journal of Nanomaterials* 2013: 1–11. https://doi.org/10.1155/2013/863951

27. De Jong, Wim H., and Paul JA Borm. 2008. "Drug delivery and nanoparticles: Applications and hazards." *International Journal of Nanomedicine* 3 (2): 133. https://doi.org/10.2147/ijn.s596'

28. Wang, Yu, Qianmei Wang, Wei Feng, Qian Yuan, Xiaowei Qi, Sheng Chen, Pu Yao et al. 2021. "Folic acid-modified ROS-responsive nanoparticles encapsulating luteolin for targeted breast cancer treatment." *Drug Delivery* 28 (1): 1695–1708. https://doi.org/10.1080/10717544.2021.1963351

29. Téllez, Jair, Maria Clara Echeverry, Ibeth Romero, Andrea Guatibonza, Guilherme Santos Ramos, Ana Carolina Borges De Oliveira, Frédéric Frézard, and Cynthia Demicheli. 2021. "Use of liposomal nanoformulations in antileishmania therapy: Challenges and perspectives." *Journal of Liposome Research* 31 (2): 169–176. https://doi.org/10.1080/08982104.2020.1749067

30. Hema, Soundararajan, Selvarathinam Thambiraj, and Dhesingh Ravi Shankaran. 2018. "Nanoformulations for targeted drug delivery to prostate cancer: An overview." *Journal of Nanoscience and Nanotechnology* 18 (8): 5171–5191. https://doi.org/10.1166/jnn.2018.15420

31. Olusanya, Temidayo OB, Rita Rushdi Haj Ahmad, Daniel M. Ibegbu, James R. Smith, and Amal Ali Elkordy. 2008. "Liposomal drug delivery systems and anticancer drugs." *Molecules* 23 (4): 907. https://doi.org/10.3390/molecules23040907

32. Allen, Theresa M. 1994. "Long-circulating (sterically stabilized) liposomes for targeted drug delivery." *Trends in Pharmacological Sciences* 15 (7): 215–220. https://doi.org/10.1016/0165-6147(94)90314-X

33. Porfire, Alina, Marcela Achim, Lucia Tefas, and Bianca Sylvester. 2017. "Liposomal nanoformulations as current tumour-targeting approach to cancer therapy." In *Liposomes*. IntechOpen. https://doi.org/10.5772/intechopen.68160 s

34. AlSawaftah, Nour, William G. Pitt, and Ghaleb A. Husseini. 2021. "Dual-targeting and stimuli-triggered liposomal drug delivery in cancer treatment." *ACS Pharmacology & Translational Science* 4 (3): 1028–1049. https://doi.org/10.1021/acsptsci.1c00066

35. Vila-Caballer, Marian, Gaia Codolo, Fabio Munari, Alessio Malfanti, Matteo Fassan, Massimo Rugge, Anna Balasso, Marina de Bernard, and Stefano Salmaso. 2016. "A pH-sensitive stearoyl-PEG-poly (methacryloyl sulfadimethoxine)-decorated liposome system for protein delivery: An application

for bladder cancer treatment." *Journal of Controlled Release* 238: 31–42. http://dx.doi.org/10.1016/j.jconrel.2016.07.024

36. Akbarzadeh, Abolfazl, Rogaie Rezaei-Sadabady, Soodabeh Davaran, Sang Woo Joo, Nosratollah Zarghami, Younes Hanifehpour, Mohammad Samiei, Mohammad Kouhi, and Kazem Nejati-Koshki. 2013. "Liposome: Classification, preparation, and applications." *Nanoscale Research Letters* 8 (1): 1–9. https://doi.org/10.1186/1556-276X-8-102

37. Dua, J. S., A. C. Rana, and A. K. Bhandari. 2012. "Liposome: Methods of preparation and applications." *International Journal of Pharmaceutical Sciences and Research* 3 (2): 14–20. https://dx.doi.org/10.1186%2F1556-276X-8-102

38. Cortesi, Rita. 1999. "Preparation of liposomes by reverse-phase evaporation using alternative organic solvents." *Journal of Microencapsulation* 16 (2): 251–256. https://doi.org/10.1080/026520499289220

39. Dwivedi, Chandraprakash, Roshni Sahu, Sandip Tiwari, Trilochan Satapathy, and Amit Roy. 2014. "Role of liposome in novel drug delivery SYSTEM." Journal of Drug Delivery and Therapeutics 4 (2): 116–129. https://doi.org/10.22270/jddt.v4i2.768.

40. Zielińska, Aleksandra, Filipa Carreiró, Ana M. Oliveira, Andreia Neves, Bárbara Pires, D. Nagasamy Venkatesh, Alessandra Durazzo et al. 2020. "Polymeric nanoparticles: Production, characterization, toxicology and ecotoxicology." *Molecules* 25 (16): 3731. https://doi.org/10.3390/molecules25163731

41. Ahlin Grabnar, Pegi, and Julijana Kristl. 2011. "The manufacturing techniques of drug-loaded polymeric nanoparticles from preformed polymers." *Journal of Microencapsulation* 28 (4): 323–335. https://doi.org/10.3109/02652048.2011.569763

42. Vauthier, Christine, and Kawthar Bouchemal. 2009. "Methods for the preparation and manufacture of polymeric nanoparticles." *Pharmaceutical Research* 26 (5): 1025–1058. https://doi.org/10.1007/s11095-008-9800-3

43. Sahoo, Satyajeet, Anitha Gopalan, S. Ramesh, P. Nirmala, G. Ramkumar, S. Agnes Shifani, Ram Subbiah, and J. Isaac JoshuaRamesh Lalvani. 2021. "Preparation of polymeric nanomaterials using emulsion polymerization." *Advances in Materials Science and Engineering*: 1–9. https://doi.org/10.1155/2021/1539230

44. Catarina Pinto Reis, Ronald J. Neufeld, Antonio J. Ribeiro, Francisco Veiga. 2021. Nanoencapsulation I. Methods for preparation of drug-loaded polymeric nanoparticles Nanomedicine: Nanotechnology, Biology, and Medicine 2 (1): 8–21. doi:10.1016/j.nano.2005.12.003

45. Farooq, Muhammad Asim, Md Aquib, Anum Farooq, Daulat Haleem Khan, Mily Bazezy Joelle Maviah, Mensura Sied Filli, Samuel Kesse et al. 2019. "Recent progress in nanotechnology-based novel drug delivery systems in designing of cisplatin for cancer therapy: An overview." *Artificial Cells, Nanomedicine, and Biotechnology* 47 (1): 1674–1692. https://doi.org/10.1080/21691401.2019.1604535

46. Pugazhendhi, Arivalagan, Thomas Nesakumar Jebakumar Immanuel Edison, Indira Karuppusamy, and Brindhadevi Kathirvel. 2018. "Inorganic nanoparticles: A potential cancer therapy for human welfare." *International Journal of Pharmaceutics* 539 (1–2): 104–111. https://doi.org/10.1016/j.ijpharm.2018.01.034

47. Liu, Jingquan, Liang Cui, and Dusan Losic. 2013. "Graphene and graphene oxide as new nanocarriers for drug delivery applications." *Acta Biomaterialia* 9 (12): 9243–9257. https://doi.org/10.1016/j.actbio.2013.08.016

48. Kostarelos, Kostas, Lara Lacerda, Giorgia Pastorin, Wei Wu, Sebastien Wieckowski, Jacqueline Luangsivilay, Sylvie Godefroy et al. 2007. "Cellular uptake of functionalized carbon nanotubes is independent of functional group and cell type." *Nature Nanotechnology* 2 (2): 108–113. https://doi.org/10.1038/nnano.2006.209

49. Vardharajula, Sandhya, Sk Z. Ali, Pooja M. Tiwari, Erdal Eroğlu, Komal Vig, Vida A. Dennis, and Shree R. Singh. 2012. "Functionalized carbon nanotubes: Biomedical applications." *International Journal of Nanomedicine* 7: 5361. https://doi.org/10.2147/IJN.S35832

50. Fabbro, Chiara, Hanene Ali-Boucetta, Tatiana Da Ros, Kostas Kostarelos, Alberto Bianco, and Maurizio Prato. 2012. "Targeting carbon nanotubes against cancer." *Chemical Communications* 48 (33): 3911–3926. https://doi.org/10.1039/C2CC17995D

51. Slowing, Igor I., Juan L. Vivero-Escoto, Chia-Wen Wu, and Victor S-Y. Lin. 2008. "Mesoporous silica nanoparticles as controlled release drug delivery and gene transfection carriers." *Advanced Drug Delivery Reviews* 60 (11): 1278–1288. https://doi.org/10.1016/j.addr.2008.03.012

52. Wang, Ying, Qinfu Zhao, Ning Han, Ling Bai, Jia Li, Jia Liu, Erxi Che et al. 2015. "Mesoporous silica nanoparticles in drug delivery and biomedical applications." *Nanomedicine: Nanotechnology, Biology and Medicine* 11 (2): 313–327. https://doi.org/10.1016/j.nano.2014.09.014

53. Zhang, Yang, and Juan Xu. 2018. "Mesoporous silica nanoparticle-based intelligent drug delivery system for bienzyme-responsive tumour targeting and controlled release." *Royal Society Open Science* 5 (1): 170986. https://doi.org/10.1098/rsos.170986

54. Sandoval-Yañez, Claudia, and Cristian Castro Rodriguez. 2020. "Dendrimers: Amazing platforms for bioactive molecule delivery systems." *Materials* 13 (3): 570. https://doi.org/10.3390/ma13030570

55. Mu, Weiwei, Qihui Chu, Yongjun Liu, and Na Zhang. 2020. "A review on nano-based drug delivery system for cancer chemoimmunotherapy." *Nano-Micro Letters* 12 (1): 1–24. https://doi.org/10.1007/s40820-020-00482-6

56. Wais, Ulrike, Alexander W. Jackson, Tao He, and Haifei Zhang. 2016. "Nanoformulation and encapsulation approaches for poorly water-soluble drug nanoparticles." *Nanoscale* 8 (4): 1746–1769. https://doi.org/10.1039/C5NR07161E

57. Jain, Vaibhav, and Prasad V. Bharatam. 2014. "Pharmacoinformatic approaches to understand complexation of dendrimeric nanoparticles with drugs." *Nanoscale* 6 (5): 2476–2501. https://doi.org/10.1039/C3NR05400D

58. Matthews, Owen A., Andrew N. Shipway, and J. Fraser Stoddart. 1998. "Dendrimers—branching out from curiosities into new technologies. "Progress in polymer science 23 (1): 1–56. https://doi.org/10.1016/S0079-6700(97)00025-7

59. Grayson, Scott M., and Jean MJ Frechet. 2001. "Convergent dendrons and dendrimers: From synthesis to applications." *Chemical Reviews* 101 (12): 3819–3868. https://doi.org/10.1021/cr990116h

60. Hawker, Craig, and Jean MJ Fréchet. 1990. "A new convergent approach to monodisperse dendritic macromolecules." *Journal of the Chemical Society, Chemical Communications* 15: 1010–1013. https://doi.org/10.1039/C39900001010

61. Sherje, Atul P., Mrunal Jadhav, Bhushan R. Dravyakar, and Darshana Kadam. 2018. "Dendrimers: A versatile nanocarrier for drug delivery and targeting." *International Journal of Pharmaceutics* 548 (1): 707–720. https://doi.org/10.1016/j.ijpharm.2018.07.030

62. Prajapati, Sunil Kumar, Sheo Datta Maurya, Manas Kumar Das, Vijay Kumar Tilak, Krishna Kumar Verma, and Ram C. Dhakar. 2016. "Dendrimers in drug delivery, diagnosis and therapy: Basics and potential applications." *Journal of Drug Delivery and Therapeutics* 6 (1): 67–92. https://doi.org/10.22270/jddt.v6i1.1190

63. Gaillard, Pieter J., Chantal CM Appeldoorn, Rick Dorland, Joan van Kregten, Francesca Manca, Danielle J. Vugts, Bert Windhorst et al. 2014. "Pharmacokinetics, brain delivery, and efficacy in brain tumour-bearing mice of glutathione pegylated liposomal doxorubicin (2B3–101)." *PloS One* 9 (1): e82331. https://doi.org/10.1371/journal.pone.0082331

64. Batist, Gerald, Karen A. Gelmon, Kim N. Chi, Wilson H. Miller, Stephen KL Chia, Lawrence D. Mayer, Christine E. Swenson, Andrew S. Janoff, and Arthur C. Louie. 2009. "Safety, pharmacokinetics, and efficacy of CPX-1 liposome injection in patients with advanced solid tumours." *Clinical Cancer Research* 15 (2): 692–700. https://doi.org/10.1371/journal.pone.0082331

65. He, Yukai, Qing Zeng, Stephanie D. Drenning, Mona F. Melhem, David J. Tweardy, Leaf Huang, and Jennifer Rubin Grandis. 1998. "Inhibition of human squamous cell carcinoma growth in vivo by epidermal growth factor receptor antisense RNA transcribed from the U6 promoter." *JNCI: Journal of the National Cancer Institute* 90 (14): 1080–1087. https://doi.org/10.1093/jnci/90.14.1080

66. US Food and Drug Administration available online. www.accessdata.fda.gov (Accessed on 15th April 2022).

67. Xu, Xu, Lin Wang, Huan-Qin Xu, Xin-En Huang, Ya-Dong Qian, and Jin Xiang. 2013. "Clinical comparison between paclitaxel liposome (Lipusu®) and paclitaxel for treatment of patients with metastatic gastric cancer." *Asian Pacific Journal of Cancer Prevention* 14 (4): 2591–2594. https://doi.org/10.7314/APJCP.2013.14.4.2591

68. Stathopoulos, G. P., D. Antoniou, J. Dimitroulis, J. Stathopoulos, K. Marosis, and P. Michalopoulou. 2011. "Comparison of liposomal cisplatin versus cisplatin in non-squamous cell non-small-cell lung cancer." *Cancer Chemotherapy and Pharmacology* 68 (4): 945–950. https://doi.org/10.1007/s00280-011-1572-5

69. ClinicalTrials.gov. https://clinicaltrials.gov/ (Accessed on 18th April 2022)

70. Hawkins, Michael J., Patrick Soon-Shiong, and Neil Desai. 2008. "Protein nanoparticles as drug carriers in clinical medicine." *Advanced Drug Delivery Reviews* 60 (8): 876–885. https://doi.org/10.1016/j.addr.2007.08.044

71. Ghamande, Sharad, Chia-Chi Lin, Daniel C. Cho, Geoffrey I. Shapiro, Eunice L. Kwak, Michael H. Silverman, Yunlong Tseng et al. 2014. "A phase 1 open-label, sequential dose-escalation study investigating the safety, tolerability, and pharmacokinetics of intravenous TLC388 administered to patients

with advanced solid tumours." *Investigational New Drugs* 32 (3): 445–451. https://doi.org/10.1007/s10637-013-0044-7

72. Roby, Katherine F., Fenghui Niu, Roger A. Rajewski, Charles Decedue, Bala Subramaniam, and Paul F. Terranova. 2008. "Syngeneic mouse model of epithelial ovarian cancer: Effects of nanoparticulate paclitaxel, Nanotax®." 622: 169–181. https://doi.org/10.1007/978-0-387-68969-2_14

73. Mamot, Christoph, Daryl C. Drummond, Charles O. Noble, Verena Kallab, Zexiong Guo, Keelung Hong, Dmitri B. Kirpotin, and John W. Park. 2005. "Epidermal growth factor receptor–targeted immunolipo-somes significantly enhance the efficacy of multiple anticancer drugs in vivo." *Cancer Research* 65 (24): 11631–11638. https://doi.org/10.1158/0008-5472.CAN-05-1093

74. Park, John W., Keelung Hong, Dmitri B. Kirpotin, Gail Colbern, Refaat Shalaby, Jose Baselga, Yvonne Shao et al. 2002. "Anti-HER2 immunoliposomes: Enhanced efficacy attributable to targeted delivery." *Clinical Cancer Research* 8 (4): 1172–1181.

75. Hosokawa, S., T. Tagawa, H. Niki, Y. Hirakawa, K. Nohga, and K. Nagaike. 2003. "Efficacy of immu-noliposomes on cancer models in a cell-surface-antigen-density-dependent manner." *British Journal of Cancer* 89 (8): 1545–1551. https://doi.org/10.1038/sj.bjc.6601341

76. Coelho, Teresa, David Adams, Ana Silva, Pierre Lozeron, Philip N. Hawkins, Timothy Mant, Javier Perez et al. 2013. "Safety and efficacy of RNAi therapy for transthyretin amyloidosis." *New England Journal of Medicine* 369 (9): 819–829. http://dx.doi.org/10.1056/NEJMoa1208760

77. Tabernero, Josep, Geoffrey I. Shapiro, Patricia M. LoRusso, Andres Cervantes, Gary K. Schwartz, Glen J. Weiss, Luis Paz-Ares et al. 2013. "First-in-humans trial of an RNA interference therapeutic targeting VEGF and KSP in cancer patients with liver involvement." *Cancer Discovery* 3 (4): 406–417. http://dx.doi.org/10.1158/2159-8290.CD-12-0429

78. Behzadi, Shahed, Vahid Serpooshan, Wei Tao, Majd A. Hamaly, Mahmoud Y. Alkawareek, Erik C. Dreaden, Dennis Brown, Alaaldin M. Alkilany, Omid C. Farokhzad, and Morteza Mahmoudi. 2017. "Cellular uptake of nanoparticles: Journey inside the cell." *Chemical Society Reviews* 46 (14): 4218–4244. https://doi.org/10.1039/C6CS00636A

79. Doherty, Gary J., and Harvey T. McMahon. 2009. "Mechanisms of endocytosis." *Annual Review of Biochemistry* 78: 857–902. https://doi.org/10.1146/annurev.biochem.78.081307.110540

80. Barenholz, Yechezkel Chezy. 2012. "Doxil®—the first FDA-approved nano-drug: Lessons learned." *Journal of Controlled Release* 160 (2): 117–134. https://doi.org/10.1016/j.jconrel.2012.03.020

81. Lim, Jet Phey, and Paul A. Gleeson. 2011. "Macropinocytosis: An endocytic pathway for internalising large gulps." *Immunology and Cell Biology* 89 (8): 836–843. https://doi.org/10.1038/icb.2011.20

82. Nan, Anjan, Xia Bai, Sang Jun Son, Sang Bok Lee, and Hamidreza Ghandehari. 2008. "Cellular uptake and cytotoxicity of silica nanotubes." *Nano Letters* 8 (8): 2150–2154. https://doi.org/10.1021/nl0802741

83. Sahay, Gaurav, Daria Y. Alakhova, and Alexander V. Kabanov. 2010. "Endocytosis of nanomedi-cines." *Journal of Controlled Release* 145 (3): 182–195. https://doi.org/10.1016/j.jconrel.2010.01.036

84. Kumari, Sudha, Swetha Mg, and Satyajit Mayor. 2010. "Endocytosis unplugged: Multiple ways to enter the cell." *Cell Research* 20 (3): 256–275. https://doi.org/10.1038/cr.2010.19

85. Chou, Leo Y.T., Kevin Ming, and Warren CW Chan. 2011. "Strategies for the intracellular delivery of nanoparticles." *Chemical Society Reviews* 40 (1): 233–245. https://doi.org/10.1039/C0CS00003E

86. Wang, Tiantian, Jing Bai, Xiue Jiang, and G. Ulrich Nienhaus. 2012. "Cellular uptake of nanoparticles by membrane penetration: A study combining confocal microscopy with FTIR spectroelectrochemis-try." *ACS Nano* 6 (2): 1251–1259. https://doi.org/10.1021/nn203892h

87. Eftekhari, Reza B., Niloufar Maghsoudnia, Shabnam Samimi, Ali Zamzami, and Farid Abedin Dorkoosh. 2019. "Co-delivery nanosystems for cancer treatment: A review." *Pharmaceutical Nanotechnology* 7 (2): 90–112. https://doi.org/10.2174/2211738507666190321112237

88. FREI III, E. M. I. L., Myron Karon, Robert H. Levin, EMIL J. FREIREICH, Robert J. Taylor, Juliet Hananian, Oleg Selawry et al. 1965. "The effectiveness of combinations of antileukemic agents in inducing and maintaining remission in children with acute leukemia." *Blood* 26 (5): 642–656. https://doi.org/10.1182/blood.V26.5.642.642

89. Elzoghby, Ahmed O., Shaimaa K. Mostafa, Maged W. Helmy, Maha A. ElDemellawy, and Salah A. Sheweita. 2017. "Multi-reservoir phospholipid shell encapsulating protamine nanocapsules for co-delivery of letrozole and celecoxib in breast cancer therapy." *Pharmaceutical Research* 34 (9): 1956–1969. https://doi.org/10.1007/s11095-017-2207-2

90. Wu, Haiqiu, Haojie Jin, Cun Wang, Zihao Zhang, Haoyu Ruan, Luyan Sun, Chen Yang, Yongjing Li, Wenxin Qin, and Changchun Wang. 2017. "Synergistic cisplatin/doxorubicin combination chemother-apy for multidrug-resistant cancer via polymeric nanogels targeting delivery." *ACS Applied Materials & Interfaces* 9 (11): 9426–9436. https://doi.org/10.1021/acsami.6b16844

91. Yang, Jiying, Zengjuan Ju, and Shufang Dong. 2017. "Cisplatin and paclitaxel co-delivered by folate-decorated lipid carriers for the treatment of head and neck cancer." *Drug Delivery* 24 (1): 792–799. https://doi.org/10.1080/10717544.2016.1236849

92. Mo, Jingxin, Li Wang, Xiaojia Huang, Bing Lu, Changye Zou, Lili Wei, Junjun Chu et al. 2017. "Multifunctional nanoparticles for co-delivery of paclitaxel and carboplatin against ovarian cancer by inactivating the JMJD3-HER2 axis." *Nanoscale* 9 (35): 13142–13152. https://doi.org/10.1039/C7NR04473A

93. Jeannot, Victor, Cony Gauche, Silvia Mazzaferro, Morgane Couvet, Laetitia Vanwonterghem, Maxime Henry, Chloé Didier et al. 2018. "Anti-tumour efficacy of hyaluronan-based nanoparticles for the co-delivery of drugs in lung cancer." *Journal of Controlled Release* 275: 117–128. https://doi.org/10.1016/j.jconrel.2018.02.024

94. Franco, Marina Santiago, Marjorie Coimbra Roque, André Luís Branco de Barros, Juliana de Oliveira Silva, Geovanni Dantas Cassali, and Mônica Cristina Oliveira. 2019 "Investigation of the antitumour activity and toxicity of long-circulating and fusogenic liposomes co-encapsulating paclitaxel and doxorubicin in a murine breast cancer animal model." *Biomedicine & Pharmacotherapy* 109: 1728–1739. https://doi.org/10.1016/j.biopha.2018.11.011

95. Stone, Vicki, Helinor Johnston, and Roel PF Schins. 2009. "Development of in vitro systems for nanotoxicology: Methodological considerations." *Critical Reviews in Toxicology* 39 (7): 613–626. https://doi.org/10.1080/10408440903120975

96. Sayes, Christie M., Kenneth L. Reed, and David B. Warheit. 2007. "Assessing toxicity of fine and nanoparticles: Comparing in vitro measurements to in vivo pulmonary toxicity profiles." *Toxicological Sciences* 97 (1): 163–180. https://doi.org/10.1093/toxsci/kfm018

97. Barenholz, Yechezkel Chezy. 2012. "Doxil®—the first FDA-approved nano-drug: Lessons learned." *Journal of Controlled Release* 160 (2): 117–134. https://doi.org/10.1016/j.jconrel.2012.03.020

98. Nijhara, Ruchika, and Krishna Balakrishnan. 2006. "Bringing nanomedicines to market: Regulatory challenges, opportunities, and uncertainties." *Nanomedicine: Nanotechnology, Biology and Medicine* 2 (2): 127–136. https://doi.org/10.1016/j.nano.2006.04.005

99. Van Calster, Geert, and Joel D'Silva. 2009. "Taking temperature—A review of European Union regulation in nanomedicine." *European Journal of Health Law* 16 (3): 249–269. https://doi.org/10.1163/15718 0909x453071

10 Nutrigenomics in Cancer Chemoprevention

*Kanchan Gairola, Shiv Kumar Dubey,
Ananya Bahuguna, and Shruti*

CONTENTS

10.1 Introduction...165
10.2 Cancer and Its Genetic Correlation..167
10.3 Dietary Signals in Nutrigenomics...168
10.4 Implications of Bioactive Phytochemicals in Cancer Therapy and Deciphering
the Genomic Puzzle..168
10.5 Natural Phytochemicals Targeting the Key Genes in Process of Carcinogenesis..............169
    10.5.1 Antioxidant Effect 169
    10.5.2 Effects of Dietary Phytochemicals on Genomic Stability
and Non-genotoxicity...170
    10.5.3 Inflammatory Mediators..170
    10.5.4 Apoptosis and Cell Cycle Arrest 171
    10.5.5 Angiogenesis...171
10.6 Chemoprevention and Natural Compounds...172
10.7 Controlling Epigenetic Changes and mi RNAs Involved in Carcinogenesis
and Cancer Prevention: The Role of Natural Products...172
    10.7.1 Isothiocyanates (ITC)...172
    10.7.2 Apigenin 173
    10.7.3 Curcumin 174
    10.7.4 Epigallocatechin Gallate (EGCG)..174
    10.7.5 Resveratrol 174
    10.7.6 Genistein...174
    10.7.7 Sulforaphane and Indole-3-Carbinol (I$_3$C) 175
    10.7.8 Quercetin 175
10.8 Crosstalk 175
10.9 Conclusion and Future Perspective...176
10.10 References..176

10.1 INTRODUCTION

Despite the tremendous and colossal advancement in the research area and the field of medicine, public health challenges still remain intact. Genomics and genetics together contributed toward the revolutionary diagnostic and prognostic approach and paved the way to personalize the treatments. Nutrigenomics emerges as a field of science that enables the researchers to scientifically study the interaction between specific genes and bioactive compounds of food. It is an attempt to observe the genome-wide influence of nutrition. Nutrigenomics gives a new perspective on the nutritional influence of the specific gene. The dietary compounds in nutrigenomics are considered as "dietary signatures." Nutrigenomics examines the effect of the dietary signatures in specific cells, tissues, or organisms and unravels their influence on homeostasis. Additionally, nutrigenomics aims to

DOI: 10.1201/b23311-10

comprehend the mechanisms underlying these genetic predispositions and assess the genome-wide risk of diseases associated with nutrition [1]. The five elementary principles of nutrigenomics are as follows:

1. Micro- and macronutrients available in food can affect the human genome either directly or indirectly by potentially changing the structure and expression of genes.
2. Diet may be considered as a decisive risk factor for inducing the diseases under certain circumstances.
3. The active substances present in diet can regulate gene expression and aid in the onset of disease or severity of disease.
4. The extent to which a diet can affect an individual's health might depend on their genetic makeup.
5. Information about the nutritional status and genotype of an individual can be assisted in the direction of preventing chronic disease.

A nutrient is not solely the chemical component that has its function in metabolism but also aids in the informational or signaling role in the cell. A system that transmits information in the form of a signal must have a receiver or a sensor to accept, decode, or relay the information [2]. Cellular proteins THAT receive and transmit the signal are entitled to receptors. These receptors are accountable for transmitting the information by the virtue of transducing mechanism to the cell, which can reprogram the cell to get acclimatized to new environmental conditions. Specificity plays a pivotal role in the principle of nutrients controlling gene expression. The receptor must be competent to bind with a nutrient-signaling molecule and initiate the adaptive changes. Earlier, it was considered that the changes in the nutritional environment affect the regulation of gene expression via hormones. However, shreds of evidence from the last decade indicate that the nutrients influence the gene expression in a hormone-independent manner [3]. Thus, it was concluded that the common dietary compound could alter the balance between health and diseases by affecting gene expression.

Interaction between nutrients and genes can be done based on three major concepts (**Figure 10.1**):

Direct interactions: The nutrient interacts with receptor and acts as a transcription factor that can directly bind to DNA and keenly regulate gene expression.
Epigenetic interactions: The nutrient shows epigenetic interaction as it can alter DNA structure and cause alteration of gene expression chronically.
Genetic variations: Alteration in gene expression can occur through genetic variations.

Cancer, being a prevalent cause of mortality all over the world, is considered as a complicated genetic predisposition that is associated with several genetic variations. Earlier studies reveal the

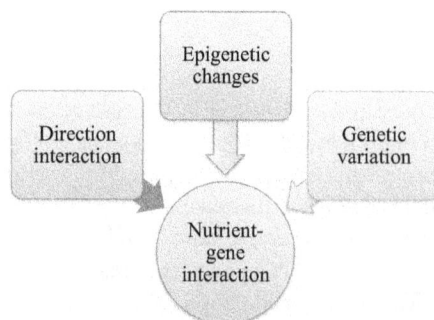

FIGURE 10.1 Three major concepts of interaction between nutrient and gene.

fact that genetic and epigenetic factors are associated with tumor progression. Several epidemiological studies stated that the dietary compound enacted a very crucial role in cancer prevention, mainly in the earlier stages of carcinogenesis [4]. It was observed that the natural dietary compound was involved in late carcinogenesis, invasion, and pro-angiogenic phase. Nutrigenomics can be utilized to personalize the diet of cancer patients as it was demonstrated in earlier studies that approximately 30–40% of cancer types are influenced by diet [5]. Fruits, vegetables, nuts, cereals, tea, wine, and other dietary compounds unveil the correlation with cancer chemoprevention. Nutrients that exhibit the tumor inhibitory mechanism, including inhibition of angiogenesis and induction of apoptosis, can be utilized for individualized medical treatment in cancer chemoprevention [6]. Dietary components, with the help of nutrigenomics, can become the plausible determinants of cancer risk and tumor behavior.

10.2   CANCER AND ITS GENETIC CORRELATION

Gene interaction is an evident cause of almost all diseases in human beings. Changes in the cellular genome have an impact on the gene's expression and functionality, which eventually control cell growth and differentiation, due to which it is considered as the leading cause of cancer. It was observed that the genetic changes promote cancer, and until today, its identification remains a challenging task for researchers [7]. In prior studies, it was discovered that the tumor and the ras gene mutation were related. K-ras gene and N-ras gene are primarily mutated ras genes in adenocarcinoma and myeloid leukemia, respectively. The mutant ras proteins are independent of any kind of incoming signal and remain in an activated state. The three ras genes—11-ras, k-ras, and N-ras—are transformed into active oncogenes by mutations in their codons 12, 13, or 61 [8]. The mutation of the tumor suppressor gene or oncogene causes gain or loss of function which leads to abnormal gene expression is quite evident in the genetic pathway of cancer. However, the epigenetic pathways to cancer are not relatively straightforward. The chromatin structure, including histone variations and modifications, DNA methylation, nucleosome remodeling, and small noncoding regulatory RNAs, determines the epigenetic pathways [9]. The epigenome is intended for multiple alterations, such as loss of DNA methylation across the whole genome, change in nucleosome occupancy, and modifications in the interim of tumor initiation and progression. It was demonstrated in earlier studies that epigenetic silencing would lead to reducing gene functionality and inducing genetic mutation. Hypermethylation in cancer is attributed to the classic tumor suppressor gene and implicated in tumorigenesis. DNA hypermethylation prevails in the expansion of progenitor cells in which there is loss of imprinting (LOI) in normal tissue, which leads to the initiation of tumor in kidney cells [10]. It was reported in earlier studies that the genes which tend to control the cell cycle or DNA repair including PTEN, RB, and BRCA1/2 have been deleted or mutated during the progression of cancer cells [11]. Several genes get mutated and remain silenced amid cancer. Promoter hypermethylation has been attributed to the predominant mechanism for functional loss. O6-methylguanine-DNA methyltransferase (MGMT), cyclin-dependent kinase inhibitor 2B (CDKN2B), and RASSF1A, which encodes a DNA repair gene and encodes a cell cycle regulator, p15, a protein that binds to the RAS oncogene, respectively, are associated with that category and have roles against tumorigenesis [12]. Critical genes, including p53 or KRAS, are subjected to genetic mutation when MGMT gets hypermethylated, which is associated with the initiation and progression of cancer [13]. Due to promoter hypermethylation, this gene's genomic stability and loss of function are both significantly impacted by MLH1, the mismatch repair gene that leads to the instability of microsatellites subsequently leading to several cancers such as endometrial and colorectal cancers [14]. Normal cells secreted frizzled-related proteins (SFRPs) are important for inhibiting WNT signaling. Due to its epigenetic silencing, WNT signaling is abnormally activated, and the gene expression that drives cell proliferation is promoted. This abnormal cell proliferation increases cancer risks. It was observed that oncogenic miRNA, such as miR-155 and miR-21, are frequently over-expressed, and tumor suppressor genes, including miR-146 or miR-15~16, are deleted at the time of cancer progression [15]. Mutation in miRNAs can lead to the disruption of its recognition ability to bind targets and result in the activation and repression of oncogenic gene and tumor suppressor gene respectively. The miRNAs such as miR-29 and miR-101 can target epigenetic modifiers, including DNMT3A/B and EZH2, respectively, and eventually lead to epigenetic alterations. Additionally, it can result in the methylation of target oncogenes and the promoters of other miRNAs [16]. Epigenomes act as the apex of the hierarchy of genome control mechanisms expedient that the mutation might have its effect on multiple pathways that are significant to the cancer phenotype. The fact that many genetic and epigenetic aberrations are present within cancer implicates that cancer therapy will be more effective when associated with epigenetic or alternative anticancer mechanisms including conventional chemotherapy [17].

10.3   DIETARY SIGNALS IN NUTRIGENOMICS

The variability and complexity of nutrients are encompassed by nutrigenomics. The body ought to process an enormous number of nutrients and food components. Each nutrient has its specificity and affinity toward numerous targets. Nutrients are not toxic even at high concentrations (mM to M) [18]. Nutritional research proves to be significant from comprehensive data on the effect of a compound at the molecular level. It has now become evident that the nutrients that function as fuels, cofactors, and micro- and macronutrients can have a crucial effect on the expression of genes and proteins and subsequently on metabolism. The specificity of nutrients to activate a specific signaling pathway is depending on its molecular structure. Sensory pathways are profoundly affected by the small changes in structure. The fine-tuned molecular specificity elucidates the varied effect of nutrients on different cellular functions. One of the illustrations is the effect of fatty acids depending on their level of saturation. It was observed in the earlier studies that ω-3 polyunsaturated fatty acid and C16–C18 saturated fatty acid have different effects on cardiac arrhythmia as the former shows a positive effect on cardiac arrhythmia whereas the latter does not show any effect [19]. Plasma levels of low-density lipoprotein are reduced by unsaturated C18 fatty acids, such as oleic and linoleic acid. It is demonstrated in the earlier observations that the nutrient signals have varied effects on different individuals [20]. A normal healthy person has a negative feedback system against the pro-inflammatory stress which is mediated by PPARs. A negative feedback system in healthy individuals acts as a nutrient sensor and can help in dealing with the variations in the free-fatty-acid level in plasma; however, in the case of individuals with obesity or diabetes, fatty acid accumulates in the form of triglycerides and accelerates toward the harmful pathway. Accumulation of TGs into non-adipose tissue increases the sensitivity of pro-inflammatory stress and might lead to severe organ dysfunction [21]. The upcoming challenge is to determine how nutrition influences the molecular pathway and analyze the downstream effect of specific nutrients. Nutrigenomics aids to identify all the nutritionally influenced pathways as it enables genome-wide characterization of genes whose expression is regulated by nutrients, it is achieved only with the absolute knowledge and information about the interlink between nutrition and genome that will enable researchers to assimilate the influence of nutrition on human health.

10.4   IMPLICATIONS OF BIOACTIVE PHYTOCHEMICALS IN CANCER THERAPY AND DECIPHERING THE GENOMIC PUZZLE

Together, omics technology and nutrigenomics enable us to understand and visualize the complete picture of the puzzle and allow researchers to shift the paradigm toward the whole new approaches to understand the nutritional agents. All types of cancer are characterized based on cellular and molecular alteration. It was observed that the diet plays an important role in the attainment of malignant phenotypes. The mystery of the nutritional genomics will be solved with the help of the integration of all genomic data and give insight into the notion of personalized nutrition. Phytochemicals, which are considered non-nutritive chemicals of the plants, are synthesized by the plant as a defense mechanism, and additionally, these phytochemicals are also responsible for the characteristics of aroma, taste, and color of fruits/vegetables [22]. The vast number of phytochemicals currently present are classified as thiols, terpenoids, and polyphenols. Flavonoids include flavones, flavonols, isoflavone, flavanones, and flavan-3-ols, such as kaempferol, quercetin, apigenin, luteolin, daidzein, genistein, naringenin, hesperidin, catechins, gallocatechin isomers, and anthocyanins [23]. It was demonstrated by the available epidemiological data that citrus consumption proves to be beneficial because of the presence of a significant amount of flavanones such as hesperetin and naringenin [24]. It was observed from earlier studies that flavonoids target the estrogen receptor and androgen receptors in human breast cancer cells and prostate cancer, respectively [25]. Flavonoids inhibit invasion and metastasis in cancer cells of breast cancer, lung cancer, ovarian cancer, and liver cancer via acting as a genomic modulator in signaling pathways [26]. Meta-analysis shows the remarkable and significant result that encompasses cancer risk and nutrigenomics. Meta-analysis studies have shown that dietary flavonoids reduce the incidence of ovarian cancer and have a preventive effect against the disease [27]. Correlation in reduction of breast cancer and exposure to flavones and flavonols in postmenopausal women was also reported in earlier studies [28].

10.5 NATURAL PHYTOCHEMICALS TARGETING THE KEY GENES IN PROCESS OF CARCINOGENESIS

Genomic approaches are utilized in clarifying several cancer-inducing mechanisms. Recently, competence to alter signal transduction mechanism and activation of transcription factors that are antagonized in the process of tumorigenesis recently attracted a lot of attention. Transcription factors bind to the response elements, causing gene expression to be triggered or inhibited. The post-genomic era's notion of the antitumor impact of phytochemicals can be speed up through the integration of all the available knowledge. The nutritional compound, in the association with standard therapies, shows beneficial effects, mainly in the direction of enhancing the chemotherapeutic efficacy. A rare case was observed in an earlier study that treatment of genistein in hepatocellular carcinoma induces the drug resistance mechanism, such as chemoresistance [29]. The natural bioactive compounds are able to target multiple pathways in cancer that will lead to carcinogenesis. The effect of the natural compounds was also observed on different mechanisms of cancer chemoprevention. These include the antioxidant activity, apoptosis activation, angiogenesis, inhibition of MAPK (mitogen-activated protein kinase), and NFκB (nuclear factor-κB) [30].

10.5.1 Antioxidant Effect

Nutrigenomics gives deep insight into the multipart view of the antioxidant mechanism and unravels the role of the bioactive compound at the initial stages of chemoprevention. In the course of time, to cope with the ROS (reactive oxygen species) imbalanced production, the human body gradually develops an all-encompassing antioxidant system, which includes antioxidant enzymes and non-enzymatic molecules like glutathione, catalase, and superoxide dismutase. The antioxidant system of the body is unbalanced and leads to oxidative damage of a broad range of biomolecules, such as DNA/RNA, lipids, and protein. Natural phytochemicals serve as a potent antioxidant and act as a scavenger of ROS and counteract the effect of ROS during the metabolic process [31]. It was observed in earlier studies that ROS exhibited a dual role. It acts as a crucial cell signaling mediator by activating the cytokine and hormone secretion and by mediating redox-responsive transcription factors [32]. It was demonstrated that the plant phytochemicals sporadically have a pro-oxidant effect based on dose-dependent manner, exposure time, and the combination with metal ions, such as Fe or Cu [33]. It was observed in an earlier study that EGCG at doses higher than 50μM shows pro-oxidant effects and induces autophagy and exhibits anticancer effects that are highly significant

[34]. EGCG doses less than 100µM inhibit transcription factors necessary for cell growth. A new class of polymeric oxidized flavanols generated by oxidizing EGCG has recently gained interest due to its significant biological action in the treatment of cancer [35]. The oxidized catechin incorporates a galloyl moiety to quinone, which is associated with the activation of electrophile responsive element (EpRE), being connected to glutathione (GSH) activation [36]. Thus, pro-oxidant mechanism induction is not consistently associated with the adverse effect. An earlier experiment showed that the catechin isomers such as quercetin act as pro-oxidants at doses over 50 µM and generate mitochondrial superoxide radicals in cell culture. Quercetin shows an ambivalent redox character to polyphenols [37].

10.5.2 EFFECTS OF DIETARY PHYTOCHEMICALS ON GENOMIC STABILITY AND NON-GENOTOXICITY

Genomic stability is one of the most common factors related to cancer etiology. Cancer cells are more susceptible to DNA damage in comparison to the normal cell, which imparts an avenue toward its therapeutic intervention. The presence of electrophiles or ROS mediates the health-threatening consequence of carcinogen exposure. The accumulation of these electrophiles leads to the initiation of cancer. There is an association between lifestyle factors, inappropriate diet or the formation of carcinogens during food processing, environmental tobacco smoke (ETS), and occupational or environmental carcinogens [38]. Nonetheless, the protective effects of natural phytochemicals against carcinogens and carcinogenesis are supported by the upsurge of pieces of evidence. Phytochemicals, isolated from the plants, are proven to be effective against malignancy that is induced by carcinogens present in the environment. Phytochemicals are also considered as health promoters. The interrelation of phytochemicals with environmental toxicology, especially in the context of genomic stability maintenance, shows a correlation with the improved phytochemicals ingestion via the ROS-detoxifying mechanism. Earlier studies demonstrate that the rise in ROS levels can lead to the early activation of carcinogenesis when it comes in contact with the pro-mutagenic lesions in DNA. Algae extracts consisting of lutein, α-carotene, β-carotene, and fatty acids have efficacy against the environmental genotoxic agents that are involved in ROS production [39]. Pelargonidin, an anthocyanide and chlorogenic, one of the most abundant polyphenols in the diet, was evaluated in HL-60 cells against 4-nitroquinoline 1-oxide (NQO) for its genotoxicity effect. It was revealed that both of them in combination show a protective effect against genotoxicity [40]. A similar result was observed in the grape seed extract, which is rich in resveratrol, ellagic acid, and lycopene [41].

10.5.3 INFLAMMATORY MEDIATORS

The natural phytochemicals have a vital role in interfering with the gene expression via transcription factor that is involved in the activation of inflammatory mechanisms [42]. According to the modern concept related to cancer biology, an apparent relation between oncogenesis and inflammation was observed. The tumor microenvironment contains a thorough list of the inflammatory mediators involved in tumor development, invasion, and metastasis [43]. The dietary nutrient targets the major mechanism of inflammation correlated with cancer by interacting and suppressing the MAPK, COX-2, iNOS, NFκB, STAT signaling, or inflammatory cytokines [44]. NFκB is a well-known transcription factor that plays a crucial role in inflammation and cancer. The pathway of NFκB is generally activated by inflammatory mediators or ROS that are repressed by polyphenolic compounds [45]. The natural phytochemical inhibits NFκB, suppresses cell proliferation, and increases cell sensitivity toward chemotherapeutic drugs, such as capecitabine [46]. It was demonstrated in earlier studies that resveratrol inhibits the expression of p-ERK and stimulates the expression of p-JNK in nigrostriatal pathway injury. Resveratrol exerts a neuroprotective effect by upregulating the expression of p-JNK and Bcl-2 and downregulating the Bax and Fluoro-Jade C (FJC) positive neurons. Resveratrol was also observed to suppress the TNF-α, IL-6, and IL-1β and show an anti-inflammatory effect [47]. The hybrid of resveratrol and caffeic acid was proven to have a good efficiency against the STAT pathway in breast cancer xenograft tumor models [48]. It was also observed that resveratrol induces the SOCS-1 and targets the STAT3 signaling pathway and acts as a sensibilizer to therapy [49].

10.5.4 Apoptosis and Cell Cycle Arrest

Apoptosis, being a programmed cell death, plays a pivotal role in the defense mechanism against cancer. Cancer often gets associated with the aberrantly regulated apoptotic cell death. The elucidation of the events that occur during apoptosis and carcinogenesis bestows the opportunity to decipher the intervention of dietary bioactive compounds [50]. Phytochemicals foster the modulation of the genomic underpinning of cell cycle regulation and apoptosis in physiological and normal status. The disabled apoptosis is more susceptible to genomic stability and can lead to tumorigenesis; therefore, it is a prime protective mechanism to prevent tumorigenesis by aiming the elimination of all the genetically damaged cells [51, 52]. It was demonstrated that the folate deficiency could lead to several cancers, including colorectal cancer, as its deficiency leads to genomic instability. The MTHFR polymorphism occurs due to the substitution of cytosine with thymine at 677 nucleotides, the most prevailing SNP variant well known in folic acid metabolism. MTHFR polymorphism is observed to have an association with apoptosis. The supplementation with folic acid prevents the teratogenic activity of MTHFR [53]. DAPK and APAF-1, which are directly associated with apoptosis and cancer, are observed to be affected by hypermethylation and can be utilized as a therapeutic target [54]. Various bioactive compounds, such as folate and B12, are involved in methylation and show epigenetic effects in the process that is linked with cancer and apoptosis. EGCG, being a natural compound, shows its effect against the cancer cell via inducing the apoptosis and cell cycle arrest without influencing the normal cell [55]. The effect of EGCG was observed on triple-negative breast cancer cells, Hs578T. EGCG proved to be significant as the proliferation of cancer cells was reduced due to activation of the anti-apoptotic genes, such as XIAP, BAG3, and RIPK2 [56].TRAF6 (TNF receptor-associated factor 6) plays a crucial role in signaling the transduction of the apoptotic mechanisms. It was observed that TRAF6 was overly expressed in the case of melanoma. It was demonstrated that the downregulation of the TRAF6 gene attenuates the tumor cell growth and metastasis. EGCG plays a pivotal role in suppressing the interaction between TRAF6 and UB13C and inhibiting the cell growth, migration, and invasion of melanoma cells. Curcumin shows a synergistic effect with resveratrol and decreases the gene expression of Bax, Bcl2, Apaf1, Fas, and FasL, simultaneously inhibiting the activated kinases, such as p38, JNK, and ERK1/2, thus playing a crucial role in cancer chemoprevention [57].

10.5.5 Angiogenesis

Angiogenesis, the generation of the new blood vessels is demonstrated as an essential mechanism in cancer progression. It proves to be an important target to inhibit tumor growth. Normal healthy cells have oxygen mechanisms, such as endothelial nitric oxide synthase (eNOS), heme-oxygenases, and oxygen-sensitive NADPH oxidases that act as a key regulating point in angiogenesis [58]. The hypoxia-inducible transcription factor (HIF) family, an important molecular interface for conveying adaptations to changes in oxygen tension, is also articulated by vascular cells as a separate class of oxygen sensors. HIF has three isoforms HIF1–3 that get heterodimerized with aryl hydrocarbon receptor nuclear translocator (HIF_/ARNT) and form an active transcriptional complex that eventually initiates the expression of many genes that are involved in the angiogenesis regulation [59]. Endothelial cells show a robust upsurge in growing cancer cells due to the release of several proteins, including EGF, VEGF, prostaglandin E1 and E2, and TNF- α, that stimulate the activation of endothelial cell growth and motility by constitutively repressing the antiangiogenic factors in cancer cells [60]. Preclinical studies show that natural phytochemicals are enabled to inhibit the

VEGF and its related receptors [61]. The inhibition of VEGF (vascular endothelial growth factor) expression via HIF-1α was observed after the treatment of resveratrol [62]. It was demonstrated in both the *in vivo* and *in vitro* studies that EGCG also shows significant activity against HIF-1α expression. It significantly reduces the HIF-1α expression, which subsequently leads to the inhibition of angiogenesis. EGCG also proves to be efficient in repressing carcinogenic molecular signals, such as MMP-2/9 and Notch1 [63]. Preclinical studies on lung cancer also reveal that EGCG is able to enhance the apoptotic activity of cancer cells [64].

10.6 CHEMOPREVENTION AND NATURAL COMPOUNDS

The intent of cancer chemoprevention is to disrupt the molecular pathways that are involved in carcinogenesis. The molecular mechanism, including increased signal transduction to NF-kB, DNA damage by reactive oxygen species, the epithelial-mesenchymal transition, and epigenomic deregulation that leads to metastasis, can be disrupted by the chemopreventive suppressing or blocking agents [65]. Cancer chemoprevention implicates the chronic administration of synthetic, natural, and/or biological agents to inhibit or reduce the malignancy. Plentiful dietary phytochemicals are involved in the inhibition of the initial phase of carcinogenesis, which enables them to be utilized as a primary chemopreventive agent [66]. It was observed that the phytochemicals are involved in the inhibition of the molecular mechanisms of metastasis. The main classes that involve in cancer chemoprevention comprise carotenoids, phytosterols, and phenolic compounds [67, 68]. The antitumor effect of phytochemicals involved in integral processes is associated with transport, biotransformation, absorption, and molecular and cellular action pathways [69]. Several aspects in which natural phytochemicals proved to be beneficial are (1) they are commonly present in consumed food and easily available to most people in their daily life, (2) they have little or no toxicity in comparison to chemotherapeutic drugs, and (3) the phytochemical compound proves to be efficient as an adjunct to chemotherapeutic drugs. Phytochemical compounds such as sulforaphane, piperine, and curcumin act as potent inhibitors of Wnt/β-catenin signaling and cancer stem cells of breast cancer at relatively low concentrations [70].

10.7 CONTROLLING EPIGENETIC CHANGES AND mi RNAs INVOLVED IN CARCINOGENESIS AND CANCER PREVENTION: THE ROLE OF NATURAL PRODUCTS

Dietary phytochemicals are involved in the alteration of DNA methylation of the essential tumor suppressor gene. Histone modification, alteration of tumor suppressive miRNAs, and HDAC activity are utilized in cancer chemoprevention. Phytochemicals, thiols, terpenoids, and polyphenols are the antecedent of taste, color, and aroma in fruits and vegetables. Flavan-3-ols (catechins and gallocatechin isomers), flavanones (naringenin and hesperidin), isoflavones (daidzein and genistein), flavonols (apigenin and luteolin), flavones (kaempferol and quercetin) and anthocyanin are classified as flavonoids (Figure 10.2). Flavonoids are also reported as a compound that targets estrogen and androgen receptors in the human breast cancer cells and prostate cancer respectively. The dietary intake of flavanones, flavonoids, and anthocyanidins effectively reduces the risk of esophageal cancer [71]. Flavonoids effectively prevent invasion and metastasis via epigenetic modification in different types of cancers, such as lung, prostate, ovarian, breast, and colon cancer. Carcinogenesis is targeted by dietary natural compounds, such as curcumin, genistein, resveratrol, indole-3-carbinol, and EGCG via regulation of miRNA [72].

10.7.1 Isothiocyanates (ITC)

Isothiocyanates are commonly present in cruciferous vegetables. Sulforaphane and glucosinolates are the dietary compounds of isothiocyanates that are mainly present in broccoli and brussels sprouts. Sulforaphane acts on the cell cycle and inhibits colon cancer cells from proliferating by

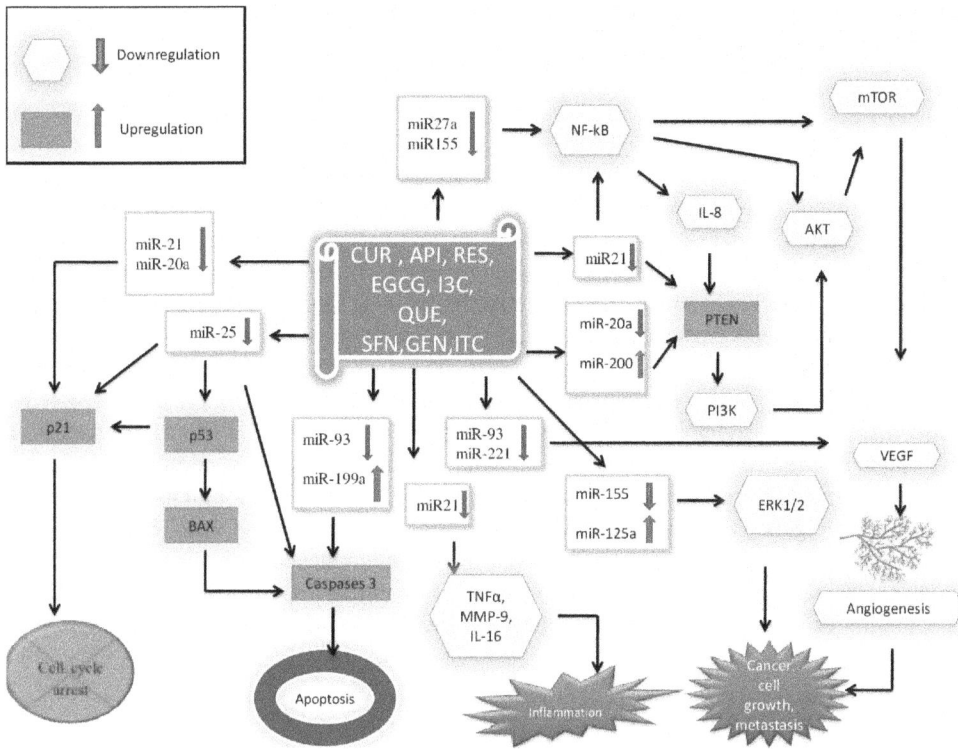

FIGURE 10.2 Role of natural products in controlling miRNAs involved in tumorigenesis of cancer.

Abbrevations: CUR—Curcumin, API—Apigenin, ITC—Isothiocyanates, EGCG—Epigallocatechin Gallate, RES—Resveratrol, GEN—Genistein, SFN—Sulforaphane, I3C—Indole-3-Carbinol, ROS—Reactive oxygen species, NF-kB—Nuclear Factor Kappa B, TNF-α—Tumor Necrosis Factor Alpha, MMP—Matrix Metallo proteinase, IL—Interleukin, mTOR—Mammalian Target of Rapamycin, VEGF—Vascular Endothelial Growth Factor, AKT—Protein kinase B, PI3K1—Phosphatidylinositol 3-kinase, PTEN—Phosphatase and Tensin Homolog, ERK—Extracellular Signal-Regulated Kinase, BAX—BCL2.

arresting the cell cycle and also suppressing the activity of DNMT and HDAC3. Recently, the activity of sulforaphane in the prevention of prostate cancer has been established. Sulforaphane involves in the prevention of prostate cancer in cell lines by reducing the expression of DNMT1 and -3b and also by reprogramming the CCND2 expression [73]. Isothiocyanates show significant activity in inducing apoptosis and cell cycle arrest by upregulation of Bax expression and overexpression of p21, respectively, in human prostate cancer epithelial cells, such as PC-3 and LNCaP cells, by preventing HDAC activity and acetylation of the level of histones [74].

10.7.2 Apigenin

Apigenin belongs to the plant flavonoid compound that is present in plenty of vegetables, beans, tea leaves, and fruits. Apigenin regulates the growth of prostate cancer cells, such as PC-3 and 22Rv1, via inhibiting the class I HDACs activity and inducing programmed cell death. Tumor suppressor gene miRNA-16, which is associated with human tumors, is observed to be downregulated by the proto-oncogene BMI1 [75]. The miR-16 is involved in suppressing NF-κB and matrix metalloproteinase (MMP-9) and reducing the degree of malignancy [76]. It was demonstrated in previous studies that apigenin exhibits a potent anticancer activity and exerts a cytotoxic effect on various cancer cells. It induces apoptotic cell death on human glioma cell U87. Additionally, it significantly increases the miR-16 levels that subsequently suppress BCL2 protein expression and NF-kB/MMP9 signaling [77].

10.7.3 Curcumin

Curcumin is responsible for hypomethylation in leukemia cells. P300/CBP HAT inhibition is observed in prostate cancer after the intervention of curcumin [78]. It is proven to be a potent chemopreventive agent. Curcumin, a bioactive compound in turmeric, shows potent anti-inflammatory effects in a variety of malignancies. By upregulating miR-22 and downregulating miR-199a, curcumin inhibits cell growth in pancreatic cancer. Additionally, it limits oncogenic miRNA expression (miR-196). It is also reported in a previous study that curcumin shows its effect on triple-negative breast cancer cells MDA-MB-231 by inducing apoptosis and reducing proliferation via overexpressing the miR-181b, which subsequently regulates the expression of matrix metalloproteinases and also the release of chemokines [79]. Curcumin abates the expression of miR-200 and miR21 and subsequently reduces the expression of PTEN, a tumor-suppressing gene in pancreatic cancer [80]. It was reported in breast adenocarcinoma cell MCF-7 that curcumin inhibits the expression of the anti-apoptotic gene Bcl-2 and induces the expression of miR-15a and miR-16 [81]. Oncogenic miRNA miR-186 suppression was also observed in A59 lung adenocarcinoma cells via induction of apoptosis by curcumin [82]. It was demonstrated in an earlier report that curcumin induces the overexpression of mi-RNA in human esophageal cancer cells, which leads to the suppression of tumors [83].

10.7.4 Epigallocatechin Gallate (EGCG)

EGCG is one of the polyphenols that act as potent anticancer agents and is present in ample amounts in green tea. It was demonstrated that EGCG exerts multiple effects on cancer cells via epigenetic changes by histone acetylation or deacetylation, DNA methyltransferases, or methylation and noncoding RNAs (microRNAs) [84]. It is reported that the re-expression of p21/Cip1 mRNA and p16 (INK4a) was observed after EGCG treatment as it reduces the DNMT1 expression at mRNA and protein levels [85]. Estrogen receptor expression in triple-negative breast cancer cells is reactivated by the treatment of EGCG. It reverses the expression of the gene RECK, which is a tumor suppressor gene, in carcinoma cells. It was observed in earlier studies that EGCG upregulates the expression of miR-16 in HepG2 cells and subsequently impedes the expression of Bcl-2, the apoptosis inhibitory gene [86]. EGCG affects neuroblastoma cell line by upregulating the tumor-suppressive miRNA, including miR-99a, miR-34a, and miR-7–1, and inhibiting the expression of oncogenic miRNAs, such as miR-106b, miR93, and miR-92 [87].

10.7.5 Resveratrol

Resveratrol shows the antiproliferative activities in the varied types of cancer, such as lung, prostate, breast, and liver cancer. It inhibits the migration of colon cancer. It is reported in earlier studies that proliferation in breast adenocarcinoma cells MCF-7 is prevented by resveratrol via upregulating the miR-774 and miR-663 [88]. Resveratrol hindered the expression of miR-21 and promote apoptosis of cancerous pancreatic cells [89]. In the triple-negative breast cancer cell line MDA-MB-231, previous investigations showed that resveratrol upregulates the tumor suppressor miRNAs miR-141 and miR-200c [90].

10.7.6 Genistein

In prostate cancer and esophageal squamous cell carcinoma cells, genistein reactivates tumor suppressor genes, such p16, RAR, and MGMT. Genistein induces the expression of miR-146a and prevents the tumor-inducing gene expression, including MTA-2, IRAK-1, NF-kB, and EGFR, in pancreatic cancer cells [91]. Genistein arrests the cell cycle at the S-phase in prostate cancer cells via inducing the expression of miR-1296 [91]. It is observed in previous studies that genistein down-regulates the expression of miR-221 and miR-222 subsequently leading to the upregulations of the ARH1 gene and inhibiting the proliferation of cancer cells in prostate cancer [92]. Genistein, in renal carcinoma, inhibits the expression of miR-23b-3p, which consequently targets the expression PTEN gene and induces apoptosis of cancer cells. It also demonstrated that it inhibits the migration and invasion of prostate cancer cells via downregulating the expression of miR-151 [93]. It was also observed that genistein inhibits the expression of miR-1260b [94].miR-34a, which is involved in the induction of programmed cell death and inhibition of cell proliferation is unregulated by genistein in the pancreatic cells [95].

10.7.7   SULFORAPHANE AND INDOLE-3-CARBINOL (I₃C)

Sulforaphane (SFN) and indole-3-carbinol (I_3C) are present in the *Brassica* and crucifers are reported as a potent chemopreventive agent in cancer cells. I_3C regulates cancer by inducing miR-26a, miR-34b, and miR-125b, respectively [96]. It was observed that miR-21 is an attractive target for cancer chemoprevention. I_3C being a chemopreventive agent downregulates VC (Vinyl carbonate), which in turn downregulates the overexpression of miR-21 and prevents cancer progression [97]. SFN induces the expression of miR-let-7a, which subsequently constrains CSC (cancer stem cell) characteristics and inhibits K-ras expression at the time of progression of pancreatic ductal adenocarcinoma [98]. Additionally, SFN also modulates the expression of mi RNAs in ductal carcinoma, such as miR-29a, miR-21, and miR-140, and induces substantial changes in the exosomal production of miRNAs that resemble non-cancer stem cells, representing a potent chemopreventive agent in the early stage of breast cancer [86].

10.7.8   QUERCETIN

Quercetin, a flavonoid that is present in leafy vegetables, apples, onions, and black tea, has the capability of inducing cytotoxic effects, including inhibition of cell proliferation and apoptosis in cancer cells. It was reported in earlier studies that the expression of miR-146a was decreased in a multitude of cancer including prostate, lung, breast, and gastric cancer [99]. It was observed that quercetin acts as an efficient chemopreventive agent against MCF-7 and MDA-MB-231 human breast cancer cells and xenograft models, respectively. Quercetin upregulates the expression of miR-146a and induces apoptosis via cleaved caspase-3 activation and mitochondrial-dependent pathway [100]. Inhibition of claudin-2 expression via overexpression of miR-16 of miR-142–3p was reported by quercetin in earlier studies [101]. It increases the expression of miR-142–3p and inhibits the proliferation of pancreatic ductal adenocarcinoma [102].

10.8   CROSSTALK

Cancer is a comprehensive and multifactorial disease that involves numerous genetic alterations as well as changes to pathways and their underlying mechanisms. Natural phytochemicals are thought to be important signaling molecules, as seen by their nutrigenomics patterns. Natural phytochemicals were found to have additive or synergistic effects in most combinatorial techniques. The natural bioactive agent shows significant results when utilized with chemotherapy as one of the most important synergistic characteristics of them is to reduce toxicity. EGCG combination with quercetin significantly inhibits the streptozotocin, induces cell damage and upsurges the glucose-stimulated insulin secretion in pancreatic β-cell. The EGCG- quercetin synergistically upregulates BCL-2 expression and reduces the level of miR-16–5p, which subsequently downregulates the apoptotic genes that cause cell damage [103]. Resveratrol, in combination with anticancer drugs, synergistically enhances its growth inhibition activity and induces apoptosis via suppressing the pathway of MAPK/ERK1/2in colon cancer DLD-1 cells **[104]**.

10.9 CONCLUSION AND FUTURE PERSPECTIVE

Nutrigenomic patterns can help the researcher to understand the key signaling pathways of natural phytochemicals. Bioavailability, routes, and dose of bioactive dietary compounds play a vital role in determining the chemopreventive strategies for cancer treatment. Nutrigenomics examines the effects of dietary substances on the expression of genes that are involved in tumor cell metastasis, DNA repair, angiogenesis, and apoptosis. Diet and supplements are critical factors and important determinants in cancer chemoprevention. Nutrigenomics is intended to understand the effect of dietary signals on cancer cells and personalize the diet of an individual based on genetic and epigenetic variations. Despite the exhilarating pioneering days of nutrigenomics, it is important to highlight the potential obstacle to its success. The nutrients and other ingredients that make up food are complex and varied in composition. The majority of nutrients have weak dietary signals. Thus, researchers still need to develop systems that can identify nutritional deficiencies by spotting these weak signals. Researchers in the field of nutrigenomics should also address the challenges related to polygenic diet-related diseases.

10.10 REFERENCES

1. Reen, Jagish Kour, Alok Kumar Yadav, and Jitendra Singh 2015. "Nutrigenomics: Concept, advances and applications." *Asian Journal of Dairy and Food Research* 34(3): 205–212.
2. Walker, W. Allan, and George Blackburn 2004. "Symposium introduction: Nutrition and gene regulation." *The Journal of Nutrition* 134(9): 2434S–2436S.
3. Chattopadhyay, Indranil 2020. "Role of nutrigenetics and nutrigenomics in cancer chemoprevention." In *Pharmacotherapeutic Botanicals for Cancer Chemoprevention*, 167–188. Springer.
4. Waltenberger, Birgit, Andrei Mocan, Karel Šmejkal, Elke H. Heiss, and Atanas G. Atanasov 2016. "Natural products to counteract the epidemic of cardiovascular and metabolic disorders." *Molecules* 21(6): 807.
5. Braicu, Cornelia, Nikolay Mehterov, Boyan Vladimirov, Victoria Sarafian, Seyed Mohammad Nabavi, Atanas G. Atanasov, and Ioana Berindan-Neagoe 2017. "Nutrigenomics in cancer: Revisiting the effects of natural compounds."*Seminars in Cancer Biology* 46: 84–106.
6. Surh, Young-Joon 2003. "Cancer chemoprevention with dietary phytochemicals." *Nature Reviews Cancer* 3(10): 768–780.
7. Barbacid, Mariano 1987. "Ras genes." *Annual Review of Biochemistry* 56(1): 779–827.
8. Bos, Johannes L 1989. "Ras oncogenes in human cancer: A review." *Cancer Research* 49(17): 4682–4689.
9. Sharma, Shikhar, Theresa K. Kelly, and Peter A. Jones 2010. "Epigenetics in cancer." *Carcinogenesis* 31(1): 27–36.
10. Baylin, Stephen B., and Peter A. Jones 2011. "A decade of exploring the cancer epigenome—biological and translational implications." *Nature Reviews Cancer* 11(10): 726–734.
11. Hatziapostolou, Maria, and Dimitrios Iliopoulos 2011. "Epigenetic aberrations during oncogenesis." *Cellular and Molecular Life Sciences* 68(10): 1681–1702.
12. Esteller, Manel 2007. "Cancer epigenomics: DNA methylomes and histone-modification maps." *Nature Reviews Genetics* 8(4): 286–298.
13. Krivtsov, Andrei V., and Scott A. Armstrong 2007. "MLL translocations, histone modifications and leukaemia stem-cell development." *Nature Reviews Cancer* 7(11): 823–833.
14. Kasinski, Andrea L., and Frank J. Slack 2011. "MicroRNAs en route to the clinic: Progress in validating and targeting microRNAs for cancer therapy." *Nature Reviews Cancer* 11(12): 849–864.
15. Fabbri, Muller, and George A. Calin 2010. "Epigenetics and miRNAs in human cancer." *Advances in Genetics* 70: 87–99.
16. You, Jueng Soo, and Peter A. Jones 2012. "Cancer genetics and epigenetics: Two sides of the same coin?" *Cancer Cell* 22(1): 9–20.
17. Evans, William E., and Howard L. McLeod 2003. "Pharmacogenomics—drug disposition, drug targets, and side effects." *New England Journal of Medicine* 348(6): 538–549.

18. Brouwer, Ingeborg A., Peter L. Zock, Ludovic GPM Van Amelsvoort, Martijn B. Katan, and Evert G. Schouten 2002. "Association between n-3 fatty acid status in blood and electrocardiographic predictors of arrhythmia risk in healthy volunteers." *American Journal of Cardiology* 89(5): 629–631.

19. Sacks, Frank M., and Martijn Katan 2002. "Randomized clinical trials on the effects of dietary fat and carbohydrate on plasma lipoproteins and cardiovascular disease." *The American Journal of Medicine* 113(9): 13–24.

20. Müller, Michael, and Sander Kersten 2003. "Nutrigenomics: Goals and strategies." *Nature Reviews Genetics* 4(4): 315–322.

21. Middleton, Frank A., Eduardo JB Ramos, Yuan Xu, Heba Diab, Xin Zhao, Undurti N. Das, and Michael Meguid 2004. "Application of genomic technologies:-DNA microarrays and metabolic profiling of obesity in the hypothalamus and in subcutaneous fat." *Nutrition-New York* 20(1):14–25.

22. Canales, Roger D., Yuling Luo, James C. Willey, Bradley Austermiller, Catalin C. Barbacioru, Cecilie Boysen, Kathryn Hunkapiller Roderick V. Jensen, Charles R. Knight, Kathleen Y. Lee, Yunqing Ma, Botoul Maqsodi, Adam Papallo, Elizabeth Herness Peters, Karen Poulter, Patricia L. Ruppel, Raymond R. Samaha, Leming Shi, Wen Yang, Lu Zhang and Federico M. Goodsaid 2006. "Evaluation of DNA microarray results with quantitative gene expression platforms." *Nature Biotechnology* 24(9): 1115–1122.

23. Tsuda, Takanori, Yuki Ueno, Toshikazu Yoshikawa, Hitoshi Kojo, and Toshihiko Osawa 2006. "Microarray profiling of gene expression in human adipocytes in response to anthocyanins." *Biochemical Pharmacology* 71(8): 1184–1197.

24. Sak, Katrin 2014. "Cytotoxicity of dietary flavonoids on different human cancer types." *Pharmacognosy Reviews* 8(16): 122–146.

25. Mc Loughlin, Patricia, Monic Roengvoraphoj, Cornelia Gissel, Jürgen Hescheler, Ulrich Certa, and Agapios Sachinidis 2004. "Transcriptional responses to epigallocatechin-3 gallate in HT 29 colon carcinoma spheroids." *Genes to Cells* 9(7): 661–669.

26. Hua, Xiaoli, Lili Yu, Ruxu You, Yu Yang, Jing Liao, Dongsheng Chen, and Lixiu Yu 2016. "Association among dietary flavonoids, flavonoid subclasses and ovarian cancer risk: A meta-analysis." *PLoS One* 11(3): e0151134.

27. Hui, Chang, Xie Qi, Zhang Qianyong, Peng Xiaoli, Zhu Jundong, and Mi Mantian 2013. "Flavonoids, flavonoid subclasses and breast cancer risk: A meta-analysis of epidemiologic studies." *PloS One* 8(1): e54318.

28. Rigalli, Juan Pablo, Nadia Ciriaci, Agostina Arias, María Paula Ceballos, Silvina Stella Maris Villanueva, Marcelo Gabriel Luquita, Aldo Domingo Mottino, Carolina Inés Ghanem, Viviana Alicia Catania, and María Laura Ruiz 2015. "Regulation of multidrug resistance proteins by genistein in a hepatocarcinoma cell line: Impact on sorafenib cytotoxicity." *PloS One* 10(3): e0119502.

29. Vittal, Ragini, Zachariah E. Selvanayagam, Yi Sun, Jungil Hong, Fang Liu, Khew-Voon Chin, and Chung S. Yang 2004. "Gene expression changes induced by green tea polyphenol (–)-epigallocatechin-3-gallate in human bronchial epithelial 21BES cells analyzed by DNA microarray." *Molecular Cancer Therapeutics* 3(9): 1091–1099.

30. Koc, Tugca Bilenler, Ebru Kuyumcu Savan, and Ihsan Karabulut 2021. "Determination of Antioxidant Properties and β-Carotene in Orange Fruits and Vegetables by an Oxidation Voltammetric Assay." *Analytical Letters* 55(6): 1–13.

31. Bjørklund, Geir, and Salvatore Chirumbolo 2017. "Role of oxidative stress and antioxidants in daily nutrition and human health." *Nutrition* 33: 311–321.

32. Lee, Ki Won, Hyong Joo Lee, Young-Joon Surh, and Chang Yong Lee 2003. "Vitamin C and cancer chemoprevention: Reappraisal." *The American Journal of Clinical Nutrition* 78(6): 1074–1078.

33. Kim, Hae-Suk, Michael J. Quon, and Jeong-A. Kim 2014. "New insights into the mechanisms of polyphenols beyond antioxidant properties; lessons from the green tea polyphenol, epigallocatechin 3-gallate." *Redox Biology* 2: 187–195.

34. Peluso, Ilaria, and Mauro Serafini 2017. "Antioxidants from black and green tea: From dietary modulation of oxidative stress to pharmacological mechanisms." *British Journal of Pharmacology* 174(11): 1195–1208.

35. Muzolf-Panek, Małgorzata, Anna Gliszczyńska-Świgło, Laura de Haan, Jac MMJG Aarts, Henryk Szymusiak, Jacques M. Vervoort, Bozena Tyrakowska, and Ivonne MCM Rietjens 2008. "Role of catechin quinones in the induction of EpRE-mediated gene expression." *Chemical Research in Toxicology* 21(12): 2352–2360.

36. De Marchi, Umberto, Lucia Biasutto, Spiridione Garbisa, Antonio Toninello, and Mario Zoratti 2009. "Quercetin can act either as an inhibitor or an inducer of the mitochondrial permeability transition pore:

A demonstration of the ambivalent redox character of polyphenols." *Biochimica et Biophysica Acta (BBA)-Bioenergetics* 1787(12): 1425–1432.

37. Liskova, Alena, Patrik Stefanicka, Marek Samec, Karel Smejkal, Pavol Zubor, Tibor Bielik, Kristina Biskupska-Bodova 2020. "Dietary phytochemicals as the potential protectors against carcinogenesis and their role in cancer chemoprevention." *Clinical and Experimental Medicine* 20(2): 173–190.

38. Bhagavathy, S., and P. Sumathi 2012. "Evaluation of antigenotoxic effects of carotenoids from green algae Chlorococcum humicola using human lymphocytes." *Asian Pacific Journal of Tropical Biomedicine* 2(2): 109–117.

39. Abraham, Suresh K., Nicole Schupp, Ursula Schmid, and Helga Stopper 2007. "Antigenotoxic effects of the phytoestrogen pelargonidin chloride and the polyphenol chlorogenic acid." *Molecular Nutrition & Food Research* 51(7): 880–887.

40. Rajendran, Praveen, Emily Ho, David E. Williams, and Roderick H. Dashwood 2011. "Dietary phytochemicals, HDAC inhibition, and DNA damage/repair defects in cancer cells." *Clinical Epigenetics* 3(1): 1–23.

41. Sales, Nevilde Maria Riselo, Patrícia Barbosa Pelegrini, and Goersch 2014. "Nutrigenomics: Definitions and advances of this new science." *Journal of Nutrition and Metabolism* 2014: 1–6.

42. Bjørklund, Geir, and Salvatore Chirumbolo 2017. "Role of oxidative stress and antioxidants in daily nutrition and human health." *Nutrition* 33: 311–321.

43. Tran, Thi Van Anh, Clemens Malainer, Stefan Schwaiger, Tran Hung, Atanas G. Atanasov, Elke H. Heiss, Verena M. Dirsch, and Hermann Stuppner 2015. "Screening of Vietnamese medicinal plants for NF-κB signaling inhibitors: Assessing the activity of flavonoids from the stem bark of Oroxylum indicum." *Journal of Ethnopharmacology* 159: 36–42.

44. Wang, Hu, Tin Oo Khor, Limin Shu, Zheng-Yuan Su, Francisco Fuentes, Jong-Hun Lee, and Ah-Ng Tony Kong 2012. "Plants vs. cancer: A review on natural phytochemicals in preventing and treating cancers and their druggability." *Anti-Cancer Agents in Medicinal Chemistry (Formerly Current Medicinal Chemistry-Anti-Cancer Agents)* 12(10): 1281–1305.

45. Tran, Thi Van Anh, Clemens Malainer, Stefan Schwaiger, Tran Hung, Atanas G. Atanasov, Elke H. Heiss, Verena M. Dirsch, and Hermann Stuppner 2015. "Screening of Vietnamese medicinal plants for NF-κB signaling inhibitors: Assessing the activity of flavonoids from the stem bark of Oroxylum indicum." *Journal of Ethnopharmacology* 159: 36–42.

46. Li, Dan, Nan Liu, Liang Zhao, Lei Tong, Hitoshi Kawano, Hong-Jing Yan, and Hong-Peng Li 2017. "Protective effect of resveratrol against nigrostriatal pathway injury in striatum via JNK pathway." *Brain Research* 1654: 1–8.

47. Li, Shanshan, Wenda Zhang, Yanwei Yang, Ting Ma, Jianpeng Guo, Shanshan Wang, Wenying Yu, and Lingyi Kong 2016. "Discovery of oral-available resveratrol-caffeic acid based hybrids inhibiting acetylated and phosphorylated STAT3 protein." *European Journal of Medicinal Chemistry* 124: 1006–1018.

48. Baek, Seung Ho, Jeong-Hyeon Ko, Hanwool Lee, Jinhong Jung, Moonkyoo Kong, Jung-woo Lee, Junhee Lee 2016. "Resveratrol inhibits STAT3 signaling pathway through the induction of SOCS-1: Role in apoptosis induction and radiosensitization in head and neck tumor cells." *Phytomedicine* 23(5): 566–577.

49. Martin, Keith R 2007. "Using nutrigenomics to evaluate apoptosis as a preemptive target in cancer prevention." *Current Cancer Drug Targets* 7(5): 438–446.

50. Ali Khan, Munawwar, Madhumitha Kedhari Sundaram, Amina Hamza, Uzma Quraishi, Dian Gunasekera, Laveena Ramesh, Payal Goala 2015. "Sulforaphane reverses the expression of various tumor suppressor genes by targeting DNMT3B and HDAC1 in human cervical cancer cells." *Evidence-based Complementary and Alternative Medicine* 2015: 1–12

51. Irimie, Alexandra Iulia, Cornelia Braicu, Oana Zanoaga, Valentina Pileczki, Claudia Gherman, Ioana Berindan-Neagoe, and Radu Septimiu Campian 2015. "Epigallocatechin-3-gallate suppresses cell proliferation and promotes apoptosis and autophagy in oral cancer SSC-4 cells." *OncoTargets and Therapy* 8: 461–470.

52. Agrelo, Ruben, Wen-Hsing Cheng, Fernando Setien, Santiago Ropero, Jesus Espada, Mario F. Fraga, Michel Herranz 2006. "Epigenetic inactivation of the premature aging Werner syndrome gene in human cancer." *Proceedings of the National Academy of Sciences* 103(23): 8822–8827.

53. Kong, Wei-Jia, Song Zhang, Chang-Kai Guo, Yan-Jun Wang, Xiong Chen, Su-Lin Zhang, Dan Zhang, Zheng Liu, and Wen Kong 2006. "Effect of methylation-associated silencing of the death-associated protein kinase gene on nasopharyngeal carcinoma." *Anti-cancer Drugs* 17(3): 251–259.

54. Khan, Naghma, Farrukh Afaq, Mohammad Saleem, Nihal Ahmad, and Hasan Mukhtar 2006. "Targeting multiple signaling pathways by green tea polyphenol (–)-epigallocatechin-3-gallate." *Cancer Research* 66(5): 2500–2505.

55. Braicu, Cornelia, and Claudia Gherman 2013. "Epigallocatechin gallate induce cell death and apoptosis in triple negative breast cancer cells Hs578T." *Journal of Drug Targeting* 21(3): 250–256.

56. Du, Qin, Bing Hu, Hong-Mei An, Ke-Ping Shen, Ling Xu, Shan Deng, and Meng-Meng Wei 2013. "Synergistic anticancer effects of curcumin and resveratrol in Hepa1–6 hepatocellular carcinoma cells." *Oncology Reports* 29(5): 1851–1858.

57. Ward, Jeremy PT 2008. "Oxygen sensors in context." *Biochimica et Biophysica Acta (BBA)-Bioenergetics* 1777(1): 1–14.

58. Rajabi, Mehdi, and Shaker A. Mousa 2017. "The role of angiogenesis in cancer treatment." *Biomedicines* 5(2): 34.

59. Pavlakovic, Helena, Werner Havers, and Lothar Schweigerer 2001. "Multiple angiogenesis stimulators in a single malignancy: Implications for anti-angiogenic tumour therapy." *Angiogenesis* 4(4): 259–262.

60. Tudoran, Oana, Olga Soritau, Ovidiu Balacescu, Loredana Balacescu, Cornelia Braicu, Meda Rus, Claudia Gherman, Piroska Virag, Florin Irimie, and Ioana Berindan-Neagoe 2012. "Early transcriptional pattern of angiogenesis induced by EGCG treatment in cervical tumour cells." *Journal of Cellular and Molecular Medicine* 16(3): 520–530.

61. Al-Ani, Bahjat 2013. "Resveratrol inhibits proteinase-activated receptor-2-induced release of soluble vascular endothelial growth factor receptor-1 from human endothelial cells." *EXCLI Journal* 12: 598–604.

62. Kwak, Tae Won, Su Bum Park, Hyun-Jung Kim, Young-IL Jeong, and Dae Hwan Kang 2017. "Anticancer activities of epigallocatechin-3-gallate against cholangiocarcinoma cells." *OncoTargets and Therapy* 10: 137–144.

63. Zhu, Jianyun, Ye Jiang, Xue Yang, Shijia Wang, Chunfeng Xie, Xiaoting Li, Yuan Li Yue Chen, Xiaoqian Wang, Yu Meng, Mingming Zhu, Rui Wu, Cong Huang, Xiao Ma, Shanshan Geng, Jieshu Wu, and Caiyun Zhong 2017. "Wnt/β-catenin pathway mediates (−)-Epigallocatechin-3-gallate (EGCG) inhibition of lung cancer stem cells." *Biochemical and Biophysical Research Communications* 482(1): 15–21.

64. Steward, W. P., and K. Brown 2013. "Cancer chemoprevention: A rapidly evolving field." *British Journal of Cancer* 109(1): 1–7.

65. Landis-Piwowar, Kristin R., and Neena R. Iyer 2014. "Cancer chemoprevention: Current state of the art." *Cancer Growth and Metastasis* 7: 1–7.

66. Li, Yanyan, G. Elif Karagöz, Young Ho Seo, Tao Zhang, Yiqun Jiang, Yanke Yu, Afonso MS Duarte Steven J. Schwartz, Rolf Boelens, Kate Carrolle, Stefan G.D. Rüdigerc, and Duxin Suna 2012. "Sulforaphane inhibits pancreatic cancer through disrupting Hsp90–p50Cdc37 complex and direct interactions with amino acids residues of Hsp90." *The Journal of Nutritional Biochemistry* 23(12): 1617–1626.

67. Kuppusamy, Palaniselvam, Mashitah M. Yusoff, Gaanty Pragas Maniam, Solachuddin Jauhari Arief Ichwan, Ilavenil Soundharrajan, and Natanamurugaraj Govindan 2014. "Nutraceuticals as potential therapeutic agents for colon cancer: A review." *Acta Pharmaceutica Sinica B* 4(3): 173–181.

68. Fenech, Michael, Ahmed El-Sohemy, Leah Cahill, Lynnette R. Ferguson, Tapaeru-Ariki C. French, E. Shyong Tai, John Milner, Woon-Puay Koh, Lin Xie, Michelle Zucker, Michael Buckley, Leah Cosgrove, Trevor Lockett, Kim Y.C. Fung, and Richard Head 2011. "Nutrigenetics and nutrigenomics: Viewpoints on the current status and applications in nutrition research and practice." *Lifestyle Genomics* 4(2): 69–89.

69. Li, Yanyan, Max S. Wicha, Steven J. Schwartz, and Duxin Sun 2011. "Implications of cancer stem cell theory for cancer chemoprevention by natural dietary compounds." *The Journal of Nutritional Biochemistry* 22(9): 799–806.

70. Manuja, Anju, R. Raguvaran, Balvinder Kumar, Anu Kalia, and B. N. Tripathi 2020. "Accelerated healing of full thickness excised skin wound in rabbits using single application of alginate/acacia based nanocomposites of ZnO nanoparticles." *International Journal of Biological Macromolecules* 155: 823–833.

71. Khan, Shabana I., Pranapda Aumsuwan, Ikhlas A. Khan, Larry A. Walker, and Asok K. Dasmahapatra 2012. "Epigenetic events associated with breast cancer and their prevention by dietary components targeting the epigenome." *Chemical Research in Toxicology* 25(1): 61–73.

72. Bishop, Karen S., and Lynnette R. Ferguson 2015. "The interaction between epigenetics, nutrition and the development of cancer." *Nutrients* 7(2): 922–947.

73. Myzak, Melinda C., Karin Hardin, Rong Wang, Roderick H. Dashwood, and Emily Ho 2006. "Sulforaphane inhibits histone deacetylase activity in BPH-1, LnCaP and PC-3 prostate epithelial cells." *Carcinogenesis* 27(4): 811–819.

74. Pandey, Mitali, Parminder Kaur, Sanjeev Shukla, Ata Abbas, Pingfu Fu, and Sanjay Gupta 2012. "Plant flavone apigenin inhibits HDAC and remodels chromatin to induce growth arrest and apoptosis in human prostate cancer cells: In vitro and in vivo study." *Molecular Carcinogenesis* 51(12): 952–962.

75. Chen, Robert W., Lynne T. Bemis, Carol M. Amato, Han Myint, Hung Tran, Diane K. Birks, S. Gail Eckhardt, and William A. Robinson 2008. "Truncation in CCND1 mRNA alters miR-16–1 regulation in mantle cell lymphoma." *Blood, The Journal of the American Society of Hematology* 112(3): 822–829.

76. Chen, Xin-Jun, Mian-Yun Wu, Deng-Hui Li, and Jin You 2016. "Apigenin inhibits glioma cell growth through promoting microRNA-16 and suppression of BCL-2 and nuclear factor-κB/MMP-9." *Molecular Medicine Reports* 14(3): 2352–2358.

77. Liu, Hong-li, Yan Chen, Guo-hui Cui, and Jian-feng Zhou 2005. "Curcumin, a potent anti-tumor reagent, is a novel histone deacetylase inhibitor regulating B-NHL cell line Raji proliferation." *Acta Pharmacologica Sinica* 26(5): 603–609.

78. Kronski, Emanuel, Micol E. Fiori, Ottavia Barbieri, Simonetta Astigiano, Valentina Mirisola, Peter H. Killian, Antonino Bruno, Arianna Pagani, Francesca Rovera, Ulrich Pfeffer, Christian P. Sommerhoff, Douglas M. Noonan, Andreas G. Nerlich, Laura Fontan, and Beatrice E. Bachmeier 2014. "miR181b is induced by the chemopreventive polyphenol curcumin and inhibits breast cancer metastasis via down-regulation of the inflammatory cytokines CXCL1 and-2." *Molecular Oncology* 8(3): 581–595.

79. Yarahmadi, Amir, Seyedeh Zahra Shahrokhi, Zohreh Mostafavi-Pour, and Negar Azarpira 2021. "MicroRNAs in diabetic nephropathy: From molecular mechanisms to new therapeutic targets of treatment." *Biochemical Pharmacology* 189: 1–39.

80. Sun, Michael, Zeev Estrov, Yuan Ji, Kevin R. Coombes, David H. Harris, and Razelle Kurzrock 2008. "Curcumin (diferuloylmethane) alters the expression profiles of microRNAs in human pancreatic cancer cells." *Molecular Cancer Therapeutics* 7(3): 464–473.

81. Zhang, Jian, Yongping Du, Changgui Wu, Xinling Ren, Xinyu Ti, Jieran Shi, Feng Zhao, and Hong Yin 2010. "Curcumin promotes apoptosis in human lung adenocarcinoma cells through miR-186 signaling pathway." *Oncology Reports* 24(5): 1217–1223.

82. Subramaniam, Dharmalingam, Sivapriya Ponnurangam, Prabhu Ramamoorthy, David Standing, Richard J. Battafarano, Shrikant Anant, and Prateek Sharma 2012. "Curcumin induces cell death in esophageal cancer cells through modulating Notch signaling." *PloS One* 7(2): 1–11

83. Zhou, Hong, Jayson X. Chen, Chung S. Yang, Mary Qu Yang, Youping Deng, and Hong Wang 2014. "Gene regulation mediated by microRNAs in response to green tea polyphenol EGCG in mouse lung cancer." *BMC Genomics* 15(11): 1–10.

84. Nandakumar, Vijayalakshmi, Mudit Vaid, and Santosh K. Katiyar 2011. "(–)-Epigallocatechin-3-gallate reactivates silenced tumor suppressor genes, Cip1/p21 and p 16 INK4a, by reducing DNA methylation and increasing histones acetylation in human skin cancer cells." *Carcinogenesis* 32(4): 537–544.

85. Li, Yiwei, Dejuan Kong, Zhiwei Wang, and Fazlul H. Sarkar 2010. "Regulation of microRNAs by natural agents: An emerging field in chemoprevention and chemotherapy research." *Pharmaceutical Research* 27(6): 1027–1041.

86. Li, Qinglin, Gabriel Eades, Yuan Yao, Yongshu Zhang, and Qun Zhou 2014. "Characterization of a stem-like subpopulation in basal-like ductal carcinoma in situ (DCIS) lesions." *Journal of Biological Chemistry* 289(3): 1303–1312.

87. Chakrabarti, Mrinmay, Mehrab Khandkar, Naren L. Banik, and Swapan K. Ray 2012. "Alterations in expression of specific microRNAs by combination of 4-HPR and EGCG inhibited growth of human malignant neuroblastoma cells." *Brain Research* 1454: 1–13.

88. Vislovukh, A., G. Kratassiouk, E. Porto, N. Gralievska, C. Beldiman, G. Pinna, A. El'skaya, A. Harel-Bellan, B. Negrutskii, and I. Groisman 2013. "Proto-oncogenic isoform A2 of eukaryotic translation elongation factor eEF1 is a target of miR-663 and miR-744." *British Journal of Cancer* 108(11): 2304–2311.

89. Liu, P., H. Liang, Q. Xia, P. Li, H. Kong, P. Lei, S. Wang, and Z. Tu 2013. "Resveratrol induces apoptosis of pancreatic cancer cells by inhibiting miR-21 regulation of BCL-2 expression." *Clinical and Translational Oncology* 15(9): 741–746.

90. Hagiwara, Keitaro, Nobuyoshi Kosaka, Yusuke Yoshioka, Ryou-U. Takahashi, Fumitaka Takeshita, and Takahiro Ochiya 2012. "Stilbene derivatives promote Ago2-dependent tumour-suppressive microRNA activity." *Scientific Reports* 2(1): 1–9.

91. Li, Yiwei, Timothy G. VandenBoom, Zhiwei Wang, Dejuan Kong, Shadan Ali, Philip A. Philip, and Fazlul H. Sarkar 2010. "miR-146a suppresses invasion of pancreatic cancer cells." *Cancer Research* 70(4): 1486–1495.

92. Chen, Yi, Mohd Saif Zaman, Guoren Deng, Shahana Majid, Shranjot Saini, Jan Liu, Yuichiro Tanaka, and Rajvir Dahiya 2011. "MicroRNAs 221/222 and genistein-mediated regulation of ARHI tumor suppressor gene in prostate cancer." *Cancer Prevention Research* 4(1): 76–86.

93. Chiyomaru, Takeshi, Soichiro Yamamura, Mohd Saif Zaman, Shahana Majid, Guoren Deng, Varahram Shahryari, Sharanjot Saini Hiroshi Hirata1, Koji Ueno, Inik Chang, Yuichiro Tanaka, Z. Laura Tabatabai, Hideki Enokida, Masayuki Nakagawa, and Rajvir Dahiya 2012. "Genistein suppresses prostate cancer growth through inhibition of oncogenic microRNA-151." *PLoS One* 7(8): 1–9

94. Hirata, H., K. Ueno, K. Nakajima, Z. L. Tabatabai, Y. Hinoda, N. Ishii, and R. Dahiya 2013. "Genistein downregulates onco-miR-1260b and inhibits Wnt-signaling in renal cancer cells." *British Journal of Cancer* 108(10): 2070–2078.

95. Xia, Jun, Qiaoling Duan, Aamir Ahmad, Bin Bao, Sanjeev Banerjee, Ying Shi, and Jia Ma 2012. "Genistein inhibits cell growth and induces apoptosis through up-regulation of miR-34a in pancreatic cancer cells." *Current Drug Targets* 13(14): 1750–1756.

96. Li, Yiwei, Timothy G. VandenBoom, Dejuan Kong, Zhiwei Wang, Shadan Ali, Philip A. Philip, and Fazlul H. Sarkar 2009. "Up-regulation of miR-200 and let-7 by natural agents leads to the reversal of epithelial-to-mesenchymal transition in gemcitabine-resistant pancreatic cancer cells." *Cancer Research* 69(16): 6704–6712.

97. Melkamu, Tamene, Xiaoxiao Zhang, Jiankang Tan, Yan Zeng, and Fekadu Kassie 2010. "Alteration of microRNA expression in vinyl carbamate-induced mouse lung tumors and modulation by the chemopreventive agent indole-3-carbinol." *Carcinogenesis* 31(2): 252–258.

98. Appari, Mahesh, Kamesh R. Babu, Adam Kaczorowski, Wolfgang Gross, and Ingrid Herr 2014. "Sulforaphane, quercetin and catechins complement each other in elimination of advanced pancreatic cancer by miR-let-7 induction and K-ras inhibition." *International Journal of Oncology* 45(4): 1391–1400.

99. Tao, Si-feng, Hai-fei He, and Qiang Chen 2015. "Quercetin inhibits proliferation and invasion acts by up-regulating miR-146a in human breast cancer cells." *Molecular and Cellular Biochemistry* 402(1): 93–100.

100. Lou, Guohua, Yanning Liu, Shanshan Wu, Jihua Xue, Fan Yang, Haijing Fu, Min Zheng, and Zhi Chen 2015. "The p53/miR-34a/SIRT1 positive feedback loop in quercetin-induced apoptosis." *Cellular Physiology and Biochemistry* 35(6): 2192–2202.

101. Sonoki, Hiroyuki, Tomonari Sato, Satoshi Endo, Toshiyuki Matsunaga, Masahiko Yamaguchi, Yasuhiro Yamazaki, Junko Sugatani, and Akira Ikari 2015. "Quercetin decreases claudin-2 expression mediated by up-regulation of microRNA miR-16 in lung adenocarcinoma A549 cells." *Nutrients* 7(6): 4578–4592.

102. Dostal, Zdenek, and Martin Modriansky 2019. "The effect of quercetin on microRNA expression: A critical review." *Biomedical Papers of the Medical Faculty of Palacky University in Olomouc* 163(2): 1–12.

103. Liu, Hui, Lu Wang, Feng Li, Yang Jiang, Hui Guan, Dan Wang, Dongxiao Sun-Waterhouse, Maoyu Wu, and Dapeng Li 2021. "The synergistic protection of EGCG and quercetin against streptozotocin (STZ)-induced NIT-1 pancreatic β cell damage via upregulation of BCL-2 expression by miR-16–5p." *The Journal of Nutritional Biochemistry* 96: 1–8.

104. Kumazaki, Minami, Shunsuke Noguchi, Yuki Yasui, Junya Iwasaki, Haruka Shinohara, Nami Yamada, and Yukihiro Akao 2013. "Anti-cancer effects of naturally occurring compounds through modulation of signal transduction and miRNA expression in human colon cancer cells." *The Journal of Nutritional Biochemistry* 24(11): 1849–1858.

11 Synergistic Effects of Natural Products in Cancer Treatment

Ritu Saini, Chitra Gupta, Shalini Pundir, Meenakshi Bajpai,
Akansha Agrwal, and Sharad Visht

CONTENTS

11.1 Introduction 184
11.2 Herbal Approaches in the Treatment of Cancer 184
    11.2.1 Examples of Various Natural Herbs Used for the Treatment
        of Cancers 185
11.3 Natural Herb–Synthetic Drug Interaction 185
    11.3.1 Potential Herb-Herb Synergistic Effect 185
        11.3.1.1 Curcumin and Resveratrol 185
        11.3.1.2 Artemisinin (ART) and Resveratrol 185
        11.3.1.3 1′-Acetoxychavicol Acetate (ACA) and Sodium Butyrate (SB) 196
        11.3.1.4 Ginsenoside Rh2 (G-Rh2) and Betulinic Acid (BA) 196
        11.3.1.5 Matrine and Resveratrol 196
        11.3.1.6 Clove, Oregano, Thyme, Walnuts, and Coffee 196
        11.3.1.7 Curcumin and Vinorelbine 196
        11.3.1.8 Ellagic Acid (EA) and Quercetin with Resveratrol 197
        11.3.1.9 Curcumin Analog Pentagamavunon-1 (PGV-1) and
                5-Flourouracil (5-FU) 197
        11.3.1.10 Thearubigin and Genistein 197
    11.3.2 Potential Herb-Drug Synergistic Effect 197
        11.3.2.1 Genistein and Arsenic Trioxide (ATO) 197
        11.3.2.2 Berberine and Rapamycin 197
        11.3.2.3 Apigenin and 5-Flourouracil (5-FU) 198
        11.3.2.4 Quercetin and Cisplatin 198
        11.3.2.5 Catechins and Doxorubicin (DOX) 198
        11.3.2.6 Gambogic Acid (GA) and Proteasome Inhibitor MS132
                or MG262 198
        11.3.2.7 Quercetin with Carboxyamidotriazole (CAI) 198
        11.3.2.8 Eicosapentaenoic Acid (EPA) on Antiproliferative Action
                of Anticancer Drugs 199
        11.3.2.9 Gemcitabine (GEM) and Impressic Acid (E12–1)
                or Acankoreanogein (E13–1) 199
11.4 Current Status of Synergistic Approach of Natural Products in the
    Treatment of Cancer 199
11.5 Future Prospects of Natural Products in the Treatment of Cancer 200
11.6 Conclusion 200
11.7 Acknowledgment 200
11.8 References 201

DOI: 10.1201/b23311-11

11.1 INTRODUCTION

Cancer is a term refers to a group of pathological conditions characterized by the uncontrollable proliferation and rapid expansion of malignant cells,[1] second biggest killer in the world. Men's cancers are mostly prevalent in the lung, prostate, bronchus, colon, rectum, and urinary bladder, while women's cancers are widespread in organs like breast, lung, bronchus, rectum, colon, uterine corpus, and thyroid[2]. During the year 2020, 19.3 million new cases were extensively documented, with a predicted increase of 47% by the year 2040. In terms of new cases (11.7%) and deaths, female breast cancer is the most reported cancer in females, outnumbering lung cancer (6.9%)[3].

For thousands of years, herbal plants have been utilized to avert and cure numerous disorders. They are good resources of functional components that have health-promoting effects, that has been an integral part of our diet. Certain functional components of plants have been shown to have anticancer properties. Approximately 50 to 60% of cancer patients in the United States use medications obtained from herbal plants, such as turmeric (curcumin), soybeans (genistein), green tea (polyphenols), grapes (resveratrol), broccoli (sulforaphane), cruciferous vegetables (isothiocyanates), milk thistle (silymarin), garlic (diallyl sulfide), tomato (lycopene), rosemary (rosmarinic acid), parsley (apigenin), and ginger (gingerol), to name a few[4]. Traditional therapeutic approaches, predominately based on land plants, remain dominant therapeutic strategies worldwide, and natural compounds responsible for a considerable share of today's pharmacological actives, notably in the area of antibiotics and cancer therapies[5].

Keep in mind Buckminster Fuller's (1968) comment: "Universe is synergetic. Life is a synergy." Synergy originates from the Greek word *synergos*, meaning "working together"[6]. Many studies are currently being conducted to study the benefits of combining anticancer treatments to overcome the limits of currently available anticancer therapies, and these drugs are also being provided to cancer patients in actual clinical practice, and some of these combinations have been proved to be successful[7].

11.2 HERBAL APPROACHES IN THE TREATMENT OF CANCER

Plants played a vital role in ancient cultures' folklore and have been used as medicines for over 5,000 years, subsequently being used as spices and foods. The nineteenth century started the trend of isolating active phytoconstituents resulted in the finding of various active constituents obtained from plants. In recent years, an increasing number of new plant-derived materials have approved as medicines, many of those with anticancer potential[8]. Toxic side effects are frequently connected with the use of synthetic anticancer drugs, as a result, the use of widely accessible and low-cost medicinal plants is a panacea for cancer patients[9]. Clinical trials and phytochemical screening have demonstrated that herbal remedies have antitumor activity against various types of cancer[10–15].

Approximately 35,000 plant species have been tested by the National Cancer Institute for anticancer activity, with approximately 3,000 plant species showing accurate results[16], and more than 114,000 plant extracts are also being tested for the same at multiple cancer institutions. Despite the reality that most of the anticancer agents are obtained plants, countless with anticancer potential are still unknown[17–19].

Presently, surgical treatments are applied for the removal of solid tumors, along with the radiation therapy and chemotherapy, but both have serious side effects and eventually affects the quality of well-being. Also, several treatments have certain limitations in practice due to their toxicity. For example, Herceptin is a medicine used in the treatment of breast cancer, but these remedies are highly expensive and effective against specific types of tumors. Resistance development to certain drugs is also observed in many cases and the patient has to switched to another medication or given with a cocktail of medications. Therefore, formulation of novel and effective anticancer medicines with lesser side effects is the necessity of the time and plants are the very much potential candidate for this. Even plants are producing around 60% of potential anticancer molecules directly or indirectly[8, 20].

11.2.1 Examples of Various Natural Herbs Used for the Treatment of Cancers

Ayurveda and Charak Samhita have always been an integral part of our health management system from ancient times and phytochemical screening of indigenous herbs has helped in the search for new drugs at some level[21]. Polyphenols, alkaloids, flavonoids, saponins, tannins, triterpenes, and quinones are the important categories of anticancer bioactive components derived from medicinal plants. These bioactive compounds inhibit angiogenesis and lower cancerous cell viability while also being antiproliferative, cytotoxic, cytostatic, antimetastatic, apoptotic, and antioxidative[19].

Podophyllotoxin, with several other lignans, isolated from *Podophyllum peltatum*, have contributed in the production of medicines used to cure testicular and lung cancer[16]. Extracts of *Chaetomium globosum*[4] and 5-methyl phenazine-1-carboxylic acid obtained from *Pseudomonas putida* were found to have cytotoxic effects on cancer cell lines. Similarly, paclitaxel and camptothecin are other plant originated drugs having anticancer potential, which have been permitted for use[8]. In another study the anticancer activity of five herbs (*Moringa oleifera, Calotropis procera, Basela alba, Millettia pinnata*, and *Euphorbia neriifolia*) that are used by the local tribals were explored. An apoptosis study with Acridine orange/ethidium bromide and 4',6-diamidino-2-phenylindole staining evidenced this work, which further indicated the maximum anticancer potential of the *M. pinnata* extract and with maximum cytotoxicity of all the other plants studied[22].

Ngyuen et al. analyzed the cytotoxic potential of chloroform extract of *Adenosma bracteosum* (Bonati). On large cell lung carcinoma (NCI-H460) and HCC (hepatocellular carcinoma) (HepG2) and found out that it inhibited their multiplication due to the bioactive constituents like xanthmicrol, 5,4'-dihydroxy-6,7,8,3'-tetramethoxyflavone, and ursolic acid[23]. The petroleum-ether extract of *Curcuma mutabilis* rhizome and a new isolated therapeutic; labdane diterpenoid,(E)-14,15-epoxylabda-8(17),12-dien-16-al (Cm epoxide), both have observed as antiproliferative against human cancer cell lines while exhibiting no cytotoxic effects on normal cells[24]. The methanolic extract of *Eclipta alba* has also opened up the possibility of developing a novel anticancer drug that could be effective against colon cancer, as it demonstrated considerable precision against HCT-116 cells while ensuring minimal damage to normal cells[25]. Some of the plant-derived substances having anticancer activity are discussed in Table 11.1.

11.3 NATURAL HERB–SYNTHETIC DRUG INTERACTION

11.3.1 Potential Herb-Herb Synergistic Effect

11.3.1.1 Curcumin and Resveratrol

Curcumin and resveratrol, respectively, are the predominant anticancer agents of *Curcuma aromatica* and *Polygonum cuspidatum*[58, 59]. In colon cancer, curcumin and resveratrol have been reported to have a synergistic anticancer response[60]. Curcumin-and-resveratrol combined treatment induce extrinsic and intrinsic apoptosis in Hepal-6 HCC cells, which is associated with the production of reactive oxygen species (ROS) as well as the suppression of X-linked inhibitor of apoptotic protein and survivin (anti-apoptosis gene)[61].

11.3.1.2 Artemisinin (ART) and Resveratrol

Resveratrol (3,4',5-trihydroxystilbene) is a phytoalexin present mostly within red wine and grape skin. Resveratrol has antitumor efficacy against a wide range of cancers[62–64]. ART is a naturally occurring

TABLE 11.1

List of Plant-Derived Substances Having Anticancer Activity Against Various Types of Cancer with Their Possible Mode of Action

Herb	Common Name	Part Used	Active Constituent	Type of Cancer/Tumor Cell Line Treated	Mode of Action	References
Acronychia baueri	Byron bay acronychia	Bark	Normelicopidine, Melicopine, Acronycine, Triterpenelupeol,	Not specified.	Antiproliferative activity	[19], [26]
Ageratum conyzoides	Billygoat weed	Leaf.	Kaempferol, Oxygenated terpenes, Sesquiterpene hydrocarbons, Monoterpene hydrocarbons	Lung, CNS, Blood, Prostate	Cytotoxic Antiproliferative activity	[19, 27, 28]
Alchornea cordifolia	Christmas bush	Leaf Bark	Flavonoids, Cardiac glycosides, Saponins, Steroids, Anthraquinone, Terpenes, Xanthones, Alkaloids, Tannins.	Blood	Apoptosis	[19, 29, 30]
Allium sativum.	Garlic	Bulb.	S-allylcysteine. S-allylmercapto-L-cysteine. Diallyl disulfide. Diallyl trisulfide. Allicin.	Skin. Colon. Lung. Prostate. Blood. Breast.	Cytotoxic. Antiproliferative activity	[29–32]
Aloe barbadensis	Indian aloe	Leaf	Aloe-emodin	Liver Lung	Cytotoxic Antioxidant effects	[31]
Alstonia boonei	God's tree	Stem/bark Leaf Root	Echitamine. Eugenol. 1, 2- Benzenedicarboxylic acid. Alstiboonine.	Pancreas. Lung. Prostate. Colon.	Antiproliferative Cytotoxic activity	[19, 33]

Plant	Common name	Part used	Compounds	Cell line/Organ	Activity	References
Anacardium occidentale.	Cashew nut	Leaf. Stem/bark.	Agathisflavone. Methyl gallate. Anacardicin. Zoapatanolide A. Tannins Alkaloids Saponins Polyphenols	Laryngeal Blood Cervical	Cytotoxic Antiproliferative activity	[34, 35]
Andrographis paniculata (Burm. F.) Nees	Bhunimba, kalmegha	Roots Leaves	Diterpene andrographolide.	KB (human epidermoid cancer cells). P388. (lymphocytic leukemia cells). MCF-7 (breast cancer cells). HCT-116. HT 29 (colon cancer cells).	Antiproliferative activity	[36, 37]
Annona atemoya, Mabb. *Annona muricata*, Linn.	Mamaphal	Fruit	Bullatacin Annomuricins (A, B)	A-549 (lung carcinoma). MCF-7. HT-29. HepG2 (human hepatoma cell line)	Apoptosis Cytotoxic	[16]
Anogeissus leiocarpus	African birch Bambara	Leaf Root	Ellagic acid. Castalagin. Flavogallonic acid.	Liver.	Antiproliferative activity	[38–40]
Astragalus membranaceus	Mongolian milkvetch	Root	Isoflavones. Calycosin. Ononin. Formononetin. Campanulin.	Breast.	Antiproliferative activity Apoptosis	[31, 41]
Attractylis lancea	Cang zhu	Root	Polyacetylenes Sesquiterpenes lactones.	Liver. Stomach.	Cell cycle arrest Antiproliferative activity Apoptosis	[42, 43]

(Continued)

TABLE 11.1 *(Continued)*
List of Plant-Derived Substances Having Anticancer Activity Against Various Types of Cancer with Their Possible Mode of Action

Herb	Common Name	Part Used	Active Constituent	Type of Cancer/Tumor Cell Line Treated	Mode of Action	References
Azadirachta indica	Neem	Leaves Seeds	Flavonoids Phenolics. Limonoids Triterpenoids.	Skin. Prostate. Breast Cervical. Blood. Liver. Colon. Lung. Stomach.	Antiproliferative activity Apoptosis	[19]
Boerhaavia diffusa	Punarnava	Leaf	Alkaloids Flavonoids Tannins. Saponins. Terpenes. Anthraquinones. Steroids.	Cervical. Breast.	Cell cycle arrest Antiproliferative activity	[44, 45]
Boswellia serrata, Roxb.	Guggul, shallaki	Gum	Isomeric Triterpenediol	Various cancer cells	Antiproliferative activity Apoptosis	[16]
Cajanus cajan	Arhar dal	Leaf	Longistylin C. Longistylin A. Stilbenoids. Flavonoids.	Colorectal. Breast Cervical. Lung. Blood. Liver.	Antiproliferative activity	[19, 31]
Camellia sinensis	Green tea	Leaf	(+)-Gallocatechin (−)-Epigallocatechingallate (−)-Epigallocatechin.	Colon. Lung. Breast Liver.	Antiproliferative activity Cytotoxic. Cell cycle arrest.	[19, 31, 32]

Plant	Common name	Part	Constituents	Cancer/Cells	Activity	References
Carica papaya.	Papaya	Leaf.	Ascorbic acid. Quercetin. Kaempferol. Tetrahydroxyflavone Kaempferol-B-D-Glucopyranoside. Papain. Lycopene. Morin. Cystatin. Fisetin Benzylisothiocyanate Luteolin-B-D-glucopyranoside Myricetin-3-O-rhamnoside.	Prostate Lung Blood Pancreas Liver	Antiproliferative activity	[46, 47]
Cassia. occidentalis	Coffee senna	Whole plant	Flavonoids Tannins. Alkaloids. Anthraquinones. Saponins.	Colon. Ovary Cervical. Breast Prostate.	Antiproliferative activity	[48, 49]
Cedrus deodara.	Deodar. cedar. Himalayan cedar. Deodar. Devdar. Xue song.	Bark Stem Wood	(–)-Wikstromal (–)-Matairesinol Dibenzylbutyrolactol	Various human cancer cell lines including leukemia cells	Cytotoxic Apoptosis	[16]
Centella asiatica, Linn.	Brahmamanduki	Whole plant Leaves	Asiaticoside. Hydrocotyline Vallerine. Pectic acid. Sterol. Stigmasterol flavonoids Thankunosides Ascorbic acid	Ehrlich ascites tumor cells. Dalton's lymphoma Ascites tumor cells.	Inhibition of proliferation of transformed cell line and directly inhibited DNA synthesis	[16]

(Continued)

TABLE 11.1 (*Continued*)
List of Plant-Derived Substances Having Anticancer Activity Against Various Types of Cancer with Their Possible Mode of Action

Herb	Common Name	Part Used	Active Constituent	Type of Cancer/Tumor Cell Line Treated	Mode of Action	References
Chromolaena odorata	Siam weed	Leaf.	n-Hexane. Ethanol 2-hydroxy-4, 4, 5,6- tetra-methoxychalcone Acacetin. Kkaempferol-3-O-rutinoside. Quercetin-3-O-rutinoside. Kaempferide. Rhamnazin.	Breast Lung. Blood.	Cytotoxic.	[19]
Citrus aurantifolia	Key lime	Not specified	Polymethoxyflavones	Colon Pancreas Breast	Apoptosis Cell cycle arrest Antimetastatic activity Immunity enhancement	[19, 31]
Croton zambesicus	Variegated laurel	Leaf	Ent-trachyloban-3-β-ol. Ent-trachyloban-3-one. Ent-18-hydroxy-trachyloban-3-one. Trans-phytol. Isopimara-7, 15-dien-3-β-ol.	Cervical	Cytotoxic	[19]
Cryptolepis sanguinolenta	Not known	Root	Cryptolepine. Ascryptolepinoic acid. Quindoline Methyl cryptolepinoate.	Lung.	Antiproliferative activity Lesser cancer cell viability	[19]
Curcuma longa, Linn.	Haridra and haldi	Rhizome	Curcumin/diferuloylmethane	Variety of tumor cells 19 strains of *H. pylori* Colon cancer Gastric cancer in rodents LNCaP prostate cancer cells	Antiproliferative activity Apoptosis	[19, 32]
Derris scandens	Tupbel, mohaguno, hog creeper, jewel vine	Not specified	Glyurallin Derriscandenon B and C Isochandaisone Derrubone	Colon	Apoptosis Mitosis inhibition	[50, 51]

Enantia chlorantha	African yellow wood	Stem/bark.	Columbamine. Saponins. Pseudocolumbamine Palmatine.	Colorectal. Liver. Lung.	Antiproliferative activity Cytotoxic.	[52, 53]
Fagara zanthoxyloides	Artar root	Root	Fagaronine	Blood Prostate	Antiproliferative activity Cytotoxic	[54]
Glochidion zeylanicum	Umbrella cheese tree	Stem/bark Leaf Root	Triterpinoids Megastigmane glycosides	Prostate Colon Liver	Antiproliferative activity Cytotoxic	[12, 19]
Glycyrrhiza glabra	Licorice	Whole plant	Licochalcone Isoliquiritigenin	Prostate Colon Breast	Antiproliferative activity Apoptosis	[19, 31]
Goniothalamus macrophyllus	Penawar hitam, chin dok diao	Root Stem	Goniothalamin	Cervical Breast Colon cancer lines	Cytotoxic Apoptosis Arrest cell cycle	[19]
Harungana madagascariensis	Dragon's blood tree	Stem/bark	Coumarins Anthraquinones Bioflavonoids Anthrone derivatives Xanthones	Blood	Cytotoxic	[19]
Khaya senegalensis	Senegal mahogany	Stem/bark	3α, 7α-dideacetylkhivorin1-O-deacetylkhayanolide.E4-Khayanolide. B2 6-dehydroxylkhayanolide. E5 1-O-acetylkhayanolide. B1 Khayano-lide. E3	Colon Cervical Liver Breast	Antiproliferative activity Apoptosis	[19]
Lophira alata	Red ironwood tree	Stem/bark.	Azobechalcone A. Flavonoids. Isolophirachalcone. Lophirone F, A, B, C Triterpenes. Sterols. Polyphenols.	Liver. Lung, Breast Skin. Colon.	Antiproliferative activity Cytotoxic	[19, 52]

(Continued)

TABLE 11.1 *(Continued)*
List of Plant-Derived Substances Having Anticancer Activity Against Various Types of Cancer with Their Possible Mode of Action

Herb	Common Name	Part Used	Active Constituent	Type of Cancer/Tumor Cell Line Treated	Mode of Action	References
Mangifera indica.	Mango	Stem/bark.	Galloyl glycosides. Lupeol. Mangiferin Gallotannins Gallic acid	Breast Ovary Kidney Colon	Antiproliferative activity	[19]
Mappia foetida, Miers	Amruta, kalgur, and narkya	Tree wood	Camptothecin Irinotecan (semisynthetic derivative) Topotecan (synthetic derivative)	Leukemia Colon Breast Lymphoma Solid mouse tumors Solid epithelial tumors Rhabdomyosarcoma cells	Inhibits DNA formation in HeLa. and L-120 cells and also nuclear enzyme type-1 DNA topoisomerase.	[16]
Milicia excelsa	African teak	Root	Neocyclomorusin. Cudraxanthone I. Betulinic acid. 6- geranylnorartocarpetin Atalantoflavone.	Cervical. Colon Liver. Brain.	Antiproliferative activity Cytotoxic.	[19]
Morinda lucida	Brimstone tree	Leaf Stem/bark	Molucidin. β-sitosterol. Stigmasterol. Oruwacin. Digitolutein. Ursolic acid. Rubiadin-1-methyl ether. Phytol. Cycloartenol. Oleanolic acid. Danna-cantha. Campesterol.	Prostate Colon Stomach Blood	Antiproliferative activity	[19]

Species	Common name	Part used	Compounds	Cancer types	Activity	References
Newbouldia laevis	Boundary tree	Root, Bark	2-Acetylfuro-1,4-naphthoquinone, Steroids, Triterpenoid, Tannins, Quinone derivatives	Pancreas, Blood	Antiproliferative activity, Cytotoxic	[19]
Ocimum gratissimum	African basil	Leaf	3,4-Dihydroxycinnamic acid, Oleanolic acid, Saponins, Linalool, Eugenol, Gerianol, Alkaloids, Thymol, Citral.	Cervical, Breast, Prostate, Lung, Colon, Pancreas, Kidney, Bone	Antiproliferative activity	[19]
Panax ginseng	Ginseng	Root, Rhizomes	Ginsenoside Rp-1	Breast	Antiproliferative activity, Apoptosis, Cell cycle arrest	[31, 32]
Phyllanthus amarus	Bhumyamalaki, jaramla, stone breaker	Whole plant, Leaves, Roots, Shoots	Flavonoids, Ellagitannins, Polyphenols, Triterpenes, Sterols, Saponins, Alkaloids, Lignans like nirtetralin, niranthrin, phyllanthin, and phyltetralin	Dalton's Lymphoma ascites, Erlich ascites carcinoma, Lung, Breast, Pancreas, Neuron, Skin, Ovarian, Prostate.	Prevent carcinogenic compound metabolic activation, cell cycle arrest, interfering with DNA repair	[19, 31]
Picrorhiza kurroa	Picrorhiza, katuka, kutki	Root	Apiocynin, Cucurbitacinesaglycone, Caffeic esters, Picrosides	Cervical, Breast.	Cytotoxic, Anti-inflammatory, Antioxidant.	[19, 23]
Psidium guajava.	Guava	Leaf.	Cryptonine, Tannins, Apigenin, Dihydrobenzop-enanthridine, Lycopene, Saponins	Breast, Blood, Cervical, Prostate	Antiproliferative activity, Apoptosis	[19, 55]

(Continued)

TABLE 11.1 *(Continued)*

List of Plant-Derived Substances Having Anticancer Activity Against Various Types of Cancer with Their Possible Mode of Action

Herb	Common Name	Part Used	Active Constituent	Type of Cancer/Tumor Cell Line Treated	Mode of Action	References
Punica granatum	Pomegranate	Fruit	Ellagic acid. Ellagitannins. Punicalagin. Tannins.	Colorectal Colon.	Cell cycle arrest. Antiproliferative activity Apoptosis	[19, 40, 56]
Rauwolfia vomitoria	Poison devil's-pepper	Root	β-Carboline. Terpenes. Tannins. Alkaloids. Steroids. Saponins. Flavonoids.	Pancreas. Ovarian. Prostate.	Cytotoxic. Apoptosis Antiproliferative activity Cell cycle arrest	[19]
Scoparia dulcis	Licorice weed, goatweed	Leaf	Dulcidiol. Iso–dulcinol. Scopadiol. 4-Epi-scopadulcic acid B. Scopanolal.	Stomach. Prostate.	Cytotoxic.	[19]
Tabebuia avellanedae	Pink trumpet	Bark	β-Lapachone	Liver	Antiproliferative activity Apoptosis	[19]
Thymus vulgaris	Garden thyme	Leaf bulb	Tannins. Triterpenes. Sterols. Flavonoids Glycosides.	Cervical.	Cytotoxic.	[30, 31, 57]
Tinospora cordifolia (wild), Miers	Giloya	Stem Roots	Choline Tinosporin Columbin Isocolumbin Palmitine Tetrahydropalmitate Magnoflorine	Ehrlich ascites Carcinoma Breast Cervical.	Cytotoxic Antiproliferative activity	[19, 32]

Epoxycleodanediterpene
Arabinogalactan
Berberine
Phenolics

Species	Common name	Part	Compounds	Cancer cell/type	Activity	Ref.
Tithonia diversifolia	Giant Mexican sunflower	Leaf	Tagitinin C. 1β-methoxydiversifolin.3-O-methyl ether 1β, 2α-epoxytagitinin C.	Colon. Blood.	Antiproliferative activity Cell cycle arrest. Cancer cell. autophagy	[19]
Uvaria chamae	Bush banana	Stem/bark Root	Dichamanetin Diuvaretin Uvaretin. Chamanetin. Pinocembrin. Isouvaretin. Pinostrobin. Isochamanetin.	Blood	Antiproliferative activity	[19]
Vernonia. amygdalina	Bitter leaf.	Leaf.	Steroid glycosides. Edotide.	Breast Cervical.	Antiproliferative activity	[19]
Withania somnifera (Linn.), Dunal	Ashwagandha and winter cherry	Leaves roots	Withaferin A	Various human cancer cell lines Ehrlich ascites tumor	Apoptosis Cytotoxic	[16, 32]
Zanthoxylum nitidum	Shiny-leaf prickly ash	Root	Nitidine Benzophenanthridine	Lung Cervical Liver	Cytotoxic Antiproliferative activity	[19]
Zingiber officinale	Ginger	Whole plant	10-Gingerol. 10-Shogaol. 6-Gingerol. 6-Shogaol.	Prostate. Liver. Breast Esophageal.	Antiproliferative activity Antimetastatic activity	[19]
Ziziphus nummularia, Wight	Harbor, wild jujube	Bark	Betulin Betulinic acid	Different type of tumor cell lines like p53 mutant cells	Apoptosis	[16]

substance derived from the sweet wormwood plant *Artemisia annualis*. ART and its derived constituents inhibit tumor cell growth in a number of tumor cell lines, such as prostate, breast, colon, and liver carcinoma[65–67]. Treatment of in vitro models of HeLa and HepG2 cells with the combination of ART and resveratrol produced a synergistic impact, showing that the 1:2 ratio of ART and resveratrol had the highest effect. Furthermore, fluorescence microscopy and cytometry analysis also revealed suppressed cancer cell growth while increased movement, necrosis, apoptosis, and ROS levels[68].

11.3.1.3   1′-Acetoxychavicol Acetate (ACA) and Sodium Butyrate (SB)

Rhizomes and seeds of *Languas galangal* and *Alpinia galangal* naturally contain ACA, which exerts anticancer activity against chemically induced cancers in the skin of mice and oral, esophageal, colonic, and pancreatic malignancies in rats[69, 70] and also induces apoptosis in numerous tumor cells, such as Ehrlich ascites tumor cells (by diminishing intracellular polyamines and increasing caspase-3 activation), human HCC cells, colon cancer cells, myeloid leukemia cells, and rat HCC cells[71]. SB prevents the proliferation of cell and stimulates apoptotic action in tumor cells cultivated in vitro[72–75]. Treatment of cells with ACA and SB produce synergistic effect on human HCC HepG2 cell by antiproliferative activity and apoptotic initiation while increased intracellular ROS levels and NADPH oxidase activity[76].

11.3.1.4   Ginsenoside Rh2 (G-Rh2) and Betulinic Acid (BA)

G-Rh2, isolated from the root of *Panax ginseng*, exhibits anticancer effects[77, 78] by causing apoptosis in MCF-7 (human breast cancer)[79], SK-HEP-1 (human hepatoma)[80], THP-1 (human leukemia)[81], and A549 (human lung adenocarcinoma)[82] cells in a number of prior investigations. BA has shown cytotoxic effects on melanoma cells only[83] and suppresses cell proliferation and causes apoptosis in number of cancer cells[84] that do not involve p53 and are not associated with the activation CD95[85]. In HeLa, A549, and HepG2 cells, G-Rh2 and BA work together to trigger apoptosis. Co-therapy of G-Rh2 with BA prompts caspase-8 stimulation and cleavage, sensitizing tumor cells, followed by caspase-9 and caspase-3 stimulation and death via a Bax-dependent mechanism[86].

11.3.1.5   Matrine and Resveratrol

Matrine is an ancient Chinese herb, *Sophora flavescens* (Ait.)[87], has recently been recognized as anticancer agent. It has antiproliferative and antimetastatic effects on malignant cells and promotes apoptosis[88] and also reduces the chances of chemotherapy-induced multidrug resistant cancerous cells and possess synergistic action with other anticancer drugs[89]. Matrine suppresses HCC cellular proliferation by eliciting apoptosis through an elevation in the Bax/Bcl-2 ratio[90], with an IC50 of 2–16 mM[91]. On HepG2 cells, researchers showed that combining resveratrol and matrine considerably decreases cell proliferation in comparison to individual drug. Matrine also dramatically increases resveratrol-induced apoptosis, which might be attributable to caspase-3 and caspase-9 activation, suppression of survivin, elevation of ROS production, and interruptions of mitochondrial membrane potential ($\Delta\Psi$m). According to the results, a combination of resveratrol with matrine is a potential new anticancer approach in treatment of liver cancer[92].

11.3.1.6   Clove, Oregano, Thyme, Walnuts, and Coffee

Inflammation and cancer are linked by NF-κB[93] that has evolved as a vital biomarker in cancer prevention and treatment. In vitro, an extract of five dietary plants (clove, oregano, thyme, walnuts, and coffee) block NF-κB activation synergistically and promote Nrf2 activity in transgenic mice. In male mice, diminished NF-κB activation was found in the liver and intestine, as well as the testis and epididymis, showing that the dietary plants have organ-specific effects[94].

11.3.1.7   Curcumin and Vinorelbine

Curcumin is a polyphenolic phytochemical obtained from the *Curcuma longa* plant[95] having chemopreventive agent that inhibits tumor instigation, proliferation, and development[96]. Vinorelbine (a semisynthetic vinca alkaloid) is a chemotherapeutic agent that is both effective and safe[97]. Curcumin pretreatment boosts vinorelbine's apoptotic activity on H520 (non-small-cell lung cancer cells) by upregulating apoptosis in Bcl-2, Bcl-XL, Bcl-xs, and Bax and activating caspase-9 and 3. Curcumin and vinorelbine both decrease NF-κB and Ap-1, which could explain their antiproliferative and apoptotic properties[95].

11.3.1.8 Ellagic Acid (EA) and Quercetin with Resveratrol

Although polyphenols are a large group of phytochemicals with anticancer properties, most studies focus on single compounds[98], occurring naturally in fruits and vegetables have been recommended as the best for cancer prevention[99, 100] like Grapes are abundant in polyphenols, muscadine species have a unique combination of three polyphenols found in fruits: EA, quercetin, and resveratrol (*Vitis rotundifolia*). Reduced proliferative activity and activation of apoptosis in cancer cells are two important mechanisms for polyphenols' anticarcinogenic actions. EA, quercetin, and resveratrol alone and in combination affects proliferation and apoptosis of human leukemic cell (MOLT-4)[98].

11.3.1.9 Curcumin Analog Pentagamavunon-1 (PGV-1) and 5-Flourouracil (5-FU)

5-FU is utilized as an adjunct to cancer therapy, post-surgical resection[101], that works by deactivating the enzyme TS (thymidylate synthase), which interfere with the production of DNA[102]. One of the curcumin analogues, PGV-1, has been investigated as a possible chemopreventive drug[103, 104]. PGV-1 initiates the blockage of S-phase and damages DNA through 5-FU mechanism. 5-FU is converted into the active metabolite 5′-dUMP in a sensitized cancer cell. It disrupts DNA synthesis and induces DNA damage-induced apoptosis. 5-FU administration enhances cyclin D1 levels in WiDr cells via promoting NF-κB activation. A high quantity of cyclin D1 in S-phase inhibits DNA synthesis and so prevents the 5-FU action. In the presence of PGV-1 (inhibitor of NF-κB), cyclin D1 levels are reduced, leading to an increase in blockage of S-phase via 5-FU. A combination of PGV-1 and 5-FU is used in the treatment of resistant colon cancer cells[102].

11.3.1.10 Thearubigin and Genistein

Genistein, a phytoestrogen derived from soy, is a potential chemotherapeutic drug capable of triggering apoptosis and reducing angiogenesis by inhibiting DNA topoisomerase or inhibiting tumor-promoting proteins like COX-2[105–107]. Catechins in black tea are oxidized for the production of the pigments theaflavins and thearubigins. According to studies, low concentrations of thearubigins individually did not prevent human prostate tumor cell growth, but in combination with genistein, they did, indicating synergy between the two natural products. Cell growth attenuation was correspondingly accompanied by G2/M phase cell cycle disruption[108].

11.3.2 Potential Herb-Drug Synergistic Effect

11.3.2.1 Genistein and Arsenic Trioxide (ATO)

Genistein, a soy-derived isoflavone[109], a promising anticancer and pro-oxidant agent, improves the efficacy of ATOs, possessing oxidation-sensitive cytotoxicity. By generating intracellular ROS and causing ER (endoplasmic reticulum) stress and mitochondrial damage, genistein causes death in HCC cells[110]. ATO's anticancer properties include cell cycle arrest and apoptosis and also tumor angiogenesis inhibition[111–112]. In vitro and in vivo, genistein combined with ATO give synergistic responses, for the reduction of ΔΨm, enhancement of cytochrome C release, activation of Bax expression, suppression of Bcl-2 expression, and activation of caspase-9 and caspase-3, eventually suppresses tumor cell proliferation and causes apoptosis in HCC cells[113].

11.3.2.2 Berberine and Rapamycin

Streptomyces hygroscopicus produces rapamycin[114] identified the mammalian target of rapamycin (mTOR) signaling pathway, which is critical for normal and malignant cell proliferation[115].

Berberine, generated from *Coptidis rhizoma*, has a powerful anticancer activity (suppression of mTOR) with no side effects[116, 117] and also causes autophagic cell death and mitochondrial apoptosis in HCC cells[118]. The usage of berberine and rapamycin together showed synergistic cytotoxic impact in HepG2 and SMMC7721 HCC cell lines by maintaining rapamycin's cytotoxic action on HCC at lower concentrations of rapamycin[119].

11.3.2.3   Apigenin and 5-Flourouracil (5-FU)

Apigenin (4′,5,7-trihydroxyflavanone) is a chemotherapeutic agent that inhibits particular signal transduction pathways and has recently been demonstrated to have antitumor characteristics in several human cancer cells by causing growth inhibition, cell cycle arrest, and cell death[120, 121]. It is also non-mutagenic and has a low toxicity[122]. The combined repressing effects of apigenin and many more chemotherapeutic drugs on cancer cell proliferation, which was greater than their individual effects. In HCC, combined incubation of apigenin with 5-FU-enhanced ROS levels is attributed to a reduction in ΔΨm. Interestingly, combined treatment stimulated the mitochondrial apoptotic pathway, as seen by suppressed Bcl-2 expression and lowered ΔΨm, as well as caspase-3 and PARP activation[123].

11.3.2.4   Quercetin and Cisplatin

Quercetin, a naturally existing polyphenolic flavonoid with anticancer potential[124], disrupts cell cycle progression in HCC cells by accelerating the levels of p21 and p27 (cyclin-dependent kinase inhibitors)[125]. Cisplatin, a widely used anticancer medication, predominantly exerts cytotoxicity by interacting with cellular DNA. Quercetin would exert synergistic suppressive effects on HCC cells, in combination with cisplatin. In HepG2 cells, they suppress cell multiplication and activate apoptotic effects, involving changes in various cell cycle and apoptotic regulators[124].

11.3.2.5   Catechins and Doxorubicin (DOX)

In HCC patients, DOX is commonly utilized as a single medication[126]. It inserts into DNA, stabilizes the topoisomerase-II protein, and induces necrosis by inhibiting topoisomerase-II and triggering redox reactions that produce ROS and free radicals[127]. (–)-Epigallocatechin-3-O-gallate (EGCG), a primary polyphenol in green tea, overpowers tumor growth by inducing apoptosis[128]. In DOX-resistant HCC cells, catechins hinder the expression of MDR1 (multidrug resistance-1) mRNA and lower the P-glycoprotein (membrane transporter) levels, which drives a wide range of xenobiotics, implying that DOX coupled with EGCG hinders P-glycoprotein efflux pump activity and massively improves intracellular DOX accretion[129].

11.3.2.6   Gambogic Acid (GA) and Proteasome Inhibitor MS132 or MG262

GA, extracted from the gamboge of the *Garcinia hanburyi* tree, has anticancer potential on a number of tumor cells, such as HCC, lung and gastric carcinoma, glioma cells, and breast cancer, through numerous cellular signaling effects[130]. Most of the poly-ubiquitinated proteins (also regulatory proteins) are degraded by the ubiquitin-proteasome system, which itself is involved in essential physiological processes, such as cell cycle procession, cell growth and distinction, programmed cell death, angiogenesis, and cell signaling pathways[131, 132]. Bortezomib is the first proteasome inhibitor to receive FDA approval for use as a first-line treatment for refractory multiple myeloma[133]. Combining GA with the proteasome inhibitors (MG 132 or MG 262) exerts a synergistic repressive effect in K562 (human leukemia cells), H22 (mouse hepatocarcinoma cells), and H22 cell allografts[130].

11.3.2.7   Quercetin with Carboxyamidotriazole (CAI)

Quercetin inhibits l-phosphatidylinositol-4-kinase and 1-phosphatidylinositol-4-phosphate 5-kinase and finally drops inositol-1,4,5-triphosphate concentration, which reduces calcium release by intracellular sources. CAI is a new anticancer drug that prevents calcium from entering cells when combined with quercetin produced synergistic effects on MDA-MB-435 (human breast cancer cells) by lowering cytosolic calcium levels[134].

11.3.2.8 Eicosapentaenoic Acid (EPA) on Antiproliferative Action of Anticancer Drugs

Several medications confirmed in the treatment of esophageal cancer, including docetaxel, paclitaxel, 5-FU, and cisdiamminedichloridoplatinum(II) (CDDP). Paclitaxel and docetaxel attach to microtubules, preventing them from separating normally during cell division and so slowing tumor development. 5-FU disrupts structure of DNA, whereas CDDP cross-linkages DNA strands and prevents them from being separated into single strands during cell division. Combination of CDDP and 5-FU is the novel drug therapy for esophageal cancer[135, 136]. EPA inhibits cell proliferation in cancer cells, like pancreatic and breast cancers[137]. On combining EPA with other anticancer medicines, it produces synergistic effects by inhibiting NF-κB activity and induces apoptotic activity (TE-1 esophageal cancer cell line)[138].

11.3.2.9 Gemcitabine (GEM) and Impressic Acid (E12–1) or Acankoreanogein (E13–1)

GEM is used as a first-line therapy for chronic cancer of pancreas, but large concentrations result in resistance and toxic effects, restricting its use[139]. E12–1 and E13–1 are two major active chemicals found in *Acanthopanax trifoliatus* (L.) Merr., having anti-inflammatory and anticancer properties. Combination of GEM through E12–1/E13–1 may drastically and synergistically suppress Panc-1 cell growth, induce apoptosis, and diminish migration[140].

11.4 CURRENT STATUS OF SYNERGISTIC APPROACH OF NATURAL PRODUCTS IN THE TREATMENT OF CANCER

When using an herbal product, it gets more complicated because its activity is often related to a combination of other ingredients rather than a single item. For example, fruits and green vegetables have been shown to significantly lower the incidence of tumor, owing to the action of a mixture of polyphenols[98]. Herbal and medication interactions have been studied for thousands of years ever since herbs were mixed together, but there is relatively little information accessible in this sector due to a lack of systematic data. In the traditional practices of Chinese medicine, Japanese Kampo, and Indian Ayurveda, several safer herbal mixtures are used.

Herb-drug interactions could either be pharmacokinetic or pharmacodynamic. Pharmacokinetic interactions like the cytochrome P-450 isoenzyme system can affect the availability of theophylline, caffeine, and other compounds, resulting in a reduction in therapeutic response. Pharmacodynamic interactions have a significant impact on drug activity, either through synergistic or antagonistic effects[141]. Curcumin inhibits cell proliferation, has pro-apoptotic effect, and inhibits COX-2, angiogenesis, oncogenesis, human epidermal growth factor receptor-2 downregulation, and NF-kB (transcription factor), as well as AP-1 inhibitor. Curcumin's effectiveness was evidenced by more than 12,000 papers published from 1924 to 2018, with 37% putting more emphasis on cancer.

Curcumin can help eliminate MDR challenges by limiting cytotoxic activity and enhancing the activity of anticancer medications, such as 5-FU, paclitaxel, GEM, cisplatin, oxaliplatin, and DOX, among several others. Curcumin does not increase the toxicity of synergistic anticancer medicines when taken with other anticancer treatments, making it an excellent choice for combination therapy. Curcumin improves the chemosensitivity of 5-FU as a first-line chemotherapy for colorectal cancer through aiming on Src, NF-kB, and NF-NB dependent regulatory gene products. Curcumin, either alone or in conjunction with 5-FU, inhibits colon polyps development, encourages apoptosis, inhibits proliferation, and suppresses colon cancer cell indicators in MMR scarce 5-FU resistant cells when detected in high-density culture[142].

11.5 FUTURE PROSPECTS OF NATURAL PRODUCTS IN THE TREATMENT OF CANCER

A common belief that a single medication is enough for the treatment of a disease is built upon the hypothesis that human ailments share a common genomic link among patient populations. But latest researches in genetics show genomic variation, including polymorphism, inferring that a certain type of patient population might require specific tailored drugs for their treatment, a concept known as personalized medicine[8]. According to the literature evidence cited in various Ayurvedic texts and latest pharmacological scholarly articles on medicinal herbs, medicinal herbs are a great resource of chemotherapeutic drugs treating various types of malignancies. Furthermore, because many herbs have chemoprotective properties, a combination of herbal remedies and conventional therapy may be recommended to inhibit cancer cell growth and alleviate symptoms of radiation and chemotherapy[31].

The use of medicinal herbs may provide a solution to the drug development crisis by using traditional herbal medicine knowledge (less expensive and more quickly), by providing a holistic perspective that accentuates the disease-targeted approach, and by the interaction between the different components of the herbs[143]. But major limitations herewith are the absence of international standardization (like how to evaluate their components, effectiveness, clinical efficacy and quality), sustained manufacturing standards, and compliance and approbation procedures. Surprisingly, the pharmaceutical industry has a wealth of knowledge and experience in drug development. As a result, uniting the merits of indigenous system of medicine and current medicine practices has previously been suggested as an effective approach for discovering and commercializing novel plant originated materials. However, only a few medicinal herbs are approved by health regulating bodies for clinical usage in recent centuries. Collaborative efforts between the WHO, the FDA, European and other regulatory bodies, and the global pharmaceutical firms may result in precise instructions regarding the formulation of herbal remedies while capitalizing on traditional medicine's enormous potential for the formulation of anticancer and several other drugs for health advancement[8].

11.6 CONCLUSION

Conventional cancer remedies are associated with number of side effects that can cause a serious problem to the body like gastrointestinal problems, kidney injury, or many other complications, but natural products are the suitable candidate to replace them as they are safest option and effective too. Natural products are constituted with numerous chemical substances like alkaloids, phenol compounds, and monoterpenes having anticancer potential against different types of cancer. Moreover, since many herbs have anticarcinogenic characteristics, a combination of herbal remedies and conventional therapy may be preferred to suppress growth of cancer cells and relieve radiation and chemotherapy side effects. There are many written evidences or studies that confirm that a combination of natural remedies and synthetic drugs is an effective and work in different manners. But the potency of these synergistic effects is dependent on various factors, like chemical nature, dose information, degree of exposure, underlying mode of action, and site of action. However, the in vitro study findings cannot be straightforwardly taken into account for clinical practice, but these findings aid in interpreting the new approaches for disease management. Amalgamation of modern remedies with conventional knowledge with help of novel scientific methodologies may help pave the way for synergistic potential of natural products in cancer treatment.

11.7 ACKNOWLEDGMENT

The authors are highly indebted to Smt. Tarawati Institute of Biomedical and Allied Sciences, Roorkee, Uttarakhand, India, for providing necessary facilities to carry out this work. Meenakshi Bajpai and Sharad Visht designed the framework of the chapter. Ritu Saini, Chitra Gupta, and

Shalini Pundir wrote the manuscript. Ritu Saini and Akansha Agrwal modified the manuscript as per journal guidelines and are involved in the submission of the work. All authors reviewed and approved the final manuscript. The authors received no funding and were not required to obtain consent in conducting this study and declares to have no competing interests.

11.8 REFERENCES

1. Mathur, G., Nain, S., and Sharma, P. K. 2015. Cancer: An overview. *Academic J Cancer Res* 8(1):1–9.
2. Hassanpour, S. H., and Dehghani, M. 2017. Review of cancer from perspective of molecular. *J Cancer Res Practice* 4:127–129.
3. Laya, A., Koubala, B. B., Pathak, K. P., and Bueno, V. 2021. Vitamins and provitamins intake as new insights to prevent and/or to treat breast cancer: A systemic review. *Int J Cancer Clin Res* 8:162.
4. Wang, H., Khor, O., Shu, L., et al. 2012. Plants against cancer: A review on natural phytochemicals in preventing and treating cancers and their druggability. *Anticancer Agents Med Chem* 12(10):1281–1305.
5. Cragg, G., and Pezzuto, J. 2015. Natural products as a vital source for the discovery of cancer chemotherapeutic and chemopreventive agents. *Med Princ Pract* 25(2):41–59.
6. Pezzani, R., Salehi, B., Vitlani, S., et al. 2019. Synergistic effects of plant derivatives and conventional chemotherapeutic agents: An update on the cancer perspective. *Medicina* 55(4):1–15.
7. Cheon, C. 2021. Synergistic effects of herbal medicines and anticancer drugs. *Cheon Medicine* 100:46.
8. Fridlender, M., Kapulnik, Y., and Koltai, H. 2015. Plant derived substances with anti-cancer activity. *Frontiers in Plant Science* 6:1–9.
9. Rashid, H., Gafur, G. M., and Sadik, R. M. A. A. 2002. Biological activities of a new derivative from Ipomoea turpithum. *Pak J Biol Sci* 5:968–969.
10. Pledgie-Tracy, A., Sobolewski, M. D., and Davidson, N. E. 2007. Sulforaphane induces cell type-specific apoptosis in human breast cancer cell lines. *Mol Cancer Ther* 6(3):1013–1021.
11. Risinger, A. L., Giles, F. J., and Mooberry, S. L. 2009. Microtubule dynamics as a target in oncology. *Cancer Treat Rev* 35(3):255–261.
12. Sharma, H., Parihar, L., and Parihar, P. 2011a. Review on anticancerous properties of some medicinal plants. *J Med Plant Res* 5(10):1818–1835.
13. Chikezie, P. C., Ojiako, O. A., and Nwufo, K.C. 2015a. Overview of anti-diabetic medicinal plants: The Nigerian research experience. *J Diabetes Metab* 6:546.
14. Solowey, E., Lichtenstein, M., Sallo, S., Paavilainen, H., Solowet, E., and Lorberboum-Galski, H. 2014. Evaluating medicinal plants for anticancer activity. *Sci World J* 721402:12.
15. Chikezie, P. C., Ekeanyanwu, R. C., Chile-Agada, A. B., and Ohiagu, F. O. 2019. Comparative FT-IR analysis of chloroform fractions of leaf extracts of Anacardium occidentale, Psidium guajava and Terminalia catappa. *J Basic Pharmacol Toxicol* 3(1):1–6.
16. Desai, A. G., Qazi, G. N., Ganju, R. K., et al. 2008. Medicinal plants and cancer chemoprevention. *Curr Drug Metab* 9(7):581–591.
17. Shoeb, M. 2006. Anti-cancer agents from medicinal plants. *Bang J Pharmacol* 1(2):35–41.
18. Sultana, S., Asif, H. M., Nazar, H. M. I., Akhtar, N., Rehman, J. U., and Rehman, R. U. 2014. Medicinal plants combating against cancer—a green anticancer approach. *Asian Pac J Cancer Prev* 15(11):4385–4394.
19. Ohiagu, F. O., Chikezie, P. C., Chikezie, C. M., and Enyoh, C. E. 2021. Anticancer activity of Nigerian medicinal plants: A review. *Future Journal of Pharmaceutical Sciences* 7(70):1–21.
20. Gordaliza, M. 2007. Natural products as leads to anticancer drugs. *Clin. Transl. Oncol* 9:767–776.
21. Bhutani, K. K., and Gohil V. 2010. Natural products drug discovery research in India: Status and appraisal. *Indian J Exp Biol* 48:199–207.
22. Kumar, G., Gupta, R., Sharan, S., Roy, P., and Pandey, D. M. 2019. Anticancer activity of plant leaves extract collected from a tribal region of India. *3 Biotech* 9(399):1–16.
23. Nguyen, V. T., Sakoff, J. A., and Scarlett, C. J. 2017. Physicochemical properties, antioxidant and cytotoxic activities of crude extracts and fractions from Phyllanthus amarus. *Med* 4(2):42
24. Soumya, T., Lakshmipriya, T., Klika, Karel D., Jayasree, P. R., and Manish Kumar, P. R. 2021. Anticancer potential of rhizome extract and a labdane diterpenoid from Curcuma mutabilis plant endemic to Western Ghats of India. *Scientific Reports* 11(1): 1–20.
25. Sahoo, N. K., Sahu, M., Pullaiah, C. P., and Muralikrishna, K. S. 2020. In vitro anticancer activity of Eclipta alba whole plant extract on colon cancer cell HCT-116. *BMC Complementary Medicine and Therapies* 20(1):1–8.

26. Svoboda, G. H., Poore, G. A., Simpson, P. J., and Boder, G. B. 1996. Alkaloids of Acronychia baueri Schott: Isolation of the alkaloids and a study of the antitumor and other biological properties of acronycine. *J Pharm Sci* 55(8):758–768.

27. Adebayo, A. H., Tan, N. H., Akindahunsi, A. A., Zeng, G. Z., and Zhang, Y. M. 2010. Anticancer and antiradical scavenging activity of Ageratum conyzoides L. (Asteraceae). *Pharmacogn Mag* 6(21):62–66.

28. Kuete, V., Tchinda, C. F., Mambe, F. T., Beng, V. P., and Efferth, T. 2016. Cytotoxicity of methanol extracts of 10 Cameroonian medicinal plants towards multifactorial drug-resistant cancer cell lines. *BMC Complement Altern Med* 16(1): 267.

29. Thomson, M., and Ali, M. 2003. Garlic (Allium sativum): A review of its potential use as an anti- cancer agent. *Curr Cancer Drug Targets* 3(1):67–81.

30. Petrovic, V., Nepal, A., Olaisen, C., Bachke, S., Hira, J., and Søgaard, C. K. 2018. Anti-cancer potential of home made fresh garlic extract is related to increased endoplasmic reticulum stress. *Nutr* 10(4):450.

31. Begum, I., Sharma, R., and Sharma, H. K. 2017. A review on plants having anti-cancer activity. *Curr Trends Pharm Res* 4(2):39–62.

32. Devi, S., Saini, R., and Kumar, P. 2020. A review on herbal drugs as anticancer agents. *JETIR* 7(10):701–714.

33. Balogun, O. S., Ajayi, O. S., and Agberotimi, B. J. 2016. A cytotoxic indole alkaloid from Alstonia boonei. *J Biol Act Prod Fr Nat* 6(4):347–351.

34. Konan, N. A., Lincopan, N., Collantes Díaz, I. E., de Fátima, J. J., Tiba, M. M., and Amarante Mendes, J. G. 2012. Cytotoxicity of cashew flavonoids towards malignant cell lines. *Exp Toxicol Pathol* 64(5):435–440.

35. Taiwo, B. J., Fatokun, A. A., Olubiyi, O. O., Bamigboye-Taiwo, O. T., van Heerden, F. R., and Wright, C. W. 2017. Identification of compounds with cytotoxic activity from the leaf of the Nigerian medicinal plant, Anacardium occidentale L. (Anacardiaceae). *Bioorg Med Chem* 25(8):2327–2335.

36. Kumar, R. A., Sridevi, K., Kumar, N. V., Nanduri, S., and Rajagopal, S. 2004. Anticancer and immunostimulatory compounds from Andrographis paninculata. *J. Ethnopharmacol* 92(2–3):291–295.

37. Jada, S. R., Subur, G. S., Matthews, C., et al. 2007. Semisynthesis and in vitro anticancer activities of andrographolide analogues. *J. Phytochemistry* 68(6):904–912.

38. Olugbami, J. O., Damoiseaux, R., France, B., Onibiyo, E. M., Gbadegesin, M. A., and Sharma, S. 2017a. A comparative assessment of antiproliferative properties of resveratrol and ethanol leaf extract of Anogeissus leiocarpus (DC) Guill and Perr against HepG2 hepatocarcinoma cells. *BMC Complement Altern Med* 17(1):381.

39. Salau, A. K., Yakubu, M., and Oladiji, A. 2013. Cytotoxic activity of aqueous extracts of Anogeissus leiocarpus and Terminalia avicennioides root barks against Ehrlich ascites carcinoma cells. *Indian J Pharm* 45(4):381–385.

40. Seeram, N. P., Adams, L. S., Henning, S. M., Niu, Y., Zhang, Y., and Nair, M. G. 2005. In vitro antiproliferative, apoptotic and antioxidant activities of punicalagin, ellagic acid and a total pomegranate tannin extract are enhanced in combination with other polyphenols as found in pomegranate juice. *J Nutr Biochem* 16(6):360–367.

41. Zhou, R., Chen, H., and Chen, J. 2018. Extract from Astragalus membranaceus inhibits breast cancer cells proliferation via PI3K/AKT/mTOR signaling pathway. *BMC Complement Altern Med* 18(83):1–8.

42. Wei, Q. G., Liang, Z. L., and Zhuo, Y. H. 2013a. Anti-proliferative effects of Atractylis lancea (Thunb.) DC. Via down-regulation of the c-myc/hTERT/telomerase pathway in Hep-G2 Cells. *Asian Pac J Cancer Prev* 14:6363–6367.

43. Zhao, M., Wang, Q., Ouyang, Z., et al. 2014a. Selective fraction of Atractylodes lancea (Thunb.) DC. And its growth inhibitory effect on human gastric cancer cells. *Cytotechnol* 66(2):201–208.

44. Muthulingam, M., and Chaithanya, K. K. 2018. In vitro anticancer activity of methanolic leaf extract of Boerhaavia diffusa Linn., against MCF-7 cell line. *Drug Invent Today* 10(2):3107–3111

45. Venkatajothi, R. 2017. In vitro anti-cancer activity of Boerhaavia diffusa Linn. *Int J Curr Res Biol Med* 2(3):20–24.

46. Pandey, S., Walpole, C., Cabot, P. J., Shaw, P. N., Batra, J., and Hewavitharana, A. K. 2017. Selective anti-proliferative activities of Carica papaya leaf juice extracts against prostate cancer. *Biomed Pharmacother* 89:515–523.

47. Nguyen, T. T., Parat, M. O., Hodson, M. P., Pan, J., Shaw, P. N., and Hewavitharana, A. K. 2015. Chemical characterization and in vitro cytotoxicity on squamous cell carcinoma cells of Carica papaya leaf extracts. *Toxins (Basel)* 8(1):7.

48. Bhagat, M., and Saxena, A. 2010. Evaluation of Cassia occidentalis for in vitro cytotoxicity against human cancer cell lines and antibacterial activity. *Indian J Pharm* 42(4):234–237.

49. Taiwo, F. O., Akinpelu, D. A., Aiyegoro, O. A., Olabiyi, S., and Adegboye, M. F. 2013. The biocidal and phytochemical properties of leaf extract of Cassia occidentalis Linn. *Afr J Microbiol Res* 7(27):3435–3441

50. Arunee, H., Kornkanok, I., Nanteetip, L., and Daniel, S. 2014. Ethanolic extract from Derris scandens Benth mediates radiosensitzation via two distinct modes of cell death in human colon cancer HT-29 cells. *Asian Pac J Cancer Prev* 15(4):1871–1877.

51. Ito, C., Matsui, T., Miyabe, K., Hasan, C. M., Rashid, M. A., Tokuda, H., and Itoigawa, M. 2020. Three isoflavones from Derris scandens (Roxb.) Benth and their cancer chemopreventive activity and in vitro antiproliferative effects. *Phytochem* 175:112376.

52. Kuete, V., Fokou, F. W., Karaosmanoglu, O., Beng, V. P., and Sivas, H. 2017. Cytotoxicity of the methanol extracts of Elephantopus mollis, Kalanchoe crenata and 4 other Cameroonian medicinal plants towards human carcinoma cells. *BMC Complement Altern Med* 17(1):280.

53. Musuyu, M. D., Fruth, B. I., Nzunzu Lami, J., Mesia, G. K., Kambu, O. K., and Tona, G. L. 2012. In vitro antiprotozoal and cytotoxic activity of 33 ethonopharmacologically selected medicinal plants from Democratic Republic of Congo. *J Ethnopharmacol* 141(1):301–308.

54. Kassim, O. O., Copeland, R. L., Kenguele, H. M., Nekhai, S., Ako-Nai, K. A., and Kanaan, Y. M. 2015. Antiproliferative activities of Fagarax anthoxyloides and Pseudocedrela kotschyi against prostate cancer cell lines. *Anticancer Res* 35(3):1453–1458.

55. Corrêa, M. G., Couto, J. S., and Teodoro, A. J. 2016. Anti-cancer properties of Psidium guajava—a mini-review. *Asian Pac Organ Cancer Prev* 17(9):4199–4204.

56. Syed, D. N., Chamcheu, J. C., Adhami, V. M., and Mukhtar, H. 2013. Pomegranate extracts and cancer prevention: Molecular and cellular activities. *Anti-Cancer Agents Med Chem* 13(8):1149–1161.

57. El-khamissi, H. A. Z., Saad, Z. H., and Rozan, H. E. 2019. Phytochemicals screening, antioxidant and anticancer activities of garlic (Allium sativum) extracts. *J Agric Chem Biotech* 10(4):79–82.

58. Li, M. and Zhang, N. 2008 To study the quality control of *Curcuma aromatica*. *J Chinese Med Mater* 31:540–543. (In Chinese)

59. Feng, L., Zhang, L. F., Yan, T., Jin, J., and Tao, W. Y. 2006. Studies on active substance of anticancer effect in *Polygonum cuspidatum*. *J Chinese Med Mater* 29:689–691. (In Chinese)

60. Majumdar, A. P., Banerjee, S., Nautiyal, J., et al. 2009. Curcumin synergizes with resveratrol to inhibit colon cancer. *Nutr. Cancer* 61:544–553.

61. Du, Q., Hu, B., and An, H. M. 2013. Synergistic anticancer effects of curcumin and resveratrol in Hepal1–6 hepatocellular carcinoma cells *Oncol. Rep* 29:1851–1858.

62. Yang, L., Tian, W. et al. 2014. Resveratrol plays dual roles in pancreatic cancer cells[J]. *J Cancer Res Clin Oncol* 140(5):749–755.

63. Zhou, S., Guo, Y., Wang, X., Wang, P.J. and Zhang, B. 2014a. Effects of resveratrol on oral squamous cell carcinoma (OSCC) cells in vitro. *J Cancer Res Clin Oncol* 140:371–374.

64. Cakir, Z., Saydam, G., Sahin, F., et al. 2011. The roles of bioactive sphingolipids in resveratrol-induced apoptosis in HL60 acute myeloid leukemia cells. *J Cancer Res Clin Oncol* 137(2):279–286.

65. Nakase, I., Gallis, B., Takatani-Nakase, T., et al. 2009. Ransferrin receptor-dependent cytotoxicity of artemisinin–transferrin conjugates on prostate cancer cells and induction of apoptosis. *Cancer Lett* 274(2):290–298.

66. Lai, H., Nakase, I., Lacoste, E., Singh, N. P., and Sasaki, T. 2009. Artemisinin transferrin conjugate retards growth of breast tumors in the rat. *Anticancer Res* 29(10):3807–3810.

67. Riganti, C., Doublier, S., Costamagna, C., et al. 2008. Activation of nuclear factor-kappa B pathway by simvastatin and RhoA silencing increases doxorubicin cytotoxicity in human colon cancer HT29 cells. *Mol Pharmacol* 74(2):476–484.

68. Li, P., Yang, S., Dou, M., Chen, Y., Zhang, J., and Zhao, X. 2014. Synergic effects of artemisinin and resveratrol in cancer cells. *J. Cancer Res. Clin. Oncol.* 140: 2065–2075.

69. Murakami, A., Ohura, S., Nakamura, Y., Koshimizu, K. and Ohigashi, H. 1996.1′-Acetoxychavicol acetate, a superoxide anion generation inhibitor, potently inhibits tumor promotion by 12-*O*-tetradecanoylphorbol-13-acetate in ICR mouse skin. *Oncology* 53:386–391.

70. Ohnishi, M., Tanaka, T., Makita, H., et al. 1997. Chemopreventive effect of a xanthine oxidase inhibitor, 1′-acetoxychavicol acetate, on rat oral carcinogenesis. *Jpn. J. Cancer Res.* 88:821–830.

71. Moffatt, J., Hashimoto, M., Kojima, A., et al. 2000. Apoptosis induced by 1′-acetoxychavicol acetate in Ehrlich ascites tumoe cells is associated with modulation of polyamine metabolism and caspase-3 activation. *Carcinogenesis* 21:2151–2157.

72. Hague, A., Manning, A. M., Hanlon, K. A., Hart, L. I., Huschtscha, D., and Paraskeva, C. 1993. Sodium butyrate induces apoptosis in human colonic tumor cell lines in a p53-independent pathway: Implication for the possible role of dietary fiber in the prevention of large-bowel cancer. *Int. J. Cancer* 55:498–505.

73. Wu, J. T., Archer, S. Y., Hinnebusch, B., Meng, S., and Hodin, R. A. 2001. Transient *vs.* prolonged histone hyperacetylation: Effects on colon cancer cell growth, differentiation and apoptosis. *Am. J. Physiol.* 280:G482–G490.

74. Augeron, C., and Laboisse, C. L. 1984. Emergence of permanently differentiated cell clones in a human colonic cancer cell line in culture after treatment with sodium butyrate. *Cancer Res.* 44:3961–3969.

75. Ho, S. B., Yan, P. S., and Dahiya, R. 1994. Stable differentiation of a human colon adenocarcinoma cell by sodium butyrate is associated with multidrug resistance. *J. Cell. Physiol.* 160:213–226.

76. Kato, R., Matsui-Yuasa, I., Azuma, H., and Kojima-Yuasa, A. 2014. The synergistic effect of 1'-acetoxychavicol acetate and sodium butyrate on the death of human hepatocellular carcinoma cells. *Chem. Biol. Interact.* 212:1–10.

77. Tatsuka, M., Maeda, M., and Ota, T. 2001. Anticarcinogenic effect and enhancement of metastatic potential of BALB/c 3T3 cell by ginsenoside Rh2. *Jpn J Cancer Res* 92:1184–1189.

78. Nakata, H., Kikuchi, Y., Tode, T., Hirata, J., and Kita, T. 1998. Inhibitory effects of ginsenoside Rh2 on tumor growth in nude mice bearing human ovarian cancer cells. *Jpn J Cancer Res* 89:733–740.

79. Oh, M., Choi, Y. H., Cho, S., et al. 1999. Anti-proliferating effects of ginsenoside Rh2 on MCF-7 human breast cancer cells. *Int J Oncol* 14:869–875.

80. Jin, Y. H., Yoo, K. J., Lee, Y. H., and Lee, S. K. 2000. Caspase 3-mediated cleavage of p21WAF1/CIP1 associated with the cyclin A-cyclindependent kinase 2 complex is a prerequisite for apoptosis in SK-HEP-1 cells. *J Biol Chem* 275:30256–30263.

81. Popovich, D. G., and Kitts, D. D. 2002. Structure-function relationship exists for ginsenosides in reducing cell proliferation and induce apoptosis in the human leukemia (THP-1) cell line. *Arch Biochem Biophys* 406:1–8.

82. Chi-Chih, C., Shu-Mei, Y., Chi-Ying, H., Jung-Chou, C., Wei-Mao, C., and Shih-Lan, H. 2005. Molecular mechanisms of ginsenoside Rh2- mediated G1 growth arrest and apoptosis in human lung adenocarcinoma A549 cells. *Cancer Chemother Pharmacol* 55:531–540.

83. Pisha, E., Chai, H., Lee, I. S., et al. 1995. Discovery of betulinic acid as a selective inhibitor of human melanoma that functions by induction of apoptosis. *Nat. Med.* 1:1046–1051.

84. Zuco, V., Supino, R., and Righetti, S. C. 2002. Selective cytotoxicity of betulinic acid on tumor cell lines, but not on normal cells. *Cancer Lett.* 175:17–25.

85. Fulda, S., Scaffidi, C., Susin, S. A., et al. 1998. Activation of mitochondria and release of mitochondrial apoptogenic factors by betulinic acid. *J. Biol. Chem.* 273:33942–33948.

86. Li, Q., Wang, X., Fang, X., et al. 2011. Co-treatment with ginsenoside Rh2 and betulinic acid synergistically induces apoptosis in human cancer cells in association with enhanced caspase-8 activation, Bax translocation, and cytochrome c release. *Mol. Carcinog.* 50:760–769.

87. Wang, Z., Zhang, J., Wang, Y., et al. 2013. Matrine, a novel autophagy inhibitor, blocks trafficking and the proteolytic activation of lysosomal proteases. *Carcinogenesis* 34:128–138.

88. Zhang, L., Wang, T., Wen, X., et al. 2007. Effect of matrine on HeLa cell adhesion and migration. *Eur J Pharmacol* 563:69–76.

89. Sun, M., Cao, H., Sun, L., et al. 2012. Antitumor activities of kushen: Literature review. *Evid Based Complement. Alternat Med* 2012:373219.

90. Zhou, H., Xu, M., Gao, Y., et al. 2014b. Matrine induces caspase-independent program cell death in hepatocellular carcinoma through bid-mediated nuclear translocation of apoptosis inducing factor. *Mol Cancer* 13:59.

91. Yu, Q., Chen, B., Zhang, X., Qian, W., Ye, B., and Zhou, Y. 2013. Arsenic trioxide-enhanced, matrine-induced apoptosis in multiple myeloma cell lines. *Planta Med* 79:775–781.

92. Ou, X., Chen, Y., Cheng, X., Zhang, X., and He, Q. 2014. Potentiation of resveratrol-induced apoptosis by matrine in human hepatoma HepG2 cells. *Oncol. Res.* 32:2803–2809.

93. Luo, J. L., Maeda, S., Hsu, L. C., Yagita, H., and Karin, M. 2004. Inhibition of NF-κB in cancer cells converts inflammation-induced tumor growth mediated by TNFα to TRAIL-mediated tumor regression. *Cancer Cell* 6:297–305.

94. Paur, I., Balstad, T. R., Kolberg, M., et al. 2010. Extract of oregano, coffee, thyme, clove, and walnuts inhibits NF-kappaB in monocytes and in transgenic reporter mice. *Cancer Prev Res (Phila)* 3:653–663.

95. Sen, S., Sharma, H., and Singh, N. 2005. Curcumin enhances Vinorelbine mediated apoptosis in NSCLC cells by the mitochondrial pathway. *Biochem Biophys Res Commun* 331:1245–1252.

96. Huang, M. T., Smart, R. C., Wong, C. Q., and Conney, A. H. 1998. Inhibitory effect of Curcumin, chlorogenic acid, caffffeic acid and ferulic acid on tumor promotion in mouse skin by 12-O-tetradecanoylphorbol-13-acetate. *Cancer Res.* 48:5941–5946.

97. Bunn Jr, P. A. 2002. Chemotherapy for advanced non-small-cell lung cancer: Who, what, when, why? *J Clin Oncol* 20:23S–33S.

98. Mertens-Talcott, S. U., and Percival, S. S. 2005. Ellagic acid and quercitin interact synergistically with reseveratol in the induction of apoptosis and cause transient cell cycle arrest in human leukemia cells. *Cancer Lett* 218:141–151.

99. Potter, J. D. 1997. Cancer prevention: Epidemiology and experiment. *Cancer Lett.* 114:7–9.

100. Boileau, T. W., Liao, Z., Kim, S., Lemeshow, S., Erdman Jr., J. W., and Clinton, S. K. 2003. Prostate carcinogenesis in N-methyl-Nnitrosourea (NMU)-testosterone-treated rats fed tomato powder, lycopene, or energy-restricted diets. *J. Natl Cancer Inst.* 95:1578–1586.

101. Sargent, D., Sobrero, A., Grothey, A., et al. 2009. Evidence for cure by adjuvant therapy in colon cancer: Observations based on individual patient data from 20,898 patients on 18 randomized trials. *J Clin Oncol* 27:872–877.

102. Meiyantol, E., Septisetyani, E. P., Larasati, Y. A., and Kawaichi, M. 2017. Curcumin Analog Pentagamavunon-1 (PGV-1) Sensitizes Widr Cells to 5-Fluorouracil through Inhibition of NF-κB Activation. *Asian Pac J Cancer Prev* 19(1):49–56.

103. Da'I, M., Supardjan, A. M., Meiyanto, E., et al. 2007. Geometric isomers and cytotoxic effect of curcumin analogs PGV-0 and PGV-1 toward T47D cells. *Indon J Pharm* 18:40–47.

104. Putri, H., Jenie, R. I., Handayani, S., et al. 2016. Combination of potassium pentagamavunon and doxorubicin induces apoptosis and cell cycle arrest and inhibits metastasis in breast cancer cells. *Asian Pac J Cancer Prev* 17:2683–2688.

105. Sarkar, F. H., and Li, Y. 2003. Soy isoflflavones and cancer prevention. *Cancer Invest* 21:744–757.

106. Nakagawa, H., Yamamoto, D., Kiyozuka, Y., et al. 2000. Effects of genistein and synergistic action in combination with eicosapentanoic acid on the growth of breast cancer cell lines. *J Cancer Res Clin Oncol* 126:448–454.

107. Ye, F., Wu, J., Dunn, T., Yi, J., Tong, X., and Zhang, D. 2004. Inhibition of cyclooxygenase-2 activity in head and neck cancer cells by genistein. *Cancer Lett* 211:39–46.

108. Sakamoto, K. 2000. Synergistic effects of thearubigin and genistein on human prostate tumor cell (PC-3) growth via cell cycle arrest. *Cancer Lett* 151:103–109.

109. Banerjee, S., Li, Y., Wang, Z., et al. 2008. Multi-targeted therapy of cancer by genistein. *Cancer Lett* 269:226–242.

110. Yeh, T. C., Chiang, P. C., Li, T. K., et al. 2007. Genistein induces apoptosis in human hepatocellular carcinomas via interaction of endoplasmic reticulum stress and mitochondrial insult. *Biochem Pharmacol* 73:782–792.

111. Kito, M., Akao, Y., Ohishi, N., et al. 2002. Arsenic trioxide-induced apoptosis and its enhancement by buthionine sulfoximine in hepatocellular carcinoma cell lines. *Biochem Biophys Res Commun* 291:861–867.

112. Kang, S. H., Song, J. H., Kang, H. K., et al. 2003. Arsenic trioxide-induced apoptosis is independent of stress-responsive signaling pathways but sensitive to inhibition of inducible nitric oxide synthase in HepG2 cells. *Exp Mol Med* 35:83–90.

113. Jiang, H., Ma, Y., Chen, X., Pan, S., Sun, B., and Krissansen, G. W. 2010. Genistein synergizes with arsenic trioxide to suppress human hepatocellular carcinoma. *Cancer Sci.* 101:975–983.

114. Saunders, R. N., Metcalfe, M. S., and Nicholson, M. L. 2001. Rapamycin in transplantation: A review of the evidence. *Kidney Int* 59:3–16.

115. Gibbons, J. J., Abraham, R. T., and Yu, K. 2009. Mammalian target of rapamycin: Discovery of rapamycin reveals a signaling pathway important for normal and cancer cell growth. *Semin Oncol* 36(Suppl 3):S3–S17.

116. Sun, Y., Xun, K., Wang, Y., and Chen, X. 2009. A systematic review of the anticancer properties of berberine, a natural product from Chinese herbs. *Anticancer Drugs* 20:757–769.

117. Tang, J., Feng, Y., Tsao, S., Wang, N., et al. 2009. Berberine and Coptidis rhizoma as novel antineoplastic agents: A review of traditional use and biomedical investigations. *J Ethnopharmacol* 126:5–17.

118. Wang, N., Feng, Y., Zhu, M., et al. 2010. Berberine induces autophagic cell death and mitochondrial apoptosis in liver cancer cells: The cellular mechanism. *J Cell Biochem* 111:1426–1436.

119. Guo, N.A., Yan, A., Gao, X., et al. 2014. Berberine sensitizes rapamycin-mediated human hepatoma cell death *in vitro*. *Molecular Medicine Reports* 10:3132–3138.

120. Choi, E. J., and Kim, G. H. 2009. Apigenin induces apoptosis through a mitochondria/caspase-pathway in human breast cancer MDA-MB-453 cells. *J Clin Biochem Nutr.* 44:260–265.

121. Turktekin, M., Konac, E., Onen, H. I., Alp, E., Yilmaz, A., and Menevse, S. 2011. Evaluation of the effects of the flavonoid apigenin on apoptotic pathway gene expression on the colon cancer cell line (HT29). *J Med Food* 14:1107–1117.

122. Gupta, S., Afaq, F., and Mukhtar, H. 2001. Selective growth-inhibitory, cell-cycle deregulatory and apoptotic response of apigenin in normal versus human prostate carcinoma cells. *Biochem Biophys Res Commun.* 287:914–920.

123. Hu, X. Y., Liang, J. Y., Guo, X. J., Liu, L., and Guo, Y. B. 2015.5-Fluorouracil combined with apigenin enhances anticancer activity through mitochondrial membrane potential-mediated apoptosis in hepatocellular carcinoma. *Clin. Exp. Pharmcol. Physiol.* 42:146–153.

124. Zhao, J. L., Zhao, J., and Jiao, H. J. 2014b. Synergistic growth-suppressive effects of quercetin and cisplatin on HepG2 human hepatocellular carcinoma cells. *Appl. Biochem. Biotechnol.* 172:784–791.

125. Mu, C., Jia, P., Yan, Z., Liu, X., Li, X., and Liu, H. 2007. Quercetin induces cell cycle G1 arrest through elevating Cdk inhibitors p21 and p27 in human hepatoma cell line (HepG2). *Methods and Findings in Experimental and Clinical Pharmacology* 29:179–183.

126. Cao, H., Phan, H., and Yang, I. X. 2012. Improved chemotherapy for hepatocellular carcinoma. *Anticancer Res.* 32:1379–1386.

127. Tsang, W. P., Chau, S. P. Y., Kong, S. K., Fung, K. P., and Kwok, T. T. 2003. Reactive oxygen species mediate doxorubicin induced p53-independent apoptosis. *Life Sci.* 73:2047–2058.

128. Kuo, P. I., and Lin, C. C. 2003. Green tea constituent (–)-epigallocatechin-3-gallate inhibits HepG2 cell proliferation and induces apoptosis through p53-dependent and Fas-mediated pathways. *J. Biomed. Sci.* 10:219–227.

129. Liang, G., Tang, A. Z., Lin, X. Z., et al. 2010. Green tea catechins augment the antitumor activity of doxorubicin in an *in vivo* mouse model for chemoresistant liver cancer. *Int. J. Oncol.* 37:111–123.

130. Huang, H., Chen, D., Li, S., et al. 2011. Gambogic acid enhances proteasome inhibitor-induced anticancer activity. *Cancer Lett.* 301:221–228.

131. Goldberg, A. L. 2003. Protein degradation and protection against misfolded or damaged proteins. *Nature* 426:895–899.

132. Hochstrasser, M. 1995. Ubiquitin, proteasomes, and the regulation of intracellular protein degradation. *Curr. Opin. Cell Biol.* 7:215–223.

133. Chen, D., Errezza, M., Schmitt, S,; Kanwar, J., and Dou, Q. P. 2011. Bortezomib as the first proteasome inhibitor anticancer drug: Current status and future perspectives. *Curr. Cancer Drug Targets* 11:239–253.

134. Yeh, Y. A., Herenyiova, M., and Weber, G. 1995. Quercitin: Synergistic action with carboxyamidotriazole in human breast carcinoma cells. *Life Sci* 57:1285–1292.

135. Scanlon, K. J., Newman, E. M., Lu, Y., and Priest, D. G. 1986. Biochemical basis for cisplatin and 5-fluorouracil synergism in human ovarian carcinoma cells. *Proc Natl Acad Sci U S A* 83:8923–8925.

136. Ogoa, A., Miyakea, S., Kubotab, H., et al. 2017. Synergistic effect of eicosapentaenoic acid on antiproliferative action of anticancer drugs in a cancer cell line model. *Ann Nutr Metab* 71:247–252.

137. Fukui, M., Kang, K. S., Okada, K., and Zhu, B. T. 2013. EPA, an omega-3 fatty acid, induces apoptosis in human pancreatic cancer cells: Role of ROS accumulation, caspase-8 activation, and autophagy induction. *J Cell Biochem* 114:192–203.

138. Kubota, H., Matsumoto, H., Higashida, M., et al. 2013. Eicosapentaenoic acid modifies cytokine activity and inhibits cell proliferation in an oesophageal cancer cell line. *Anticancer Res* 33:4319–4324.

139. Binenbaum, Y., Na'ara, S., and Gil, Z. 2015. Gemcitabine resistance in pancreatic ductal adenocarcinoma. *Drug Resist Updat.* 23:55–68.

140. Jiang, S., Li, D-L., Jie Chen, J., et al. 2020. Synergistic Anticancer Effect of Gemcitabine Combined With Impressic Acid or Acankoreanogein in Panc-1 Cells by Inhibiting NF-κB and Stat 3 Activation. *Natural Product Communications* 15(12):1–7.

141. HemaIswarya, S., and Doble, M. 2006. Potential synergism of natural products in the treatment of cancer. *Phytother. Res* 20:239–249.

142. Karthika, C., Hari, B., Rahman, M. H. et al. 2021. Multiple strategies with the synergistic approach for addressing colorectal cancer. *Biomed Pharmacother* 140(111704):1–11.

143. Li, W. 2002. Botanical drugs: The next new new thing? Available at: http://nrs.harvard.edu/urn-3:HUL.InstRepos:8965577.

12 Medicinal Chemistry, Pharmacodynamics, and Pharmacokinetics of Camptothecin and Its Derivatives in Clinical Chemotherapeutics

*Thadiyan Parambil Ijinu, Parameswaran Sasikumar,
Vandhanam Aparna, Sreejith Pongillyathundiyil Sasidharan,
Vipin Mohan Dan, Farkhodjon Tukhtaev, Varughese George,
and Palpu Pushpangadan*

CONTENTS

12.1   Introduction...207
12.2   Camptothecin-Bearing Plants...208
12.3   Biosynthesis of Camptothecin 208
12.4   Chemistry of Camptothecin..209
12.5   Synthesis of Camptothecin 210
12.6   Synthesis of Camptothecin Derivatives... 212
12.7   Chemotherapeutic Potential of Camptothecin and Its Derivatives..................................... 215
     12.7.1   Topotecan.. 215
     12.7.2   Irinotecan.. 216
     12.7.3   Belotecan 217
12.8   Conclusion and Prospects 217
12.9   Acknowledgments... 217
12.10 References...218

12.1   INTRODUCTION

Camptothecin (CPT) is a tryptophan-derived quinoline alkaloid with a unique planar pentacyclic structure [1–3], which was originally obtained from the stem wood and bark of *Camptotheca acuminata*, Decne. (*xi shu* tree, or happy tree; family Nyssaceae) by the botanists from the US Department of Agriculture in the mid-1950s [1, 4, 5]. The plant, *C. acuminata*, is native to China and is used in traditional Chinese medicine for the treatment of psoriasis, leukemia, and diseases of the abdominal organs [6]. CPT effectively inhibited a number of cancer cells derived from leukemia, small-cell lung cancer, and colorectal cancer. Chemical synthesis of CPT and follow-up preclinical and clinical studies were actively conducted in the late 1950s and mid to late 1960s [7]. CPT was studied in the US in cancer patients in both the first and second clinical phases [8–10]. In China, CPT was

DOI: 10.1201/b23311-12

clinically used until the mid-1970s to treat specific types of leukemia, stomach, and bladder cancer, frequently in conjunction with corticosteroids [11].

Studies revealed that the water-soluble sodium salt of CPT (carboxylate form) has some promising effects against bladder or neck cancer [12]. The uncertain clinical efficacy of the carboxylate form prompted scientists to shift their attention to the lactone form for further research and development. Nevertheless, due to its water-insoluble property in the lactone form, unclear mechanism of action, low response rates [8–10], and side effects [13, 14], clinical experiments with CPT were abandoned in the 1970s. Later, several researchers studied the biochemical activity of CPT and found inhibitory activity both in DNA and RNA synthesis [15–22]. Subsequently, researchers confirmed that the action of the CPT was due to its lactone form and also proved that the lactone and carboxylate forms were in an equilibrium that was pH-dependent [23].

Again, in the late 1980s, the DNA topoisomerase I (Top-I) enzyme was discovered to be the therapeutic target for CPT, which brought more attention to CPT in anticancer drug research [24–29]. It is now clear that the toxicity of CPT is mainly due to the ring-open form (carboxylate or sodium salt form), which is inactive against its molecular target Top-I. Later, it was replaced by more effective and safer derivatives, such as topotecan, irinotecan, and belotecan [30, 31]. Also, hundreds of other CPT derivatives have been synthesized, and many of these have been actively studied by the research community over the past six decades.

12.2   CAMPTOTHECIN-BEARING PLANTS

The growing demand for CPT have encouraged scientists to search for alternate CPT-bearing plants besides the first reported *Camptotheca acuminata* (syn. *C. yunnanensis*) [1, 32]. More than 42 CPT-bearing plants belonging to ten families have been reported, which include the following: *C. lowreyana* (Nyssaceae) [33], *Alnus nepalensis* (Betulaceae) [34, 35], *Apodytes dimidiate* (Metteniusaceae) [36, 37], *Codiocarpus andamanicus* (Stemonuraceae), *Gomphandra comosa* (Stemonuraceae), *G. coriacea*, *G. tetrandra* (syn. *G. polymorpha*), *Iodes cirrhosa* (Icacinaceae), *Mappianthus hookerianus* (syn. *Iodes hookeriana*; Icacinaceae) [36, 38], *Merrilliodendron megacarpum* (Icacinaceae) [39], *Miquelia dentata* (Icacinaceae), *M. assamica* (syn. *M. kleinii*; Icacinaceae) [36], *Mappia nimmoniana* (syn. *Nothapodytes nimmoniana*; Icacinaceae) [40], *M. pittosporoides* (syn. *N. pittosporoides*) [41, 42], *M. obscura* (syn. *N. obscura*), *M. obtusifolia* (syn. *N. obtusifolia*), *M. tomentosa* (syn. *N. tomentosa*), *M. collina* (syn. *N. collina*) [43], *Natsiatum herpeticum* (Icacinaceae) [36], *Pyrenacantha klaineana* (Icacinaceae) [44], *P. volubilis* [45], *Sarcostigma kleinii* (Icacinaceae) [36], *Tabernaemontana alternifolia* (syn. *Ervatamia heyneana*; Apocynaceae) [46], *Chonemorpha fragrans* (syn. *C. grandiflora*; Apocynaceae) [47], *Mostuea brunonis* (Gelsemiaceae) [48], *Ixora chinensis* (syn. *I. coccinea*; Rubiaceae) [49], *Ophiorrhiza alata* (Rubiaceae) [50], *O. filistipula* [51], *O. fucosa*, *O. japonica* [52], *O. kuroiwae* (syn. *O. liukiuensis*) [53], *O. oblonga* (syn. *O. plumbea*), *O. pectinate* (syn. *O. mungos*) [54], *O. rugosa* (syn. *O. harrisiana*, *O. prostrata*) [55, 56], *O. pumila* [57, 58], *O. ridleyana*, *O. trichocarpos*, *O. wattii* [59–61], *Dysoxylum gotadhora* (syn. *D. binectariferum*; Meliaceae) [62], and *Rinorea anguifera* (Violaceae) [63], among others. In addition to CPT being produced in plant species, several biotechnological methods for the commercial production of CPT have also been reported [64].

12.3   BIOSYNTHESIS OF CAMPTOTHECIN

During the last six decades, various research groups around the globe have worked to build a thorough picture of CPT biosynthesis. Wenkert and co-workers proposed the first biosynthetic pathway of CPT based on the synthesis of corynantheidine (indole alkaloid) precursor [65]. Winterfledt et al. (1971) proposed another biosynthetic pathway, where the plausible precursor is geissoschizine [66]. Later, several research groups explained the biosynthetic pathway of CPT [38]. From the research findings, it is clear that, like any other monoterpene indole alkaloid, CPT is formed from

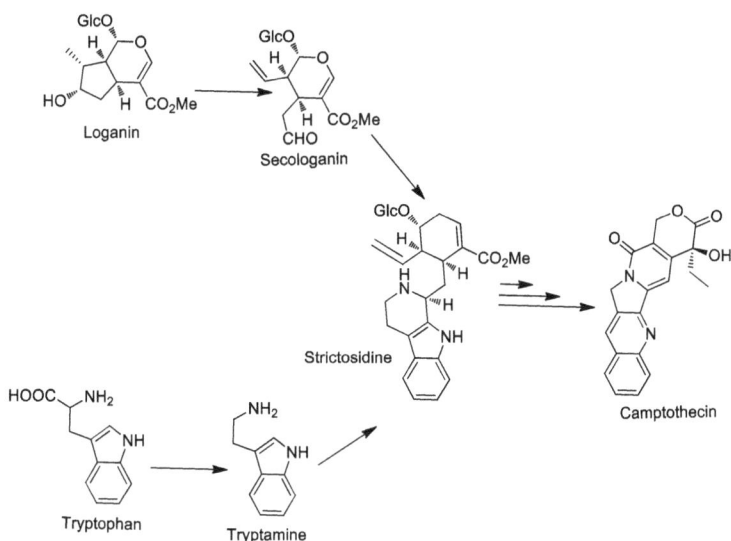

SCHEME 12.1 Biosynthesis of camptothecin.

strictosidine precursor. The shikimate pathway and the mevalonate and methylerythritol phosphate pathways are well described in the formation of strictosidine. The tryptamine is produced via the shikimate pathway [67, 68], with chorismite as a precursor, while the secologanin is produced by the mevalonate (occurring in the cytosol) and methylerythritol phosphate (occurring in the plastid) pathways, with isopentenyl diphosphate as a precursor [64, 69].

Strictosidine is formed when the condensation reaction between an indole, tryptamine (decarboxylation product tryptophan), and a terpenoid, secologanin (formed from the monoterpene loganin), occurs in the presence of an enzyme, strictosidine synthase [70]. Later, an intermolecular cyclization reaction transforms strictosidine into strictosamide. Strictosamide is transformed to 3(S)-pumiloside, which is further converted to 3(S)-deoxypumiloside and finally to CPT (Scheme 12.1). The formation of CPT from strictosamide can be achieved by oxidative cleavage of the ring intramolecular recyclization, reduction of the quinolone part (which forms double bond rearrangement), and hydrolytic cleavage of the glucose moiety [64].

12.4 CHEMISTRY OF CAMPTOTHECIN

Camptothecin ($C_{20}H_{16}N_2O_4$; molecular weight = 348.4) is a monoterpene indole alkaloid with a pentacyclic structure consisting of a pyrrolo(3,4-β)-quinoline moiety (rings A, B, C, and D) and one asymmetric center within the α-hydroxy lactone ring with 20(S) configuration (ring E) [1, 71]. It contains only a single chiral center at C20. Unfortunately, the labile E-ring lactone form opens at physiological pH to transform into the inactive carboxylate form (Figure 12.1), and this can easily be reversed upon acidification [1]. Thus, the E-ring experiences an equilibrium reaction between an active lactone form (water-insoluble) and a less active carboxylate form (water-soluble).

CPT is a pale-yellow needle-like crystalline compound that dissolves slightly in ethanol and chloroform but poorly in water. CPT can produce sodium salt (Figure 12.2) by heating with sodium hydroxide solution and is soluble in water. The melting point is 264–267°C [74]. Due to its low solubility in water, the sodium salt of CPT was initially used in clinical trials. However, it induced severe toxicity. Many CPT derivatives with enhanced pharmacological properties, such as topotecan, irinotecan, and belotecan, were later approved as chemotherapeutic medications by the authorities, and a number of other new-generation CPT derivatives are currently in preclinical and clinical trials.

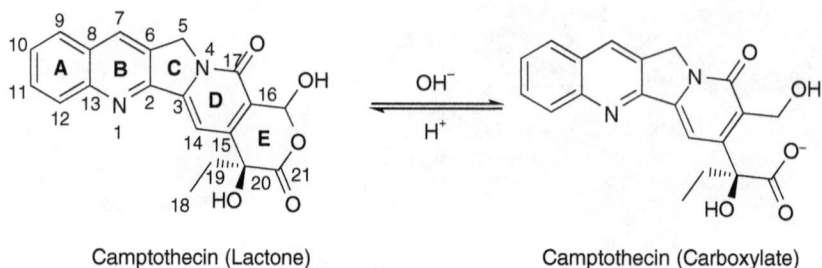

Camptothecin (Lactone)   Camptothecin (Carboxylate)

FIGURE 12.1   Chemical structure of the camptothecin lactone and carboxylate forms.

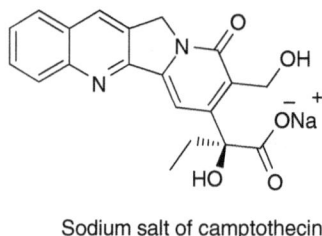

Sodium salt of camptothecin

FIGURE 12.2   Structure of sodium salt of camptothecin.

12.5   SYNTHESIS OF CAMPTOTHECIN

CPT total synthesis has advanced dramatically as a result of restricted raw material supply, rising demand, and environmental concerns. In 1971, Stork and Schultz successfully carried out the first total synthesis of CPT racemic mixture with o-aminobenzaldehyde and pyrrolidine as the preparatory materials [75]. The mechanism is shown in Scheme 12.2.

Corey et al. (1975) first reported that the chiral intermediary might be utilized in the synthesis of 20(S)-CPT [76]. Ejima et al. (1989) published the first report on the enantioselective synthesis of 20(S)-CPT [77]. The chiral auxiliary used in the synthesis was a derivative of N-tosyl-R-proline. Imura et al. (1998) reported the enzyme-mediated enantioselective synthesis of an important chiral intermediary [78]. Comins et al. (1992) determined the stereochemistry of the 20(S)-hydroxyl group using chiral auxiliaries [79]. CPT was formed when the C-ring of the tetracyclic intermediate formed by the Mitsunobu reaction of the DE-ring intermediate (formed from 2-methoxypyridine), and the AB-ring intermediate (formed from 2-bromoquinoline) was closed by the Heck reaction. The DE-ring intermediate was formed in three steps, and the AB-ring intermediate in a single step (Scheme 12.3) [80–82].

The 4 + 1 radical cascade annulation-based synthesis of (S)-CPT has been described by Curran et al. (1995) [83]. Fortunak et al. (1996) proposed an alternative way for the formation of B- and C-rings using the intramolecular Diels Alder reaction [84]. Chavan and Venkatraman (1998) constructed D-ring through cascade oxidative cyclization [85]. Ciufolini and Roschangar (1996, 1997, 2000) constructed D-ring through intermolecular radical cyclization [86–88]. Bennasar et al. (2000, 2002) established C20 asymmetry and thereby afforded access to 20(S)-CPT from (2R,5R)-2-tert-butyl-5-ethyl-1,3-dioxolan-4-one [89, 90]. Blagg and Boger (2002) described the formation of a CPT precursor by the Diels-Alder cycloaddition reaction of N-sulfonyl-1-aza-1,3-butadiene with dienophile [91]. Bennasar et al. (2002) formed the C-ring using intramolecular radical cyclization and then used enolate chemistry to asymmetrically construct the E-ring [90]. Chavan and Sivappa (2004) proposed another method of synthesis of the D-ring by using an intramolecular ring-closing metathesis process [92]. Tam and Wicki (2004) prepared CPT by ortho-methylation using

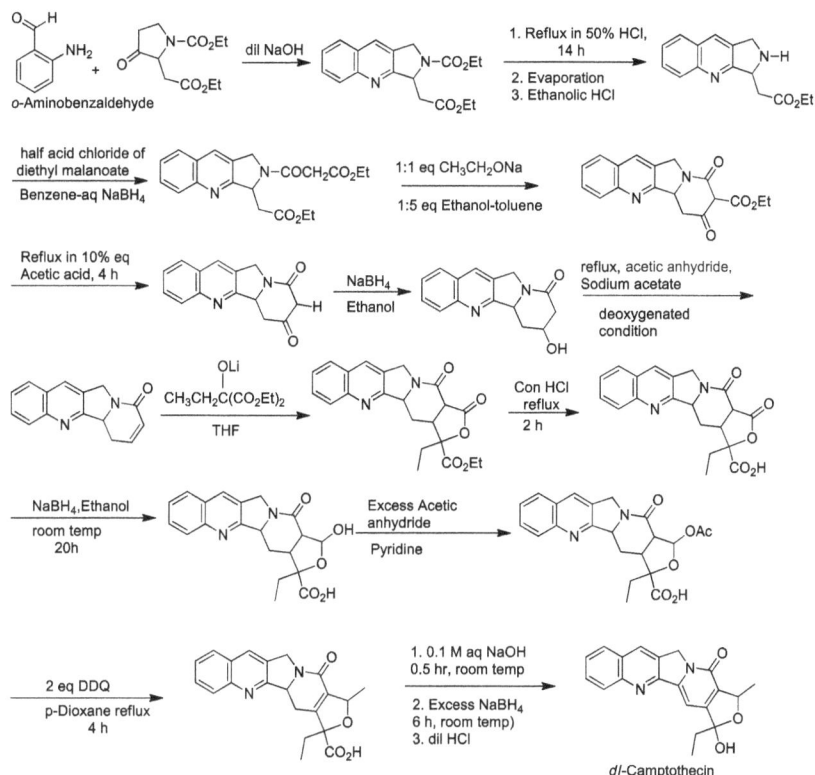

SCHEME 12.2 Total synthesis of camptothecin racemic form.

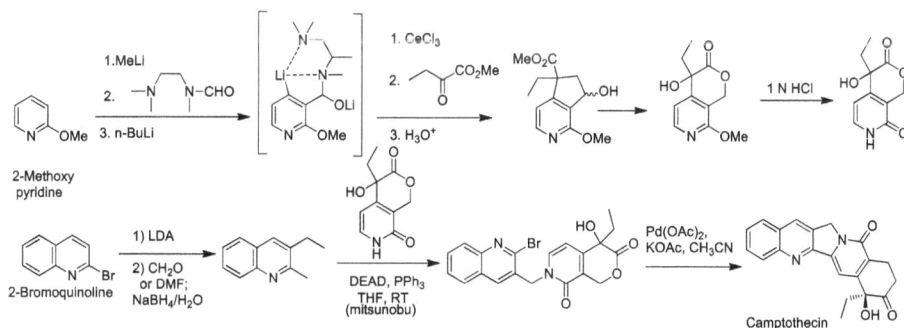

SCHEME 12.3 Synthesis of 20(S)-camptothecin.

a transition metal-mediated cross-coupling from quinolone [93]. The organozinc compound thus formed was subjected to cross-coupling using palladium.

Anderson et al. (2005) reported the synthesis of CPT racemic mixture (Scheme 12.4) [94]. Diazo compounds undergo isomunchnone cycloaddition to give a hydroxyl pyridine molecule. It undergoes esterification with (Z)-methyl-4-chloro-2-methoxybut-2-enoate. Further Claisen rearrangement introduces hydroxyl group in the beta position after hydrogenation followed by reaction with ozone, and the reduction with sodium borohydride yields lactone. Further reaction with selenium dioxide gives a hydroxyl derivative, and further Friedlander condensation reaction with o-amino benzaldehyde yields quinoline. Further reaction of quinoline with hot aqueous hydrogen bromide gives CPT.

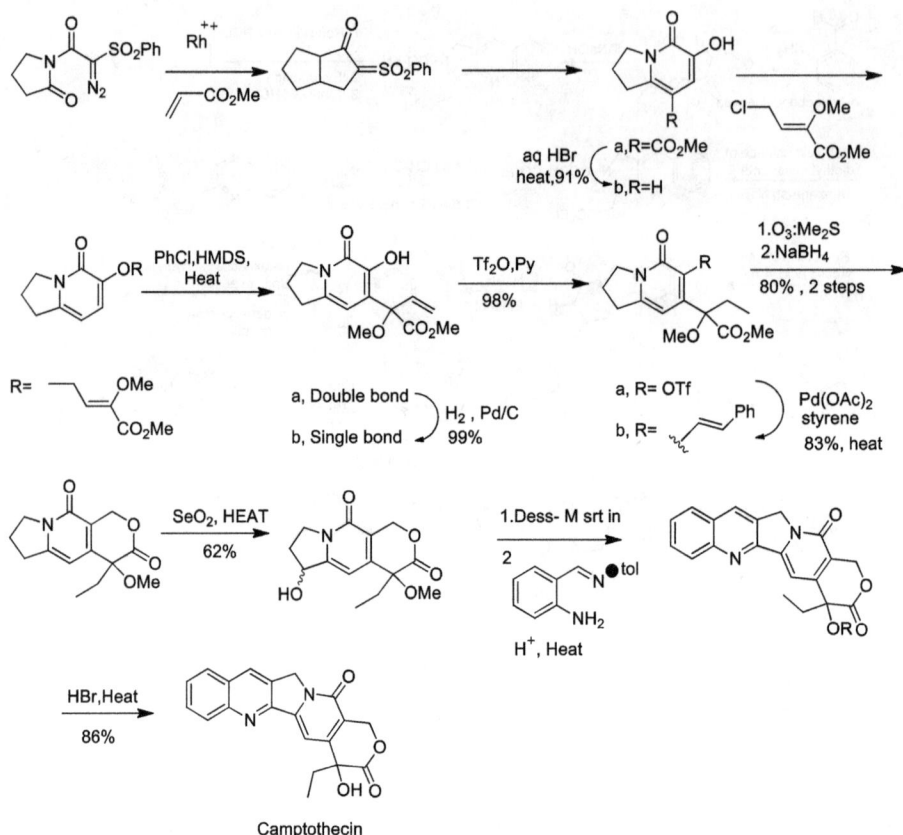

SCHEME 12.4   Synthesis of racemic mixture of camptothecin.

Yu et al. (2012) developed the total synthesis of enantiopure CPT from inexpensive and commercially available starting materials, 2-methoxynicotinic acid, and glyceraldehyde derivatives, without any column chromatography separation [95]. This method offers an economical industrial process for the CPT and its derivatives. Li et al. (2016) developed a versatile approach for building skeletons that includes indolizinone or quinolizinone [96]. In this method, they achieved the total synthesis of CPT in nine steps in a unique ring-forming method under mild reaction conditions and without the use of any transition metals. Recently, Liu et al. (2019) developed a very simplistic approach for the synthesis of racemic CPT via domino Heck/aza-Michael addition reactions [97]. This approach offers a novel and condensed path to the total synthesis of CPT and its derivatives (Scheme 12.5).

12.6   SYNTHESIS OF CAMPTOTHECIN DERIVATIVES

Along with various total syntheses of CPT, a number of synthetic attempts are focused on developing more effective derivatives of CPT with low or no toxicity. The majority of these initiatives have focused on manipulating CPT in a semisynthetic manner. Topotecan, irinotecan, and belotecan are the most important water-soluble and clinically approved derivatives of CPT, and many are now under preclinical and clinical trials (Figure 12.3).

Topotecan was initially made semisynthetically by Kingsbury et al. (1991) from SmithKline Beecham Pharmaceuticals [98]. The CPT was converted to 10-hydroxy CPT, which further underwent the Mannich reaction. The regioselective products formed include water-soluble and clinically active topotecan (Scheme 12.6).

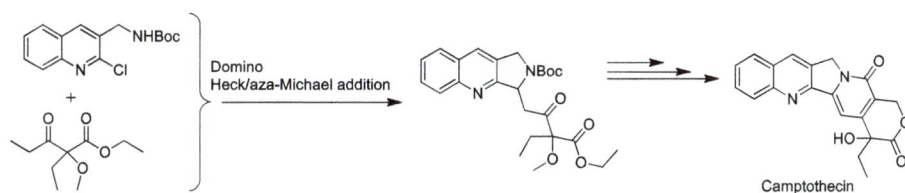

SCHEME 12.5 Total synthesis of racemic camptothecin.

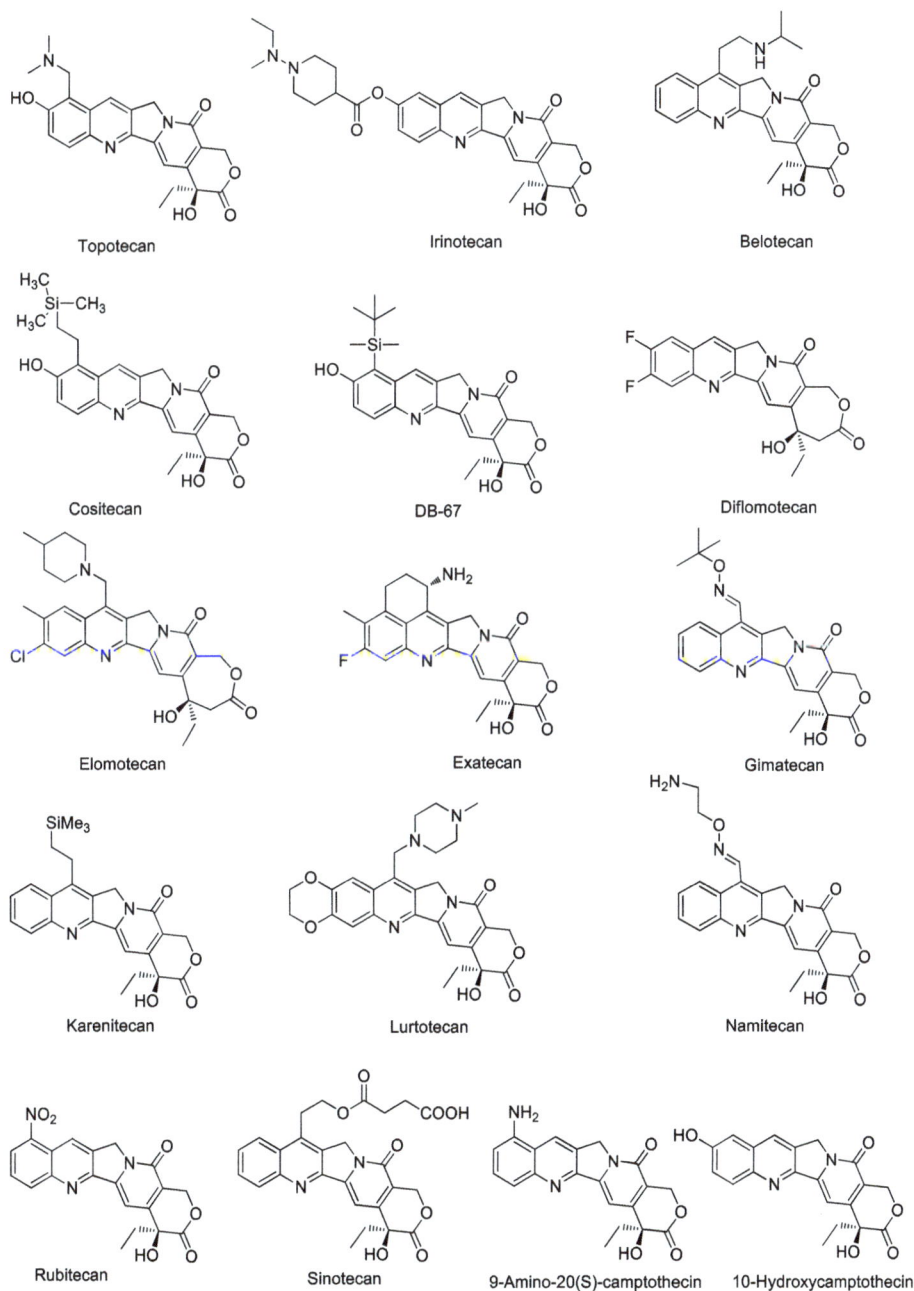

FIGURE 12.3 Structures of some camptothecin derivatives.

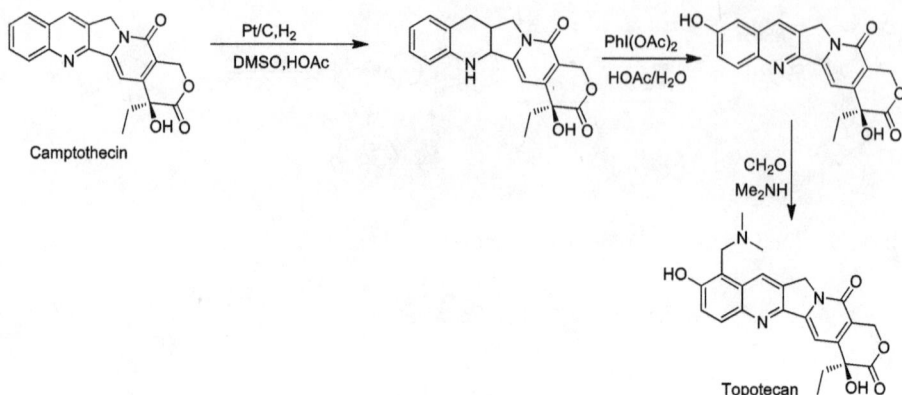

SCHEME 12.6   Semisynthetic pathway of topotecan.

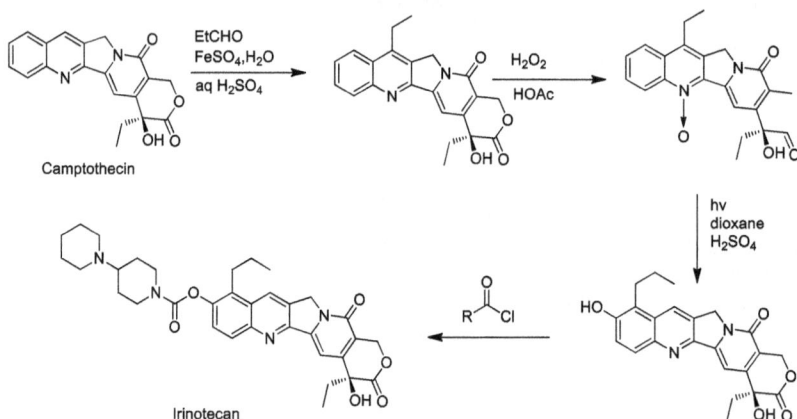

SCHEME 12.7   Semisynthetic pathway of irinotecan.

Irinotecan was first formed by pharmaceutical companies Yakult Honsha and Daiichi Pharmaceutical in the year 1984 (Scheme 12.7). Irinotecan has two main methods for its synthesis using C11 phosphene and by C11 carbon dioxide fixation [99, 100].

A highly stable belotecan-7-(2-[N-isopropylamine] ethyl)-(20S)-CPT was synthesised by Ahn et al. (2000) (Scheme 12.8) [101]. CPT undergoes a Minisci type reaction to give 7-methyl CPT, which further undergoes a Mannich type reaction to give belotecan.

Rubitecan (9-nitro CPT) was prepared by nitration reaction using sulfuric acid and nitric acid as nitrating agent, or by using potassium nitrate and thallium nitrate as nitrating agents [102, 103], gimatecan was prepared from 7-hydroxy methyl CPT via Minisci type reaction, then by reflux reaction using acetic acid, the alcohol part was converted to aldehyde, and by further reaction with o-tert-butylhydroxyamine [104], karenitecan was synthesized using Menisci alkylation of CPT [105], and 7-tert-butyldimethylsilyl-10-hydroxy-camptothecin (DB-67 or AR-67) is prepared by oxidation of 7-silyl CPT and further irradiation in the presence of dioxane and sulfuric acid [106]. All of the synthetic and semisynthetic methods thus stated have made it possible to identify novel derivatives with enhanced properties. In addition to these semisynthetic methods for the production of CPT derivatives, several hundred CPT derivatives have been reported through diverse methods, and many are now undergoing different phases of preclinical and clinical research.

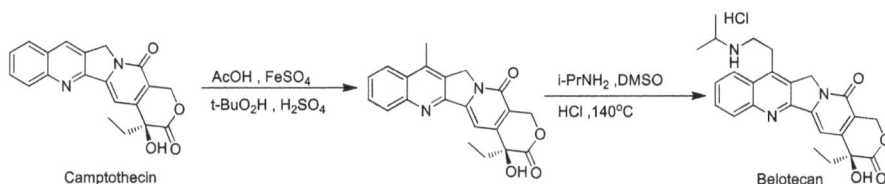

SCHEME 12.8 Semisynthetic pathway of belotecan.

12.7 CHEMOTHERAPEUTIC POTENTIAL OF CAMPTOTHECIN AND ITS DERIVATIVES

In the late 1980s, scientists at Johns Hopkins Medical School and Smith, Kline & French revealed that CPT selectively inhibits Top-I [24]. DNA topoisomerases are the enzymes that regulate the topological structure of DNA and perform catalytic functions during replication and transcription—overwinding or underwinding of DNA. These enzymes are highly expressed in tumor cells. Topoisomerases are currently a target for many chemotherapeutic drugs and drug leads, with CPTs being one such special class of selective S-phase-specific Top-I inhibitors [107]. CPT functions as a noncompetitive inhibitor that exclusively binds the Top-I-DNA transitory covalent binary complex. The specific mechanism by which it stabilizes the Top-I is still not completely understood. However, a number of structural models have been put forth, commencing with the x-ray crystal structure of the human Top-I-DNA cleavable complex [108–113]. Because of the toxicity and low water solubility of the CPT molecule, derivatives such as topotecan, irinotecan, and belotecan are now employed in clinical applications.

12.7.1 TOPOTECAN

Topotecan, or TPT (9-[(dimethylamino)methyl]-10-hydroxycamptothecin; Hycamtin, GlaxoSmith-Kline) is a semisynthetic CPT derivative, developed with the intention of lowering the adverse effects and improving bioavailability in comparison to those of the precursor molecule, and has been licensed by the Food and Drug Administration (FDA) as a broad-spectrum anticancer drug to treat ovarian (May 1996), cervical (June 2006), and small-cell lung cancers (October 2007) and tumors of the central nervous system [114–116]. TPT has a potency that is usually 100–1,000-fold higher than that of the drug precursor CPT [117]. TPT poisons the Topo I enzyme by inhibiting the cut-and-repair mechanism of the enzyme, and thereby, it delays the reversal of the covalent Topo I–DNA intermediate (the cleavable complex) and the relegation of DNA single-strand breaks [118–120]. It has more detrimental effects during the synthesis phase of the cell cycle, in which TPT induces DNA double-strand breaks, causing cell cycle arrest and death [121, 122]. Unfortunately, at pH 7 or even at a higher pH, the lactone ring of TPT opens to form carboxylate form, resulting in lower activity [115, 123].

TPT is administered intravenously along with fluids or orally as a capsule during the clinical chemotherapeutic process [124]. Intravenous administration has the advantage of having very high bioavailability but also has the drawback of raising the risk of systemic side effects and invasive procedures [125]. TPT is typically administered at a dose of 1.5 mg/m^2/day on days 1–5 of a 21-day cycle, with response rates varying between 13 and 33% [126]. TPT-lactone is broadly dispersed in the periphery, with a mean volume of dispersion at a steady state of 75 L/m^2. The renal clearance of the lactone form accounts for around 40% of the administered dose with significant interindividual variability, and the mean total body clearance is 30 L/h/m^2 with a mean elimination half-life (t$^1/_2$β) of three hours. TPT has approximately 35% oral bioavailability. This low bioavailability may be caused by the poorly absorbed carboxylate form produced when TPT-lactone hydrolyses in the gut [124, 127].

Neutropenia (61%), thrombocytopenia (38%), and anemia (25%) are the major dose-limiting hematological toxicities of TPT. These toxicities may usually be controlled with dose reduction and dose delay since they are typically predictable, short-lived, and non-cumulative. The majority of non-hematological toxicity is typically mild, which includes diarrhea (6%), hair loss, and fatigue [128–130]. As second-line therapy for recurrent small-cell lung cancer (SCLC), oral and intravenous administration of TPT showed nearly comparable response rates (18% vs. 22%) and 1-year survival rates (33% vs. 29%), but the prevalence of grade 4 neutropenia was higher for intravenous than for oral (64% vs. 47%) [131]. TPT is increasingly being combined with other standard chemotherapeutic agents (e.g., cisplatin, carboplatin, amrubicin, cyclophosphamide, etc.) for improved therapy [132–138].

12.7.2 Irinotecan

Irinotecan, also known as CPT-11 or Camptosar (Pharmacia, now Pfizer), is a water-soluble yellow crystalline derivative of CPT. Compared to the precursor CPT, CPT-11 often has a pharmacological potency that is 1,00–1,000 times higher [117, 139]. CPT-11 gained more attention after getting FDA clearance for clinical trial in June 1996 and full endorsement in 1998 as a medication to treat colorectal cancer [140, 141]. Numerous cancers, including pulmonary [142–145], pancreatic, gastric [146, 147], ovarian [148, 149], cervical [150], colorectal [151], gastric, ovarian, and others [152, 153], have been treated with CPT-11. In the human body, CPT-11 breaks down into its active metabolite, 7-ethyl-10-hydroxycamptothecin (SN-38), which is a hundred to thousand times more potent than CPT-11 [154]. This reaction is carried out by the liver and intestinal carboxylesterases. Uridine diphosphate glucuronosyltransferase 1A1 (UGT1A1) in the liver further breaks down SN-38 into SN-38G, an inactive substance. In human plasma, both CPT-11 and SN-38 achieve equilibrium, with CPT-11 predominately in carboxylate form and SN-38 predominately in lactone form [155].

In contrast to TPT, CPT-11 and SN-38 bind to blood proteins at a rate of roughly 80% and 100%, respectively [156]. SN-38 has a terminal half-life of roughly 47 hours [157]. Regarding the use of CPT-11 in individuals with changed UGT1A1 activity, the FDA issued a caution in 2005. There are major regional variations in CPT-11 dosage and administration. In Europe, a starting dosage of 350 mg/m^2 of CPT-11 administered as an intravenous infusion lasting 30 to 90 minutes once every three weeks was advised. In the US, the first dosage of CPT-11 was recommended to be 125 mg/m^2, administered as a 90-minute intravenous infusion once a week for four weeks, followed by a two-week respite. In Japan, the dosage and frequency were 100 mg/m^2 once a week for every two weeks [139, 158].

CPT-11 is often used as first or second-line chemotherapy to treat advanced or recurrent colorectal cancer. It can be administered in a conventional regimen with 5-fluorouracil and leucovorin to possibly increase survival time [159]. For advanced colorectal cancer, the stated response rate to single-use CPT-11 is 20–30%, and when coupled with additional medications, it can reach 50% [160]. In the first and second phases of the clinical investigations of CPT-11 in advanced colorectal cancer, the median overall survival times were 6.6–16.1 and 9.1–10.8 months, respectively [161]. In the second and third phases of clinical studies, severe adverse effects of CPT-11, such as diarrhea, asthenia, nausea, vomiting, and treatment-related death, have also been reported [161]. Numerous clinical investigations have demonstrated the advantages of combining CPT-11 with other anticancer medications in addition to its single-use use [162]. For instance, CPT-11, 5-fluorouracil, and leucovorin (folinic acid) administered as first-line therapy for metastatic colorectal cancer led to significantly higher remission rates and longer progression-free survival and median survival [163–166]. Similar to this, the first-line CPT-11 with bevacizumab regimen had a 67% overall response rate, as well as median and overall survival times of 1 year, 3 days and 1 year, 11 months, 7 days, respectively (95% confidence interval) [167].

12.7.3 BELOTECAN

The CPT-derivative belotecan ([7-(2-isopropylamino) ethyl]-CPT) was made by introducing a water-solubilizing group to the seventh position of the B-ring [168]. Studies in six human tumor xenografts revealed that belotecan inhibited topoisomerase I more potently than CPT and TPT [169]. In individuals with previously untreated extensive disease SCLC, intravenous administration of belotecan at a dose of 0.5 mg/m^2/day, days 1–5, every three weeks, demonstrated a median overall survival time of 10.4 and 11.9 months, respectively, in a phase 2 clinical investigation. The overall response rates were 53.2% and 63.6%, respectively. The most prevalent effect, neutropenia, was easily manageable [170, 171].

Belotecan therapy for advanced SCLC in the elderly resulted in a median overall survival time of 6.4 months and a median overall response rate of 35%, indicating that belotecan alone may be a viable therapeutic option for patients who cannot tolerate etoposide or cisplatin [172]. The results of belotecan therapy for older people with progressive SCLC showed a median overall survival time of 6.4 months and a median overall response rate of 35%, suggesting that belotecan alone may be an effective treatment option for those who cannot tolerate etoposide or cisplatin. In patients with recurrent severe disease SCLC, belotecan demonstrated a 22–24% overall response rate and a 9.9–13.1-month overall median survival time [173, 174].

In addition, three phase II trials in individuals with earlier untreated extensive disease SCLC, treated with the combination of cisplatin (60 mg/m^2/day, intravenous, on day 1) and belotecan (0.5 mg/m^2/day, intravenous, days 1–4, every three weeks), exhibited a median overall survival time of ten months and an overall response rate of 70% [175, 176]. Patients with recurrent ovarian cancer responded well to intravenous belotecan 0.5 mg/m^2/day for five days every three weeks, with an overall response rate of 45% and minimal toxicity. Using the same regimen, a thorough study revealed that the platinum-sensitive group had a greater overall response rate (53.3%) than the platinum-resistant group (20%) [177, 178]. A retrospective analysis showed that in recurrent ovarian cancer patients, particularly those who were platinum-sensitive, belotecan-based chemotherapy was more effective than TPT-based chemotherapy (overall response rate of 45.7% vs. 24.4%) [179]. In individuals with platinum-resistant or extensively pretreated ovarian cancer, oral administration of a combination regimen of etoposide (50, 75 mg/day, days 6–10, every three weeks) and belotecan (0.5 mg/m^2/day, days 1–5, every three weeks) resulted in an overall response rate of 44% [180]. Patients with platinum-sensitive or resistant recurrent ovarian cancer were treated successfully and safely with belotecan (0.3 mg/m^2/day, days 1–5, every three weeks) in combination with carboplatin (on day 5, every three weeks) or cisplatin (50 mg on day 1, every three weeks) [181, 182].

12.8 CONCLUSION AND PROSPECTS

The success of camptothecin and its three clinically approved derivatives, such as topotecan, irinotecan, and belotecan, highlights the potential of naturally derived molecules as a source of pharmaceutical leads. Camptothecin was the starting point for DNA topoisomerase I inhibitors in chemotherapeutic research. Despite great advances in total asymmetric camptothecin syntheses and semisynthetic approaches to camptothecin derivatives, further research is required to develop cost-effective methods for large-scale production. Current camptothecin-based chemotherapeutic research has focused on developing novel formulations to improve drug delivery, stability, and toxicity, particularly at the nanoscale. There are high hopes for the next generation of camptothecin derivatives, and research is underway on multiple fronts.

12.9 ACKNOWLEDGMENTS

TPI, VG, and PP express their sincere thanks to Dr. Ashok K. Chauhan, Founder President, Ritnand Balved Education Foundation and Amity Group of Institutions, and Dr. Atul Chauhan, Chancellor, Amity University Uttar Pradesh, India, for facilitating this work. PS is very thankful to the

Principal and Head (Drug Testing Laboratory, Department of Rasasastra and Bhaishajyakalpana), Government Ayurveda College, Thiruvananthapuram, India, for providing necessary support and encouragement. Ms. Sulochana Priji of the Department of Botany, University of Kerala, India, provided technical assistance to the authors. TPI received the Young Scientist Fellowship from the Department of Science and Technology, Government of India (SP/YO/413/2018).

12.10   REFERENCES

1. Wall, Monroe E., M. C. Wani, C. E. Cook, Keith H. Palmer, A. T. McPhail, and G. A. Sim. 1966. "Plant Antitumor Agents. I. the Isolation and Structure of Camptothecin, a Novel Alkaloidal Leukemia and Tumor Inhibitor from *Camptotheca acuminata*." *Journal of the American Chemical Society* 88 (16): 3888–3890.
2. Wyk, Ben-Erik van, and Michael Wink. 2004. *Medicinal Plants of the World: An Illustrated Scientific Guide to Important Medicinal Plants and Their Uses*. Portland, OR: Timber Press.
3. Afzal, Obaid, Suresh Kumar, Md Rafi Haider, Md Rahmat Ali, Rajiv Kumar, Manu Jaggi, and Sandhya Bawa. 2015. "A Review on Anticancer Potential of Bioactive Heterocycle Quinoline." *European Journal of Medicinal Chemistry* 97: 871–910.
4. Wall, Monroe E., and Mansukh C. Wani. 1996. "Camptothecin and Taxol: From Discovery to Clinic." *Journal of Ethnopharmacology* 51 (1–3): 239–254.
5. Perdue, R. E., R. L. Smith, M. E. Wall, J. L. Hartwell, and B. J. Abbott. 1970. *Camptotheca acuminata* Decaisne (Nyssaceae), Source of Camptothecin, An Antileukemic Alkaloid. *Technical Bulletin, No. 1415, U.S. Department of Agriculture*, Agricultural Research Service, Washington, DC.
6. Huang, S. Y. 1986. *Seven Hundred Herbal Prescriptions for Cancer Medicine Treatment (in Chinese)*. Taipei, Chinese Rep.: Bada Educational and Cultural Publishers.
7. Schultz, Arthur G. 1973. "Camptothecin." *Chemical Reviews* 73 (4): 385–405.
8. Gottlieb, J. A., and J. K. Luce. 1972. "Treatment of Malignant Melanoma with Camptothecin (NSC-100880)." *Cancer Chemotherapy Reports. Part 1* 56 (1): 103–105.
9. Moertel, C. G., A. J. Schutt, R. J. Reitemeier, and R. G. Hahn. 1972. "Phase II Study of Camptothecin (NSC-100880) in the Treatment of Advanced Gastrointestinal Cancer." *Cancer Chemotherapy Reports. Part 1* 56 (1): 95–101.
10. Muggia, F. M., P. J. Creaven, H. H. Hansen, M. H. Cohen, and O. S. Selawry. 1972. "Phase I Clinical Trial of Weekly and Daily Treatment with Camptothecin (NSC-100880): Correlation with Preclinical Studies." *Cancer Chemotherapy Reports. Part 1* 56 (4): 515–521.
11. Pettit, George R. 1976. "A View of Cancer Treatment in the People's Republic of China." *The China Quarterly* 68: 789–796.
12. Xu, B. 1980. "Clinical studies with camptothecin sodium." In *U.S.-China Pharmacology Symposium*, eds. J. J. Burns and P. J. Tsuchiatani, 156. Washington, DC: National Academy of Sciences.
13. Horwitz, Susan B. 1975. "Camptothecin." In *Mechanism of Action of Antimicrobial and Antitumor Agents*, eds. J. W. Corcoran, F. E. Hahn, J. F. Snell, and K. L. Arora, 48–57. Berlin, Heidelberg: Springer.
14. Rozencweig, M., M. Slavik, F. M. Muggia, and S. K. Carter. 1976. "Overview of Early and Investigational Chemotherapeutic Agents in Solid Tumors." *Medical and Pediatric Oncology* 2 (4): 417–432.
15. Bosmann, H. B. 1970. "Camptothecin Inhibits Macromolecular Synthesis in Mammalian Cells but Not in Isolated Mitochondria of *E. coli*." *Biochemical and Biophysical Research Communications* 41 (6): 1412–1420.
16. Gallo, R. C., J. Whang-Peng, and R. H. Adamson. 1971. "Studies on the Antitumor Activity, Mechanism of Action, and Cell Cycle Effects of Camptothecin." *Journal of the National Cancer Institute* 46: 789–795.
17. Horwitz, M. S., and S. B. Horwitz. 1971. "Intracellular Degradation of HeLa and Adenovirus Type 2 DNA Induced by Camptothecin." *Biochemical and Biophysical Research Communications* 45 (3): 723–727.
18. Horwitz, S. B., C. K. Chang, and A. P. Grollman. 1971. "Studies on Camptothecin. I. Effects of Nucleic Acid and Protein Synthesis." *Molecular Pharmacology* 7 (6): 632–644.
19. Kessel, David. 1971. "Effects of Camptothecin on RNA Synthesis in Leukemia L1210 Cells." *Biochimica et Biophysica Acta* 246 (2): 225–232.

20. Wu, R. S., A. Kumar, and J. R. Warner. 1971. "Ribosome Formation Is Blocked by Camptothecin, a Reversible Inhibitor of RNA Synthesis." *Proceedings of the National Academy of Sciences of the United States of America* 68 (12): 3009–3014.

21. Abelson, H. T., and S. Penman. 1972. "Selective Interruption of High Molecular Weight RNA Synthesis in HeLa Cells by Camptothecin." *Nature: New Biology* 237 (74): 144–146.

22. Kessel, David, H. Bruce Bosmann, and Kristine Lohr. 1972. "Camptothecin Effects on DNA Synthesis in Murine Leukemia Cells." *Biochimica et Biophysica Acta* 269 (2): 210–216.

23. Muggia, F. 1995. "Twenty Years Later: Review of Clinical Trials with Camptothecin Sodium (NSC-100880)." In *Camptothecins: New Anticancer Agents*, eds. M. Potmesil and H. Pinedo, 43–50. Boca Raton: CRC Press.

24. Hsiang, Y. H., R. Hertzberg, S. Hecht, and L. F. Liu. 1985. "Camptothecin Induces Protein-Linked DNA Breaks via Mammalian DNA Topoisomerase I." *The Journal of Biological Chemistry* 260 (27): 14873–14878.

25. Pommier, Y., J. M. Covey, D. Kerrigan, J. Markovits, and R. Pham. 1987. "DNA Unwinding and Inhibition of Mouse Leukemia L1210 DNA Topoisomerase I by Intercalators." *Nucleic Acids Research* 15 (16): 6713–6731.

26. Thomsen, B., S. Mollerup, B. J. Bonven, R. Frank, H. Blöcker, O. F. Nielsen, and O. Westergaard. 1987. "Sequence Specificity of DNA Topoisomerase I in the Presence and Absence of Camptothecin." *The EMBO Journal* 6 (6): 1817–1823.

27. Jaxel, C., K. W. Kohn, and Y. Pommier. 1988. "Topoisomerase I Interactions with SV40 DNA in the Presence and Absence of Camptothecin." *Nucleic Acids Research* 16: 11157–11170.

28. Jaxel, C., K. W. Kohn, M. C. Wani, M. E. Wall, and Y. Pommier. 1989. "Structure-Activity Study of the Actions of Camptothecin Derivatives on Mammalian Topoisomerase I: Evidence for a Specific Receptor Site and a Relation to Antitumor Activity." *Cancer Research* 49 (6): 1465–1469.

29. Kjeldsen, E., S. Mollerup, B. Thomsen, B. J. Bonven, L. Bolund, and O. Westergaard. 1988. "Sequence dependent Effect of Camptothecin on Human Topoisomerase I DNA Cleavage." *Journal of Molecular Biology* 202: 333–342.

30. Chazin, Eliza de Lucas, Raisa da Rocha Reis, Walcimar Trindade Vellasco Junior, Lucas Fajardo Elmor Moor, and Thatyana Rocha Alves Vasconcelos. 2014. "An Overview on the Development of New Potentially Active Camptothecin Analogs against Cancer." *Mini Reviews in Medicinal Chemistry* 14 (12): 953–962.

31. Du, Hongzhi, Yue Huang, Xiaoying Hou, Xingping Quan, Jingwei Jiang, Xiaohui Wei, Yang Liu, et al. 2018. "Two Novel Camptothecin Derivatives Inhibit Colorectal Cancer Proliferation via Induction of Cell Cycle Arrest and Apoptosis in Vitro and in Vivo." *European Journal of Pharmaceutical Sciences: Official Journal of the European Federation for Pharmaceutical Sciences* 123: 546–559.

32. Khan, N., E. T. Tamboli, V. K. Sharma, and S. Kumar. 2013. "Phytochemical and Pharmacological Aspects of Nothapodytes Nimmoniana, an Overview." *Herba Polonica* 59: 53–66.

33. Li, S. Y. 1997. "*Camptotheca lowreyana*, a New Species of Anti-Cancer Happy Trees, Bull." *Bulletin of Botanical Research* 17: 348–352.

34. Chen, R., Y. G. Luo, X. Y. Hu, X. Z. Chen, and G. L. Zhang. 2008. "Chemical Study on *Alus nepalensis*." *Natural Product Research and Development* 20: 578–581.

35. Phan, Minh Giang, Thi To Chinh Truong, Tong Son Phan, Katsuyoshi Matsunami, and Hideaki Otsuka. 2010. "Mangiferonic Acid, 22-Hydroxyhopan-3-one, and Physcion as Specific Chemical Markers for *Alnus nepalensis*." *Biochemical Systematics and Ecology* 38 (5): 1065–1068.

36. Ramesha, B. T., H. K. Suma, U. Senthilkumar, V. Priti, G. Ravikanth, R. Vasudeva, T. R. Santhosh Kumar, K. N. Ganeshaiah, and R. Uma Shaanker. 2013. "New Plant Sources of the Anti-Cancer Alkaloid, Camptothecine from the Icacinaceae Taxa, India." *Phytomedicine* 20 (6): 521–527.

37. Foubert, Kenn, Filip Cuyckens, Ann Matheeussen, Arnold Vlietinck, Sandra Apers, Louis Maes, and Luc Pieters. 2011. "Antiprotozoal and Antiangiogenic Saponins from *Apodytes dimidiata*." *Phytochemistry* 72 (11–12): 1414–1423.

38. Pu, Xiang, Cheng-Rui Zhang, Lin Zhu, Qi-Long Li, Qian-Ming Huang, Li Zhang, and Ying-Gang Luo. 2019. "Possible Clues for Camptothecin Biosynthesis from the Metabolites in Camptothecin-Producing Plants." *Fitoterapia* 134: 113–128.

39. Gunasekera, S. P., M. M. Badawi, G. A. Cordell, N. R. Farnsworth, and M. Chitnis. 1979. "Plant Anticancer Agents X. Isolation of Camptothecin and 9-Methoxycamptothecin from *Ervatamia heyneana*." *Journal of Natural Products* 42 (5): 475–477.

40. Shweta, S., S. Zuehlke, B. T. Ramesha, V. Priti, P. Mohana Kumar, G. Ravikanth, M. Spiteller, R. Vasudeva, and R. Uma Shaanker. 2010. "Endophytic Fungal Strains of *Fusarium solani*, from *Apodytes dimidiata* E. Mey. Ex Arn (Icacinaceae) Produce Camptothecin, 10-Hydroxycamptothecin and 9-Methoxycamptothecin." *Phytochemistry* 71 (1): 117–122.

41. Longze, Lin, Shen Jihui, Zhou Tong, Shen Chunyi, and K. E. Minmin. 1989. "New Alkaloid: 10-Hydroxydeoxycamptothecin." *Hua Xue Xue Bao. Acta Chimica Sinica* 47 (5): 506.

42. Ma, Xiaohua, Lili Song, Weiwu Yu, Yuanyuan Hu, Yang Liu, Jiasheng Wu, and Yeqing Ying. 2015. "Growth, Physiological, and Biochemical Responses of *Camptotheca acuminata* Seedlings to Different Light Environments." *Frontiers in Plant Science* 6: 321.

43. Arisawa, M., S. P. Gunasekera, G. A. Cordell, and N. R. Farnsworth. 1981. "Plant Anticancer Agents XXI. Constituents of *Merrilliodendron megacarpum*." *Planta Medica* 43 (4): 404–407.

44. Roja, G. 2006. "Comparative Studies on the Camptothecin Content from *Nothapodytes foetida* and *Ophiorrhiza* Species." *Natural Product Research* 20 (1): 85–88.

45. Ramesha, B. T., T. Amna, G. Ravikanth, Rajesh P. Gunaga, R. Vasudeva, K. N. Ganeshaiah, R. Uma Shaanker, R. K. Khajuria, S. C. Puri, and G. N. Qazi. 2008. "Prospecting for Camptothecines from *Nothapodytes nimmoniana* in the Western Ghats, South India: Identification of High-Yielding Sources of Camptothecin and New Families of Camptothecines." *Journal of Chromatographic Science* 46 (4): 362–368.

46. Uday, P., M. Maheshwari, P. Sharanappa, Z. Nafeesa, V. H. Kameshwar, B. S. Priya, and S. N. Swamy. 2017. "Exploring Hemostatic and Thrombolytic Potential of Heynein-a Cysteine Protease from *Ervatamia heyneana* Latex." *Journal of Ethnopharmacology* 199: 316–322.

47. Kedari, P., and N. Malpahak. 2013. "Subcellular Localization and Quantification of Camptothecin in Different Plant Parts of *Chonemorpha fragrans*." *Advances in Zoology and Botany* 1: 34–38.

48. Dai, Jin-Rui, Yali F. Hallock, John H. Cardellina, and Michael R. Boyd. 1999. "20-O-β-Glucopyranosyl Camptothecin from *Mostuea brunonis*: A Potential Camptothecin pro-Drug with Improved Solubility." *Journal of Natural Products* 62 (10): 1427–1429.

49. Idowu, Thomas Oyebode, Abiodun Oguntuga Ogundaini, Abiola Oladimeji Salau, Efere Martins Obuotor, Merhatibeb Bezabih, and Berhanu Molla Abegaz. 2010. "Doubly Linked, A-Type Proanthocyanidin Trimer and Other Constituents of *Ixora coccinea* Leaves and Their Antioxidant and Antibacterial Properties." *Phytochemistry* 71 (17–18): 2092–2098.

50. Ya-ut, Pornwilai, Piyarat Chareonsap, and Suchada Sukrong. 2011. "Micropropagation and Hairy Root Culture of *Ophiorrhiza alata* Craib for Camptothecin Production." *Biotechnology Letters* 33 (12): 2519–2526.

51. Arbain, D., D. P. Putra, and M. V. Sargent. 1993. "The Alkaloids of *Ophiorrhiza filistipula*." *Australian Journal of Chemistry* 46 (7): 977.

52. Feng, Tao, Kai-Ting Duan, Shi-Jun He, Bin Wu, Yong-Sheng Zheng, Hong-Lian Ai, Zheng-Hui Li, Juan He, Jian-Ping Zuo, and Ji-Kai Liu. 2018. "Ophiorrhines A and B, Two Immunosuppressive Monoterpenoid Indole Alkaloids from *Ophiorrhiza japonica*." *Organic Letters* 20 (24): 7926–7928.

53. Asano, T., I. Watase, M. Kitajima, H. Takayama, N. Aimi, M. Yamazaki, and K. Saito. 2004. "Camptothecin Production by in Vitro Cultures of *Ophiorrhiza liukiuensis* and *O. kuroiwai*." *Plant Biotechnology* 21: 275–281.

54. Tafur, S., J. D. Nelson, D. C. DeLong, and G. H. Svoboda. 1976. "Antiviral Components of *Ophiorrhiza mungos*. Isolation of Camptothecin and 10-Methoxycamptothecin." *Lloydia* 39 (4): 261–262.

55. Beegum, A. S., K. P. Martin, C. L. Zhang, I. K. Nishitha, S. A. Ligimol, and P. V. Madhusoodanan. 2007. "Organogenesis from Leaf and Inernode Explants of *Ophiorrhiza prostrata*, an Anticancer Drug (Camptothecin) Producing Plant, Electron." *Electronic Journal of Biotechnology* 10: 114–123.

56. Gharpure, G., B. Chavan, U. Lele, A. Hastak, A. Bhave, N. Malpure, R. Vasudeva, and A. Patwardhan. 2010. "Camptothecin Accumulation in *Ophiorrhiza rugosa* var. *prostrata* from Northern Western Ghats." *Current Science* 98: 302–304.

57. Saito, K., H. Sudo, M. Yamazaki, M. Koseki-Nakamura, M. Kitajima, H. Takayama, and N. Aimi. 2001. "Feasible Production of Camptothecin by Hairy Root Culture of *Ophiorrhiza pumila*." *Plant Cell Reports* 20 (3): 267–271.

58. Asano, Takashi, Hiroshi Sudo, Mami Yamazaki, and Kazuki Saito. 2009. "Camptothecin Production by in Vitro Cultures and Plant Regeneration in *Ophiorrhiza* Species." In *Methods in Molecular Biology*, 337–345. Totowa, NJ: Humana Press.

59. Sibi, C. V., K. P. Dintu, R. Renjith, M. V. Krishnaraj, G. Roja, and K. Satheeshkumar. 2012. "A New Record of *Ophiorrhiza trichocarpon* Blume (Rubiaceae: Ophiorrhizeae) from Western Ghats, India: Another Source Plant of Camptothecin." *Journal of Scientific Research* 4: 529.

60. Sibi, C. V., R. Renjith, K. P. Dintu, P. Ravichandran, and K. Satheeshkumar. 2015. "A New Record of *Ophiorrhiza wattii* (Rubiaceae) for Western Ghats, India-a Source of an Anticancer Drug, Camptothecin." *Indian Journal of Tropical Biodiversity*. 23: 246–249.

61. Rajan, Renjith, Sibi Chirakkadamoolayil Varghese, Rajani Kurup, Roja Gopalakrishnan, Ramaswamy Venkataraman, Krishnan Satheeshkumar, and Sabulal Baby. 2016. "HPTLC-Based Quantification of

Camptothecin in *Ophiorrhiza* Species of the Southern Western Ghats in India." *Cogent Chemistry* 2 (1): 1275408.

62. Jain, Shreyans K., Samdarshi Meena, Ajai P. Gupta, Manoj Kushwaha, R. Uma Shaanker, Sundeep Jaglan, Sandip B. Bharate, and Ram A. Vishwakarma. 2014. "*Dysoxylum binectariferum* Bark as a New Source of Anticancer Drug Camptothecin: Bioactivity-Guided Isolation and LCMS-Based Quantification." *Bioorganic & Medicinal Chemistry Letters* 24 (14): 3146–3149.

63. Ma, Ji, Shannon H. Jones, Rebekah Marshall, Xihan Wu, and Sidney M. Hecht. 2005. "DNA Topoisomerase I Inhibitors from *Rinorea anguifera.*" *Bioorganic & Medicinal Chemistry Letters* 15 (3): 813–816.

64. Sirikantaramas, Supaart, Takashi Asano, Hiroshi Sudo, Mami Yamazaki, and Kazuki Saito. 2007. "Camptothecin: Therapeutic Potential and Biotechnology." *Current Pharmaceutical Biotechnology* 8 (4): 196–202.

65. Wenkert, E., K. G. Dave, R. G. Lewis, and P. W. Sprague. 1967. "General Methods of Synthesis of Indole Alkaloids. VI. Syntheses of dl-Corynantheidine and a Camptothecin Model." *Journal of the American Chemical Society* 89: 6741–6745.

66. Winterfeldt, Ekkehard. 1971. "Reaktionen an Indolderivaten, XIII. Chinolon-Derivate durch Autoxydation." *Justus Liebigs Annalen der Chemie* 745 (1): 23–30.

67. Hutchinson, C. Richard, Amos H. Heckendorf, John L. Straughn, Peter E. Daddona, and David E. Cane. 1979. "Biosynthesis of Camptothecin. 3. Definition of Strictosamide as the Penultimate Biosynthetic Precursor Assisted by Carbon-13 and Deuterium NMR Spectroscopy." *Journal of the American Chemical Society* 101 (12): 3358–3369.

68. Lu, Hua, Elizabeth Gorman, and Thomas D. McKnight. 2005. "Molecular Characterization of Two Anthranilate Synthase Alpha Subunit Genes in Camptotheca Acuminata." *Planta* 221 (3): 352–360.

69. Eisenreich, W., F. Rohdich, and A. Bacher. 2001. "Deoxyxylulose Phosphate Pathway to Terpenoids." *Trends in Plant Science* 6 (2): 78–84.

70. Stöckigt, Joachim, Andrey P. Antonchick, Fangrui Wu, and Herbert Waldmann. 2011. "The Pictet-Spengler Reaction in Nature and in Organic Chemistry." *Angewandte Chemie* 50 (37): 8538–8564.

71. Shrivastava, Vanshika, Naveen Sharma, Vikas Shrivastava, and Ajay Sharma. 2021. "Review on Camptothecin Producing Medicinal Plant: *Nothapodytes nimmoniana.*" *Biomedical & Pharmacology Journal* 14 (4): 1799–1813.

72. Gabr, A., A. Kuin, M. Aalders, H. El-Gawly, and L. A. Smets. 1997. "Cellular Pharmacokinetics and Cytotoxicity of Camptothecin and Topotecan at Normal and Acidic pH." *Cancer Research* 57 (21): 4811–4816.

73. Lorence, Argelia, and Craig L. Nessler. 2004. "Camptothecin, over Four Decades of Surprising Findings." *Phytochemistry* 65 (20): 2735–2749.

74. Kang, De, Ai-Lin Liu, Jin-Hua Wang, and Guan-Hua Du. 2018. "Camptothecin." In *Natural Small Molecule Drugs from Plants*, ed. G. H. Du, 491–496. Singapore: Springer Singapore.

75. Stork, Gilbert, and Arthur G. Schultz. 1971. "Total Synthesis of Dl-Camptothecin." *Journal of the American Chemical Society* 93 (16): 4074–4075.

76. Corey, E. J., Dennis N. Crouse, and Jerome E. Anderson. 1975. "Total Synthesis of Natural 20(S)-Camptothecin." *The Journal of Organic Chemistry* 40 (14): 2140–2141.

77. Ejima, Akio, Hirofumi Terasawa, Masamichi Sugimori, and Hiroaki Tagawa. 1989. "Asymmetric Synthesis of (S)-Camptothecin." *Tetrahedron Letters* 30 (20): 2639–2640.

78. Imura, Akihiro, Motohiro Itoh, and Akihiko Miyadera. 1998. "Enantioselective Synthesis of 20(S)-Camptothecin Using an Enzyme-Catalyzed Resolution." *Tetrahedron, Asymmetry* 9 (13): 2285–2291.

79. Comins, Daniel L., Matthew F. Baevsky, and Hao Hong. 1992. "A 10-Step, Asymmetric Synthesis of (S)-Camptothecin." *Journal of the American Chemical Society* 114 (27): 10971–10972.

80. Comins, Daniel L., Hao Hong, and Gao Jianhua. 1994. "Asymmetric Synthesis of Camptothecin Alkaloids: A Nine-Step Synthesis of (S)-Camptothecin." *Tetrahedron Letters* 35 (30): 5331–5334.

81. Comins, Daniel L., and Jayanta K. Saha. 1995. "Asymmetric Synthesis of a Key Camptothecin Intermediate from 2-Fluoropyridine." *Tetrahedron Letters* 36 (44): 7995–7998.

82. Comins, D. L., and J. M. Nolan. 2001. "A Practical Six-Step Synthesis of (S)-Camptothecin." *Organic Letters* 3 (26): 4255–4257.

83. Curran, Dennis P., Sung-Bo Ko, and Hubert Josien. 1996. "Cascade Radical Reactions of Isonitriles: A Second-Generation Synthesis of(20S)-Camptothecin, Topotecan, Irinotecan, and GI-147211C." *Angewandte Chemie* 34 (2324): 2683–2684.

84. Fortunak, Joseph M. D., Antonietta R. Mastrocola, Mark Mellinger, Nicolas J. Sisti, Jeffery L. Wood, and Zhi-Ping Zhuang. 1996. "Novel Syntheses of Camptothecin Alkaloids, Part I. Intramolecular [4+2]

Cycloadditions of N-Arylimidates and 4H-3,1-Benzoxazin-4-Ones as 2-Aza-1,3-Dienes." *Tetrahedron Letters* 37 (32): 5679–5682.

85. Chavan, Subhash P., and M. S. Venkatraman. 1998. "A Practical and Efficient Synthesis of (±)-Camptothecin." *Tetrahedron Letters* 39 (37): 6745–6748.

86. Ciufolini, Marco A., and Frank Roschangar. 1996. "Total Synthesis of(+)-Camptothecin." *Angewandte Chemie* 35 (15): 1692–1694.

87. Ciufolini, Marco A., and Frank Roschangar. 1997. "Practical Total Synthesis of (+)-Camptothecin: The Full Story." *Tetrahedron* 53 (32): 11049–11060.

88. Ciufolini, M. A., and F. Roschangar. 2000. "Practical Synthesis of (20S)-(+)-Camptothecin: The Progenitor of a Promising Group of Anticancer Agents." *Targets Heterocyclic Systems* 4: 25–55.

89. Bennasar, M-Lluïsa, Cecília Juan, and Joan Bosch. 2000. "A Short Synthesis of Camptothecin via a 2-Fluoro-1,4-Dihydropyridine." *Chemical Communications* 24: 2459–2460.

90. Bennasar, M-Lluïsa, Ester Zulaica, Cecília Juan, Yolanda Alonso, and Joan Bosch. 2002. "Addition of Ester Enolates to N-Alkyl-2-Fluoropyridinium Salts: Total Synthesis of (±)-20-Deoxycamptothecin and (+)-Camptothecin." *The Journal of Organic Chemistry* 67 (21): 7465–7474.

91. Blagg, Brian S. J., and Dale L. Boger. 2002. "Total Synthesis of (+)-Camptothecin." *Tetrahedron* 58 (32): 6343–6349.

92. Chavan, Subhash P., and Rasapalli Sivappa. 2004. "A Synthesis of Camptothecin." *Tetrahedron Letters* 45 (15): 3113–3115.

93. Tam, Markus A., and Victor Wicki. 2004. "Combined Directed Ortho Metalation/Cross-Coupling Strategies: Synthesis of the Tetracyclic A/B/C/D Ring Core of the Antitumor Agent Camptothecin." *The Journal of Organic Chemistry* 69 (23): 7816–7821.

94. Anderson, Regan J., Gajendra B. Raolji, Alice Kanazawa, and Andrew E. Greene. 2005. "A Novel, Expeditious Synthesis of Racemic Camptothecin." *Organic Letters* 7 (14): 2989–2991.

95. Yu, Shanbao, Qing-Qing Huang, Yu Luo, and Wei Lu. 2012. "Total Synthesis of Camptothecin and SN-38." *The Journal of Organic Chemistry* 77 (1): 713–717.

96. Li, Ke, Jinjie Ou, and Shuanhu Gao. 2016. "Total Synthesis of Camptothecin and Related Natural Products by a Flexible Strategy." *Angewandte Chemie* 55 (47): 14778–14783.

97. Liu, Xiaoxi, Eyob Adane, Fei Tang, and Markos Leggas. 2019. "Pharmacokinetic Modeling of the Blood-Stable Camptothecin Analog AR-67 in Two Different Formulations." *Biopharmaceutics & Drug Disposition* 40 (8): 265–275.

98. Kingsbury, William D., Jeffrey C. Boehm, Dalia R. Jakas, Kenneth G. Holden, Sidney M. Hecht, Gregory Gallagher, Mary Jo Caranfa, et al. 1991. "Synthesis of Water-Soluble (Aminoalkyl) Camptothecin Analogs: Inhibition of Topoisomerase I and Antitumor Activity." *Journal of Medicinal Chemistry* 34 (1): 98–107.

99. Nitta K., Yokokura T., Sawada S., Kunimoto T., Tanaka T., Uehara N., Baba H., Takeuchi M., Miyasaka T., and Mutai M. 1987. "Antitumor Activity of New Derivatives of Camptothecin." *Gan to kagaku ryoho. Cancer & Chemotherapy* 14 (3 Pt 2): 850–857.

100. Sawada, S., S. Okajima, R. Aiyama, K-I Nokata, T. Furata, T. Yokpkura, E. Sugimo, K. Yamaguchi, and T. Miyasaka. 1991. "Synthesis and Antitumor Activity of 20(S)-Camptothecin Derivatives: Carbamate-Linked, Water-soluble Derivatives of 7-Ethyl-10-Hydroxycamptothecin." *Chemical & Pharmaceutical Bulletin* 39 (6): 1446–1450.

101. Ahn, Soon Kil, Nam Song Choi, Byeong Seon Jeong, Kye Kwang Kim, Duck Jin Journ, Joon Kyum Kim, Sang Joon Lee, Jung Woo Kim, Chung Hong II, and Sang-Sup Jew. 2000. "Practical Synthesis of (S)-7-(2-Isopropylamino) Ethylcamptothecin Hydrochloride, Potent Topoisomerase I Inhibitor." *Journal of Heterocyclic Chemistry* 37 (5): 1141–1144.

102. Wani, M. C., A. W. Nicholas, M. E. Wall, and M. E. Wall. 1986. "Plant Antitumor Agents: 23. Synthesis and Antileukemic Activity of Camptothecin Analogues." *Journal of Medicinal Chemistry* 29: 2358–2363.

103. Cao, Zhisong, Kim Armstrong, Marcus Shaw, Eddie Petry, and Nick Harris. 1998. "Nitration of Camptothecin with Various Inorganic Nitrate Salts in Concentrated Sulfuric Acid: A New Preparation of Anticancer Drug 9-Nitrocamptothecin." *Synthesis* (12): 1724–1730.

104. De Cesare, M., G. Pratesi, P. Perego, N. Carenini, S. Tinelli, L. Merlini, S. Penco, et al. 2001. "Potent Antitumor Activity and Improved Pharmacological Profile of ST1481, a Novel 7-Substituted Camptothecin." *Cancer Research* 61 (19): 7189–7195.

105. Haridas, Kochat, and Frederick H. Hausheer. 2000. Highly lipophilic Camptothecin derivatives. 6057303. *US Patent*, filed October 20, 1998, and issued May 2, 2000.

106. Du, W. 2003. "Semisynthesis of DB-67 and Other Silatecans from Camptothecin by Thiol-Promoted Addition of Silyl Radicals." *Bioorganic & Medicinal Chemistry* 11 (3): 451–458.

107. Pommier, Yves, and Christophe Marchand. 2012. "Erratum: Interfacial Inhibitors: Targeting Macromolecular Complexes." *Nature Reviews. Drug Discovery* 11 (3): 250.
108. Wang, X., L. K. Wang, W. D. Kingsbury, R. K. Johnson, and S. M. Hecht. 1998. "Differential Effects of Camptothecin Derivatives on Topoisomerase I-Mediated DNA Structure Modification." *Biochemistry* 37 (26): 9399–9408.
109. Redinbo, M. R., L. Stewart, P. Kuhn, J. J. Champoux, and W. G. Hol. 1998. "Crystal Structures of Human Topoisomerase I in Covalent and Noncovalent Complexes with DNA." *Science (New York, N.Y.)* 279 (5356): 1504–1513.
110. Padlan, E. A., and E. A. Kabat. 1991. "Modeling of Antibody Combining Sites." *Methods in Enzymology* 203: 3–21.
111. Laco, Gary S., Jack R. Collins, Brian T. Luke, Heiko Kroth, Jane M. Sayer, Donald M. Jerina, and Yves Pommier. 2002. "Human Topoisomerase I Inhibition: Docking Camptothecin and Derivatives into a Structure-Based Active Site Model." *Biochemistry* 41 (5): 1428–1435.
112. Kerrigan, J. E., and D. S. Pilch. 2001. "A Structural Model for the Ternary Cleavable Complex Formed between Human Topoisomerase I, DNA, and Camptothecin." *Biochemistry* 40 (33): 9792–9798.
113. Giannini, Giuseppe. 2014. "Camptothecin and Analogs." In *Methods and Principles in Medicinal Chemistry*, ed. S. Hanessian, 181–224. Weinheim, Germany: Wiley-VCH Verlag GmbH & Co. KGaA.
114. Wong, Eric T., and Anna Berkenblit. 2004. "The Role of Topotecan in the Treatment of Brain Metastases." *The Oncologist* 9 (1): 68–79.
115. Fugit, Kyle D., and Bradley D. Anderson. 2014. "The Role of pH and Ring-Opening Hydrolysis Kinetics on Liposomal Release of Topotecan." *Journal of Controlled Release* 174: 88–97.
116. Buzun, Kamila, Anna Bielawska, Krzysztof Bielawski, and Agnieszka Gornowicz. 2020. "DNA Topoisomerases as Molecular Targets for Anticancer Drugs." *Journal of Enzyme Inhibition and Medicinal Chemistry* 35 (1): 1781–1799.
117. Tanizawa, A., A. Fujimori, Y. Fujimori, and Y. Pommier. 1994. "Comparison of Topoisomerase-I Inhibition, DNA-Damage, and Cytotoxicity of Camptothecin Derivatives Presently in Clinical-Trials." *Journal of the National Cancer Institute* 86: 836–842.
118. Kollmannsberger, C., K. Mross, A. Jakob, L. Kanz, and C. Bokemeyer. 1999. "Topotecan—A Novel Topoisomerase I Inhibitor: Pharmacology and Clinical Experience." *Oncology* 56 (1): 1–12.
119. Arun, Banu, and Eugene P. Frenkel. 2001. "Topoisomerase I Inhibition with Topotecan: Pharmacologic and Clinical Issues." *Expert Opinion on Pharmacotherapy* 2 (3): 491–505.
120. Staker, Bart L., Kathryn Hjerrild, Michael D. Feese, Craig A. Behnke, Alex B. Burgin Jr, and Lance Stewart. 2002. "The Mechanism of Topoisomerase I Poisoning by a Camptothecin Analog." *Proceedings of the National Academy of Sciences of the United States of America* 99 (24): 15387–15392.
121. Feeney, G. P., R. J. Errington, M. Wiltshire, N. Marquez, S. C. Chappell, and P. J. Smith. 2003. "Tracking the Cell Cycle Origins for Escape from Topotecan Action by Breast Cancer Cells." *British Journal of Cancer* 88 (8): 1310–1317.
122. Carol, Hernan, Peter J. Houghton, Christopher L. Morton, E. Anders Kolb, Richard Gorlick, C. Patrick Reynolds, Min H. Kang, et al. 2010. "Initial Testing of Topotecan by the Pediatric Preclinical Testing Program: Topotecan Pediatric Preclinical Testing." *Pediatric Blood & Cancer* 54 (5): 707–715.
123. Fassberg, J., and V. J. Stella. 1992. "A Kinetic and Mechanistic Study of the Hydrolysis of Camptothecin and Some Analogues." *Journal of Pharmaceutical Sciences* 81 (7): 676–684.
124. Herben, V. M., W. W. ten Bokkel Huinink, and J. H. Beijnen. 1996. "Clinical Pharmacokinetics of Topotecan." *Clinical Pharmacokinetics* 31 (2): 85–102.
125. Madhav, N. V. Satheesh, Ashok K. Shakya, Pragati Shakya, and Kuldeep Singh. 2009. "Orotransmucosal Drug Delivery Systems: A Review." *Journal of Controlled Release* 140 (1): 2–11.
126. Morris, R., and A. Munkarah. 2002. "Alternative Dosing Schedules for Topotecan in the Treatment of Recurrent Ovarian Cancer." *Oncologist* 7: 29–35.
127. Dennis, M. J., J. H. Beijnen, L. B. Grochow, and L. J. van Warmerdam. 1997. "An Overview of the Clinical Pharmacology of Topotecan." *Seminars in Oncology* 24 (1 Suppl 5): S5–12-S5–18.
128. Bokkel, H. W., J. Carmicheal, D. Armstrong, A. Gordon, and J. Malfetano. 1997. "Efficacy and Safety of Topotecan in the Treatment of Advanced Ovarian Carcinoma." *Journal of Clinical Oncology* 15 (1): 177–186.
129. Herzog, T. J. 2002. "Update on the Role of Topotecan in the Treatment of Recurrent Ovarian Cancer." *The Oncologist* 7 (90005): 3–10.
130. Pirker, Robert, Peter Berzinec, Stephen Brincat, Peter Kasan, Gyula Ostoros, Milos Pesek, Signe Plāte, et al. 2010. "Therapy of Small Cell Lung Cancer with Emphasis on Oral Topotecan." *Lung Cancer (Amsterdam, Netherlands)* 70 (1): 7–13.

131. Eckardt, John R., Joachim von Pawel, Jean-Louis Pujol, Zsolt Papai, Elisabeth Quoix, Andrea Ardizzoni, Ruth Poulin, Alaknanda J. Preston, Graham Dane, and Graham Ross. 2007. "Phase III Study of Oral Compared with Intravenous Topotecan as Second-Line Therapy in Small-Cell Lung Cancer." *Journal of Clinical Oncology* 25 (15): 2086–2092.

132. Eckardt, John R., Joachim von Pawel, Zsolt Papai, Antoaneta Tomova, Valentina Tzekova, Theresa E. Crofts, Sarah Brannon, Paul Wissel, and Graham Ross. 2006. "Open-Label, Multicenter, Randomized, Phase III Study Comparing Oral Topotecan/Cisplatin versus Etoposide/Cisplatin as Treatment for Chemotherapy-Naive Patients with Extensive-Disease Small-Cell Lung Cancer." *Journal of Clinical Oncology* 24 (13): 2044–2051.

133. Vecchione, F., R. Fruscio, T. Dell'Anna, A. Garbi, R. Garcia Parra, S. Corso, and A. A. Lissoni. 2007. "A Phase II Clinical Trial of Topotecan and Carboplatin in Patients with Newly Diagnosed Advanced Epithelial Ovarian Cancer." *International Journal of Gynecological Cancer* 17 (2): 367–372.

134. Gupta, D., R. L. Owers, M. Kim, D. Y. Kuo, G. S. Huang, S. Shahabi, G. L. Goldberg, and M. H. Einstein. 2009. "A Phase II Study of Weekly Topotecan and Docetaxel in Heavily Pretreated Patients with Recurrent Uterine and Ovarian Cancers." *Gynecologic Oncology* 113 (3): 327–330.

135. Nogami, Naoyuki, Katsuyuki Hotta, Shoichi Kuyama, Katsuyuki Kiura, Nagio Takigawa, Kenichi Chikamori, Takuo Shibayama, et al. 2011. "A Phase II Study of Amrubicin and Topotecan Combination Therapy in Patients with Relapsed or Extensive-Disease Small-Cell Lung Cancer: Okayama Lung Cancer Study Group Trial 0401." *Lung Cancer (Amsterdam, Netherlands)* 74 (1): 80–84.

136. Kretschmar, Cynthia S., Morris Kletzel, Kevin Murray, Paul Thorner, Vijay Joshi, Robert Marcus, E. Ide Smith, Wendy B. London, and Robert Castleberry. 2004. "Response to Paclitaxel, Topotecan, and Topotecan-Cyclophosphamide in Children with Untreated Disseminated Neuroblastoma Treated in an Upfront Phase II Investigational Window: A Pediatric Oncology Group Study." *Journal of Clinical Oncology* 22 (20): 4119–4126.

137. Kacprzak, Karol Michał. 2013. "Chemistry and Biology of Camptothecin and Its Derivatives." In *Natural Products*, 643–682. Berlin, Heidelberg: Springer.

138. Frumovitz, M., M. F. Munsell, J. K. Burzawa, L. A. Byers, P. Ramalingam, J. Brown, and R. L. Coleman. 2017. "Combination Therapy with Topotecan, Paclitaxel, and Bevacizumab Improves Progression-Free Survival in Recurrent Small Cell Neuroendocrine Carcinoma of the Cervix." *Gynecologic Oncology* 144 (1): 46–50.

139. Lee, Wai-Leng, Jeng-Yuan Shiau, and Lie-Fen Shyur. 2012. "Taxol, Camptothecin and Beyond for Cancer Therapy." In *Advances in Botanical Research*, eds. Lie-Fen Shyur and Allan S. Y. Lau, 62:133–178. San Diego, CA: Elsevier.

140. Kawato, Y., M. Aonuma, Y. Hirota, H. Kuga, and K. Sato. 1991. "Intracellular Roles of SN-38, a Metabolite of the Camptothecin Derivative CPT-11, in the Antitumor Effect of CPT-11." *Cancer Research* 51 (16): 4187–4191.

141. Negoro, S., M. Fukuoka, N. Masuda, M. Takada, Y. Kusunoki, K. Matsui, N. Takifuji, S. Kudoh, H. Niitani, and T. Taguchi. 1991. "Phase I Study of Weekly Intravenous Infusions of CPT-11, a New Derivative of Camptothecin, in the Treatment of Advanced Non-Smallcell Lung Cancer." *Journal of the National Cancer Institute* 83 (16): 1164–1168.

142. Langer, C. J. 2001. "The Emerging World Role of Irinotecan in Lung Cancer." *Oncology (Williston Park, N.Y.)* 15 (7 Suppl 8): 15–21.

143. Sandler, Alan. 2002. "Irinotecan plus Cisplatin in Small-Cell Lung Cancer." *Oncology (Williston Park, N.Y.)* 16 (9 Suppl 9): 39–43.

144. Pectasides, Dimitrios, Nikolaos Mylonakis, Dimitrios Farmakis, Maria Nikolaou, Maria Koumpou, Ioannis Katselis, Asimina Gaglia, Vassiliki Kostopoulou, Athanasios Karabelis, and Christos Kosmas. 2003. "Irinotecan and Gemcitabine in Patients with Advanced Non-Small Cell Lung Cancer, Previously Treated with Cisplatin-Based Chemotherapy. A Phase II Study." *Anticancer Research* 23 (5b): 4205–4211.

145. Yang, Xue-Qin, Chong-Yi Li, Ming-Fang Xu, Hong Zhao, and Dong Wang. 2015. "Comparison of First-Line Chemotherapy Based on Irinotecan or Other Drugs to Treat Non-Small Cell Lung Cancer in Stage IIIB/IV: A Systematic Review and Meta-Analysis." *BMC Cancer* 15 (1): 949.

146. Enzinger, Peter C., Matthew H. Kulke, Jeffrey W. Clark, David P. Ryan, Haesook Kim, Craig C. Earle, Michele M. Vincitore, Ann L. Michelini, Robert J. Mayer, and Charles S. Fuchs. 2005. "A Phase II Trial of Irinotecan in Patients with Previously Untreated Advanced Esophageal and Gastric Adenocarcinoma." *Digestive Diseases and Sciences* 50 (12): 2218–2223.

147. Makiyama, Akitaka, Kohei Arimizu, Gen Hirano, Chinatsu Makiyama, Yuzo Matsushita, Tsuyoshi Shirakawa, Hirofumi Ohmura, et al. 2018. "Irinotecan Monotherapy as Third-Line or Later Treatment in Advanced Gastric Cancer." *Gastric Cancer* 21 (3): 464–472.

148. Gershenson, David M. 2002. "Irinotecan in Epithelial Ovarian Cancer." *Oncology (Williston Park, N.Y.)* 16 (5 Suppl 5): 29–31.

149. Musa, Fernanda, Bhavana Pothuri, Stephanie V. Blank, Huichung T. Ling, James L. Speyer, John Curtin, Leslie Boyd, et al. 2017. "Phase II Study of Irinotecan in Combination with Bevacizumab in Recurrent Ovarian Cancer." *Gynecologic Oncology* 144 (2): 279–284.

150. Verschraegen, Claire F. 2002. "Irinotecan for the Treatment of Cervical Cancer." *Oncology (Williston Park, N.Y.)* 16 (5 Suppl 5): 32–34.

151. Fuchs, Charles, Edith P. Mitchell, and Paulo M. Hoff. 2006. "Irinotecan in the Treatment of Colorectal Cancer." *Cancer Treatment Reviews* 32 (7): 491–503.

152. Rothenberg, M. L. 2001. "Irinotecan (CPT-11): Recent Developments and Future Directions-Colorectal Cancer and Beyond." *The Oncologist* 6 (1): 66–80.

153. Kciuk, Mateusz, Beata Marciniak, and Renata Kontek. 2020. "Irinotecan-Still an Important Player in Cancer Chemotherapy: A Comprehensive Overview." *International Journal of Molecular Sciences* 21 (14): 4919.

154. Khanna, R., C. L. Morton, M. K. Danks, and P. M. Potter. 2000. "Proficient Metabolism of Irinotecan by a Human Intestinal Carboxylesterase." *Cancer Research* 60 (17): 4725–4728.

155. Sparreboom, A., M. J. de Jonge, P. de Bruijn, E. Brouwer, K. Nooter, W. J. Loos, R. J. van Alphen, R. H. Mathijssen, G. Stoter, and J. Verweij. 1998. "Irinotecan (CPT-11) Metabolism and Disposition in Cancer Patients." *Clinical Cancer Research* 4 (11): 2747–2754.

156. Combes, O., J. Barré, J. C. Duché, L. Vernillet, Y. Archimbaud, M. P. Marietta, J. P. Tillement, and S. Urien. 2000. "In Vitro Binding and Partitioning of Irinotecan (CPT-11) and Its Metabolite, SN-38, in Human Blood." *Investigational New Drugs* 18 (1): 1–5.

157. Kehrer, D. F., W. Yamamoto, J. Verweij, M. J. de Jonge, P. de Bruijn, and A. Sparreboom. 2000. "Factors Involved in Prolongation of the Terminal Disposition Phase of SN-38: Clinical and Experimental Studies." *Clinical Cancer Research* 6 (9): 3451–3458.

158. Herben, Virginie M. M., Jos H. Beijnen, Wim W. ten Bokkel Huinink, and Jan H. M. Schellens. 1998. "Clinical Pharmacokinetics of Camptothecin Topoisomerase I Inhibitors." *Pharmacy World & Science* 20 (4): 161–172.

159. Ikeguchi, Masahide, Yosuke Arai, Yoshihiko Maeta, Keigo Ashida, Kuniyuki Katano, and Toshiro Wakatsuki. 2011. "Topoisomerase I Expression in Tumors as a Biological Marker for CPT-11 Chemosensitivity in Patients with Colorectal Cancer." *Surgery Today* 41 (9): 1196–1199.

160. Douillard, J. Y., D. Cunningham, A. D. Roth, M. Navarro, R. D. James, P. Karasek, P. Jandik, et al. 2000. "Irinotecan Combined with Fluorouracil Compared with Fluorouracil Alone as First-Line Treatment for Metastatic Colorectal Cancer: A Multicentre Randomised Trial." *Lancet* 355 (9209): 1041–1047.

161. Oostendorp, Linda J. M., Peep F. Stalmeier, Pieternel C. Pasker-de Jong, Winette T. Van der Graaf, and Petronella B. Ottevanger. 2010. "Systematic Review of Benefits and Risks of Second-Line Irinotecan Monotherapy for Advanced Colorectal Cancer." *Anti-Cancer Drugs* 21 (8): 749–758.

162. Vanhoefer, U., P. Rougier, M. Borner, A. Muñoz, J-L Van Laethem, and A. Sobrero. 2004. "Irinotecan in Combination with New Agents." *European Journal of Cancer Supplements* 2 (7): 14–20.

163. Vanhoefer, U., A. Harstrick, W. Achterrath, S. Cao, S. Seeber, and Y. M. Rustum. 2001. "Irinotecan in the Treatment of Colorectal Cancer: Clinical Overview." *Journal of Clinical Oncology* 19 (5): 1501–1518.

164. Fuchs, Charles S., John Marshall, Edith Mitchell, Rafal Wierzbicki, Vinod Ganju, Mark Jeffery, Joseph Schulz, et al. 2007. "Randomized, Controlled Trial of Irinotecan plus Infusional, Bolus, or Oral Fluoropyrimidines in First-Line Treatment of Metastatic Colorectal Cancer: Results from the BICC-C Study." *Journal of Clinical Oncology* 25 (30): 4779–4786.

165. Peeters, Marc, Timothy Jay Price, Andrés Cervantes, Alberto F. Sobrero, Michel Ducreux, Yevhen Hotko, Thierry André, et al. 2010. "Randomized Phase III Study of Panitumumab with Fluorouracil, Leucovorin, and Irinotecan (FOLFIRI) Compared with FOLFIRI Alone as Second-Line Treatment in Patients with Metastatic Colorectal Cancer." *Journal of Clinical Oncology* 28 (31): 4706–4713.

166. Moehler, M., A. Mueller, T. Trarbach, F. Lordick, T. Seufferlein, S. Kubicka, M. Geißler, et al. 2011. "Cetuximab with Irinotecan, Folinic Acid and 5-Fluorouracil as First-Line Treatment in Advanced Gastroesophageal Cancer: A Prospective Multi-Center Biomarker-Oriented Phase II Study." *Annals of Oncology* 22 (6): 1358–1366.

167. García-Alfonso, P., A. J. Muñoz-Martin, S. Alvarez-Suarez, Y. Jerez-Gilarranz, M. Riesco-Martinez, P. Khosravi, and M. Martin. 2010. "Bevacizumab in Combination with Biweekly Capecitabine and Irinotecan, as First-Line Treatment for Patients with Metastatic Colorectal Cancer." *British Journal of Cancer* 103 (10): 1524–1528.

168. Crul, Mirjam. 2003. "CKD-602. Chong Kun Dang." *Current Opinion in Investigational Drugs (London, England: 2000)* 4 (12): 1455–1459.

169. Lee, J. H., J. M. Lee, J. K. Kim, S. K. Ahn, S. J. Lee, M. Y. Kim, S. S. Jew, J. G. Park, and C. I. Hong. 1998. "Antitumor Activity of 7- 2-(N-Isopropylamino)Ethyl]-(20S)-Camptothecin, CKD602, as a Potent DNA Topoisomerase I Inhibitor." *Archives of Pharmacal Research* 21 (5): 581.

170. Lee, D. H., S-W Kim, C. Suh, J-S Lee, J. H. Lee, S-J Lee, B. Y. Ryoo, et al. 2008. "Belotecan, New Camptothecin Analogue, Is Active in Patients with Small-Cell Lung Cancer: Results of a Multicenter Early Phase II Study." *Annals of Oncology* 19 (1): 123–127.

171. Kim, S. J., J. S. Kim, S. C. Kim, Y. K. Kim, Y. K. Kim, J. Y. Kang, H. K. Yoon, et al. 2010. "A Multicenter Phase II Study of Belotecan, New Camptothecin Analogue, in Patients with Previously Untreated Extensive Stage Disease Small Cell Lung Cancer." *Lung Cancer (Amsterdam, Netherlands)* 68 (3): 446–449.

172. Rhee, Chin Kook, Sang Haak Lee, Ju Sang Kim, Seung Joon Kim, Seok Chan Kim, Young Kyoon Kim, Hyun Hee Kang, et al. 2011. "A Multicenter Phase II Study of Belotecan, a New Camptothecin Analogue, as a Second-Line Therapy in Patients with Small Cell Lung Cancer." *Lung Cancer (Amsterdam, Netherlands)* 72 (1): 64–67.

173. Jeong, Jaeheon, Byoung Chul Cho, Joo Hyuk Sohn, Hye Jin Choi, Se Hyun Kim, Young Joo Lee, Min Kyu Jung, et al. 2010. "Belotecan for Relapsing Small-Cell Lung Cancer Patients Initially Treated with an Irinotecan-Containing Chemotherapy: A Phase II Trial." *Lung Cancer (Amsterdam, Netherlands)* 70 (1): 77–81.

174. Lee, Dae Ho, Sang-We Kim, Cheolwon Suh, Jung-Shin Lee, Jin Seok Ahn, Myung-Ju Ahn, Keunchil Park, et al. 2010. "Multicenter Phase 2 Study of Belotecan, a New Camptothecin Analog, and Cisplatin for Chemotherapy-Naive Patients with Extensive-Disease Small Cell Lung Cancer." *Cancer* 116 (1): 132–136.

175. Hong, Junshik, Minkyu Jung, Yu Jin Kim, Sun Jin Sym, Sun Young Kyung, Jinny Park, Sang Pyo Lee, et al. 2012. "Phase II Study of Combined Belotecan and Cisplatin as First-Line Chemotherapy in Patients with Extensive Disease of Small Cell Lung Cancer." *Cancer Chemotherapy and Pharmacology* 69 (1): 215–220.

176. Lim, Seungtaek, Byoung Chul Cho, Ji Ye Jung, Gun Min Kim, Se Hyun Kim, Hye Ryun Kim, Han Sang Kim, et al. 2013. "Phase II Study of Camtobell Inj. (Belotecan) in Combination with Cisplatin in Patients with Previously Untreated, Extensive Stage Small Cell Lung Cancer." *Lung Cancer (Amsterdam, Netherlands)* 80 (3): 313–318.

177. Lee, Hyo-Pyo, Sang-Soo Seo, Sang-Young Ryu, Jong-Hyeok Kim, Yung-Jue Bang, Sang-Yoon Park, Joo-Hyun Nam, Soon-Beom Kang, Kyung-Hee Lee, and Yong Sang Song. 2008. "Phase II Evaluation of CKD-602, a Camptothecin Analog, Administered on a 5-Day Schedule to Patients with Platinum-Sensitive or -Resistant Ovarian Cancer." *Gynecologic Oncology* 109 (3): 359–363.

178. Kim, Yong-Man, Shin Wha Lee, Dae-Yeon Kim, Jong-Hyeok Kim, Joo-Hyun Nam, and Young-Tak Kim. 2010. "The Efficacy and Toxicity of Belotecan (CKD-602), a Camptothericin Analogue Topoisomerase I Inhibitor, in Patients with Recurrent or Refractory Epithelial Ovarian Cancer." *Journal of Chemotherapy (Florence, Italy)* 22 (3): 197–200.

179. Kim, Hee Seung, Noh Hyun Park, Sokbom Kang, Sang-Soo Seo, Hyun Hoon Chung, Jae Weon Kim, Yong Sang Song, and Soon-Beom Kang. 2010. "Comparison of the Efficacy between Topotecan- and Belotecan-, a New Camptothecin Analog, Based Chemotherapies for Recurrent Epithelial Ovarian Cancer: A Single Institutional Experience." *The Journal of Obstetrics and Gynaecology Research* 36 (1): 86–93.

180. Hwang, Jong Ha, Heon Jong Yoo, Myong Cheol Lim, Sang-Soo Seo, Sang-Yoon Park, and Sokbom Kang. 2012. "Phase I Clinical Trial of Alternating Belotecan and Oral Etoposide in Patients with Platinum-Resistant or Heavily Treated Ovarian Cancer." *Anti-Cancer Drugs* 23 (3): 321–325.

181. Nam, Eun Ji, Jae Wook Kim, Jae Hoon Kim, Sunghoon Kim, Sang Wun Kim, Si Young Jang, Dae Woo Lee, Yong Wook Jung, and Young Tae Kim. 2010. "Efficacy and Toxicity of Belotecan with and without Cisplatin in Patients with Recurrent Ovarian Cancer." *American Journal of Clinical Oncology* 33 (3): 233–237.

182. Choi, Chel Hun, Yoo-Young Lee, Tae-Jong Song, Hwang-Shin Park, Min Kyu Kim, Tae-Joong Kim, Jeong-Won Lee, Je-Ho Lee, Duk-Soo Bae, and Byoung-Gie Kim. 2011. "Phase II Study of Belotecan, a Camptothecin Analogue, in Combination with Carboplatin for the Treatment of Recurrent Ovarian Cancer." *Cancer* 117 (10): 2104–2111.

13 *Brassica* Phytochemicals
A Potential Source of Cancer Prevention and Treatment

Himanshu Punetha and Shivanshu Garg

CONTENTS

13.1 Introduction 227
13.2 The Caesar Experiment with *Brassica* 230
13.3 Diversity of *Brassica* 230
13.4 *Brassica* Interacting with Carcinogens..230
13.5 The Molecular Framework of the Human Body and Its Response to Cancer...................231
13.6 The Universality and Uniqueness of Organosulfur Compounds (OSC) 232
    13.6.1 OSC and Enzymatic Interaction 232
    13.6.2 Molecular Insights of Brassicaceae Metabolites 232
13.7 An Example the World Needs to Learn 233
13.8 The Pathway Hindrances and Open Doors 233
    13.8.1 Integrators of Metabolic Pathway..234
        13.8.1.1 Direct and Indirect Interaction....................................234
        13.8.1.2 The Concentration Regime 234
        13.8.1.3 Integration of Collective Data......................................235
13.9 The Chemistry of Indole-3-Carbinol in the Human Body 235
    13.9.1 The Important Condensate 236
    13.9.2 Diversity Within I3C 236
    13.9.3 Chemosensitization's Companionship 237
    13.9.4 Mode of Action by I3C/DIM 238
13.10 Conclusion 238
13.11 References..238

13.1 INTRODUCTION

Asia welcomed *Brassica* about three million years ago from the Northeastern Mediterranean region of the globe. In Europe, *Brassica* was quite a famous, taste-rich consumable of great kings. No one has predicted in those ancient times that the main cause of morbidity and mortality around the world will be the increase in cases of different types of cancer. The enormous growth of melanomas in different tissue types, even after a vast array of therapeutic technologies have been made available by the advent of 2022, is attributed to the side effects of currently available medicines [1]. Hippocrates made a reasonable statement more than two thousand years ago, "Let food be medicine and medicine be food," which proved to be valid for human civilization [2]. The metabolites present in *Brassica* consumed either as vegetable or oilseed, as food or component of food, harboring rich diversity of metabolites, may act as new weapons in anticancer therapeutics [**Table 13.1**]. Pertaining to its biochemical properties, the *Brassica* has its documented life-saving roles in the prevention of obesity, diabetes (type I and type II), and cardiovascular ailments. The presence of organosulfur compounds such as indole-3-carbinol [**Figure 13.1**] in *Brassica* worldwide is chiefly

DOI: 10.1201/b23311-13

TABLE 13.1

Anticarcinogenic Properties of Brassica

S. No.	Anticarcinogenic Components of Brassica	Target Site	Molecular Mechanism	Reference
1.	Flavanols (quercetin, kaempferol)	Liver, kidney	• Controls cell cycle phases • Involved in pro-apoptosis of developing tumor cells	[10]
2.	Indole-3-carbinol (I3C)	Heart, lung, liver, kidney, cell lines (breast)	• Depolarizes the mitochondrial membrane which leads to cytochrome C release, ultimately helping in detoxifying mechanisms activation • Inhibits the growth of the estrogen-dependent human MCF-7 breast cancer cell line	[11, 12]
3.	Sulphorafane (SFN)	Liver, heart	• Inhibits the NF-κB DNA binding factor • Inhibits other genes involved in transactivation of κB-dependent genes	[13]
4.	Allyl isothiocyanate (ITC)	Cell lines (prostrate)	• Inhibits human prostate cancer cell lines LNCaP by causing G2/M arrest • Induction of apoptosis	[14]
5.	Benzyl isothiocyanate (BTIC)	Cell lines (head and neck)	• Inhibits head and neck squamous cell carcinoma (HNSCC) cells by activation of caspase-3	[15]
6.	Diallyl sulfide	Cell lines (colon, gastric)	• Induces differentiation and apoptosis in cancer cells • Arrests G2/M phase of cell cycle	[16, 17]

responsible for inducing G1 cell cycle arrest and programmed cell death (PCD) of melanomas by upregulating the *p23*, a tumor suppressor protein, and, simultaneously, downregulating the *pRb*, a check protein in the cell cycle [3]. Basically, isothiocyanates do not quench free radicals, but being present in *Brassica* has a feature of inducing phase II enzymes N-acetyltransferase (NAT), UDP-glucuronosyltransferase (UGT), and glutathione S-transferase (GST), which guard the cells against the damage done to structure and sequence of DNA arising from free radicals, teratogens, and carcinogens. Luteolin, zeaxanthin [**Figure 13.2**], flavonoids, and β-carotene [**Figure 13.3**] are universally present in all cultivars of *Brassica*, imparting antioxidant properties. Biochemically, the phenylpropanoid pathway is highly active in *Brassica*, converting the amino acid phenylalanine into tetrahydroxychalcone, which generates flavonols like quercetin [**Figure 13.4**], apigenin, lutein [**Figure 13.5**], and kaempferol [**Figure 13.6**][4]. The World Cancer Research Fund has delivered information governing that *Brassica*-rich diets act specifically against melanomas of the rectum, thyroid, and colon [88]. Mizuna, bok choi, Siberian kale, daikon, tatsoi, wasabi, and arugula are the cultivars of *Brassica* with a worldwide occurrence and have been in use by humans for consumption since the medieval period, imparting flavor and anticarcinogenic health benefits [5].

Glucoraphanin, gluconasturtiin, glucotropaeolin, and sinigrin are different glucosinolates present in *Brassica*'s water-soluble entities, which, in their way into the intestine, neutralize pathogenic bacteria, viruses, and even fungi. Fat-soluble vitamins A, D, E, and K present in *Brassica* protect against numerous carcinomas. The importance of vegetable *Brassica* can be judged by the fact that despite insolubility, the fiber present act as a check to control the occurrence of bowel carcinoma, stop diverticulitis, and avoid irritable bowel syndrome. Oxygen radical absorbance capacity (ORAC), measured in *μmol, Trolox Equivalent (TE)/100 g fresh weight*, of *Brassica* ranges from 352 to 3,602; the values serve themselves as an indicator of high scavenging of free radicals [6]. The bioavailability of the

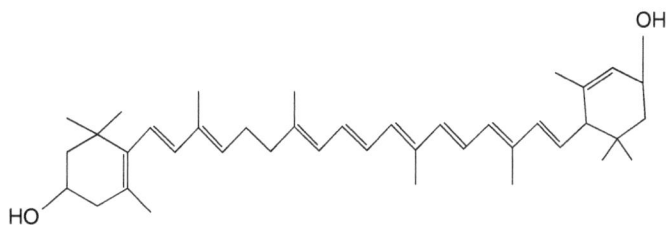

FIGURE 13.1   Zeaxanthin.

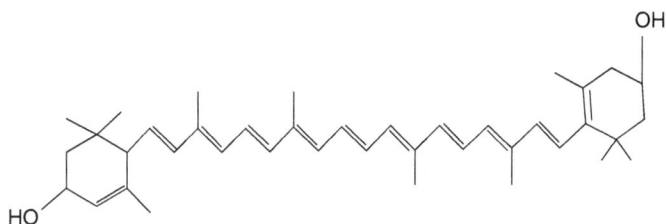

FIGURE 13.2   Lutein.

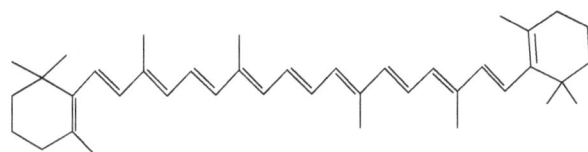

FIGURE 13.3   Beta-carotene.

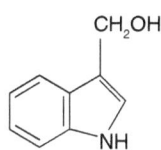

FIGURE 13.4   Indole-3-carbinol.

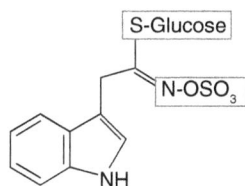

FIGURE 13.5   Glucobrassicin.

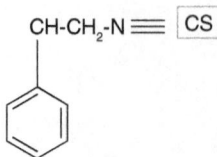

$$CH\text{-}CH_2\text{-}N \equiv \boxed{CS}$$

FIGURE 13.6   Phenylisothiocyanate.

important carotenoids involved in cancerous inactivation may be enhanced by formulating them into water-dispersible beads as they are fat-soluble in nature. The definite role of the growing season on the nutritional attributes of *Brassica* has been illustrated by the fact that winter or summer lowers the antioxidant activity and autumn or spring boosts up the levels of different antioxidants [7]. The antioxidative properties in the seed meal of *Brassica* are attributed to the presence of polyphenolic compounds having a hydroxyl group responsible for scavenging reactive oxygen species [8, 9].

13.2   THE CAESAR EXPERIMENT WITH *BRASSICA*

When numerous methods were not available for phytochemicals screening in ancient history, even then, *Brassica* has been listed by Julius Caesar for treating indigestion and, in biochemical terms, the prevention of free radical generation. Going to royal banquets and having a protein-rich meal in terms of animal-derived diet, Julius Caesar happened to be a sufferer of indigestion, acidity, and constipation. By constantly monitoring his diseased condition derived from high-animal-based diet, he decided to have ingestion experiment of cruciferous vegetables, which were then believed by many to alleviate the indigestion symptoms. Julius Caesar monitored his health condition in next few weeks by incorporating cruciferous vegetables in his diet and found himself to be free from indigestion derived ailments. Hence, the experiment done by Julius Caesar onto himself proved successful, and later, he promoted the inclusion of cruciferous vegetables like broccoli to stay healthy [18].

It can be said, "Eat Brassica, and Live Long and Stay Healthy."

13.3   DIVERSITY OF *BRASSICA*

With 3,709 species, 320 genera, and 49 tribes, Brassicaceae is a large family in the plant kingdom, comprising vegetables, and ornamentals (*Iberis, Matthiola, Erysimum, Aurbrieta, Arabis*, and many others), as well as oilseed crops like rapeseed and mustard. Not only these notable palatable crops, molecular and breeding-based experimental work plants serving as model organisms like *Arabis alpine, Cardamine hirsute, Arabidopsis thaliana*, and *A. suecica* also fall under Brassicaceae [19]. Although not a single member of the species is present in Antarctica, the diversification of the group is centered in the Mediterranean and Iran-Turanian regions [20, 21].

13.4   *BRASSICA* INTERACTING WITH CARCINOGENS

The interaction of the metabolome of *Brassica* governs a change in the molecular process of tumor formation. Tumor formation is influenced by hereditary and environmental exposure to an organism. The buildup of metastatic events, as they are defined, is brought about by deviation from a normal metabolic pathway, say, into extreme division till cell bursts, automatic death triggered by nuclear factors, or rapid cell death. Initiation, promotion, and progression mark the hallmark of carcinogenic events in a cell [22].

To begin with, initiation is the pioneering state governed by mutagenic DNA, damaged by physical (x-ray, UV ray), chemical (carcinogens), or viral exposure (adenovirus). The preneoplastic cell population results from epigenetic alterations depending upon their genetic framework. Whatsoever, the chronic exposure is ultimate in the development of metastasis. The critical stage required for a cell to be in metastasis is progression. Referring this to the final stage of carcinogenesis, the event of converting preneoplastic cells into an invasive and metastatic cell population are repeated again and again till the cell starves to death and bursts finally [23].

The achievement of cell death through carcinogenesis is credited to exogenous teratogens and carcinogens and endogenous carcinogens. The endogenous carcinogens owe their development and destruction ability to alter the oxygen to a different form: the reactive oxygen species (ROS). The ROS either are originated by electrophilic derivatives of the environment, say, smoke and air pollution, or are generated metabolically by the body itself, in vivo, otherwise destined for destruction by helpful enzymatic quenching. Any sort of imbalance, impeding the promotion led by the initiation of the tumor genesis process, results in carcinogenesis [10].

Hitting at molecular targets that lead to the development of metastatic cells, by virtue of chemical metabolite mediators, is chemoprevention. As such, the chemoprevention may be natural, if agents are naturally derived, say, from plants or microbes, and may be synthetic, say, by chemical synthesis designed by robotics and computer-generated software in laboratories. They all prevent cell metastasis [24].

The metastatic proliferation of different tissues is metabolically stopped when such tissues counteract metabolites provided to them, being sourced as the consumption of edibles, from origin Brassicaceae, by humans. Brassicaceae chemicals mediate their bioactive properties by virtue of a class of compounds called isothiocyanates (ISC), which are end products of enzymatic hydrolysis of glucosinolates (GSL). This enzymatic hydrolysis is brought about by myrosinase. Before humans can benefit from these substances, the test of whether getting a benefit or not was done on mice, in vivo. These benefits turned to be not only anticarcinogenic but anti-inflammatory too. The common reason for isothiocyanates being both anticarcinogenic and anti-inflammatory to mammals owed directly to the function of nuclear erythroid-2-related factor *Nrf2* found in the nucleus. The *Nrf2* is activated by a redox reaction involving the simultaneous activation of phase II enzymes [25].

13.5 THE MOLECULAR FRAMEWORK OF THE HUMAN BODY AND ITS RESPONSE TO CANCER

The human body encompasses metabolic routes to all sorts of carcinogens, whether dietary or environmentally driven. These metabolic routes are specified by an enzymatic process mediated through extracting electrons from agents or delivering hydrogen ions in a way supported by the dynamic transition of oxidation and reduction. This imparts a hydrophilic nature to the agent, making it accessible to cytochrome P450 (CYP), and the whole process is designated as phase I metabolism, a physicochemical event. Here comes the binding of the foreign agents, existing in dynamic oxidized and reduced state, to the cell's macromolecules, primarily proteins, lipids, and DNA and RNA. This binding is facilitated by yet another physiological but specifically chemically defined phase II enzyme. Ultimately, due to hydrophilic particle generation, out of foreign agents, they are destined to be secreted in urinary or biliary tracts [23].

A dynamic yet balanced gene action required for the generation of enzymatic proteins involved in phase I and phase II metabolism is an essential condition to meet the inactivation of carcinogens, which is further supported by xenobiotic factors found in certain vegetables and fruits [10, 26].

Phase I enzymes fall under the superfamily of microsomal enzymes and cytochrome P450 (CYP). The phase II enzymes act as conjugators of adjoining the foreign metabolites to those of the body's own defense-generating precursors. Some of them are glutathione S-transferase (GST), N-acetyltransferases (NAT), sulfotransferases (SULT), epoxide hydrolase (EPH), UDP-glucuronosyltransferases (UGT), diaphorase (DP), NAD quinone oxidoreductase (NQO) or NAD (P) H, menadione oxidoreductase (NMO), and quinone reductase (QR) [13].

On common phylogenetic analysis of phase II enzymes, the consensus region conferring antioxidant response element/electrophile responsive element (ARE/EpRE) is located at the 5′ region of genes. The fate of this consensus region is to make an advance in xenobiotics resolving chemoprevention by certain fruits and vegetables [22].

So to regulate chemoprevention under cancer-induced stress in cells, the enzymes organize themselves in variable expression levels, as the induction was seen in detoxifying enzymes of phase II (GST and QR) and the elevation of glutathione (GSH) against ROS and chemical carcinogens [9].

13.6 THE UNIVERSALITY AND UNIQUENESS OF ORGANOSULFUR COMPOUNDS (OSC)

Broccoli, cauliflower, brussels sprouts, kale, garlic, and onion are rich sources of organosulfur compounds (OSCs). OSCs have sulfur atoms bound to a cyanate group or the same sulfur atom is bound to another carbon atom in a non-cyclic or cyclic double bond formation.

13.6.1 OSC AND ENZYMATIC INTERACTION

Once the cells of Brassicaceae plants are disrupted in a mechanical way of cutting, OSCs are released. There takes place the formation of isothiocyanates from glucosinolates by the enzymatic involvement of the myrosinase enzyme. Glucosinolates (GSLs) are water-soluble entities, with the hydroxylation of sulfate residues, which are sulfur-linked β-D-glucopyranose units, and the variable part is of an amino side chain (R). This variable side chain defines the kind or class of isothiocyanates [27]. The biological activation of ITC is brought about by diminishing of cell compartmentalization, which otherwise separates myrosin cell content, myrosinase, a thioglucohydrolase (E.C. 3.2.1.147), from GSLs placed in vacuoles of the plant cell. The decompartmentalization of cells may arise from the chewing on leaves by insects during food localization. Overall, in this process, the thioglucosidic bond is hydrolyzed [28].

Besides *Brassica* genus, the *Allium* genus possesses an organosulfur compound, such as allicin, chemically cleaved from the organosulfur metabolite alliin by virtue of the enzyme alliinase, in which the metabolite alliin has a very short half-life, making it unstable kinetically and converted into to diallyl disulfide, diallyl trisulfide, or diallyl sulfide. Interestingly and fortunately, all these end products have beneficial impacts on health; the major ones are anticancer and anti-inflammatory [29, 30].

Of the several OSCs, diallyl sulfides are the ones with the potency of devastating the cancerous cells. OSCs are not limited in their action by the presence of polyphenols and carotenoids. Rather, their activity of being anticancerous is enhanced in such a biochemical diverse environment. A synergistic kind of response is generated, which wards off reactive oxygen species (ROS) and thus the free radicals generated from the ROS. It has been reported that garlic juice extract mixed with water has shown keen potential in lowering down platelet aggregation and reduction in metastasis of the lung specifically [27, 31]. Broccoli, garlic, and onion are the best examples to be given for in this kind of warding off process, which lowers down the low-density lipoprotein and simultaneously increases high-density lipoprotein [32]. In a true sense, the concentration regimen for OSCs has been different as OSC extract of garlic at 200 mg/mL caused inflammation to the liver in rats, while doses below it did not show any beneficial effect on the organisms [33, 34].

13.6.2 MOLECULAR INSIGHTS OF BRASSICACEAE METABOLITES

Chronic disorders are associated with complex metabolic pathways whose intermediates are enacted upon by several phytochemicals halting the development of disease [35]. This multiverse association is universally found in humans, making them easy prey to heart ailments and stroke [36]. To escape this, humans have to incorporate dietary phytochemicals that are present in Brassicaceae and ultimately live healthy lives [35]. These food-derived bioactive chemicals are generated by different biosynthetic routes, and within such routes, the metastatic proliferation pathways are encountered, which are ultimately targeted by bioactive compounds, improving human health by saving cell's life from chronic metabolic disorders. The best-studied example is of interleukin-6 (IL-6), released by macrophages and T-cells upon infection by reverse transcribing viruses in order to reduce inflammation-mediated metastasis in humans. The diverse health effects are attributed to OSCs for their anticancer, anti-inflammatory, age-enhancing effects and relieving oxidative stress [37].

The metabolic activities involved in oxidative stress are such that tissue injury takes place resulting in inflammation and thereby the release of cytokines activating the cascade pathways. In this context, the role of sulforaphane present in Brassicaceae comes into play, which suppresses the NF-κB by binding itself to thiol groups of NFκB—p65 and p50. This biochemical interaction serves as a biochemical marker to identify the kind of oxidative stress involved [38]. Beneficially speaking, OSCs' inclusion in the diet reduces pathological implications of oxidation reactions that are otherwise involved in chronic diseases like cancer.

13.7 AN EXAMPLE THE WORLD NEEDS TO LEARN

A disease-free healthy life can be lived only by ensuring optimal nutritional intake. To mark this statement in context to cruciferous vegetables, the government of Japan conducted a survey in 2011–2013, which highlighted the fact that people in Japan consume the cruciferous vegetables most (21% of all the vegetables) and live a less disease-prone life as a compared global level [39].

One thing is consumption and the other is the bioavailability of consumed palatable. For cruciferous vegetables, the bioavailability is determined by the analysis of metabolites and their conjugates that appear in the urine. S-allyl cysteine is one of the water-soluble OSCs that appear in urine in form of N-acetyl-s allyl cysteine, as its acetylation reaction is catalyzed by N-acetyltransferase. The bioavailability of cruciferous-derived OSCs is 103.0% in *Rattus norvegicus* and 87.2% in *Canis familaris* [40]. It is also a fact that the bioavailability of OSCs is a mere reflection of their types, depending on the kind of cruciferous vegetables. As the data suggested, sulforaphane (SFN), whose concentration is highest in broccoli only, is checked for its bioavailability by urinary metabolite glutathione excretion [41, 42] and not by S-allyl cysteine present in all cruciferous vegetables [43].

13.8 THE PATHWAY HINDRANCES AND OPEN DOORS

The downregulation of pro-inflammation-causing cytokines is governed by the SFN existing in significantly larger amounts in cabbage and broccoli, among all other cruciferous vegetables. The proportion of inactivation of genes related to pro-inflammatory and activation of anti-inflammatory cytokines is regulated in an inverse manner in presence of SFN. During downregulation of pro-inflammatory cytokines, these SFNs block the NF-κB or Nrf2 pathway. The transcription factor of NF-κB is ARE (autosomal repressing element), which represses p38 MAPK (mitogen-activated protein kinase), which in turn represses inflammation and finally reduces it [44], whereas COX-2 (cyclooxygenase-2) enzyme is regulated as molecular an on/off switch in the p38 MAPK pathway by the presence/absence of SFN [45]. As the complex gene action is related to inflammation response, a number of other genes and proteins have been characterized that invoke inflammation in tissue. Examples are tumor necrosis factor (TNF), MCP-1 (mono chemoattractant protein-1), and vascular cell adhesion molecule, all involved in the p39 MAPK pathway [15, 46]. The ability of manipulation exists in very large proteins and its complexity; for example, the dystrophin-deficient mdx mouse model has revealed the fact that SFN dose administered directly leads to downregulation of mRNA of NF-kB (p65), which in turn leads to the diminished release of pro-inflammatory cytokines from immune cells [47].

A simple scholastic view can be inferred from these proven facts: OSC, although dose-dependent, shares strong communications of molecular interaction with diminishing inflammation.

13.8.1 INTEGRATORS OF METABOLIC PATHWAY

Isothiocyanates (ITC) arrive at their molecular existence through enzymatic degradation of glucosinolates (GSL), which Brassicaceae plants secrete through their metabolic routes, and lead to the diminution of metastasis-like properties by virtue of inducing the cell skeleton dynamic state maintenance genes. In such a dynamic state, the molecular mechanisms include very long cascade pathways of apoptosis, activation and deactivation, cell cycle blockages, and cessation of angiogenesis. The energy hub of the cell, mitochondria, makes sure the release of MAPK signaling cascades to the cytoplasm, with Bcl-2 family-related proteins as activators of numerous transcription factors and cleavage of caspases in order to activate them [47].

Regulation of apoptosis by ITC is accomplished primarily through the mitochondrial release of cytochrome C, Bcl-2 family regulation, and subsequent activation of caspases. To bring all these kinds of events in action, cytochrome P450 senses ITC and stops the carcinogens from following the metabolic route to metastasis by the body. All the travel done by all molecules involved into action halts to a common point of action, such as the upregulation of phase II detoxifying genes, the majority of which are enzymes [48]. The protection from cancer can be achieved by not coming into contact with carcinogens or, if they are entered, making the diet rich with those nutritionally beneficial chemicals that stop the process of metastasis.

13.8.1.1 Direct and Indirect Interaction

Phytoalexins and phytoanticipins are key secondary metabolites of non-resourceful origin to the Brassicaceae family that protect the plants themselves. Besides these, glucosinolates (GSL) mark their presence limited to only 15 botanical families of the Capparales order but are abundant in Brassicaceae family [44]. The inclusion of GSL-containing cruciferous vegetables in farming, along with staple crops like rice, maize, and wheat, controls or delimits the pathogens in the soil on which the plants are planted. The result is biofumigation, free of synthetic fumigation and also free of carcinogenic hazards of chemical fumigation [45]. Thus, we should not rely only on the direct benefits of cruciferous vegetables as their consumption but use them as a biofumigants too.

The biological reaction existing behind the breakdown of GSL is done by myrosinase, an enzyme that breaks it down into toxic compounds, but as the reaction is hydrolytic, the involvement of water molecules is there [46]. The biological approach is based on the release of toxic compounds derived from GSL with the destruction of tissue in the presence of water. The task of biofumigation is assigned naturally to GSLs, which ooze out from tissues by mechanical damage to tissue themselves [49]. The origin of the anticancer compound ITC from GSL also produces oxazolidones, thiocyanates, nitrites, and epithionitriles [10]. The characteristic unique odor and taste of Brassicaceae is also due to ITC presence and has anticancer properties in different metabolic pathways [50, 51]. Not only cruciferous vegetable secondary metabolites halt the growth of cancer cells, but they have proven to be very effective in diminishing the growth of pathogenic bacteria and fungi, as well as in removing unwanted weeds and insects [52]. To further extend the beneficial of natural ITC derived from cruciferous vegetables, erucin and SFN are the ones for whom the inhibition of virulence has been achieved against *Pseudomonas aeruginosa*, [53] an oil-degrading microbe, to stop the reaction of oil degradation if spilled in the sea.

Apart from all these qualities, the major noted properties of cruciferous vegetables are proven to be anticancer [54]. This is because GSLs are majorly involved in the regulation of oxidative stress [55]. Alkylation, thiolation, and mitochondrial damage resulting in the elevation of levels of ROS (reactive oxygen species) are events that increase metabolic oxidative stress, and to counteract such stress, GSL-derived ITC acts as natural antioxidants by stimulating many antioxidant enzymes and proteins of non-enzymatic nature [51].

13.8.1.2 The Concentration Regime

Anticarcinogenic activity is based upon the concentration regime of phytochemicals present in *Brassica*. At low concentrations, there is upregulation in cell skeleton maintenance genes and downregulation in cytochrome P450 genes. As the higher concentration of ITC is reached, the cell cycle arrest and apoptosis take place in the host cell's microenvironment [49]. The relation between dose-dependent manners of effect displayed over the ITC concentration proved the dynamic state of maintenance of the cell's microenvironment in response to oxidative stress induction by chemo-preventive methods [51].

The process of extrinsic apoptosis mediated by cell signaling factors is enacted upon by media-tors of signal transduction. In the case of *Brassica*, these are allyl isothiocyanates (AITC) and phenyl isothiocyanates (PITC) [**Figure 13.8**], which inhibit a protein TNF essential to extrinsic apoptosis [56]. The presence of benzylisothiocyanate (BITC) in cruciferous vegetables like broc-coli involves the energy hub of the cell's cytochrome oxidase c release [57], which in turn activates intrinsic apoptosis [58]. Comparative experimental shreds of evidence of ITCs have revealed the fact that AITC and SFN are less potent signaling factors for the activation of apoptotic cascade mechanisms than PITC and BITC [59].

13.8.1.3 Integration of Collective Data

All the collective data indicates the experiment-generated fact that carcinogens are directly metabo-lized to inactivation by phase I enzymes. This metabolic response leads to a cascade involving phase II enzymes, which are partly involved in doing detoxification and providing cytoskeleton defense against ROS. Here comes the role of antioxidant responsive element (ARE), which is located upstream of the transcription factor, Nrf2 [23]. The responsiveness of phase I and phase II enzymes depends on their intracellular concentrations [60] and the generation of ROS [61].

An enzymatic framework fits better with a molecular system when the substrate and enzyme concentration fits optimally. As the glutathione-S-transferase metabolizes ITC, at the same time the mere presence of ITC does not make it available to be anticarcinogenic [60]. Additional evidence suggested that ITCs induced directly phase II enzymes, which in turn suppressed the maturation of carcinogenic cells by inducing apoptotic events in the cell [62]. The stage of metastasis from where it could be reverted is accelerated by BITC, thus counteracting carcinogenesis [61].

In the earlier instances where ARE was shown to direct ITC expression, the molecular targets are heme oxygenase 1 (HO1), NAD (P) H: quinine oxidoreductase (NQO1) and gamma-glutamyl cysteine synthetase (γ-GCS). This gene expression comes into action when phase II enzymes are induced by SFN [63]. The final framework raised is detoxification of carcinogens and removal of ROS by nuclear factor Nrf2 (nuclear factor E2-related factor), a member of the NF-E2 family. Nrf2 binding takes place at ARE upstream elements in a cis-activation fashion [64].

Another pathway for induction of phase II enzymes is mediated by ARE/EpRE, which is also located at upstream region, but its activation is mediated in a trans fashion. The cis and trans fashion binding studies of Nrf2 and EpRE, respectively, have demonstrated that PITC, AITC, BITC, and SFN elevate the levels of γ-GCS and numerous cytokine factors [10].

13.9 THE CHEMISTRY OF INDOLE-3-CARBINOL IN THE HUMAN BODY

A wide perspective of cohort and case studies has been drafted in breast cancer, with the conclusion that consuming cruciferous vegetables in a significant amount slows down the metabolic routes of attaining metastasis [65].

Indole-3-carbinol engineers itself within the metabolism of the breakdown of the glucosinolate named glucobrassicin [**Figure 13.7**]. The glucobrassicin is unique to cruciferous plants only. With a single glucose unit, a single sulfur atom, a single sulfate group, and a group of aglucon when attached in a manner of minimum steric hindrance of molecules, the existence of glucosinolates comes into action. By taking up variable chemical groups in different numbers and modes of attach-ment, the glucosinolates reach out to their diversity in around 150. The chemical activity varies with respect to the change in aglucon moieties and sulfate groups. Of these, aglucon moieties exist in any

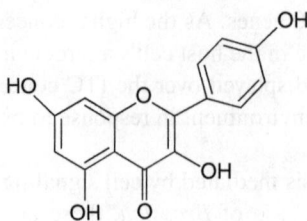

FIGURE 13.7 Kaempferol.

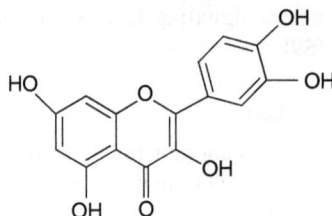

FIGURE 13.8 Quercetin.

of the bonding patterns, like alkenyl, alkyl, and indolyl structures. This bonding pattern is random and fits true to its diversity in producing the diverse physiological effects in glucosinolates [49].

For every key possible natural chemical molecule to form, a series of steps are involved. Likewise, for indole-3-carbinol (I3C) to form, first glucose is detached, leading to the formation of an unstable intermediate. Then from this unstable intermediate, the sulfur cyanate is detached, leading to formation of an indole ring of indole-3-carbinol. The key difference between aliphatic isothiocyanates like SFN and I3C is presence of indole ring in I3C. Cruciferous vegetables like kale, broccoli, and cabbage account for a major dietary source of I3C for humans. The maximum levels of I3C are attained in seed by the plant [66],

13.9.1 THE IMPORTANT CONDENSATE

By the time I3C in dietary form reaches the stomach, the gastric juice, due to its pH value of 2.2, acidifies I3C by dimerizing it to 3,3′-diindolylmethane (DIM). This dimerized condensate holds account for the biological activity of I3C, such as its anticarcinogenic potential. Current molecular techniques utilize I3C and DIM against cancerous metabolic routes. A connecting link has been established between the old orthodox medicinal system and new experimentally enriched proven anticancer therapies where mustard is used to slow down the pace of cancer and inflammation. CD-1 mice were given an oral dose of radiolabeled DIM, the trace of which was tracked by its distribution in the lung, liver, heart, and kidney. This proved their tissue localization in these important organs [67].

13.9.2 DIVERSITY WITHIN I3C

In humans the I3C dose of 700 mg was traced, the condensed DIM in serum with profound blockages created in carcinogens metabolic routes [68]. However, the I3C supplied was in purified form. The natural intake of such a higher amount of I3C will require ingestion of 0.6 kg sprout material, which, if taken, would lead to laxative effect with high chances of toxicity. Therefore, chemically synthesized I3C was used for anticancer studies. From such an instance, it is apparent that I3C has potential role in halting metastasis with antioxidant, anti-inflammatory, and detoxification effects **[69, 70]**.

The most important prerequisite that matter for cancer therapy is the natural killing of cancerous cells **[71]**. This task is accomplished by I3C/DIM, which interferes with signaling cascades at many intermetabolite levels. DIM condensates were found to have induced the apoptosis of cancerous cells from target tissues of the heart, liver, and kidney **[72, 73]**. The enhancement of effectiveness of a drug is also attributed to I3C/DIM, or the I3C/DIM carries out effective chemosensitization **[74]**—namely, with doxorubicin, cisplatin, and paclitaxel (isolated from *Taxus baccata*) **[75]**.

13.9.3 CHEMOSENSITIZATION'S COMPANIONSHIP

The programmed cell death of metastatic breast tissue cells is known to be accelerated by the *Bcl2* gene inhibition present in NF-kB pathway. DIM condensates alter the promoter elements of this gene and overexpress it. The chemosensitization effect of I3C/DIM has been widely studied and demonstrated in the case of breast cancer therapeutic drugs—Taxol and Taxotere **[76, 77]**. The same level of chemosensitization has been noticed in drugs involved in halting the metastatic stage—gemcitabine, oxaliplatin, cisplatin, doxorubicin, vincristine, and vinblastine **[78, 79]**. Herceptin, a monoclonal antibody used as a drug to treat stomach cancer; Tamoxifen, an estrogen receptor modulator used as drug to treat breast cancer; and Bortezomib, used to treat mantle cell melanoma, are all regulated by chemosensitization modulated by I3C/DIM in action **[80, 81]**.

The breakthroughs in chemosensitization mechanisms have provided a promising approach to combination therapy for which, I3C and DIM are ideal candidates. Not only are these benefits acquainted with I3C and DIM, but right from the progression of cancer, they are involved in a way of hindering the conversion of pro-carcinogen to the carcinogen. Even, in the case of drug-metabolizing machinery, cyp450 enzymes face competitive inhibition from I3C and DIM **[66]**.

Division in cells is inflicted by an increase in mass and number. This is also a hallmark property of the metastatic cell. Revolving around the repetitive phases of the cell cycle (G1, S, G2, and M) for the division to occur protein kinases cyclins and cyclin-dependent kinases (CDK) are involved. But the presence of I3C/DIM metabolite takes over the control of cell cycle as they inhibit CDK2, cyclin D&E1 **[82]** while transcriptionally regulating the cell division inhibitory factors like (phosphorylated) p21, p27, and KIP 1 (Kinase Inhibitory Protein 1) in a way **[83]** that they directly downregulate the growth of tumor. These instances lie at the same conclusion that cell division is terminated, G1 phase is interrupted, chromatin duplication comes to a halt, and mitosis is inactivated when I3C/DIM controls cell cycle **[73]**. Even the nuclear factor kappa B (NF-κB) is inhibited by I3C/DIM. Reducing the mitochondrial membrane potential by regulating the p53 gene and induction of apoptosis are yet other valuable functions of I3C/DIM **[75]**.

Besides the carcinogens' diversity, there are some tumor-promoter agents whose mere presence proliferates the tumor to such an extent that the cell bursts. One such is 16-alpha hydroxyl estrone, a conversion product of estradiol hormone. The I3C/DIM interaction with estradiol results in negligible levels of 16-alpha hydroxyl estrone production, leading to less risk of breast cancer **[84]**. Higher levels of 2-hydroxyestrone and lower levels of 16-alpha-hydroxyestrone are indicators of reduced risk of breast cancer in women. This metabolite-to-metabolite ratio of estrogen, a determining figure in metastatic event, is maintained in opposition to cancer by I3C/DIM **[85]**. Another approach that was used earlier to lower the risk of breast cancer originating through estrogen metabolism is anti-estrogen drug tamoxifen. Here is the interesting fact that was found later: I3C/DIM has a similar metabolic mechanism to lower estrogen level with more effectiveness. I3C/DIM lowers estrogen levels down to 90%, compared to 60% with tamoxifen **[86]**.

The use of combination therapy involving tamoxifen and I3C on MCF-7, a breast cancer cell line, inhibits estrogen receptors and finally less estrogen production. About 50% estrogen receptors were blocked by I3C alone, whereas tamoxifen has shown no effect on p21 **[87]**.

Women's Healthy Eating and Living (WHEL) was a trial study conducted with the participation of more than 3,000 breast cancer patients to generate the proof after molecular insights were achieved by combination therapy. The reduction in estrogen receptors was noticed when cruciferous plants were included in the diet along with the tamoxifen dose [88].

13.9.4 Mode of Action by I3C/DIM

Principles of the I3C action include protection from deleterious effect of epigenesis, scavenging of free radicals, detoxification, DNA repair mechanisms, inhibition of hormone-dependent metastasis, tumor growth inhibition of cell division (cell cycle arrest), inhibition of neo-angiogenesis, reducing the metastasis by induction of apoptosis, and increase of chemosensitivity. The modes of action employed by I3C/DIM are deactivation of oncogenes, activation of tumor suppressors by acetylation, breakdown in conversion of pro-carcinogens to carcinogens [89], activation of detoxification systems, compensation for DNA damages, suppression of hormone intake in the cell and transport into the cell nucleus, no activation of hormone-dependent genes, inhibition of protein kinase, induction of inhibitors of kinases, halting of regular organization of tubulin and cell division, reduced organization of epithelial cells and formation of capillaries, inactivation of growth factors, mitochondrial membrane depolarization, cytochrome C release, and invalidation of metastatic cellular mechanisms [90].

13.10 CONCLUSION

The metastasis progression involves complex molecular machinery—namely, DNA, RNA, and proteins. Moreover, a number of cascade mechanisms accelerate the carcinogenesis. The Brassicaceae family possesses the kind of chemicals that stops the progression of metastasis at earlier stages. In earlier times, when *Brassica* phytochemicals were not analyzed for their functional attributes, even then the cruciferous vegetables were consumed to get rid of ailments like indigestion, constipation, and allergies. By the time different chemical compounds of *Brassica* came into light by virtue of modern screening technologies, anticancer properties were tested and proven to be true in these chemical compounds. The notable ones are I3C/DIM, ITC, BITC, PITC, kaempferol, and diallyl disulfide. These *Brassica*-derived phytochemicals have been tested in various cancer cell lines, which proved effective in controlling cancer by modulating the cell's genetic machinery. Not only do they interfere in the carcinomas cell cycle but also mask that gene action which otherwise helps in cancer development. The activation of tumor suppressor genes and inactivation of tumor proliferation genes are the events triggered in presence of *Brassica* phytochemicals. Not only do they protect the DNA from damage by carcinogens but also maintain the cell's cytoskeleton to modulate all the biological activities in a normal manner. The Japanese population survey in 2011–2013 and the WHEL trial are the classical illustrations of cruciferous vegetables' anticarcinogenic effect. The World Cancer Research Fund has also reflected the beneficial role of *Brassica* in the prevention of cancer, which makes its cultivation essential and consumption useful.

13.11 REFERENCES

1. Özüdoǧru, B., Akaydın, G., Erik, S., Al-Shehbaz, I.A., and Mummenho, K. 2015. Phylogeny, diversification and biogeographic implications of the eastern Mediterranean genus Ricotia (Brassicaceae). *Taxon.* 64(04): 727–740.
2. Chishti, Hakim. 1988. *The Traditional Healer's Handbook*, Vermont: Healing Arts Press, ISBN 978-0-89281-438-1.
3. Lee, C. S., Baek, J., and Han, S.Y. 2017. The role of kinase modulators in cellular senescence for use in cancer treatment. *Molecule.* 22(09): E1411.
4. Fridlender, M., Kapulnik, Y., and Koltai, H. 2015. Plant derived substances with anti-cancer activity: From folklore to practice. *Frontiers in Plant Science.* 6: 799.

5. Stewart, A.V. 2002. A review of Brassica species, cross pollination and implications for pure seed production in New Zealand. *Agronomy*.19: 12–24.

6. Mithen, R.F., Dekker, M., Verkerk, R., Rabot, S., and Johnson, I.T. 2000. The nutritional significance, biosynthesis and bioavailability of glucosinolates in human foods. *Journal of the Science of Food and Agriculture*. 80: 967–984.

7. Holst, B., and Williamson, G. 2004. A critical review of the bioavailability of glucosinolates and related compounds. *Natural Products Report*. 21(03): 425–447.

8. Punetha, H., Chandra, M., Bhutia, S., Triphati, S., and Praksh, O. 2015. Antioxidative properties and mineral composition of defatted meal of oileferous Brassica germplasm. *Journal of Biologically Active Products from Nature*. 5(1): 43–51.

9. Punetha, H., Kumar, S, Mudila, H., and Prakash, O. 2018. Brassica meal as source of health protecting neutraceutical and its antioxidative properties. *IGI Global,* 1: 108–131.

10. Ye, L., and Zhang, Y. 2001. Total intracellular accumulation levels of dietary isothiocyanates determine their activity in elevation of cellular glutathione and induction of Phase 2 detoxification enzymes. *Carcinogenesis*, 22(12): 1987–1992.

11. Yang, J., and Liu, R.H. 2009. Induction of phase II enzyme, quinine reductase, in murine hepatoma cells in vitro by grape extracts and selected phytochemicals. *Food Chemistry*, 114(03): 898–904.

12. Özüdoˇgru, B., Akaydın, G., Erik, S., Al-Shehbaz, I.A., and Mummenho, K. 2015. Phylogeny, diversification and biogeographic implications of the eastern Mediterranean endemic genus Ricotia (Brassicaceae). *Taxon*. 64(04): 727–740.

13. Xiao, D., Srivastava, S. K., Lew, K. L., Zeng, Y., Hershberger, P. and Johnson, C. S. 2003. Allyl isothiocyanate, a constituent of cruciferous vegetables, Inhibits proliferation of human prostate cancer cells by causing G2/M arrest and inducing apoptosis. *Carcinogenesis*. 24: 891–897.

14. Xu, C., Li, Y.-T., and Kong, A.-N. 2005. Induction of phase I, II and III drug metabolism/transport by xenobiotics. *Archives of Pharmacal Research*. 28(3): 249–268.

15. Xu, W., Yan, M., Lu, L., Sun, L., Theze, J., and Zheng, Z. 2001. The p38 MAPK pathway is involved in the IL-2 induction of TNF-beta gene via the EBS element. *Biochemical and Biophysical Research Communications*. 289(5): 979–986.

16. Ling, H., Zhang, L.Y., and Su, Q. 2006. Erk is involved in the differentiation induced by diallyl disulfide in the human gastric cancer cell line MGC803. *Cell &Molecular Biology Letters*. 11: 408–423.

17. Ling, H., Wen, L., and Ji, X.X. 2010. Growth inhibitory effect and Chk1-dependent signalinginvolved in G2/M arrest on human gastric cancer cells induced by diallyldisulfide. *Brazilian Journal of Medical and Biological Research*. 43: 271–278

18. Hedges, L.J., and Lister, C.E. 2006. Nutritional attributes of Brassica vegetables. *Crop and Food Research*. Report no 1618.

19. Al-Shehbaz, I.A. 2012. A generic and tribal synopsis of the Brassicaceae (Cruciferae). *Taxon*. 61: 931–954.

20. Kang, L., Qian, L., and Zheng, M. 2021. Genomic insights into the origin, domestication and diversification of *Brassica juncea*. *Nature Genetics*. 53: 1392–1402.

21. Franzke, A., German, D., Al-Shehbaz, I.A., and Mummenho, K. 2008. Arabidopsis family ties: Molecular phylogeny and age estimates in the Brassicaceae. *Taxon*. 58(2): 425–437.

22. Kong, A.N., Yu R., Hebbar, V., Chen, C., Owuor, E., Hu, R., Ee, R., and Mandlekar, S. 2001. Signal transduction events elicited by cancer prevention compounds. *Mutation Research*. 480–481: 231–241.

23. Keum, Y.S., Jeong, W.S., and Kong, A.N.T. 2004. Chemoprevention by isothiocyanates and their underlying molecular signaling mechanisms. *Mutation Research*. 555(1–2): 191–202.

24. Fimognari, C., and Hrelia, P. 2007. Sulforaphane as a promising molecule for fighting cancer. *Mutation Research*. 635(2–3): 90–104.

25. Zhang, M., An, C., Gao, Y., Leak, R. K., Chen, J., and Zhang, F. 2013. Emerging roles of Nrf2 and phase II antioxidant enzymes in neuroprotection. *Progress in Neurobiology*. 100: 30–47.

26. Belgiorno, V., Rizzo, L., Fatta, D., Rocca, C.D., Lofrano, G., Nikolaou, A., Naddeo, and Meric, S. 2007. Review on endocrine disrupting-emerging compounds in urban wastewater: Occurrence and removal by photocatalysis and ultrasonic irradiation for wastewater reuse. *Desalination*. 215(1–3): 166–176.

27. Srividya, A.R., Nagasamy, V., and Vishnuvarthan, V.J. 2010. Nutraceutical as medicine: A review. *Pharmanest*. 1(2): 132–145.

28. Fenwick, G.R., Heaney, R.K., and Mullin, W.J. 1983. Glucosinolates and their breakdown products in food and food plants. *Critical Reviews in Food Science and Nutrition*.18(02): 123–201.

29. Swanson, J.E. 2003. Bioactive food components. 2001. In: Wildman, R.E.C. (Ed.), *Encyclopedia of Food Culture*, first ed. Boca Raton: Handbook Nutraceuticals Functional Foods, The Gale Group Incorporation. ISBN: 0-8493-8734-5. CRC Series.

30. Abuajah, C.I., Ogbonna, A.C., and Osuji, C.M. 2015. Functional components and medicinal properties of food: A review. *Journal of Food Science and Technology*. 52(5): 2522–2529.

31. Crandell, K., and Duren, S. 2007. Nutraceuticals: What are they and they work? *Journal of Biotechnology*. 34(3): 29–36.

32. Dureja, H.D., and Kaushik, K.V. 2003. Development of nutraceuticals. *Indian Journal of Pharmacology*. 35(06): 363–372.

33. Egen-Schwind, C., Eckard, R., and Kemper, F. 2008. Metabolism of garlic constituents in the isolated perfused rat liver. *Planta Medica*. 58(04): 301–305.

34. Banerjee, S.K., Maulik, M., Mancahanda, S.C., Dinda, A.K., Gupta, S.K., and Maulik, S.K. 2002. Dose-dependent induction of endogenous antioxidants in rat heart by chronic administration of garlic. *Life Sciences*. 70(30):1509–1518.

35. Craig, W.J. 1997. Phytochemicals: Guardians of our health. *Journal of American Dietic Association*. 97(10): 199–204.

36. WHO. 2018. *The Top 10 Causes of Death*. Available online at: www.who.int/news-room/fact-sheets/detail/the-top-10-causes-of-death (accessed December 02, 2020).

37. Ouchi, N., Parker, J.L., Lugus, J.J., and Walsh, K. 2011. Adipokines in inflammation and metabolic disease. *Nature Reviews Immunology*. 11(02):85–97.

38. Ruhee, R.T., Roberts, L.A., Ma, S., and Suzuki, K. 2020. Organosulfur compounds: A review of their anti-inflammatory effects in human health. *Frontiers in Nutrition*. 7: 64.

39. What Vegetable is Most Consumed in Japan? 2014. Available online at: http://nbakki.hatenablog.com/entry/2014/03/31/124219 (accessed March 20, 2019).

40. Nagae, S., Ushijima, M., Hatono, S., Imai, J., Kasuga, S., and Matsuura, H. 1994. Pharmacokinetics of the garlic compound S-allylcysteine. *Planta Medica*. 60(03): 214–217.

41. Kassahun, K., Davis, M., Hu, P., Martin, B., and Baillie, T. 1997. Biotransformation of the naturally occurring isothiocyanate sulforaphane in the rat: Identification of phase I metabolites and glutathione conjugates. *Chemical Research in Toxicology*. 10(11): 1228–1233.

42. Egner, P.A., Chen, J.G., Wang, J.B., Wu, Y., Sun, Y., and Lu, J.H. 2011. Bioavailability of Sulforaphane from two broccoli sprout beverages: Results of a short term, cross-over clinical trial in Qidong, China. *Cancer Prevention Research*. 4(3): 384–395.

43. Shuruq, A., Noura, A., Ghada, A., Alammari, G., Kavita, M.S., and Maha, A.T. 2017. Role of phytochemicals in health and nutrition. *BAOJ Nutrition*. 3: 28.

44. Fridlender, M., Kapulnik, Y., and Koltai, H. 2015. Plant derived substances with anti-cancer activity: From folklore to practice. *Frontiers in Plant Science*. 6: 799.

45. Galletti, S., E. Sala, O. Leoni, S. Cinti, and C. Cerato. 2008. *Trichoderma* spp. tolerance to *Brassica carinata* seed meal for a combined use in biofumigation. *Biological Control*. 45(3): 319–327.

46. Sun, C.C., Li, S.J., Yang, C.L., Xue, R.L., Xi, Y.Y., and Wang, L. 2015. Sulforaphane attenuates muscle inflammation in dystrophin-deficient Mdx mice via Nrf2-mediated inhibition of NF-kB signaling pathway. *Journal of Biological Chemistry*. 290: 17784–17795.

47. Grubb, C.D., and Abel, S. 2006. Glucosinolate metabolism and its control. *Trends in Plant Science*, 11(2): 89–100.

48. Haefner, B. 2002. NF-κB: Arresting a major culprit in cancer. *Drug Discovery Today*. 7(12): 653–663.

49. Hayes, J.D., Kelleher, M.O., and Eggleston, I.M. 2008. The cancer chemopreventive actions of phytochemicals derived from glucosinolates. *European Journal of Nutrition*. 47(2): 73–88.

50. Pasini, F.V., Verardo, M.F., Caboni and D.'Antuono, L.F. 2012. Determination of glucosinolates and phenolic compounds in rocket salad by HPLC-DAD–MS: Evaluation of *Eruca sativa* Mill and *Diplotaxis tenuifolia* L. genetic resources. *FoodChemistry*. 133: 1025–1033.

51. Zhang, Y., Li, J., and Tang, L. 2005. Cancer-preventive isothiocyanates: Dichotomous modulators of oxidative stress. *Free Radical Biology and Medicine*. 38(1): 70–77.

52. Vig, A.P., Rampal, L., Thind, T.S., and Arora, S. 2009. Bio-protective effects of glucosinolates—A review. LWT—*Journal of Food Science and Technology*. 42: 1561–1572

53. Ganin, H., Rayo, J., Amara, N., Levy, N., Krief, P., and Meijler, M. 2013. Sulforaphane and erucin, natural isothiocyanates from broccoli, inhibit bacterial quorum sensing. *Medicinal Chemistry Communications*. 4: 175–179.

54. Blazevic, I., and Mastelic, J. 2009. Glucosinolate degradation products and other bound and free volatiles in the leaves and roots of radish (*Raphanus sativus* L.). *Food Chemistry*. 113: 96–102.

55. Fimognari, C., Nusse, M., Cesari, R., Iori, R., Cantelli-Forti, G., and Hrelia, P. 2002. Growth inhibition, cell-cycle arrest and apoptosis in human T-cell leukemia by the isothiocyanate sulforaphane. *Carcinogenesis*. 23: 581–586.

56. Thejass, P., and Kuttan, G. 2007. Allyl isothiocyanate (AITC) and phenyl isothiocyanate (PITC) inhibit tumor-specific angiogenesis by downregulating nitric oxide (NO) and tumor necrosis factor (TNF) production. *Nitric Oxide.* 16: 247–257.

57. Shawn, L., Straszewski-Chavez, V., Abrahams, M., and Mor, G. 2005. The role of apoptosis in the regulation of trophoblast survival and differentiation during pregnancy. *Endocrine Reviews.* 26(7): 877–897.

58. Miyoshi, N., Uchida, K., Osawa, T., and Nakamura, Y. 2004. A link between benzyl isothiocy-anate-induced cell cycle arrest and apoptosis: Involvement of mitogen-activated protein kinases in the Bcl-2 phosphorylation. *Cancer Research.* 64(21): 34–42.

59. Tang, L., and Zhang, Y. 2005. Mitochondria are the primary target in isothiocyanate induced apoptosis in human bladder cancer cells. *Molecular Cancer Therapeutics.* 5: 1250–1259.

60. Zhang, Y., and Talalay, P. 1998. Mechanism of differential potencies of isothiocyanates as inducers of anticarcinogenic phase 2 enzymes. *Cancer Research.* 58(46): 32–39.

61. Seow, A., Jian-Min, Y., Can-Lan, S., Den Berg, V., Hin-Peng, L., and Mimi, Y. 2002. Dietary isothio-cyanates, glutathione S-transferase polymorphisms and colorectal cancer risk in the Singapore Chinese Health Study. *Carcinogenesis.* 23(12): 2055–2061.

62. Bonnesen, C., Eggleston, I.M., and Hayes, J.D. 2001. Dietary indoles and isothiocyanates that are generated from cruciferous vegetables can both stimulate apoptosis and confer protection against DNA damage in human colon cell lines. *Cancer Research.* 61: 6120–6130.

63. Clarke, J.D., Dashwood, R.H., and Ho, E. 2008. Multi-targeted prevention of cancer by sulforaphane. *Cancer Letters.* 269: 291–304.

64. Yeh, C.T., and Yen, G. C. 2009. Chemopreventive functions of sulforaphane: A potent inducer of antioxidant enzymes and apoptosis. *Journal of Functional Foods.* 1: 23–32.

65. Ambrosone, C.B., McCann, S.E., and Freudenheim, J.L. 2004. Breast cancer risk in premenopausal women is inversely associated with consumption of broccoli, a source of isothiocyanates, but is not modified by GST genotype. *Journal of Nutrition.* 134: 1134–1138.

66. Wortelboer, H.M., de Kruif, C.A., and van Iersel, A.A. 1992. Acid reaction products of indole-3- carbinol and their effects on cytochrome P450 and phase II enzymes in rat and monkey hepatocytes. *Biochemical Pharmacology.* 43: 1439–1447.

67. Anderton, M.J., Manson, M.M., and Verschoyle R. 2004. Physiological modeling of formulated and crystalline 3,3-diindolylmethane pharmacokinetics following oral administration in mice. *Drug Metabolism and Disposition.* 32: 632–638.

68. Reed, G.A., Arneson, D.W., and Putman, W.C. 2006. Single-dose and multiple-dose administration of indole-3-carbinol to women: Pharmacokinetics based on 3,3′-diindolylmethane. *Cancer Epidemiology, Biomarkers & Prevention.* 15: 2477 2481.

69. Haefner, B. 2002. NF-κB: Arresting a major culprit in cancer. *Drug Discovery Today.* 7(12): 653–663.

70. Karin, M., and Greten, F. R. 2005. NF-κB: Linking inflammation and immunity to cancer development and progression. *Nature Reviews Immunology.* 5(10): 749–759.

71. Karin, M. 2006. NF-κB in cancer development and progression. *Nature.* 441: 431–436.

72. Aggarwal, B.B., and Ichikawa, H. 2005. Molecular targets and anticancer potential of indole-3-carbinol and its derivatives. *Cell Cycle.* 4: 1201–1215.

73. Chinni, S.R., Li, Y., and Upadhyay, S. 2001. Indole- 3-carbinol (I3C) induced cell growth inhibition, G1 cell cycle arrest and apoptosis in prostate cancer cells. *Oncogene.* 20: 2927–2936.

74. Ahmad, A., Sakr, W.A., and Rahman, K.M. 2010. Anticancer properties of indole compounds: Mechanism of apoptosis induction and role in chemotherapy. *Current Drug Targets.* 11: 652–666.

75. Rahman, K. M., Ali, S., and Aboukameel, A. 2007. Inactivation of NF-kappaB by 3,3′-diindolylmethane contributes to increased apoptosis induced by chemotherapeutic agent in breast cancer cells. *Molecular Cancer Therapeutics.* 6: 2757–2765.

76. Ahmad A., Ali, S., and Wang, Z. 2011.3,3-diindolylmethane enhances taxotere-induced growth inhibition of breast cancer cells through down-regulation of FoxM1. *International Journal of Cancer.* 129: 1781–1791.

77. McGuire, K.P., Ngoubilly, N., and Neavyn, M, 2006.3,3-diindolylmethane and paclitaxel act synergistically to promote apoptosis in HER2/Neu human breast cancer cells. *Journal of Surgical Research.* 132: 208–213.

78. Christensen, J.G., and Le Blanc, G.A. 1996. Reversal of multidrug resistance in vivo by dietary administration of the phytochemical indole-3- carbinol. *Cancer Research.* 56: 574–581.

79. Wang, H., Word, B.R., and Lyn-Cook, B.D. 2011. Enhanced efficacy of gemcitabine by indole-3-carbinol in pancreatic cell lines: The role of human equilibrative nucleoside transporter 1. *Anticancer Research.* 31: 3171–3180.

80. Malejka-Giganti, D., Parkin, D.R., and Bennett, K.K. 2007. Suppression of mammary gland carcino-genesis by post-initiation treatment of rats with tamoxifen or indole-3-carbinol or their combination. *European Journal of Cancer Prevention*. 16: 130–141.

81. Taylor-Harding, B., Agadjanian, H., and Nassanian, H. 2012. Indole-3-carbinol synergistically sensitises ovarian cancer cells to bortezomib treatment. *British Journal of Cancer*. 106: 333–343.

82. Stewart, Z.A., Westfall, M.D., and Pietenpol, J.A. 2003. Cell cycle dysregulation and anticancer therapy. *Trends in Pharmacological Science*. 24: 139–145.

83. Gong, Y., Sohn, H., and Xue, L. 2006. 3,3-diindolylmethane is a novel mitochondrial H(+)-ATP syn-thase inhibitor that can induce p21 (Cip/ Waf1) expression by induction of oxidative stress in human breast cancer cells. *Cancer Research*.66: 4880–4887.

84. Fowke, J.H., Longcope, C., and Hebert, J.R. 2009. Brassica vegetable consumption shifts estrogen metabolism in healthy postmenopausal women. *Cancer Epidemiology & Biomarkers Prevention*. 9: 773–779.

85. Michnovicz, J.J., and Bradlow, H.L.1991. Altered estrogen metabolism and excretion in humans follow-ing consumption of indole-3-carbinol. *Nutrition and Cancer*. 16: 59–66.

86. Cover, C.M., Hsieh, S.J., Cram, E.J., Hong, C., Riby, J.E., and Bjeldanes, L.F. 1999. Indole-3-carbinol and tamoxifen cooperate to arrest the cell cycle of MCF-7 human breast cancer cells. *Cancer Research*. 59: 1244–1251.

87. Cover, C.M., Hsieh, S.J., and Tran, S.H., 1998. Indole-3- carbinol inhibits the expression of cyclindepen-dent kinase-6 and induces a G1 cell cycle arrest of human breast cancer cells independent of estrogen receptor signaling. *Journal of Biological Chemistry*. 273: 3838–3847.

88. Vivar, O., Lin, C.L., and Firestone G.L. 2009. 3,3- Diindolylmethane induces a G(1) arrest in human prostate cancer cells irrespective of androgen receptor and p53 status. *Biochemical Pharmacology*. 78: 469–476.

89. Mao, C.G., Tao, Z.Z., and Chen, Z. 2014. Indole-3- carbinol inhibits nasopharyngeal carcinoma cell growth in vivo and in vitro through inhibition of the PI3K/Akt pathway. *Experimental and Therapeutic Medicine*. 8: 207–212.

90. Cho, H.J., Seon, M.R, and Lee, Y.M. 2008. 3,3'-Diindolylmethane suppresses the inflammatory response to lipopolysaccharide in murine macrophages. *Journal of Nutrition*. 138: 17–23.

14 Opportunities with Nano-Formulations in Cancer Chemoprevention

Akansha Agrwal and Sheetal Mittal

CONTENTS

14.1 Introduction 243
14.2 Nano-Formulations for Cancer Chemoprevention: Opportunities and Challenges 245
14.3 Chemoprevention Opportunities Based on Nano-Formulations 246
    14.3.1 Carbon-Based Nano-Formulations for Drug Delivery and Cancer Treatment 246
        14.3.1.1 Carbon Nanotubes (CNTs) 247
        14.3.1.2 Graphene 247
    14.3.2 Polymeric Nano-Formulations for Cancer Therapy 247
    14.3.3 Liposome-Based Formulations for Cancer Prevention 248
    14.3.4 Lipid-Polymer Hybrid Nano-Formulations for Cancer Therapy 248
    14.3.5 Nutraceuticals and Phytochemical-Based Nano-Formulations
        for Cancer Treatment 249
    14.3.6 Dendrimer-Based Nano-Formulations for Cancer Chemoprevention 250
    14.3.7 Metal Nano-Formulations for Drug Delivery and Cancer Treatment 250
    14.3.8 Quantum-Dot-Based Nano-Formulations for Cancer Prevention 251
14.4 Conclusion and Future Prospects 251
14.5 References 252

14.1 INTRODUCTION

Cancer is a foremost community health issue that is the world's second largest cause of death. The worldwide problem of cancer remains to rise, with the number of cases anticipated to double by 2040 as compared to 2012 [1]. Cancer is a fatal disease triggered by both internal factors, like congenital mutations, biological messengers, and insusceptible situations, and ecological/acquired factors, like food, radioactivity, tobacco, and pathogens. It is uncontrolled tissue proliferation that can lead to invasion of other organs if it is not properly regulated or differentiated.

Currently, more than 30% of all cancer-related deaths arise in low- and middle-income nations due to environmental risk factors and poor health management (World Health Organization, 2020). The most common cancer types diagnosed (Figure 14.1) are lung (2.21 million, 22.21% of the total), breast (2.26 million, 22.6%), colorectal (1.93 million, 19.3%), prostate (1.41 million, 14.10%), skin (non-melanoma) (1.20 million, 12.0%), and stomach (1.09 million, 10.9%) (International Agency for Research on Cancer, 2020)[2].

In 2020, the major causes of cancer death were due to lung cancer, which caused 1.8 million deaths; colorectal cancer, which caused 916,000 deaths; liver cancer, which caused 830,000 deaths; stomach cancer, which caused 769,000 deaths; and breast cancer, which also caused 769,000 deaths[2].

Cancer grows once normal cells get converted into bump-like structures in a composite process, which typically develops after a premalignant abrasion to a hostile tumor. The whole variations

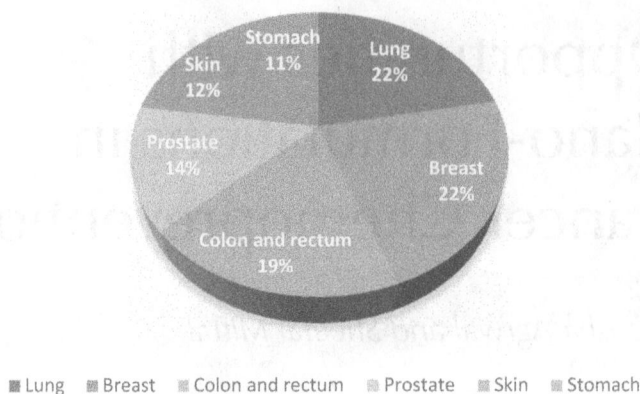

FIGURE 14.1 Statistics of cancer cases diagnosed in 2020, data released by World Health Organization on February 3, 2022.

are the outcome of an individual's hereditary characteristics cooperating through three sorts of peripheral stimuli: corporeal cancer-causing agents, such as UV and ionization-causing radiation; chemical cancer-causing agent, such as cigarette products, alcohol, food contaminants, and arsenic, which is a drinking water pollutant; and biological carcinogens, such as viruses, bacteria, and other parasites [3].

Usage of tobacco and alcohol, poor diet, sedentary lifestyle, and different types of pollutions are the risk factors for cancer. According to a recent study directed by researchers of International Agency for Research on Cancer (IARC), alcohol intake would be linked to an estimated 741,000 additional cancer cases worldwide in 2020. Although perilous and substantial drinking habits accounted for the major cancer problem, light to moderate drinking is reported in 1 in 7 alcohol-attributable cases and described for more than 100,000 new cancer cases globally, according to the latest data published in *The Lancet Oncology* [4, 5].

Traditional cancer therapy options include surgery, radiation, and chemotherapeutic medications, all of which cause harm to healthy cells and consequently cause toxicity in patients. Furthermore, most powerful anticancer medicines have low solubility in the biological milieu, reducing their pharmacokinetic qualities significantly [6]. As a result, not only has the research of new medications been prioritized but so, too, are the creation of unique therapeutic strategies. In this framework, nanoscale drug distribution schemes suggest noteworthy potential due to their varied behaviors for precise targeting to tumors and improved defense of drugs. They can grasp the target site, ensuring enhancements in biodistribution and pharmacokinetics of drugs in the complete flow, skillful release of the drugs over a longer period of time at anticipated amounts, co-delivery of several kinds of drugs, or indicative agents to fight multidrug resistance (MDR).

Nanotechnology is a division of science that works through particles ranging in size from a few nanometers (nm) to several hundred nanometers, depending on their intended application. Over the last decade, there has been a lot of interest in evolving precise drug delivery systems because they have a lot of advantages over traditional formulations. Since drugs can reach tissues at the molecular level, they hold great promise in both cancer diagnosis and treatment. Cancer nanotechnology is being eagerly examined and utilized in cancer treatment, representing a significant advancement in disease discovery, finding, and treatment [7, 8]. This chapter discusses the benefits, accomplishments, and applications of nano-formulations in cancer detection and treatment.

14.2   NANO-FORMULATIONS FOR CANCER CHEMOPREVENTION: OPPORTUNITIES AND CHALLENGES

Chemoprevention of cancer is described by means of the usage of ordinary substances to reduce, converse, or avoid the cancer-causing progression from continuing to violent cancer. Several natural food components with varied chemical assemblies, like tannins, curcumins, flavonoids, and poly-phenols, have been projected as chemopreventive mediators throughout the last two decades [9].

Nanotechnology-based products broadly refer to nano-formulations containing particles smaller than 100 nm and those smaller than 1,000 nm from a literature standpoint [10]. These drug carrier systems have superior properties in terms of absorption, targeting, and safety due to their nanoscale size. Nanotechnology has provided significant advantages to healthcare over time and has also addressed many of the flaws and obstacles that come with traditional therapy. Nano-formulations have become extremely popular in practically all scientific domains. Nanoscale materials have very different surface chemistry, reactivity, and other features than their macro/micro counterparts. As a result, they have applications in engineering, biological sciences, cosmetology, environmental sciences, and other fields [11]. Drug solubility is considerably improved by nano-formulations, such as nanocrystals, nanoparticles, nano-emulsions, and micellar encapsulation. Nano-formulation-based targeted therapy is a benefit to healthcare since it enhances overall efficacy and lowers unfavorable effects on normal cells [12]. Targeting can be done in two ways: passive or active (Figure 14.2).

Particle size and surface chemistry are crucial factors in passive targeting. Enhanced negative charge or hydrophobicity of nanocarriers can effectively target medicines to the reticuloendothelial system (RES), resulting in opsonization and enlarged permission by macrophages [13]. At the targeted region, active targeting uses a variety of overexpressed receptors, proteins, and enzymes. On the external of the nanocarriers that attach to the targeted cells, monoclonal antibodies or ligands are coated. Greater drug acceptance, enhanced passage time, enlarged drug competence, and less unfavorable effects on usual cells are all advantages of this method [14]. Drug delivery techniques based on nanoparticles have proven to be effective. Because of their ability to target, nanoparticles are the preferred medication carriers. Nanoparticles can both passively and actively target the active medicinal ingredient to the target region. Nano-formulations have shown to be beneficial in the treatment of a variety of life-threatening disorders. Nanocarriers have the potential to eliminate the limitations of traditional therapy, such as unfavorable effects on normal cells, excessive dose size and frequency, lower patient compliance, treatment length, non-selective targeting, lack of tailored

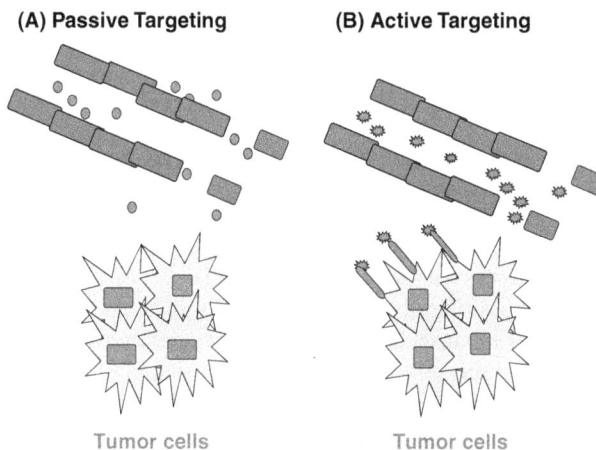

FIGURE 14.2   (a) Passive targeting is based on nanoparticle extravasation through a leaky tumor vasculature; (b) active targeting is based on surface modified nanoparticles.

medication, and so on [15]. Nanoparticles are also employed in biological tissue imaging (known as bioimaging). The importance of bioimaging lies in disease diagnosis and therapeutic monitoring. Nanoparticles can be utilized to identify diseases as serious as tumor at an initial phase. This use of nanoparticles can be attributed to their superior contrasting qualities. Nanoparticles are also favored as imaging agents since they have a longer retention time in vivo and can be tailored to target a specific region while also improving their imaging properties [16].

It is unusual to find something that has only great benefits and no drawbacks or undesirable side effects. Nano-formulations are not any different. They have had a significant and favorable impact on the health business in recent decades; nevertheless, despite their spectacular success, they still confront a few problems. The successful application of nano-formulations necessitates their large-scale synthesis in order to benefit a larger segment of the health business. Although various new synthesis technologies have been created, scaling them up remains a hurdle in the majority of cases. Scaling up necessitates highly skilled labor and costly chemicals. Running expenses rise dramatically, potentially affecting the nano-formulations' risk-to-benefit ratio [17]. The largest hurdle for scientists working on successful utilization of nanocarriers for better disease management with nano-formulations is the safety concerns related with nano-formulations. Nanoscale, while proving to be a blessing to the therapeutic sector, also offers a safety risk due to nonspecific interactions with bodily cells.

Nano-formulations are deserving of all the attention they have received in recent decades. There are many investigations are going on currently to figure out how to make the finest usage of active NPs. NPs have demonstrated their potential in the biomedical area across the board, from drug delivery to bioimaging. Nanoparticles have features that make them suitable for application in a variety of diseases, from their innate capacity to passively target to their ability to be desirable adjusted. Even severe diseases such as cancer have benefited from the use of numerous types of nanoparticles.

14.3 CHEMOPREVENTION OPPORTUNITIES BASED ON NANO-FORMULATIONS

Nanotechnology's application in cancer chemoprevention has demonstrated its ability to deliver medications in an additional active, harmless, and tailored manner. This field of study is progressing toward the creation of nanovaccines for cancer inhibition. Early detection strategies based on nanoplatforms accomplished at detecting premalignant indicators are also acquisition traction as a preventative intervention, with nano-formulations such as carbon-based and polymer-based nano-formulation, formulation of nucleic acids, lipid-polymer hybrid nano-formulations, nutraceutical-based nano-formulations, bio-inspired nano-formulations, phytochemical-based nano-formulations, metal nano-formulations, quantum-dot-based nano-formulations, and surface-decorated nano-formulations having been reported [18].

14.3.1 CARBON-BASED NANO-FORMULATIONS FOR DRUG DELIVERY AND CANCER TREATMENT

Since their exclusive structural dimensions and physicochemical qualities, carbon-based nanoparticles (CBNs) have piqued attention in a variety of fields [19]. CNTs, graphene, mesoporous carbon, and fullerenes are all types of carbon materials. Because of their high visual movement and huge multifunctional external expanse, these materials have demonstrated improved drug-loading size, biocompatibility, and immunogenicity [20]. The construction of biocompatible scaffolds and nanomedicines has been intensively investigated using functionalized CBNs. CBNs with certain moieties, such as functional groups, molecules, and polymers discovered to be useful for biomedical applications, have been chemically modified. Because of their outstanding supramolecular stacking, high adsorption capacity, and photothermal conversion ability, most of the materials under CBNs have been intensively researched for cancer therapy [21].

14.3.1.1 Carbon Nanotubes (CNTs)

Carbon nanotubes (CNTs) are one of the most studied carbon-related nanostructures and have gotten a lot of attention. CNTs are not only which can be used as eccentric drug trailers, but they also play a vital role in phototherapy-built cancer handling due to their distinct physicochemical qualities and exclusive construction, such as huge shallow areas, amusing surface alteration potentials, high drug load ability, and outstanding effects. The characteristic boundaries of CNTs, such as inacceptable solubility, accumulation, and cell poisonousness, can be overwhelmed through optimal functionalization, which recovers their relaxing concert in cancer treatment owing to improved biocompatibility and biodegradability, increased tumor dissemination capability, and attained dynamic targeting competence. Furthermore, numerous studies have established the multifunctional claims of carbon nanotubes in antitumor drug transport, ranging from intracellular targeting to TME mechanisms, from solitary therapeutic modality to combinational treatment strategies, all pointing to CNTs' bright future in cancer treatment. CNTs are significant participants in cancer detection because of their unique properties, in addition to their therapeutic applications. CNTs can be used in a variety of cancer imaging techniques, including MRI, Raman imaging, radionuclide imaging, PA imaging, and near-infrared fluorescence imaging. Single-walled carbon nanotubes (SWCNTs) and multiple-walled carbon nanotubes (MWCNTs) are two types of carbon nanotubes. CNTs can also be used as nano biosensors aimed at initial discovery of many tumor categories, such as prostate cancer, cervical cancer, and pancreatic cancer, by excellent specificity when combined with other diagnostic agents. As a result, CNTs are a versatile nanomaterial with a wide range of biomedical applications in cancer treatment and diagnosis [22–24].

14.3.1.2 Graphene

It is a cage-like structure made up of carbon atoms used in imaging, medication administration, photosensitization, and immune response stimulation. Gd-metallo fullerenol was found to decrease breast cancer stem cells, prevent epithelial-to-mesenchymal transition, and be nontoxic to normal mammary epithelial cells. Both tumor initiation and metastasis were reduced by their targeted activities [25]. Various drugs formulated with carbon material have been previously reported for the prevention of different types of cancer (Table 14.1).

14.3.2 POLYMERIC NANO-FORMULATIONS FOR CANCER THERAPY

Colloidal organizations comprising natural or synthetic polymers make up polymer-based nanoparticles. They have a number of advantages over additional nano-transporters, like liposomes, inorganic nanosystems, and micelles, including the ability to gauge up and the trade procedure being compliant with GMPs (GMP) [31, 32]. Polymeric nanoparticles are distinguished by their significant stability in biological fluids, as well as their extensive obtainability of numerous polymers, the ability to functionalize their exteriors and moderate polymer deprivation, and the leakage of the

---

TABLE 14.1

Carbon-Based Nano-Formulation Used for the Prevention of Different Types of Cancer

S. No.	Material	Cancer Type	Drug Used	Reference
1.	Single-walled carbon nanotubes (SWCNTs)	Tumor	Paclitaxel (PTX)	[26]
2.	Single-walled carbon nanotubes (SWCNTs) + Congo red	Breast cancer	Doxorubicin (DOX)	[27]
3.	Multiple-walled carbon nanotubes (MWCNTs) + ethylenediamine and phenylboronic acid	Colon cancer	Paclitaxel (PTX)	[28]
4.	Multiple-walled carbon nanotubes (MWCNTs) + TAT peptide and chitosan	All types of cancer	Doxorubicin (DOX)	[29]
5.	Graphene oxide + folic acid	Bone tumor	Temozolomide	[30]

ensnared complexes as a purpose of precise incentives [33, 34]. Numerous chemotherapeutics were captured in polymeric transport methods to improve antitumor activity, limit metastasis, and reduce actual dosage and side effects. Polymers can capture or adsorb a dynamic substance into their assembly or on their exteriors [35]. Various drugs formulated with polymeric material have been previously reported for the prevention of different type of cancers (Table 14.2).

14.3.3   LIPOSOME-BASED FORMULATIONS FOR CANCER PREVENTION

Alec D. Bangham revealed liposomes in 1965, and he approved them as a first type of beneficial NPs for tumor management [46]. Liposomes are the globular vesicles which are made up of phospholipids and cholesterol that form a minimum of one lipid bilayer around an aqueous core in water. They can capture mutually hydrophilic medications (e.g., Doxil, encapsulated doxorubicin in the aqueous core) and hydrophobic complexes (e.g., AmBisome, trapped amphotericin B) that are engrossed in the gills by Van der Waals forces [47]. Doxorubicin HCL liposome (Doxil), the first nanosized liposomal medication approved in the United States, was accepted in 1995 aimed at the handling ovarian tumors. Liposomes have proven to be a highly effective and safe drug delivery technology that is both biocompatible and biodegradable, with no immunogenicity or toxicity risks [48]. Table 14.3 contains a complete list of phytochemicals that have been nano-formulated and tested in cancer models.

14.3.4   LIPID-POLYMER HYBRID NANO-FORMULATIONS FOR CANCER THERAPY

Polymers and lipids are the most often employed matrices in the creation of nanocarriers for the transport of anticancer drugs, each having their own set of advantages [54]. Lipid-based nanocarriers have various advantages, including a low production cost and a higher therapeutic drug trapping efficiency; however, they have a lower constancy, a faster freight proclamation, and a higher polydispersity. Polymeric NPs also have additional rewards, such as the capacity to make NPs with minor sizes and low polydispersity, as well as a variety of preparation processes, a modest and repeatable combination procedure, and improved stability [55, 56]. The construction of a hybrid scheme composed of a polymeric fundamental covered through a lipid shell, dubbed lipid-polymer hybrid

---

TABLE 14.2

Polymer-Based Nano-Formulation Used for the Prevention of Several Types of Cancer

S. No.	Polymeric Formulation	Cancer Type	Name of Brand	Reference
1.	Polymeric NP micelle formulation of mono methoxy poly(ethylene glycol)-block-poly(D,L-lactide) (mPEG-PDLLA) and paclitaxel	Lung cancer	Genexol-PM	[36, 37]
2.	Biodegradable polymer poly(octylcyanoacrylate) (POCA)	Pancreatic cancer	Gemcitabine	[38, 39]
3.	Poly-DL-lactide-co glycolide (PLGA) and polyethylene glycol (PEG)	Prostate cancer	Docetaxel-bicalutamide	[40]
4.	Bovine serum albumin	Breast cancer	Curcumin	[41, 42]
5.	PEG-docetaxel	Solid tumor	NKTR-105	[43]
6.	Amphiphilic linear–dendritic copolymer (named telodendrimer) composed of polyethylene glycol (PEG), cholic acid (CA, a facial amphiphilic molecule), and lysine	Ovarian cancer	Paclitaxel	[44]
7.	Pluronic-based micellar formulation of doxorubicin (Dox)	Leukemia	SP1049C	[45]

TABLE 14.3

Liposome-Based Nano-Formulation Used for the Prevention of Several Types of Cancer

S.No.	Liposome and Drug Formulation	Cancer Type	Name of Brand	Reference
1.	DOPC, cholesterol, cardiolipin (90:5:5) Lipid, PTX (33:1)	Ovarian cancer	LEP-ETU	[49]
2.	72 g PC, 10.8 cholesterol in ethanol + paclitaxel	Gastric, ovarian and lung cancer	Lipusu	[50]
3.	DSPC, cholesterol, PEG 2000-DSPE (56:39:5) + doxorubicin	Breast and ovarian cancer	Lipo-Dox	[51]
4.	DSPC, cholesterol, daunorubicin (10:5:1) + daunorubicin	Leukemia	DaunoXome	[52]
5.	DSPE, HER2, PEG + doxorubicin	Breast cancer	MM-302	[53]

TABLE 14.4

LPHNPs That Have Been Nano-Formulated and Tested in Cancer Models

S.No.	Formulation	Polymer	Lipid	Drug	Cancer Type	Reference
1.	Polymer-lipid hybrid nanoparticles (PLN)	HPESO (hydrolyzed polymer of epoxidized soybean oil)	Stearic acid	Doxorubicin (DOX) and GG918	Breast cancer	[58]
2.	Transferrin (Tf)-conjugated lipid-coated poly(d,l-lactide-coglycolide) (PLGA) nanoparticles	PLGA	DOPE, Tf-DOPE, MB-DOPE	7α-(4′-amino) phenylthio-1,4-androstadiene-3,17-dione (7α-APTADD)	Breast cancer	[59]
3.	Nanocells	PLGA	DSPE-PEG	DOX, combretastatin A4	Lung cancer	[60]
4.	Psoralen polymer–lipid hybrid nanoparticles	PSO	PLN	DOX	Breast cancer	[61]

nanoparticles, has resulted in a new generation of nanosystems with the benefits of both polymeric and lipid-based systems (LPHNPs) [57]. LPHNPs are made up of a hydrophilic or hydrophobic polymer core encasing the therapeutic drug of interest, surrounded by a lipid layer that increases the biocompatibility and stability of the nanoparticles after systemic delivery. Table 14.4 contains list of LPHNPs that have been nano-formulated and tested in cancer models.

LPHNPs can be categorized into the following groups based on their structure: (1) monolithic hybrid nanosystems, (2) core-shell nanosystems, (3) hollow-core-shell nanoparticles, (4) biomimetic lipid–polymer hybrid nanosystems, and (5) polymer-caged liposomes are all examples of hybrid nanosystems.

14.3.5 NUTRACEUTICALS AND PHYTOCHEMICAL-BASED NANO-FORMULATIONS FOR CANCER TREATMENT

Nutraceuticals are bioactive chemical substances derived from natural sources that have significant pharmacological effects. Curcumin, resveratrol, quercetin, genistein, and epigallocatechin gallate are some of the well-known phyto-ingredients having great anticancer potential [62]. Phytochemicals are naturally occurring compounds found in a wide range of plants that have biologically active properties, such as plant growth and defense. For ages, several of these phytochemicals have been investigated for their diverse effects and medical significance [25, 63]. The newer, less-explored sector of nano nutraceuticals may prove to be the lowest hanging fruit of all, as the trend toward natural alternatives to pharmaceutical medications continues to develop. Nutraceuticals made from phytochemicals that have been researched for their anticancer effects have shown promise. Table 14.5 contains a complete list of phytochemicals that have been nano-formulated and tested in cancer models.

14.3.6 DENDRIMER-BASED NANO-FORMULATIONS FOR CANCER CHEMOPREVENTION

Dendrimers are spherical macromolecules through an extremely branched construction and a large quantity of marginal collections that support encapsulate hydrophobic drugs. They are made up of a central core, branching units, and functional groups at the end. The central core determines the environment of the nanocavities and their solubilizing characteristics [63]. Dendrimers, like liposomes, can improve the solubility and bioavailability of water-insoluble substances by encapsulating them in their internal cavities or adhering them to their surface via electrostatic or hydrophobic interactions. Dendrimers can also be used to transport nucleic acid-based chemotherapies, which are hydrophilic molecules that are difficult to pass through the cell membrane [73]. Polymeric dendrimers, such as polyamidoamine (PAMAM) dendrimers, are of tremendous interest in biological applications because to their low toxicity [74]. Inorganic dendrimers by biomedical uses comprise phosphorus and carbosilane dendrimers. These dendrimers have shown to be effective against a variety of cancers and can be loaded with various anticancer medicines for medication delivery [75].

14.3.7 METAL NANO-FORMULATIONS FOR DRUG DELIVERY AND CANCER TREATMENT

Due to the potential landscapes and therapeutic implications in cancer treatment, metal nanoparticles have gotten a lot of interest. Gold and iron oxides, for example, have completely changed the way cancer is treated. Gold nanoparticles (GNPs) are engrained nanostructures that fascinate a lot of light and can generate thermal energy, which is typically used to destroy malignant tissue [76]. Moreover, iron oxides such as magnetite (Fe_3O_4) have magnetic characteristics that can be used for tumor thermotherapy guiding. The emission of harmful hydroxyl radicals is a serious problem when employing this substance. As a result, chemically functionalized maghemite (γ-Fe_2O_3) is employed

---

TABLE 14.5

Phytochemical-/Nutraceutical-Based Nano-Formulation Used for the Prevention of Several Types of Cancer

S. No.	Phytochemical/ Nutraceutical	Formulation	Cancer Type	Reference
1.	β-Lapachone	β-Lapachone—PEG-PLA polymer micelles	Lung cancer, breast cancer	[64, 65]
2.	Eugenol	Eugenol—nanoemulsions, magnetic NP	Colon and liver cancer	[66, 67]
3.	Resveratrol	Resveratrol—gold NP	Skin cancer, breast cancer, prostate cancer, pancreatic cancer	[68]
4.	Ellagic acid	Ellagic acid—PLGA NP, PEG, mesoporous silica NP	Bladder cancer	[69]
5.	Ursolic acid	Ursolic acid—poly(DL-lactide-co-glycolide) nanoparticles	Cervical cancer	[70]
6.	Quercetin	Quercetin—poly(DL-lactide-co-glycolide) nanoparticles	Liver cancer	[71]
7.	Genistein	Genistein—pegylated silica hybrid nanomaterials	Breast cancer, prostate cancer, colon cancer	[72]

as a substitute in drug delivery arrangements that use material surface alterations to target tumor cells [77]. Inorganic transporters made of C or Si couples like silicon dioxide have been developed as a result of efforts to develop diverse and dependable delivery mechanisms. Mesoporous silica NPs, for example, have been projected as effective therapeutic drug transport architectures with properties such as improved drug solubility, higher lading competence, multifunctionality, and stimuli-responsive proclamation regulation. Silica NPs are particularly appealing for GI cancer treatment because their exterior can be effortlessly changed to penetrate GI barriers or target malignant cells, among other advantages [78, 79]. The metal organic framework (MOF), which is designed to accomplish a range of roles, is another inorganic NP delivery device. MOFs are hybrid nano-motifs made up of metal ions entrenched in a web of organic linkers that have recently been shown to be beneficial in photodynamic therapy and enhanced immunotherapy. MOFs also have a high capacity for drug loading, biocompatibility, and multifunctionality [80, 81].

14.3.8 QUANTUM-DOT-BASED NANO-FORMULATIONS FOR CANCER PREVENTION

QD (quantum dot) nanoparticles are semiconductors that can be used to portray live cells and animals in fluorescent form. They can be a decent foundation of light ranging from UV to IR, contingent on their conformation and size. There are approximately 100–100,000 atoms in the crystal core of a QD. The diameter of a QD is typically between 2 and 10 nm. QDs' dimensions, on the other hand, are determined by the substance used to create them [82, 83]. When determining whether a nanoparticle is a quantum dot or not just grounded on its size, there is no rational line to draw. Metals or semiconductor materials can be used to make quantum dots (Ni, Co, Pt, and Au). Aside from that, some work has been done using metalloid QDs, such as silicon (Table 14.6). QDs fabricated with various legends or anticancer agents/genes can vastly improve fluorescence bio imaging and target delivery proficiency by simultaneously imaging tumor cells and delivering malignant growth therapy via specific authoritative to receptors over communicated on tumor cells and tissue surface [84, 85].

14.4 CONCLUSION AND FUTURE PROSPECTS

Over the years, nanotechnology has revolutionized the cancer therapy radically by changing the treatment process. Major limitations like lack of solubility and drug resistance of chemotherapeutic drugs have been effectively overcome by a number of nanocarriers, such as micelles, liposomes, dendrimers, carbon nanotubes, nanocapsules, and nanospheres. These nanocarriers either entrap, covalently bind, encapsulate, or adsorb the hydrophobic therapeutic agents and transport them to the desired area after administration, thus restraining the drug resistance. Nanomedicines like Doxil and its generics, as well as Abraxane, continue to grow in popularity, with sales approaching $1 billion a year. However, the field's expectations have yet to be met. Using the term "nanoparticle" as a search term on clinicaltrials.gov yields 240 trials, 11 of which are at the phase III stage, representing a reasonable and promising growth. Nanomaterials have helped to improve cancer detection and treatment by improving pharmacokinetic and pharmacodynamic qualities. Because of their

TABLE 14.6

Quantum-Dot-Based Nano-Formulation Used for the Prevention of Several Types of Cancer

S. No.	Formulation	Drug	Cancer Type	Reference
1.	Quantum dot + dox	Doxorubicin	Lung cancer	[86]
2.	QD-PSMA	Streptavidin	Prostate cancer	[87]

specificities, nanotechnology allows for the medical management of injured organs with minimal systemic toxicity. Nano-formulation techniques have been proposed as possible alternatives to traditional drug delivery systems because they provide unique drug delivery features. Nanocarriers have shown a significant boost in medication therapeutic efficacy when compared to standard cancer chemotherapy, with only a few side effects. However, nanotechnology, like additional therapeutic techniques, is not without side effects and postures a variety of claim issues, with as universal and organ-specific poisonousness, which have delayed scientific utilization. Given the limitations of nanotechnology, more development is required to optimize pharmaceutical delivery, increase efficacy, and minimize side effects. By increasing the connections among the physicochemical effects of the nanomaterials, harmless and additional operative derivatives for cancer diagnosis and therapy can be made available. Finally, we want to underline both the underlying benefits of nanotechnology and the limitations of its applicability to cancer treatment needs.

Additionally, nanotechnology's therapeutic results and upcoming developments could make it a therapeutic option for numerous infections. Ischemic stroke and rheumatoid arthritis, for example, would require the administration of a suitable pharmacologic drug to the affected region. Nanotechnology can further deliver a revolutionary development in other potential areas of cancer applications like immunotherapy, combination therapies, and intraoperative imaging.

14.5 REFERENCES

1. International Agency of Research on Cancer, 2022. World Health Organization. 2022. Sixth edition of World Cancer Research Day: Driving progress against cancer, together. www.iarc.who.int/wp-content/uploads/2021/09/pr302_E.pdf. Accessed March 28.
2. World Health Organization, 2020. 2022. Cancer the problem. www.who.int/en/news-room/fact-sheets/detail/cancer. Accessed March 28.
3. Fan, Anna M. 2009. Cancer. *Information Resources in Toxicology*: 103–121. https://doi.org/10.1016/B978-0-12-373593-5.00011-2.
4. Rumgay, Harriet, Kevin Shield, Hadrien Charvat, Pietro Ferrari, Bundit Sornpaisarn, Isidore Obot, Farhad Islami, Valery E.P.P. Lemmens, Jürgen Rehm, and Isabelle Soerjomataram. 2021. Global burden of cancer in 2020 attributable to alcohol consumption: A population-based study. *The Lancet Oncology* 22. Lancet Publishing Group: 1071–1080. https://doi.org/10.1016/S1470-2045(21)00279-5/ATTACHMENT/F8F564ED-5B9B-4E60-974A-F71B4884888B/MMC1.PDF.
5. International Agency of Research on Cancer, 2021. World Health Organization. 2022. Latest global data on cancer burden and alcohol consumption: More than 740 000 new cases of cancer in 2020 attributed to alcohol. www.iarc.who.int/wp-content/uploads/2021/07/pr299_E.pdf. Accessed March 28.
6. Din, Fakhar Ud, Waqar Aman, Izhar Ullah, Omer Salman Qureshi, Omer Mustapha, Shumaila Shafique, and Alam Zeb. 2017. Effective use of nanocarriers as drug delivery systems for the treatment of selected tumors. *International Journal of Nanomedicine* 12. Dove Press: 7291. https://doi.org/10.2147/IJN.S146315.
7. Sutradhar, Kumar Bishwajit, and Md. Lutful Amin. 2014. Nanotechnology in cancer drug delivery and selective targeting. *ISRN Nanotechnology* 2014: 1–12. https://doi.org/10.1155/2014/939378.
8. Desai, Preshita, Naga Jyothi Thumma, Pushkaraj Rajendra Wagh, Shuyu Zhan, David Ann, Jeffrey Wang, and Sunil Prabhu. 2020. Cancer chemoprevention using nanotechnology-based approaches. *Frontiers in Pharmacology* 11: 1–9. https://doi.org/10.3389/fphar.2020.00323.
9. Muqbil, Irfana, Ashiq Masood, Fazlul H. Sarkar, Ramzi M. Mohammad, and Asfar S. Azmi. 2011. Progress in nanotechnology based approaches to enhance the potential of chemopreventive agents. *Cancers* 3: 428–445. https://doi.org/10.3390/cancers3010428.
10. Jeevanandam, Jaison, Ahmed Barhoum, Yen S. Chan, Alain Dufresne, and Michael K. Danquah. 2018. Review on nanoparticles and nanostructured materials: History, sources, toxicity and regulations. *Beilstein Journal of Nanotechnology 9:98* 9. Beilstein-Institut: 1050–1074. https://doi.org/10.3762/BJNANO.9.98.
11. Martin, Charles R. 1994. Nanomaterials: A membrane-based synthetic approach. *Science* 266. American Association for the Advancement of Science: 1961–1966. https://doi.org/10.1126/SCIENCE.266.5193.1961.

12. Torchilin, Vladimir P. 2000. Drug targeting. *European Journal of Pharmaceutical Sciences* 11. Elsevier: S81–S91. https://doi.org/10.1016/S0928-0987(00)00166-4.

13. Alexis, Frank, Eric Pridgen, Linda K. Molnar, and Omid C. Farokhzad. 2008. Factors affecting the clearance and biodistribution of polymeric nanoparticles. *Molecular Pharmaceutics* 5. American Chemical Society: 505. https://doi.org/10.1021/MP800051M.

14. Siddiqui, Lubna, Harshita Mishra, Pawan Kumar Mishra, Zeenat Iqbal, and Sushama Talegaonkar. 2018. Novel 4-in-1 strategy to combat colon cancer, drug resistance and cancer relapse utilizing functionalized bioinspiring lignin nanoparticle. *Medical Hypotheses* 121. Churchill Livingstone: 10–14. https://doi.org/10.1016/J.MEHY.2018.09.003.

15. Obyn, B., R. J. Arst, L. J. Ewis, R. Ubin, W. A. Alker, L. Ong, M. D. Ichael, M. C. G. Oon, S. R. Tuart Ich, D. B. Avid, B. Adesch, B. M. Ertron, G. Roves, et al. 2009. A comparison of continuous intravenous epoprostenol (prostacyclin) with conventional therapy for primary pulmonary hypertension. *http://dx.doi.org/10.1056/NEJM199602013340504* 51. Massachusetts Medical Society: 993. https://doi.org/10.1056/NEJM199602013340504.

16. Singh, Ravina, and Hari Singh Nalwa. 2011. Medical applications of nanoparticles in biological imaging, cell labeling, antimicrobial agents, and anticancer nanodrugs. *Journal of Biomedical Nanotechnology* 7. J Biomed Nanotechnol: 489–503. https://doi.org/10.1166/JBN.2011.1324.

17. Maniam, Geetha, Chun Wai Mai, Mohd Zulkefeli, Christine Dufès, Doryn Meam Yee Tan, and Ju Yen Fu. 2018. Challenges and opportunities of nanotechnology as delivery platform for tocotrienols in cancer therapy. *Frontiers in Pharmacology* 9. Frontiers Media S.A.: 1358. https://doi.org/10.3389/FPHAR.2018.01358/BIBTEX.

18. Jin, Cancan, Kankai Wang, Anthony Oppong-Gyebi, and Jiangnan Hu. 2020. Application of nanotechnology in cancer diagnosis and therapy—A mini-review. *International Journal of Medical Sciences* 17: 2964–2973. https://doi.org/10.7150/ijms.49801.

19. Rauti, Rossana, Mattia Musto, Susanna Bosi, Maurizio Prato, and Laura Ballerini. 2019. Properties and behavior of carbon nanomaterials when interfacing neuronal cells: How far have we come? *Carbon* 143. Pergamon: 430–446. https://doi.org/10.1016/J.CARBON.2018.11.026.

20. Mohajeri, Mohammad, Behzad Behnam, and Amirhossein Sahebkar. 2019. Biomedical applications of carbon nanomaterials: Drug and gene delivery potentials. *Journal of Cellular Physiology* 234. John Wiley & Sons, Ltd: 298–319. https://doi.org/10.1002/JCP.26899.

21. Wang, Shang Yu, Hong Zhi Hu, Xiang Cheng Qing, Zhi Cai Zhang, and Zeng Wu Shao. 2020. Recent advances of drug delivery nanocarriers in osteosarcoma treatment. *Journal of Cancer* 11. Ivyspring International Publisher: 69. https://doi.org/10.7150/JCA.36588.

22. Sinha, Niraj, and John T.W. Yeow. 2005. Carbon nanotubes for biomedical applications. *IEEE Transactions on Nanobioscience* 4: 180–195. https://doi.org/10.1109/TNB.2005.850478.

23. Liu, Zhuang, Joshua T. Robinson, Scott M. Tabakman, Kai Yang, and Hongjie Dai. 2011. Carbon materials for drug delivery & cancer therapy. *Materials Today* 14. Elsevier: 316–323. https://doi.org/10.1016/S1369-7021(11)70161-4.

24. Mahor, Alok, Prem Prakash Singh, Peeyush Bharadwaj, Neeraj Sharma, Surabhi Yadav, Jessica M. Rosenholm, and Kuldeep K. Bansal. 2021. Carbon-based nanomaterials for delivery of biologicals and therapeutics: A cutting-edge technology. *C* 7: 19. https://doi.org/10.3390/c7010019.

25. Salama, Lavinia, Elizabeth R. Pastor, Tyler Stone, and Shaker A. Mousa. 2020. Emerging nanopharmaceuticals and nanonutraceuticals in cancer management. *Biomedicines* 8: 1–22. https://doi.org/10.3390/BIOMEDICINES8090347.

26. Al Garalleh, Hakim, and Ali Algarni. 2020. Modelling of paclitaxel conjugated with carbon nanotubes as an antitumor agent for cancer therapy. *Journal of Biomedical Nanotechnology* 16. NLM (Medline): 224–234. https://doi.org/10.1166/JBN.2020.2886.

27. Jagusiak, Anna, Katarzyna Chłopaś, Grzegorz Zemanek, Małgorzata Jemioła-Rzemińska, Barbara Piekarska, Barbara Stopa, and Tomasz Pańczyk. 2019. Self-assembled supramolecular ribbon-like structures complexed to single walled carbon nanotubes as possible anticancer drug delivery systems. *International Journal of Molecular Sciences* 20. Multidisciplinary Digital Publishing Institute (MDPI). https://doi.org/10.3390/IJMS20092064.

28. Rathod, Vishakha, Rahul Tripathi, Parth Joshi, Prafulla K. Jha, Pratap Bahadur, and Sanjay Tiwari. 2019. Paclitaxel encapsulation into dual-functionalized multi-walled carbon nanotubes. *AAPS Pharmscitech* 20. Springer New York LLC: 51–51. https://doi.org/10.1208/S12249-018-1218-6.

29. Dong, Xia, Zhiting Sun, Xiaoxiao Wang, and Xigang Leng. 2017. An innovative MWCNTs/DOX/TC nanosystem for chemo-photothermal combination therapy of cancer. *Nanomedicine: Nanotechnology, Biology and Medicine* 13. Elsevier: 2271–2280. https://doi.org/10.1016/J.NANO.2017.07.002.

30. Wu, Kunzhe, Beibei Yu, Di Li, Yangyang Tian, Yan Liu, and Jinlan Jiang. 2022. Recent advances in nanoplatforms for the treatment of osteosarcoma. *Frontiers in Oncology* 12. Frontiers Media SA: 241. https://doi.org/10.3389/FONC.2022.805978/BIBTEX.

31. Van Vlerken, Lilian E., Tushar K. Vyas, and Mansoor M. Amiji. 2007. Poly(ethylene glycol)-modified nanocarriers for tumor-targeted and intracellular delivery. *Pharmaceutical Research 2007 24:8* 24. Springer: 1405–1414. https://doi.org/10.1007/S11095-007-9284-6.

32. Rezvantalab, Sima, Natascha Ingrid Drude, Mostafa Keshavarz Moraveji, Nihan Güvener, Emily Kate Koons, Yang Shi, Twan Lammers, and Fabian Kiessling. 2018. PLGA-based nanoparticles in cancer treatment. *Frontiers in Pharmacology* 9. Frontiers Media S.A.: 1260. https://doi.org/10.3389/FPHAR.2018.01260/BIBTEX.

33. Goodall, Stephen, Martina L. Jones, and Stephen Mahler. 2015. Monoclonal antibody-targeted polymeric nanoparticles for cancer therapy—future prospects. *Journal of Chemical Technology & Biotechnology* 90. John Wiley & Sons, Ltd: 1169–1176. https://doi.org/10.1002/JCTB.4555.

34. Sarcan, E. Tugce, Mine Silindir-Gunay, and A. Yekta Ozer. 2018. Theranostic polymeric nanoparticles for NIR imaging and photodynamic therapy. *International Journal of Pharmaceutics* 551. Elsevier: 329–338. https://doi.org/10.1016/J.IJPHARM.2018.09.019.

35. Masood, Farha. 2016. Polymeric nanoparticles for targeted drug delivery system for cancer therapy. *Materials Science and Engineering: C* 60. Elsevier: 569–578. https://doi.org/10.1016/J.MSEC.2015.11.067.

36. Ahn, Hee Kyung, Minkyu Jung, Sun Jin Sym, Dong Bok Shin, Shin Myung Kang, Sun Young Kyung, Jeong Woong Park, Sung Hwan Jeong, and Eun Kyung Cho. 2014. A phase II trial of Cremorphor EL-free paclitaxel (Genexol-PM) and gemcitabine in patients with advanced non-small cell lung cancer. *Cancer Chemotherapy and Pharmacology* 74. Springer Verlag: 277–282. https://doi.org/10.1007/S00280-014-2498-5/TABLES/3.

37. Kim, D. W., S. Y. Kim, H. K. Kim, S. W. Kim, S. W. Shin, J. S. Kim, K. Park, M. Y. Lee, and D. S. Heo. 2007. Multicenter phase II trial of Genexol-PM, a novel Cremophor-free, polymeric micelle formulation of paclitaxel, with cisplatin in patients with advanced non-small-cell lung cancer. *Annals of Oncology* 18. Elsevier: 2009–2014. https://doi.org/10.1093/ANNONC/MDM374.

38. Arias, José L., L. Harivardhan Reddy, and Patrick Couvreur. 2009. Polymeric nanoparticulate system augmented the anticancer therapeutic efficacy of gemcitabine. *Journal of Drug Targeting* 17. J Drug Target: 586–598. https://doi.org/10.1080/10611860903105739.

39. Zhang, Xiaoping, and Maoquan Li. 2021. Molecular-targeted therapy of pancreatic carcinoma and its progress. *Integrative Pancreatic Intervention Therapy*. Elsevier: 487–503. https://doi.org/10.1016/B978-0-12-819402-7.00020-6.

40. Kesch, Claudia, Veronika Schmitt, Samir Bidnur, Marisa Thi, Eliana Beraldi, Igor Moskalev, Virginia Yago, et al. 2019. A polymeric paste-drug formulation for intratumoral treatment of prostate cancer. *Prostate Cancer and Prostatic Diseases 2019 23:2* 23. Nature Publishing Group: 324–332. https://doi.org/10.1038/s41391-019-0190-x.

41. Jithan, A. V., K. Madhavi, M. Madhavi, and K. Prabhakar. 2011. Preparation and characterization of albumin nanoparticles encapsulating curcumin intended for the treatment of breast cancer. *International Journal of Pharmaceutical Investigation* 1. Wolters Kluwer—Medknow Publications: 119. https://doi.org/10.4103/2230-973X.82432.

42. Grewal, Ikmeet Kaur, Sukhbir Singh, Sandeep Arora, and Neelam Sharma. 2021. Polymeric nanoparticles for breast cancer therapy: A comprehensive review. *Biointerface Research in Applied Chemistry* 11: 11151–11171. https://doi.org/10.33263/BRIAC114.1115111171.

43. Calvo, E., U. Hoch, D. J. Maslyar, and A. W. Tolcher. 2010. Dose-escalation phase I study of NKTR-105, a novel pegylated form of docetaxel. *https://doi.org/10.1200/jco.2010.28.15_suppl.tps160* 28. American Society of Clinical Oncology: TPS160–TPS160. https://doi.org/10.1200/JCO.2010.28.15_SUPPL.TPS160.

44. Xiao, Kai, Juntao Luo, Wiley L. Fowler, Yuanpei Li, Joyce S. Lee, Li Xing, R. Holland Cheng, Li Wang, and Kit S. Lam. 2009. A self-assembling nanoparticle for paclitaxel delivery in ovarian cancer. *Biomaterials* 30. NIH Public Access: 6006. https://doi.org/10.1016/J.BIOMATERIALS.2009.07.015.

45. Alakhova, Daria Y., Yi Zhao, Shu Li, and Alexander V. Kabanov. 2013. Effect of doxorubicin/pluronic SP1049C on tumorigenicity, aggressiveness, DNA methylation and stem cell markers in murine leukemia. *PLoS ONE* 8. https://doi.org/10.1371/journal.pone.0072238.

46. Allen, Theresa M., and Pieter R. Cullis. 2013. Liposomal drug delivery systems: From concept to clinical applications. *Advanced Drug Delivery Reviews* 65. Elsevier: 36–48. https://doi.org/10.1016/J.ADDR.2012.09.037.

47. Amber Gonda, Nanxia Zhao, Jay V. Shah, Hannah R. Calvelli, Harini, and Vidya Ganapathy Kantamneni, Nicola L. Francis. 2019. Engineering tumor-targeting nanoparticles as vehicles for precision nanomedicine. *Med One*. Hapres. https://doi.org/10.20900/MO.20190021.

48. Lungu, Iulia Ioana, Alexandru Mihai Grumezescu, Adrian Volceanov, Ecaterina Andronescu, Alejandro Baeza, Fernando Novio, and Juan Luis Paris. 2019. Nanobiomaterials used in cancer therapy: An up-to-date overview. *Molecules 2019, Vol. 24, Page 3547* 24. Multidisciplinary Digital Publishing Institute: 3547. https://doi.org/10.3390/MOLECULES24193547.

49. Bulbake, Upendra, Sindhu Doppalapudi, Nagavendra Kommineni, and Wahid Khan. 2017. Liposomal formulations in clinical use: An updated review. *Pharmaceutics* 9. Pharmaceutics. https://doi.org/10.3390/PHARMACEUTICS9020012.

50. Xu, Xu, Lin Wang, Huan Qin Xu, Xin En Huang, Ya Dong Qian, and Jin Xiang. 2013. Clinical comparison between paclitaxel liposome (Lipusu®) and paclitaxel for treatment of patients with metastatic gastric cancer. *Asian Pacific Journal of Cancer Prevention* 14. Asian Pacific Journal of Cancer Prevention: 2591–2594. https://doi.org/10.7314/APJCP.2013.14.4.2591.

51. Smith, Judith A., Lata Mathew, Maryam Burney, Pranavanand Nyshadham, and Robert L. Coleman. 2016. Equivalency challenge: Evaluation of Lipodox® as the generic equivalent for Doxil® in a human ovarian cancer orthotropic mouse model. *Gynecologic Oncology* 141. Elsevier: 357–363. https://doi.org/10.1016/J.YGYNO.2016.02.033.

52. Olusanya, Temidayo O.B., Rita Rushdi Haj Ahmad, Daniel M. Ibegbu, James R. Smith, and Amal Ali Elkordy. 2018. Liposomal drug delivery systems and anticancer drugs. *Molecules 2018, Vol. 23, Page 907* 23. Multidisciplinary Digital Publishing Institute: 907. https://doi.org/10.3390/MOLECULES23040907.

53. Miller, Kathy, Javier Cortes, Sara A. Hurvitz, Ian E. Krop, Debu Tripathy, Sunil Verma, Kaveh Riahi, et al. 2016. HERMIONE: A randomized Phase 2 trial of MM-302 plus trastuzumab versus chemotherapy of physician's choice plus trastuzumab in patients with previously treated, anthracycline-naïve, HER2-positive, locally advanced/metastatic breast cancer. *BMC Cancer* 16. BioMed Central Ltd: 1–11. https://doi.org/10.1186/S12885-016-2385-Z/FIGURES/5.

54. Martinelli, Chiara, Carlotta Pucci, and Gianni Ciofani. 2019. Nanostructured carriers as innovative tools for cancer diagnosis and therapy. *APL Bioengineering* 3. American Institute of Physics. https://doi.org/10.1063/1.5079943.

55. Ghitman, Jana, Elena Iuliana Biru, Raluca Stan, and Horia Iovu. 2020. Review of hybrid PLGA nanoparticles: Future of smart drug delivery and theranostics medicine. *Materials and Design* 193. Elsevier Ltd. https://doi.org/10.1016/J.MATDES.2020.108805.

56. Martínez Rivas, Claudia Janeth, Mohamad Tarhini, Waisudin Badri, Karim Miladi, Hélène Greige-Gerges, Qand Agha Nazari, Sergio Arturo Galindo Rodríguez, Rocío Álvarez Román, Hatem Fessi, and Abdelhamid Elaissari. 2017. Nanoprecipitation process: From encapsulation to drug delivery. *International Journal of Pharmaceutics* 532. Elsevier: 66–81. https://doi.org/10.1016/J.IJPHARM.2017.08.064.

57. Rao, Shasha, and Clive A. Prestidge. 2016. Polymer-lipid hybrid systems: Merging the benefits of polymeric and lipid-based nanocarriers to improve oral drug delivery. *https://doi.org/10.1517/17425247.2016.1151872* 13. Taylor & Francis: 691–707. https://doi.org/10.1517/17425247.2016.1151872.

58. Wong, Ho Lun, Reina Bendayan, Andrew Mike Rauth, and Xiao Yu Wu. 2006. Simultaneous delivery of doxorubicin and GG918 (Elacridar) by new Polymer-Lipid Hybrid Nanoparticles (PLN) for enhanced treatment of multidrug-resistant breast cancer. *Journal of Controlled Release* 116. Elsevier: 275–284. https://doi.org/10.1016/J.JCONREL.2006.09.007.

59. Zheng, Yu, Bo Yu, Wanlop Weecharangsan, Longzhu Piao, Michael Darby, Yicheng Mao, Rumiana Koynova, et al. 2010. Transferrin-conjugated lipid-coated PLGA nanoparticles for targeted delivery of aromatase inhibitor 7α-APTADD to breast cancer cells. *International Journal of Pharmaceutics* 390. Elsevier: 234–241. https://doi.org/10.1016/J.IJPHARM.2010.02.008.

60. Sengupta, Shiladitya, David Eavarone, Ishan Capila, Ganlin Zhao, Nicki Watson, Tanyel Kiziltepe, and Ram Sasisekharan. 2005. Temporal targeting of tumour cells and neovasculature with a nanoscale delivery system. *Nature* 436. Nature: 568–572. https://doi.org/10.1038/NATURE03794.

61. Huang, Qingqing, Tiange Cai, Qianwen Li, Yinghong Huang, Qian Liu, Bingyue Wang, Xi Xia, et al. 2018. Preparation of psoralen polymer-lipid hybrid nanoparticles and their reversal of multidrug resistance in MCF-7/ADR cells. *Drug Delivery* 25. Informa Healthcare USA, Inc: 1056–1066. https://doi.org/10.1080/10717544.2018.1464084.

62. Illahi, Aroosa Fazal, Faqir Muhammad, and Bushra Akhtar. 2019. Nanoformulations of Nutraceuticals for Cancer Treatment. *Critical Reviews™ in Eukaryotic Gene Expression* 29. Begel House Inc.: 449–460. https://doi.org/10.1615/CRITREVEUKARYOTGENEEXPR.2019025957.

63. Khan, Tabassum, and Pranav Gurav. 2018. PhytoNanotechnology: Enhancing delivery of plant based anti-cancer drugs. *Frontiers in Pharmacology* 8. Frontiers Media S.A.: 1002. https://doi.org/10.3389/FPHAR.2017.01002/BIBTEX.

64. Jeong, Seong Yun, Sung Jin Park, Sang Min Yoon, Joohee Jung, Ha Na Woo, So Lyoung Yi, Si Yeol Song, et al. 2009. Systemic delivery and preclinical evaluation of Au nanoparticle containing β-lapachone for radiosensitization. *Journal of Controlled Release* 139. Elsevier: 239–245. https://doi.org/10.1016/J.JCONREL.2009.07.007.

65. Yang, Yang, Xianchun Zhou, Ming Xu, Junjie Piao, Yuan Zhang, Zhenhua Lin, and Liyan Chen. 2017. β-lapachone suppresses tumour progression by inhibiting epithelial-to-mesenchymal transition in NQO1-positive breast cancers. *Scientific Reports 2017 7:1* 7. Nature Publishing Group: 1–13. https://doi.org/10.1038/s41598-017-02937-0.

66. Majeed, Hamid, John Antoniou, and Zhong Fang. 2014. Apoptotic effects of eugenol-loaded nanoemulsions in human colon and liver cancer cell lines. *Asian Pacific Journal of Cancer Prevention* 15. Asian Pacific Journal of Cancer Prevention: 9159–9164. https://doi.org/10.7314/APJCP.2014.15.21.9159.

67. Abdullah, Mashan L., Mohamed M. Hafez, Ali Al-Hoshani, and Othman Al-Shabanah. 2018. Antimetastatic and anti-proliferative activity of eugenol against triple negative and HER2 positive breast cancer cells 11 Medical and Health Sciences 1112 Oncology and Carcinogenesis. *BMC Complementary and Alternative Medicine* 18. BioMed Central Ltd.: 1–11. https://doi.org/10.1186/S12906-018-2392-5/FIGURES/6.

68. Thipe, Velaphi C., Kiandohkt Panjtan Amiri, Pierce Bloebaum, Alice Raphael Karikachery, Menka Khoobchandani, Kavita K. Katti, Silvia S. Jurisson, and Kattesh V. Katti. 2019. Development of resveratrol-conjugated gold nanoparticles: Interrelationship of increased resveratrol corona on anti-tumor efficacy against breast, pancreatic and prostate cancers. *International Journal of Nanomedicine* 14. Dove Press: 4413–4428. https://doi.org/10.2147/IJN.S204443.

69. Neamatallah, Thikryat, Nagla El-Shitany, Aymn Abbas, Basma G. Eid, Steve Harakeh, Soad Ali, and Shaker Mousa. 2020. Nano ellagic acid counteracts cisplatin-induced upregulation in OAT1 and OAT3: A possible nephroprotection mechanism. *Molecules 2020, Vol. 25, Page 3031* 25. Multidisciplinary Digital Publishing Institute: 3031. https://doi.org/10.3390/MOLECULES25133031.

70. Wang, Shaoguang, Xiaomei Meng, and Yaozhong Dong. 2017. Ursolic acid nanoparticles inhibit cervical cancer growth in vitro and in vivo via apoptosis induction. *International Journal of Oncology* 50. Spandidos Publications: 1330–1340. https://doi.org/10.3892/IJO.2017.3890/HTML.

71. Ren, Ke Wei, Ya Hua Li, Gang Wu, Jian Zhuang Ren, Hui Bin Lu, Zong Ming Li, and Xin Wei Han. 2017. Quercetin nanoparticles display antitumor activity via proliferation inhibition and apoptosis induction in liver cancer cells. *International Journal of Oncology* 50. Spandidos Publications: 1299–1311. https://doi.org/10.3892/IJO.2017.3886/HTML.

72. Pool, Héctor, Rocio Campos-Vega, María Guadalupe Herrera-Hernández, Pablo García-Solis, Teresa García-Gasca, Isaac Cornelius Sánchez, Gabriel Luna-Bárcenas, and Haydé Vergara-Castañeda. 2018. Development of genistein-PEGylated silica hybrid nanomaterials with enhanced antioxidant and antiproliferative properties on HT29 human colon cancer cells. *American Journal of Translational Research* 10. e-Century Publishing Corporation: 2306.

73. Mendes, Livia Palmerston, Jiayi Pan, and Vladimir P. Torchilin. 2017. Dendrimers as nanocarriers for nucleic acid and drug delivery in cancer therapy. *Molecules 2017, Vol. 22, Page 1401* 22. Multidisciplinary Digital Publishing Institute: 1401. https://doi.org/10.3390/MOLECULES22091401.

74. Uram, Łukasz, Aleksandra Filipowicz, Maria Misiorek, Natalia Pieńkowska, Joanna Markowicz, Elżbieta Wałajtys-Rode, and Stanisław Wołowiec. 2018. Biotinylated PAMAM G3 dendrimer conjugated with celecoxib and/or Fmoc-l-Leucine and its cytotoxicity for normal and cancer human cell lines. *European Journal of Pharmaceutical Sciences* 124. Elsevier: 1–9. https://doi.org/10.1016/J.EJPS.2018.08.019.

75. Caminade, Anne Marie. 2020. Phosphorus dendrimers as nanotools against cancers. *Molecules 2020, Vol. 25, Page 3333* 25. Multidisciplinary Digital Publishing Institute: 3333. https://doi.org/10.3390/MOLECULES25153333.

76. Wang, Chungang, Jiji Chen, Tom Talavage, and Joseph Irudayaraj. 2009. Gold nanorod/Fe3O4 nanoparticle "nano-pearl-necklaces" for simultaneous targeting, dual-mode imaging, and photothermal ablation of cancer cells. *Angewandte Chemie* 121. John Wiley & Sons, Ltd: 2797–2801. https://doi.org/10.1002/ANGE.200805282.

77. Vangijzegem, Thomas, Dimitri Stanicki, and Sophie Laurent. 2018. Magnetic iron oxide nanoparticles for drug delivery: Applications and characteristics. *https://doi.org/10.1080/17425247.2019.1554647* 16. Taylor & Francis: 69–78. https://doi.org/10.1080/17425247.2019.1554647.

78. Wang, Ying, Yating Zhao, Yu Cui, Qinfu Zhao, Qiang Zhang, Sara Musetti, Karina A. Kinghorn, and Siling Wang. 2018. Overcoming multiple gastrointestinal barriers by bilayer modified hollow mesoporous silica nanocarriers. *Acta Biomaterialia* 65. Elsevier: 405–416. https://doi.org/10.1016/J. ACTBIO.2017.10.025.

79. Zhou, Yixian, Guilan Quan, Qiaoli Wu, Xiaoxu Zhang, Boyi Niu, Biyuan Wu, Ying Huang, Xin Pan, and Chuanbin Wu. 2018. Mesoporous silica nanoparticles for drug and gene delivery. *Acta Pharmaceutica Sinica B* 8. Elsevier: 165–177. https://doi.org/10.1016/J.APSB.2018.01.007.

80. Zhou, Jingrong, Gan Tian, Lijuan Zeng, Xueer Song, and Xiu Wu Bian. 2018. Nanoscaled metal-organic frameworks for biosensing, imaging, and cancer therapy. *Advanced Healthcare Materials* 7. John Wiley & Sons, Ltd: 1800022. https://doi.org/10.1002/ADHM.201800022.

81. He, Liangcan, Yuan Liu, Joseph Lau, Wenpei Fan, Qunying Li, Chao Zhang, Pintong Huang, and Xiaoyuan Chen. 2019. Recent progress in nanoscale metal-organic frameworks for drug release and cancer therapy. *https://doi.org/10.2217/nnm-2018-0347* 14. Future Medicine Ltd London, UK: 1343–1365. https://doi.org/10.2217/NNM-2018-0347.

82. Chen, Lu, and Jiangong Liang. 2020. An overview of functional nanoparticles as novel emerging antiviral therapeutic agents. *Materials Science and Engineering: C* 112. Elsevier: 110924. https://doi.org/10.1016/J.MSEC.2020.110924.

83. Zhao, Mei Xia, and Bing Jie Zhu. 2016. The research and applications of quantum dots as nano-carriers for targeted drug delivery and cancer therapy. *Nanoscale Research Letters* 11. Springer New York LLC: 1–9. https://doi.org/10.1186/S11671-016-1394-9/FIGURES/8.

84. Ye, Fei, Åsa Barrefelt, Heba Asem, Manuchehr Abedi-Valugerdi, Ibrahim El-Serafi, Maryam Saghafian, Khalid Abu-Salah, Salman Alrokayan, Mamoun Muhammed, and Moustapha Hassan. 2014. Biodegradable polymeric vesicles containing magnetic nanoparticles, quantum dots and anticancer drugs for drug delivery and imaging. *Biomaterials* 35. Elsevier: 3885–3894. https://doi.org/10.1016/J. BIOMATERIALS.2014.01.041.

85. Devi, Sheetal, Manish Kumar, Abhishek Tiwari, Varsha Tiwari, Deepak Kaushik, Ravinder Verma, Shailendra Bhatt, et al. 2022. Quantum dots: An emerging approach for cancer therapy. *Frontiers in Materials* 8. Frontiers Media S.A.: 585. https://doi.org/10.3389/FMATS.2021.798440/BIBTEX.

86. Ruzycka-Ayoush, Monika, Patrycja Kowalik, Agata Kowalczyk, Piotr Bujak, Anna M. Nowicka, Maria Wojewodzka, Marcin Kruszewski, and Ireneusz P. Grudzinski. 2021. Quantum dots as targeted doxo-rubicin drug delivery nanosystems. *Cancer Nanotechnology* 12. BioMed Central Ltd: 1–27. https://doi.org/10.1186/S12645-021-00077-9/FIGURES/11.

87. Gao, Xiaohu, Yuanyuan Cui, Richard M. Levenson, Leland W.K. Chung, and Shuming Nie. 2004. In vivo cancer targeting and imaging with semiconductor quantum dots. *Nature Biotechnology* 22: 969–976. https://doi.org/10.1038/nbt994.

15 Obstacles in Utilizing Nanomedicine for Cancer Management

Kamana Singh, Vineeta Kashyap, and Addanki P Kumar

CONTENTS

15.1 Introduction..259
15.2 Limitations Crossing the Physiological Barriers of Body 260
    15.2.1 Activation of Complement Proteins with Nanomedicines...................................261
    15.2.2 Nanomedicine Uptake and Clearance by Macrophage Phagocytic System (MPS)..261
15.3 Tumor Environment and Vasculature as a Barrier to Delivery of Nanoparticles to the Tumor Site...262
15.4 Different Properties of Colloids Affecting the EPR Effect in Tumor Cells 263
    15.4.1 Size of Nanoparticles 263
    15.4.2 Charge of Nanoparticles 263
    15.4.3 Shape of Nanoparticles 264
    15.4.4 Effect of EPR of NM in Cancerous Cells in Humans 264
15.5 Targeted Delivery of NM by Nanocarriers to the Tumor Site 265
    15.5.1 Passive Targeting of NMs...265
    15.5.2 Active Targeting of NMs 265
    15.5.3 Triggered Release of NMs 265
15.6 Eliminating Multidrug Resistance (MDR) by Nanoparticles...266
15.7 Government Rules and Regulations on the Manufacturing and Commercialization of Nanomedicines 267
15.8 Challenges Associated with the Translation of NM from Laboratory to Clinic 268
15.9 Conclusions 269
15.10 References..269

15.1 INTRODUCTION

Cancer remains a major life-threatening disease among people in India and worldwide, despite significant improvements in our understanding of the molecular mechanism and pathology of cancer and the ongoing development of chemotherapy in cancer treatment. Though chemotherapy is the most popular for the treatment of cancer cells as it reasonably cures malignant tissues and increases patient survival, it has several drawbacks. Anticancer drugs are often associated with adverse side effects that create a lot of discomfort in cancer patients. They have low tumor selectivity and significant toxicity in the healthy tissue, and so a challenge remains to overcome these drawbacks. Even if the treatment is successful, multidrug resistance (MDR), while initially successful, needs to be overcome as it may reduce the efficacy of chemotherapy and hinders the inhibition of tumor growth [1].

Nanomedicine (NM) is a newly emerging field of nanotechnology that can overcome many of the drawbacks of chemotherapeutic drugs. Its nano size, composition, and capacity to modify the

DOI: 10.1201/b23311-15

surface give unique characteristics that can enhance the pharmacokinetics of integrated medicinal medicines considerably. Modifications of the drugs in nanoparticles (NPs) can increase solubility, lengthen circulation duration, prevent immune system clearance, and allow for controlled drug release. Targeted delivery of nanomedicines (NMs) to a tumor site by either passive or active targeting methods makes them more advantageous over the anticancer drugs and results in minimizing the toxicity on the surrounding healthy normal tissues. The specific targeting of these NMs to the cancer tissues led to the possibility of the development of personalized cancer therapies.

Over the past few years, recent advancements in the field of nanotechnology have led to the generation of an arsenal of NPs, like organic NPs, liposomes, niosomes, polymeric [2] NPs, dendrimer NPs, lipid NPs, carbon-based NPs, inorganic NPs (quantum dots, silica, gold), and magnetic NPs. NMs are preferred over conventional drugs because of their higher surface-area-to-volume ratio, improved loading capacity, bioavailability, higher reactivity, and efficacy. These micro-sized particles can specifically deliver the drug at the tumor site and prove to be beneficial in the treatment of many types of cancer and cardiovascular and neurodegenerative diseases. Despite different types of NPs and NMs available, still, a relatively small number of NMs are available in market to be used clinically for treatment of cancer. Thus, in this chapter, an attempt is made to understand all the problems underlying the poor clinical translation of NMs.

15.2 LIMITATIONS CROSSING THE PHYSIOLOGICAL BARRIERS OF BODY

The innate immune system is the body's immune response that provide the first line of defense against the invading pathogens. To defend itself, human body comprise many layers of physical, chemical, physiological, and biological barriers. Hairs covering the skin and most parts of our body, the outer layer of skin that is the epidermis, the air-blood barrier in the lungs, the mucous membrane lining the gastrointestinal tract and the reproductive system, and the blood-brain barrier all fall under the examples of physical barriers that the nanomaterials have to cross to reach the bloodstream [3, 4]. To bypass these surface barriers and utilize the nanotherapeutic approach at the illness site, nanomaterials must be properly tailored. The first physiological site/environment that the NP meets after it is intravenously injected is the blood. Blood has thousands of proteins that may interact with these NPs, affecting their bioavailability and penetrance into the other specific cellular tissues or aid in removal by the macrophages or may [5] promote undesirable allergic reactions and inflammatory reactions [2]. Further, as these NMs travel through the body, they interact with different proteins, like transferrin and albumin, present in the blood, which will compete for binding to the surface of NMs, forming a layer known as protein corona [6]. The formation of this layer of protein corona is influenced by the physicochemical properties of NPs, like size, charge, density, shape, conformation, porosity, colloidal stability, density, chemical composition, affinity, and surface area, that influence its activity [7].

On one hand, nanomaterials can have positive impacts on proteins, such as immobilizing enzymes and enhancing protein activity and stability, and, on other hand, have a negative impact by being harmful to cells and organs. The presence of large number of serum protein molecules on the surface of NP forming the protein corona gives the NP its unique biological identity that helps it to targeted selectively to the tumor tissues, but if this interaction happens in an uncontrolled manner, the corona can no longer protect the NPs, and it can be cleared by tissue macrophages and may lead to undesirable inflammatory reactions. Positively charged NPs interact with negatively charged biological membranes with more precision, enhancing cellular uptake, and improving lysosomal escape without producing toxicity. Apart from the charge on the NP, different types of carbon nanomaterials may have different effects on protein structure and function. Graphene sheets, for example, have a greater capacity for adsorption of serum proteins than carbon nanotubes [8]. Carbon nanotubes, on the other hand, are substantially more difficult to disintegrate enzymatically than graphene and fullerene derivatives [9]. Thus, nanomaterial characteristics, both individually and in combination, have a critical role in defining their interaction with the different immune cells affecting their cellular uptake and biological function.

15.2.1 ACTIVATION OF COMPLEMENT PROTEINS WITH NANOMEDICINES

A group of 30 serum proteins known as complement proteins, present in both soluble and membrane-bound forms, constitute an important arm of innate immunity. These proteins recognize and neutralize the foreign invading microbes that are picked up immediately by phagocytic macrophages, B cells, and dendritic cells. Likewise, when NPs are injected intravenously into the bloodstream, they are immediately recognized as foreign and activates the complement pathway. During this process, many proinflammatory split products of complement proteins known as anaphylatoxins (C3a, C4a, and C5a) are also produced that can either lead to adverse reactions and tissue damage in individuals hypersensitive to them, while some complement proteins, such as C3b, iC3, and C4b, can aid in binding and coating of the NPs, resulting in attracting phagocytic cells to absorb them [10]. These complement proteins and antibodies are referred to as *opsonins*, and the process is known as *opsonization*, which helps the phagocytic cells to bind to the surface of these NPs/nanodrugs and clear them from the blood before they can even reach the target tissue (e.g., tumor). This was demonstrated in many studies conducted both in vitro and in vivo where the adsorption of the IgG antibodies on NP surface was shown to improve macrophage detection and absorption of a variety of nanomaterials [11–13]. For example, in vitro and in vivo study shows about sixfold enhanced absorption of IgG coated liposome by isolated liver macrophages [14]. Likewise, the coating of IgG opsonin on surface of NPs made up of iron oxide boosts uptake by macrophages by about tenfold [15].

15.2.2 NANOMEDICINE UPTAKE AND CLEARANCE BY MACROPHAGE PHAGOCYTIC SYSTEM (MPS)

Peripheral lymphocytes and resident macrophages are known to effectively remove microorganisms in organs like the liver, spleen, and lymph nodes. These macrophages are also responsible for transport of NPs from these organs to the tumor site. However, as more than 90% of these NMs are cleared early, this has an impact on their transport to the tumor site, thereby resulting in a poor prognosis for cancer patients. NMs are also eliminated via autophagosomes, which are vesicles fused with lysosomes and contain hydrolytic enzymes that help destroy intracellular contents and old/damaged cells while maintaining the integrity of the cell. Some of the NPs, such as Fe_3O_4 NPs, have been found to trigger autophagy in human cervical cancer cells. The autophagic response of the cell is decreased when more high-molecular-weight/large-size proteins are adsorbed on the surface of Fe_3O_4 NPs, according to Kong H. et al. [16]. However, while this prevents NPs clearance by MPS, the substantial dosage of these nanomaterials elevates the risk of toxicity owing to particle buildup in the liver and spleen. The chemicals used during the manufacturing of NPs may lead to cytotoxicity. In addition, interactions between nanocarriers and leukocytes have the potential to cause inflammatory responses or change the usual course of an immune response, both of which might have an influence on tumor growth. The tumor microenvironment (TME) has an impact on the phenotype of macrophages in the solid tumors. TME can either activate the macrophages to secrete chemicals/interleukins to help the in the progression of the tumor by promoting angiogenesis and cell proliferation, or the presence of nano-based drugs may divert the macrophages to secrete interleukins that lead to inflammation and tissue damage. Thus, strategies that avoid premature clearance by MPS, reduce the toxicity, and avoid the unwanted immune response are much needed.

To achieve this, four strategies that involve modification of NPs' surface, physicochemical properties of NPs, biomimetic architecture, and regulation of the MPS are followed to counteract MPS absorption, reduce off-target rates, and improve cancer therapy efficacy [17]. To escape early detection and clearance by the macrophage phagocyte system, one can modify the surface of NMs by attaching polyethylene glycol (PEG), dysopsonic proteins, and other carbohydrate moieties that help NMs stay for a longer time in the blood improving their efficacy [18]. Using cell membrane extracts from blood or tumor cells as a defense against phagocytosis is another intriguing method. The extracts can be utilized to encapsulate NPs to prevent MPS absorption. Thus, the number of processes and their respective contributions are determined not only by the type of cell but also by the biological identity that is different from synthetic identity of NPs. However, the process responsible for a certain nanomaterial's uptake is unknown.

15.3 TUMOR ENVIRONMENT AND VASCULATURE AS A BARRIER TO DELIVERY OF NANOPARTICLES TO THE TUMOR SITE

Normal and cancerous cells have significantly diverse morphologies. Tumor cells are disorganized cells that divide in an erratic manner. When a solid tumor reaches a particular size, more oxygen is required to allow the massive mass of cells to flourish to their full potential. Tumor cells secrete growth factors that promote angiogenesis by producing aberrant blood vessels with fenestrations and an improper vascular bed that allows NPs to flow through more readily [19]. This illustrates the improved permeability component of tumor cells referred as enhanced permeation and retention (EPR) effect.

Furthermore, in these cells, limited interstitial fluid absorption is caused by insufficient lymphatic capillary activity. Hence, NPs cannot diffuse back to the circulation because of large hydrodynamic radii and end up being accumulated in the tumor tissues [20] illustrating the enhanced retention component of the EPR process.

The mechanism of EPR effect must be well understood since it varies based on the kind of tumor, as well as across patients and over time within the same patient's tumor progression. Although several attempts at noninvasive delivery of NPs via oral/nasal/transdermal routes have recently been made, many nanodrugs used in cancer immunotherapy are still delivered intravenously and accumulate in cancerous tissues, indicating the prominent role of EPR in the success of clinical trials by NMs.

M2-tumor-associated macrophages are specialized macrophages that gather and destroy functional nanomaterials introduced into the tumor [21]. Nanomaterials will be forced back into circulation due to the high tumor interstitial fluid pressure, preventing them from reaching tumor areas. Intratumoral delivery of nanomaterials is particularly problematic due to these pathophysiological characteristics. To circumvent the stromal barrier, researchers have tried drugs to deplete stromal barrier and to block transforming growth factor-beta (TGF-β) cytokine pathways and interleukin-6 (IL-6) pathway [4].

Pei et al. (2019) [22] observed an improved therapeutic efficiency in pancreatic cancer, when TGF signaling and the KRAS mutation were targeted at the same time.

Therefore, it is critical to understand the association between the immune cells present in tumor cells and nanomaterials so that nanomaterials that may help in overcoming these obstacles can be synthesized accordingly.

The cancer cells release angiogenic agents for that dictates the design and shape of blood vessels [23]. These new vessels are disorganized and discontinuous and have tiny pores [24] that make them leaky, allowing movement of endothelial cells and exchange of substances like nanomaterials and macromolecules to pass through the surrounding cells. The permeability of substances though these fenestrated blood vessel is influenced by a number of factors, including type and stage of cancer, and site of implantation in xenograft [19, 25].

The amount of collagen and glycosaminoglycans (GAG) in the ECM and its degree of organization are also related to the resistance to interstitial transport [26]. This can be overcome by the enzyme collagenase that will help in restoring the mobility of slow diffusing species by breaking the entangled protein [26]. The consequences of GAG-disrupting enzymes, on the other hand, are less evident. Heparinases, another enzyme, can prevent the interaction of colloids with the heparin sulfate moieties of ECM by facilitating the movement of macromolecules having positively charged, in some circumstances [27]. Finally, MPS cells in tumor interstitium through their phagocytes can obstruct the passage of NPs toward the cancer cells and retain them in the tumor interstitium [28].

15.4 DIFFERENT PROPERTIES OF COLLOIDS AFFECTING THE EPR EFFECT IN TUMOR CELLS

As the physicochemical properties of colloids have an influence on their circulation time in blood, interaction with the cells and proteins of ECM in tumor, extravasation through the vessel, and their diffusion rate, thereby reaching the tumor tissue, it is difficult to correctly assess the impact of each attribute on the EPR [29, 30].

15.4.1 SIZE OF NANOPARTICLES

As the design and functionalization choices for NPs are so diverse, and the list of possible uses keeps growing, the current trend in NM is toward custom-made devices [31]. It is crucial to note, however, that treatment based on nanodrugs are not magical cures as the manufacturing of these NM are still gripped by many obstacles. Protein adsorption leading to forming corona on surface of NPs can modify not only particle size, stability, and surface characteristics but also their behavior and distribution inside the body affecting cellular uptake [32–34]. In terms of circulation time, interaction of NM nonspecifically with serum proteins is one limiting factor. Another problem is that due to the coating of NM with opsonin protein in the liver and spleen, they are easily identified by the MPS and are quickly cleared from the blood. This would be critical if these organs are the targeted for drug delivery [2]. The most common method for avoiding the quick clearance and to promote longer stay of NMs in the blood and avoid liver accumulation is to coat them with PEG polymer. The hydrophilic ether oxygen of PEG polymer on the surface of NPs helps in reducing the steric effects between them NPs, thereby reducing their adsorption [2, 35]. Furthermore, it was observed that there was a reduced uptake of gold NPs in vitro by the macrophages when a large number of PEG molecules (i.e., density of PEGylation) was increased on the surface of gold NPs [13]. For tissues with high blood flow, a short blood circulation half-life may be adequate. To gradually extravasate in weakly perfused tissues, however, longer circulation half-lives are required [36]. Furthermore, the size of a nanosystems has a significant impact on its biodistribution inside the body. Renal filtration, thus, clears nanodevices smaller than 10 nm [2]. On the other hand, nanodevices having the size of 100 nm or more than that concentrate in the liver, and that is why the liver is most affected organ due to the side effects of these drugs [37]. For overcoming biological obstacles and attaining their goal, several sizes of nanodevices are therefore recommended. It was noticed that in lung tumors or inflammatory tissue, there was a specific and precise delivery of nanodevices having diameter in the range of 31–80 nm (Souris et al., 2010). Therefore, nanosystems with a diameter of 10 to 100 nm are the preferrable size used for delivery of drugs to specific tissues.

15.4.2 CHARGE OF NANOPARTICLES

The EPR effect can be influenced by the charge of macromolecules and nanomaterials via altering systemic circulation timings and intratumoral processes. Again, dealing with these phenomena separately is frequently challenging. The charge on the surface of nanomaterial can affect the material's opsonization, cell recognition in MPS organs, and overall plasma circulation profile [29, 30, 38, 39]. For example, positive charges on the surface of nanomaterial are usually regarded as having a detrimental influence on circulation kinetics of the nanomaterial in the plasma [38, 39], whereas negative charges on their surface can either enhance, reduce, or have no impact on the clearance or circulation time of these NPs [40, 41].

Despite the shorter blood circulation times in tumor-bearing animals, non-PEGylated, liposomes containing the lipid 1,2-diacyl-trimethylammonium propane (DOTAP) that are positively charged are found to be accumulated in tumor tissues with poor diffusion to interstitium compared to the liposomes that have negative charges or neutral charges [42]. Therefore, positive charges may favor NP interactions with tumor blood vessels, preventing their intratumoral diffusion. Recently, in humans and preclinical models [42, 43], this nanomaterials have been used for therapeutic purposes by targeting the tumor vascular endothelium with antiangiogenic medicines.

Charges had a detrimental influence on the migration of NPs through the matrix in ex vivo tests on ECM extracted from mice sarcoma. Indeed, the presence of charge beyond a particular threshold—that is, zeta potential in the range of +10 mV to −30 mV, respectively—stopped diffusion of the NPs in the interstitium [27]. This is consistent with the findings of studies related to administration of NPs that are neutral and charged, which revealed that colloids that have charge on their surface had a longer interaction with the tumor than neutral colloids [44].

15.4.3 SHAPE OF NANOPARTICLES

Besides the size and charge, the shape of NPs also impacts the extravasation and its blood circulation kinetics of NPs [30, 45, 46]. Furthermore, in spite of the high aspect ratio of single-wall carbon nanotubes (i.e., 100:1 to 500:1) and being 10–20 times greater than the threshold size (i.e., 100–500 nm) that are permissible and filtered through glomerulus of kidneys, these nanotubes are cleared and filtered efficiently through the kidneys because of their elongated shapes [47]. These concerns motivated researchers to investigate the possibility of varied aspect ratio nanomaterials to accumulate in malignancies.

Likewise, when tumor dissemination kinetics of nanorod and nanosphere having the same hydrodynamic radius, identical blood circulation kinetics but with different lengths were compared, it was observed that nanorods with a length of 44 nm were able to extravasate four times quicker to the interstitium and diffuse deeper into tumor engaging 1.7-fold bigger volume of tumor cells compared to 35 nm nanospheres [48]. Similarly, viral nanofilaments produced from plant viruses that have elongated shape demonstrated greater accumulation and diffuse deeper into the tumor cell compared to other viral spherical constructions [49]. Overall, our data show that elongated design may have advantageous EPR features.

15.4.4 EFFECT OF EPR OF NM IN CANCEROUS CELLS IN HUMANS

Despite the well-known mechanism of EPR in tumor cells, further research is needed to understand it fully in newly developed rodent and orthotopic xenografts models to determine its occurrence in human primary and metastatic cancers [50, 51]. This is especially crucial, if we want to enhance overall survival in patient cohorts by administration of these few available first-generation anticancer NMs in the market. In this patient cohort, there could be some non-responsive individuals that may hide subpopulations with higher susceptibility to NPs [28]. Thus, a careful selection of patients that are impacted by EPR process needs to be done that might help to improve the efficacy of NMs in clinical practice [28].

The lack of evidence regarding the EPR effect in humans can be attributed to several factors, the most prominent of which being the difficulty in acquiring relevant biodistribution data in humans.

Because main and metastatic lesions might behave differently [50, 51], it is still unclear if NPs deposition in metastases follows the same pattern as original tumors [52]. Patients in early clinical trials are seldom naive to therapeutic drugs, and the majority had received many therapies that may have influenced the outcome.

The tumors microenvironment is complicated. Finally, analyzing the true impact of EPR process in human cancer cells is complicated by the fact that the ideal settings for maximizing dispersion and retention in people may differ from those necessary in animal models as animals and humans have different clearance mechanisms and tumor biology [53].

15.5 TARGETED DELIVERY OF NM BY NANOCARRIERS TO THE TUMOR SITE

15.5.1 PASSIVE TARGETING OF NMs

The EPR effect [54] allows NPs with sufficiently long circulation periods to collect passively within malignant tissues and is known as passive targeting [55]. However, a recent literature review on NMs showed that hardly less than 1% of the injected dose of NM was only able to reach the malignant tissue in many situations. Whether 1% is enough to have a therapeutic impact was a point of contention. In tiny tumors, a concentration of 1% of the entire administered dosage may be sufficient.

Furthermore, many people believe that the important question is that it is not the dose of NM that enters the tumor but how much of the active medicine finally reaches the tumor cells and is absorbed. However, to treat tumors effectively with NPs, it is important that the active pharmacological ingredients (APIs) of NPs concentrate at the targeted tissue, and the release of the encapsulated drug from the NP must be enhanced. It is also critical that NMs carrying drug cargo circulate long enough in the bloodstream and in order to achieve this polymer like PEG can be conjugated to NMs' surface. Theek et al. (2014) discovered a link between tumor vascularization and NM passive targeting. Doxil (pegylated liposomal doxorubicin), for example, is the first FDA-approved NM, and it has shown improved effectiveness in ovarian cancer and AIDS-related Kaposi's sarcoma as compared to typical conventional therapy [56]. When doxorubicin is encapsulated in PEGylated liposomes, PEGylation prevents clearance of these NPs by the reticuloendothelial system (RES), extending circulation half-life and to accumulate in the tumor tissues [57]. Furthermore, by inhibiting doxorubicin release into the cardiac vasculature, these NPs will prevent the concentration of free drug in the plasma [58] and considerably lowers the risk of cardiotoxicity [57].

15.5.2 ACTIVE TARGETING OF NMs

The physically or chemically attachment of ligands, such as antibodies or peptides, onto the surface of NMs to facilitate uptake by target cells is known as active targeting [55, 59]. In vivo studies by Willis and Forssen (1998) [60] and Hua (2013) [61] found that ligands bound to NMs can bind to receptors present on the tumor cells and can deliver anticancer drugs to these cancerous cells [62–64]. The question of whether ligand-targeted NMs may considerably increase accumulation of NM at target locations over non-targeted NM (passive targeting) is currently being debated [65]. Even though enhanced accumulation of ligand-targeted NMs have been proven in tumor cells, there were no variations in accumulation of non-targeted NMs in target tissues. For instance, in HER2-positive breast cancer xenograft model, active (liposomes coupled with monoclonal antibodies to HER2 antibody fragments in tumor tissue; 7–8% administered dose/g tumor tissue) or passive targeting of liposome leads to comparable high quantities of both the liposomes in the tumor tissue [66]. However, anti-HER2 immunoliposomes loaded with doxorubicin outperformed all other control groups, including recombinant Mab trastuzumab against HER2 ligand, non-targeted liposomes loaded with doxorubicin, and free doxorubicin [67]. Thus, these in vivo pharmacodynamic differences of the targeted NM formulation compared to other control groups forms the basis for the increased antitumor efficacy of these NPs [68].

15.5.3 TRIGGERED RELEASE OF NMs

A third targeting technique uses NMs that respond to stimuli and is known as triggered drug release. Here, an endogenous or external stimulation was used for the increased release of drug in tumors.

The NM susceptible to endogenous stimuli involves use of factors, such as low pH, specific enzymes, and redox gradient, in the microenvironment of tumor cells. Exogenous-responsive NMs release drugs in response to external stimuli, like temperature, ultrasound, magnetic fields, or light. Thus, presently, the most promising strategy appears to be the deployment of external temperature changes to trigger release of therapeutic chemicals from NMs, such as thermosensitive liposomal doxorubicin (ThermoDox) [69]. ThermoDox was found to be superior to Doxil in treatment of non-resectable hepatocellular carcinoma. Lipids like distearoyl phosphocholine (DSPC) and/ or polymers such as poly(N-isopropylacrylamide) are commonly used to modify thermosensitive liposomes. This composition ensures stability of these NMs while keeping their contents stable at physiologic temperatures. On the other hand, some NPs maintain a phase transition temperature that keeps their surface more permeable, allowing the cargo to be released on application of heat [70]. With the addition of imaging moieties, the benefits of these NMs can be further enhanced, allowing for monitoring of biodistribution, target accumulation, and effectiveness.

15.6 ELIMINATING MULTIDRUG RESISTANCE (MDR) BY NANOPARTICLES

MDR is a major phenomenon observed in cancerous cells where they develop resistance to a wide range of chemotherapeutic drugs. These cells overexpress transmembrane transporters that leads to increased efflux of these drugs and so these drugs not being absorbed by them. NPs can prevent this MDR by either distributing several medicines or siRNAs. If several siRNAs are administered at the same time, many genes, including drSug resistance genes, may be silenced. For example, NPs containing drug doxorubicin and siRNA were used to silence resistant Bcl-2 gene by targeting the epidermal growth factor receptor (EGFR) on cancer cells. Also, co-delivery of doxorubicin and siRNA lead to more effective suppression of the size (twofold decrease) of lung tumors in mice compared to their treatment with doxorubicin or siRNA alone [71]. Short single-stranded sequences of DNA known as aptamers have showed potential in cancer cells for bypassing this resistance mechanism [72]. Liu et al. [73] created an aptamer against the target protein nucleolin. In vivo finding shows showed an 88% tumor inhibition rate with the aptamer functionalized liposomes compared to a 22% inhibition rate following free doxorubicin injection [74]. NM's higher efficacy in MDR is thought to be the result of its unique enhanced absorption mechanism. This enhanced cellular absorption is brought about by these targeted NPs. The NPs can easily fuse inside the membrane of target cell and deliver the drug specifically.

Another major problem leading to chemoresistance is the presence of active oxidative phosphorylation (OXPHOS) present in mitochondria of cancer cells [75, 76]. As a result, mitochondrial targeting NPs have been developed to cause cell death by selective drug accumulation. Jiang et al. [77] created paclitaxel-containing liposomes by attaching 2,3-dimethylmaleic anhydride (DMA) to these NPs that could be used for treatment of cancer cells. DMA promotes cellular absorption by reversing the charge of NPs. Similarly, death of MDR lung cancer cells in humans and in xenografted mice was promoted by inserting mitochondria specific peptide lysine-leucine-alanine into the surface of NPs.

A recent study found that following mitochondrial drug accumulation, treatment results in MDR cancers improved [78]. For example, when doxorubicin loaded dendrigraft poly-L-lysine (DGL) NPs with aptamers used to target nucleolin on tumor cells and cytochrome C in mitochondria were released from the mitochondria, it resulted in apoptosis of MDR resistant cancer cells in vitro and in vivo, demonstrating that DGL NPs successfully circumvented acquired and intrinsic MDR. Another way to overcome MDR mechanism is to modulate ceramide concentration in the cell membrane [79]. Ceramides are membrane lipids that govern many cellular signaling pathways including cell differentiation, proliferation, and programmed cell death. Overexpression of the enzyme glucosylceramide synthase is linked to MDR [79].

Intravenous injection of ceramide with the chemotherapy drug paclitaxel restored apoptotic signaling in MDR human ovarian cancer cells [80]. It was observed that inhibition of glucosylceramide synthase enzyme by addition of the drug tamoxifen enhanced the antitumor effect by tenfold of paclitaxel in MDR human ovarian adenocarcinoma cells as compared to threefold in non-MDR ovarian cancer cells [81]. This was further supported by findings in xenograft mouse model where in the same enhanced antitumor effect was observed in tumor cells to paclitaxel-loaded NPs in presence of tamoxifen. Therefore, in addition to specific targeting of NPs to tumor cells, many drugs can be co-delivered with the NP to repair the disrupted signaling pathway to overcome the MDR [81]. Thus, it is critical to have a deep understanding of MDR mechanism in patients such that an appropriately designed NP can be chosen for the therapy of chemo-resistant cancers

15.7 GOVERNMENT RULES AND REGULATIONS ON THE MANUFACTURING AND COMMERCIALIZATION OF NANOMEDICINES

NMs have the potential to boost pharmaceutical industry growth and enhance health outcomes. Manufacturers have to follow government rules regulation for the commercial production of good quality NMs. They also have to comply with government rules set for patenting of these NMs. The absence of clear government laws pertaining to manufacturing of safe NMs in a regulated manner has hampered the development of NMs and prevented them from reaching the clinic in a timely and effective manner [82–84]. Polymers, for example, have been studied extensively as a platform for NM methods; nevertheless, their safety and efficiency are largely reliant on the molecular weight, conjugation chemistry, molecular structure, and polydispersity of the polymer [82, 85]. For the evaluation of newly innovative complicated API containing NM formulation, an adequate regulatory framework is required on an urgent basis [82]. Because each polymer-based NM is unique, dosages, administration methods, dosing frequency, and intended therapeutic application must all be considered separately. This is true for most NM systems. NMs are now managed by the traditional regulatory framework, which is overseen by each country's major regulatory authority (e.g., FDA). First-generation NM products were simply required to fulfill broad requirements applicable to pharmaceutical substances in order to gain regulatory clearance, but the same regulatory rules do not apply to the newly emerging NM products prepared for clinical use because of the complicated chemical nature of these NMs, the type of nanomaterials used and their unclear interaction with proteins in the tumor cell. Both pharmaceutical and medical device laws require regulatory standards and methods that are verified particularly for newly developed NPs. These laws should consider all the factors starting from manufacturing of NMs to their safety profile, their mode of delivery, pharmacokinetic profile and their preclinical and clinical trial studies [84]. To maintain the safety and quality of NMs, a delicate balance must be struck without much restriction, which might stifle progress by increasing the price of bringing new items to the market, obtaining regulatory permission, and/or using a considerable amount of resources a period of time during which a patent is valid. Global regulatory guidelines for NMs should be developed and be formed in conjunction with significant countries with vested interests.

Although significant progress has been made in the previous five years, a meaningful partnership is required between the academia, regulatory agencies, and the industry [86]. This is especially important given the scarcity of contract manufacturing companies worldwide that specialize in creating NM goods in line with GMP [86]. It is worth noting that this small group of manufacturers might be further split based on their infrastructural capabilities for generating various NM platforms. All the countries should follow the same rules and regulation pertaining to the manufacturing of these NMs.

NM manufacturing procedures will need to be thoroughly evaluated for their suitable industrial standards, quality control, and issues related to environment [87]. Newly manufactured NMs must still fulfill certain requirements regarding their purity, stability, sterility, and other general pharmaceutical criteria manufacturing activities and associated standards of control industrial operations. Furthermore, new analytical tools, toxicity testing methods, and established processes need to be documented in order to assess critical physical properties of NMs and their in vivo performance [83, 84, 87].

15.8   CHALLENGES ASSOCIATED WITH THE TRANSLATION OF NM FROM LABORATORY TO CLINIC

Nanotechnology is a fast-evolving field that offers prospects for delivering tiny compounds for cancer detection, diagnosis, therapy, and MDR avoidance. However, there are still a slew of challenges and concerns concerning their clinical development, stability, and toxicity. Nanodrug development is problematic due to a lack of adequate in vitro models capable of accurately reproducing the in vivo situation. Current therapeutic applications of nano-formulations are based on in vitro research with cell lines, which fail to reflect the complexities of NP-cell interactions in vivo.

To discover and overcome the hazards associated with nanotechnology and to limit the negative impacts of nanotechnology on human health and the environment, preliminary in vivo performance and toxicity testing of NPs using appropriate animal models must be carefully carried out. This is a new field, and its yet unclear which type of NPs will be most effective in treating cancer. An improved knowledge of the pharmacological, cellular, and physiological variables that regulate nanotech-based medication delivery is essential for rational nanotechnology design. Before therapeutic NPs can be created, the optimal physicochemical features must be discovered. Difficulties in developing NP libraries with distinct characteristics in a timely, accurate, and repeatable manner along with screening the quality of several NPs remain a challenge.

The burden of producing complex NPs with adjustable physicochemical properties and large-scale, repeatable, and quick development of NPs [78, 88–90], which was not feasible with the traditional, antiquated methods used to produce polydisperse NPs on a bulk scale, has been lessened by the recently developed Microfluidic technology.

Another technique called the particle replication in non-wetting template (PRINT) technology [91, 92] can produce NPs in a reproducible and regulated way with the required chemical composition, drug loading, surface characteristics, size, and shape.

Preliminary in vivo performance and toxicity testing of NPs using suitable animal models must be carried out carefully to overcome the risks associated and to minimize the adverse effects of nanotechnology on human health and the environment. In contrast to *in vitro* models, recent attempts to construct biomimetic organ-/tumor-on-a-chip tools [93, 94] will help to provide a detailed insight of the mechanism of NPs accumulation and the factors (particle size, flow rate of interstitial fluid, and interaction with cells), leading to their diffusion into the tumor cell. By contrasting the behaviors of NPs in such chip tools to those of animal models, the potential of these biomimetic microdevices may be demonstrated.

One major challenge is the disparity between preclinical efficacies and clinical trial outcomes, which is due to unavailability of proper tumor model that can accurately mimic human cancers [95, 96]. Many cell line-derived xenografts (CDX), genetically engineered murine models (GEMMs) [95, 96], and patient-derived xenografts (PDXs) [97] are recently developed animal and human tumor metastasis models [98] that closely resemble the anatomy and heterogeneous nature of a tumor cell. However, there is not a single animal or human xenograft model that can purely mimic all the characteristic of a cancerous cell in human and can avoid the EPR effect. Furthermore, because tumor metastases play such a large role in cancer mortality, we need a human tumor metastasis model that can be used to compare EPR effects and which can be targeted for NP delivery. Thus, for clinical translation of nanotherapeutics, one may need to develop more such humanized mouse models and GEMMs with aggressive metastasis [99].

During the manufacturing of NPs in industries, one needs to follow standard chemistry, manufacturing, and controls (CMC) and good manufacturing practices (GMP). Pharmaceutical companies have been so far using standard CMC and GMP protocols for manufacturing simple NPs containing APIs at commercial level, but with the development of new types of NMs at the preclinical stage, industries are being forced to adapt their production processes and use new CMC and GMP protocols. Another obstacle to the clinical development of NMs is the pharmaceutical companies' compliance with these additional CMC and GMP processes as a result of the complexity of NMs.

The most difficult task is producing NPs on a large scale and in a repeatable manner when new technologies or several phases are involved in the manufacturing process. Even while the PRINT approach makes it possible to produce NPs repeatedly [91], scaling to kilograms has not yet been demonstrated.

A coaxial turbulent jet mixer system has recently been created [100] to manufacture 3 kg/day of polymeric NPs per channel, and it offers the advantages of homogeneity, repeatability, and tunability that can only be achieved with methods like microfluidics. Therefore, sufficient numbers of NMs must be produced employing these reliable and flexible approaches, including PRINT and turbulent jet mixer system, to hasten the clinical translation of NMs.

15.9   CONCLUSIONS

In summary, this review helps us to understand various challenges involved in the manufacturing and development of NMs and their translation from lab to clinic for cancer therapy. A better understanding of tumor microenvironment, EPR effect, tumor markers/receptors, the interaction of NMs with cell, and specific targeting of these NMs to the tumor tissues may lead to the possibility of the development of personalized NMs for cancer patients. Despite numerous hurdles faced by researchers in translating nanodevices into clinically authorized medicines, their prospective benefits should contribute to their successful development. Furthermore, the ongoing demand for innovative anti-cancer medicines necessitates the need for scientific studies to develop nano-based techniques that will help to enhance patient survival in near future.

15.10   REFERENCES

1. M. M. Gottesman, "Mechanisms of Cancer Drug Resistance," *Annual Review of Medicine*, vol. 53, no. 1, pp. 615–627, Feb. 2002, doi: 10.1146/annurev.med.53.082901.103929.
2. D. Owensiii and N. Peppas, "Opsonization, biodistribution, and pharmacokinetics of polymeric nanoparticles," *International Journal of Pharmaceutics*, vol. 307, no. 1, pp. 93–102, Jan. 2006, doi: 10.1016/j.ijpharm.2005.10.010.
3. G. Nabil, K. Bhise, S. Sau, M. Atef, H. A. El-Banna, and A. K. Iyer, "Nano-engineered delivery systems for cancer imaging and therapy: Recent advances, future direction and patent evaluation," *Drug Discovery Today*, vol. 24, no. 2, pp. 462–491, Feb. 2019, doi: 10.1016/j.drudis.2018.08.009.
4. H. Meng, W. Leong, K. W. Leong, C. Chen, and Y. Zhao, "Walking the line: The fate of nanomaterials at biological barriers," *Biomaterials*, vol. 174, pp. 41–53, Aug. 2018, doi: 10.1016/j.biomaterials.2018.04.056.
5. I. I. Lungu, A. M. Grumezescu, A. Volceanov, and E. Andronescu, "Nanobiomaterials used in cancer therapy: An up-to-date overview," *Molecules*, vol. 24, no. 19, p. 3547, Sep. 2019, doi: 10.3390/molecules24193547.
6. I. Lynch and K. A. Dawson, "Protein-nanoparticle interactions," *Nano Today*, vol. 3, no. 1–2, pp. 40–47, Feb. 2008, doi: 10.1016/S1748-0132(08)70014-8.
7. T. Stylianopoulos and R. K. Jain, "Design considerations for nanotherapeutics in oncology," *Nanomedicine: Nanotechnology, Biology and Medicine*, vol. 11, no. 8, pp. 1893–1907, Nov. 2015, doi: 10.1016/j.nano.2015.07.015.
8. L. S. Franqui *et al.*, "Interaction of graphene oxide with cell culture medium: Evaluating the fetal bovine serum protein corona formation towards in vitro nanotoxicity assessment and nanobiointeractions," *Materials Science and Engineering: C*, vol. 100, pp. 363–377, Jul. 2019, doi: 10.1016/j.msec.2019.02.066.
9. X. Wang, Y. Zhu, M. Chen, M. Yan, G. Zeng, and D. Huang, "How do proteins 'response' to common carbon nanomaterials?" *Advances in Colloid and Interface Science*, vol. 270, pp. 101–107, Aug. 2019, doi: 10.1016/j.cis.2019.06.002.
10. S. Inturi *et al.*, "Modulatory role of surface coating of superparamagnetic iron oxide nanoworms in complement opsonization and leukocyte uptake," *ACS Nano*, vol. 9, no. 11, pp. 10758–10768, Nov. 2015, doi: 10.1021/acsnano.5b05061.
11. G. Borchard and J. Kreuter, " No title found]," *Pharmaceutical Research*, vol. 13, no. 7, pp. 1055–1058, 1996, doi: 10.1023/A:1016010808522.
12. O. Lunov *et al.*, "Differential uptake of functionalized polystyrene nanoparticles by human macrophages and a monocytic cell line," *ACS Nano*, vol. 5, no. 3, pp. 1657–1669, Mar. 2011, doi: 10.1021/nn2000756.

13. C. D. Walkey and W. C. W. Chan, "Understanding and controlling the interaction of nanomaterials with proteins in a physiological environment," *Chemical Society Reviews*, vol. 41, no. 7, pp. 2780–2799, 2012, doi: 10.1039/C1CS15233E.

14. J. T. P. Derksen, H. W. M. Morselt, and G. L. Scherphof, "Uptake and processing of immunoglobulin-coated liposomes by subpopulations of rat liver macrophages," *Biochimica et Biophysica Acta (BBA)—Molecular Cell Research*, vol. 971, no. 2, pp. 127–136, Sep. 1988, doi: 10.1016/0167-4889(88)90184-X.

15. N. Rama *et al.*, "Study of magnetic properties of A_2B^'NbO_6 (A=Ba,Sr, (BaSr): And B^'=Fe and Mn) double perovskites," *Journal of Applied Physics*, vol. 95, no. 11, pp. 7528–7530, Jun. 2004, doi: 10.1063/1.1682952.

16. H. Kong *et al.*, "Serum protein corona-responsive autophagy tuning in cells," *Nanoscale*, vol. 10, no. 37, pp. 18055–18063, Sep. 2018, doi: 10.1039/C8NR05770B.

17. Y. Zhou and Z. Dai, "New strategies in the design of nanomedicines to oppose uptake by the mononuclear phagocyte system and enhance cancer therapeutic efficacy," *Chemistry—An Asian Journal*, vol. 13, no. 22, pp. 3333–3340, 2018, doi: 10.1002/asia.201800149.

18. J. Y. Oh *et al.*, "Cloaking nanoparticles with protein corona shield for targeted drug delivery," *Nature Communications*, vol. 9, no. 1, p. 4548, Dec. 2018, doi: 10.1038/s41467-018-06979-4.

19. S. K. Hobbs *et al.*, "Regulation of transport pathways in tumor vessels: Role of tumor type and microenvironment," *Proceedings of National Academy of Science U S A*, vol. 95, no. 8, pp. 4607–4612, Apr. 1998.

20. T. P. Padera, B. R. Stoll, J. B. Tooredman, D. Capen, E. di Tomaso, and R. K. Jain, "Pathology: Cancer cells compress intratumour vessels," *Nature*, vol. 427, no. 6976, p. 695, Feb. 2004, doi: 10.1038/427695a.

21. L. D. Leserman, J. Barbet, F. Kourilsky, and J. N. Weinstein, "Targeting to cells of fluorescent liposomes covalently coupled with monoclonal antibody or protein A," *Nature*, vol. 288, no. 5791, pp. 602–604, Dec. 1980, doi: 10.1038/288602a0.

22. Y. Pei *et al.*, "Sequential Targeting TGF-β signaling and KRAS mutation increases therapeutic efficacy in pancreatic cancer," *Small*, vol. 15, no. 24, p. 1900631, 2019, doi: 10.1002/smll.201900631.

23. D. O. Bates, N. J. Hillman, B. Williams, C. R. Neal, and T. M. Pocock, "Regulation of microvascular permeability by vascular endothelial growth factors," *Journal of Anatomy*, vol. 200, no. 6, pp. 581–597, Jun. 2002, doi: 10.1046/j.1469-7580.2002.00066.x.

24. P. Carmeliet and R. K. Jain, "Principles and mechanisms of vessel normalization for cancer and other angiogenic diseases," *Nature Reviews Drug Discovery*, vol. 10, no. 6, pp. 417–427, Jun. 2011, doi: 10.1038/nrd3455.

25. R. K. Jain and T. Stylianopoulos, "Delivering nanomedicine to solid tumors," *Nature Reviews Clinical Oncology*, vol. 7, no. 11, pp. 653–664, Nov. 2010, doi: 10.1038/nrclinonc.2010.139.

26. P. A. Netti, D. A. Berk, M. A. Swartz, A. J. Grodzinsky, and R. K. Jain, "Role of extracellular matrix assembly in interstitial transport in solid tumors," *Cancer Research*, vol. 60, no. 9, pp. 2497–2503, May 2000.

27. O. Lieleg, R. M. Baumgärtel, and A. R. Bausch, "Selective filtering of particles by the extracellular matrix: An electrostatic bandpass," *Biophysical Journal*, vol. 97, no. 6, pp. 1569–1577, Sep. 2009, doi: 10.1016/j.bpj.2009.07.009.

28. U. Prabhakar *et al.*, "Challenges and key considerations of the enhanced permeability and retention effect for nanomedicine drug delivery in oncology," *Cancer Research*, vol. 73, no. 8, pp. 2412–2417, Apr. 2013, doi: 10.1158/0008-5472.CAN-12-4561.

29. F. Alexis, E. Pridgen, L. K. Molnar, and O. C. Farokhzad, "Factors affecting the clearance and biodistribution of polymeric nanoparticles," *Molecular Pharmaceutics*, vol. 5, no. 4, pp. 505–515, Aug. 2008, doi: 10.1021/mp800051m.

30. N. Bertrand and J.-C. Leroux, "The journey of a drug-carrier in the body: An anatomo-physiological perspective," *Journal of Controlled Release*, vol. 161, no. 2, pp. 152–163, Jul. 2012, doi: 10.1016/j.jconrel.2011.09.098.

31. S. Tran, P.-J. DeGiovanni, B. Piel, and P. Rai, "Cancer nanomedicine: A review of recent success in drug delivery," *Clinical and Translational Medicine*, vol. 6, no. 1, p. e44, 2017, doi: 10.1186/s40169-017-0175-0.

32. D. Hühn *et al.*, "Polymer-coated nanoparticles interacting with proteins and cells: Focusing on the sign of the net charge," *ACS Nano*, vol. 7, no. 4, pp. 3253–3263, Apr. 2013, doi: 10.1021/nn3059295.

33. P. P. Karmali and D. Simberg, "Interactions of nanoparticles with plasma proteins: Implication on clearance and toxicity of drug delivery systems," *Expert Opinion on Drug Delivery*, vol. 8, no. 3, pp. 343–357, Mar. 2011, doi: 10.1517/17425247.2011.554818.

34. M. P. Monopoli, C. Åberg, A. Salvati, and K. A. Dawson, "Biomolecular coronas provide the biological identity of nanosized materials," *Nature Nanotechnology*, vol. 7, no. 12, pp. 779–786, Dec. 2012, doi: 10.1038/nnano.2012.207.

35. R. Gref *et al.*, "'Stealth' corona-core nanoparticles surface modified by polyethylene glycol (PEG): Influences of the corona (PEG chain length and surface density) and of the core composition on phago-cytic uptake and plasma protein adsorption," *Colloids and Surfaces B: Biointerfaces*, vol. 18, no. 3, pp. 301–313, Oct. 2000, doi: 10.1016/S0927-7765(99)00156-3.

36. J. Shi, P. W. Kantoff, R. Wooster, and O. C. Farokhzad, "Cancer nanomedicine: Progress, challenges and opportunities," *Nature Reviews Cancer*, vol. 17, no. 1, pp. 20–37, Jan. 2017, doi: 10.1038/nrc.2016.108.

37. J. Shi, Z. Xiao, N. Kamaly, and O. C. Farokhzad, "Self-assembled targeted nanoparticles: Evolution of technologies and bench to bedside translation," *Accounts of Chemical Research*, vol. 44, no. 10, pp. 1123–1134, Oct. 2011, doi: 10.1021/ar200054n.

38. K. Xiao *et al.*, "The effect of surface charge on in vivo biodistribution of PEG-oligocholic acid based micellar nanoparticles," *Biomaterials*, vol. 32, no. 13, pp. 3435–3446, May 2011, doi: 10.1016/j.biomaterials.2011.01.021.

39. C. He, Y. Hu, L. Yin, C. Tang, and C. Yin, "Effects of particle size and surface charge on cellular uptake and biodistribution of polymeric nanoparticles," *Biomaterials*, vol. 31, no. 13, pp. 3657–3666, May 2010, doi: 10.1016/j.biomaterials.2010.01.065.

40. R. R. Arvizo *et al.*, "Modulating Pharmacokinetics, Tumor Uptake and Biodistribution by Engineered Nanoparticles," *PLoS ONE*, vol. 6, no. 9, p. e24374, Sep. 2011, doi: 10.1371/journal.pone.0024374.

41. D. Peer and R. Margalit, "Tumor-targeted hyaluronan nanoliposomes increase the antitumor activity of liposomal Doxorubicin in syngeneic and human xenograft mouse tumor models," *Neoplasia*, vol. 6, no. 4, pp. 343–353, Aug. 2004, doi: 10.1593/neo.03460.

42. M. Schmitt-Sody *et al.*, "Neovascular targeting therapy: Paclitaxel encapsulated in cationic liposomes improves antitumoral efficacy," *Clinical Cancer Research*, vol. 9, no. 6, pp. 2335–2341, Jun. 2003.

43. R. B. Campbell *et al.*, "Cationic charge determines the distribution of liposomes between the vascular and extravascular compartments of tumors," *Cancer Research*, vol. 62, no. 23, pp. 6831–6836, Dec. 2002.

44. T. Nomura, N. Koreeda, F. Yamashita, Y. Takakura, and M. Hashida, "Effect of particle size and charge on the disposition of lipid carriers after intratumoral injection into tissue-isolated tumors," *Pharmaceutical Research*, vol. 15, no. 1, pp. 128–132, Jan. 1998, doi: 10.1023/a:1011921324952.

45. J. A. Champion and S. Mitragotri, "Shape induced inhibition of phagocytosis of polymer particles," *Pharmaceutical Research*, vol. 26, no. 1, pp. 244–249, Jan. 2009, doi: 10.1007/s11095-008-9626-z.

46. Y. Geng *et al.*, "Shape effects of filaments versus spherical particles in flow and drug delivery," *Nature Nanotechnology*, vol. 2, no. 4, pp. 249–255, Apr. 2007, doi: 10.1038/nnano.2007.70.

47. A. Ruggiero *et al.*, "Paradoxical glomerular filtration of carbon nanotubes," *Proceedings of National Academy of Science U S A*, vol. 107, no. 27, pp. 12369–12374, Jul. 2010, doi: 10.1073/pnas.0913667107.

48. V. P. Chauhan *et al.*, "Fluorescent nanorods and nanospheres for real-time in vivo probing of nanopar-ticle shape-dependent tumor penetration," *Angewandte Chemie International Edition in English*, vol. 50, no. 48, pp. 11417–11420, Nov. 2011, doi: 10.1002/anie.201104449.

49. S. Shukla, A. L. Ablack, A. M. Wen, K. L. Lee, J. D. Lewis, and N. F. Steinmetz, "Increased tumor homing and tissue penetration of the filamentous plant viral nanoparticle Potato virus X," *Molecular Pharmaceutics*, vol. 10, no. 1, pp. 33–42, Jan. 2013, doi: 10.1021/mp300240m.

50. T. Lammers, F. Kiessling, W. E. Hennink, and G. Storm, "Drug targeting to tumors: Principles, pitfalls and (pre-) clinical progress," *Journal of Controlled Release*, vol. 161, no. 2, pp. 175–187, Jul. 2012, doi: 10.1016/j.jconrel.2011.09.063.

51. S. Taurin, H. Nehoff, and K. Greish, "Anticancer nanomedicine and tumor vascular permeability; Where is the missing link?," *Journal of Controlled Release*, vol. 164, no. 3, pp. 265–275, Dec. 2012, doi: 10.1016/j.jconrel.2012.07.013.

52. A. Schroeder *et al.*, "Treating metastatic cancer with nanotechnology," *Nature Reviews Cancer*, vol. 12, no. 1, pp. 39–50, Dec. 2011, doi: 10.1038/nrc3180.

53. A. Gabizon, H. Shmeeda, and Y. Barenholz, "Pharmacokinetics of pegylated liposomal Doxorubicin: Review of animal and human studies," *Clinical Pharmacokinetics*, vol. 42, no. 5, pp. 419–436, 2003, doi: 10.2165/00003088-200342050-00002.

54. Y. Matsumura and H. Maeda, "A new concept for macromolecular therapeutics in cancer chemother-apy: Mechanism of tumoritropic accumulation of proteins and the antitumor agent smancs," *Cancer Reserach*, vol. 46, no. 12 Pt 1, pp. 6387–6392, Dec. 1986.

55. F. Danhier, "To exploit the tumor microenvironment: Since the EPR effect fails in the clinic, what is the future of nanomedicine?" *Journal of Controlled Release*, vol. 244, pp. 108–121, Dec. 2016, doi: 10.1016/j.jconrel.2016.11.015.

56. J. W. Nichols and Y. H. Bae, "EPR: Evidence and fallacy," *Journal of Controlled Release*, vol. 190, pp. 451–464, Sep. 2014, doi: 10.1016/j.jconrel.2014.03.057.

57. A. M. Rahman, S. W. Yusuf, and M. S. Ewer, "Anthracycline-induced cardiotoxicity and the cardiac-sparing effect of liposomal formulation," *International Journal of Nanomedicine*, vol. 2, no. 4, pp. 567–583, 2007.

58. O. Lyass *et al.*, "Correlation of toxicity with pharmacokinetics of pegylated liposomal doxorubicin (Doxil) in metastatic breast carcinoma," *Cancer*, vol. 89, no. 5, pp. 1037–1047, Sep. 2000, doi: 10.1002/1097-0142(20000901)89:5<1037::AID-CNCR13>3.0.CO;2-Z.

59. R. van der Meel, L. J. C. Vehmeijer, R. J. Kok, G. Storm, and E. V. B. van Gaal, "Ligand-targeted particulate nanomedicines undergoing clinical evaluation: Current status," *Advanced Drug Delivery Reviews*, vol. 65, no. 10, pp. 1284–1298, Oct. 2013, doi: 10.1016/j.addr.2013.08.012.

60. E. Forssen and M. Willis, "Ligand-targeted liposomes," *Advanced Drug Delivery Reviews*, vol. 29, no. 3, pp. 249–271, Feb. 1998, doi: 10.1016/S0169-409X(97)00083-5.

61. S. Hua, "Targeting sites of inflammation: Intercellular adhesion molecule-1 as a target for novel inflammatory therapies," *Frontiers in Pharmacology*, vol. 4, 2013, doi: 10.3389/fphar.2013.00127.

62. M. Coimbra *et al.*, "Liposomal pravastatin inhibits tumor growth by targeting cancer-related inflammation," *Journal of Controlled Release*, vol. 148, no. 3, pp. 303–310, Dec. 2010, doi: 10.1016/j.jconrel.2010.09.011.

63. F. Danhier, O. Feron, and V. Préat, "To exploit the tumor microenvironment: Passive and active tumor targeting of nanocarriers for anti-cancer drug delivery," *Journal of Controlled Release*, vol. 148, no. 2, pp. 135–146, Dec. 2010, doi: 10.1016/j.jconrel.2010.08.027.

64. S. A. Kuijpers, M. J. Coimbra, G. Storm, and R. M. Schiffelers, "Liposomes targeting tumour stromal cells," *Molecular Membrane Biology*, vol. 27, no. 7, pp. 328–340, Oct. 2010, doi: 10.3109/09687688.2010.522204.

65. A. Puri *et al.*, "Lipid-based nanoparticles as pharmaceutical drug carriers: From concepts to clinic," *Critical Reviews in Therapeutic Drug Carrier Systems*, vol. 26, no. 6, pp. 523–580, 2009, doi: 10.1615/critrevtherdrugcarriersyst.v26.i6.10.

66. D. Kirpotin *et al.*, "Sterically stabilized anti-HER2 immunoliposomes: Design and targeting to human breast cancer cells in vitro," *Biochemistry*, vol. 36, no. 1, pp. 66–75, Jan. 1997, doi: 10.1021/bi962148u.

67. J. W. Park *et al.*, "Anti-HER2 immunoliposomes: Enhanced efficacy attributable to targeted delivery," *Clinical Cancer Research*, vol. 8, no. 4, pp. 1172–1181, Apr. 2002.

68. D. B. Kirpotin *et al.*, "Antibody targeting of long-circulating lipidic nanoparticles does not increase tumor localization but does increase internalization in animal models," *Cancer Research*, vol. 66, no. 13, pp. 6732–6740, Jul. 2006, doi: 10.1158/0008-5472.CAN-05-4199.

69. D. Needham, G. Anyarambhatla, G. Kong, and M. W. Dewhirst, "A new temperature-sensitive liposome for use with mild hyperthermia: Characterization and testing in a human tumor xenograft model," *Cancer Research*, vol. 60, no. 5, pp. 1197–1201, Mar. 2000.

70. K. Kono, "Thermosensitive polymer-modified liposomes," *Advanced Drug Delivery Review*, vol. 53, no. 3, pp. 307–319, Dec. 2001, doi: 10.1016/s0169-409x(01)00204-6.

71. W. Zhang, Y. Yu, F. Xie, X. Gu, J. Wu, and Z. Wang, "High pressure homogenization versus ultrasound treatment of tomato juice: Effects on stability and in vitro bioaccessibility of carotenoids," *Lebensmittel-Wissenschaft + [i.e. und] Technologie*, 2019, Accessed: Jan. 15, 2022. [Online]. Available: https://10.1016/j.lwt.2019.108597

72. M. Bar-Zeev, Y. D. Livney, and Y. G. Assaraf, "Targeted nanomedicine for cancer therapeutics: Towards precision medicine overcoming drug resistance," *Drug Resistance Update*, vol. 31, pp. 15–30, Mar. 2017, doi: 10.1016/j.drup.2017.05.002.

73. J. Liu *et al.*, "Multifunctional aptamer-based nanoparticles for targeted drug delivery to circumvent cancer resistance," *Biomaterials*, vol. 91, pp. 44–56, Jun. 2016, doi: 10.1016/j.biomaterials.2016.03.013.

74. S. Yu *et al.*, "Co-delivery of paclitaxel and PLK1-targeted siRNA using aptamer-functionalized cationic liposome for synergistic anti-breast cancer effects in vivo," *J Biomedical Nanotechnology*, vol. 15, no. 6, pp. 1135–1148, Jun. 2019, doi: 10.1166/jbn.2019.2751.

75. N. Guaragnella, S. Giannattasio, and L. Moro, "Mitochondrial dysfunction in cancer chemoresistance," *Biochemical Pharmacolology*, vol. 92, no. 1, pp. 62–72, Nov. 2014, doi: 10.1016/j.bcp.2014.07.027.

76. T. Farge *et al.*, "Chemotherapy-resistant human acute myeloid leukemia cells are not enriched for leukemic stem cells but require oxidative metabolism," *Cancer Discovery*, vol. 7, no. 7, pp. 716–735, Jul. 2017, doi: 10.1158/2159-8290.CD-16-0441.

77. L. Jiang *et al.*, "Overcoming drug-resistant lung cancer by paclitaxel loaded dual-functional liposomes with mitochondria targeting and pH-response," *Biomaterials*, vol. 52, pp. 126–139, Jun. 2015, doi: 10.1016/j.biomaterials.2015.02.004.

78. D. Chen *et al.*, "Rapid discovery of potent siRNA-containing lipid nanoparticles enabled by controlled microfluidic formulation," *Journal of American Chemical Society*, vol. 134, no. 16, pp. 6948–6951, Apr. 2012, doi: 10.1021/ja301621z.

79. M. Itoh *et al.*, "Possible role of ceramide as an indicator of chemoresistance: Decrease of the ceramide content via activation of glucosylceramide synthase and sphingomyelin synthase in chemoresistant leukemia," *Clinical Cancer Research*, vol. 9, no. 1, pp. 415–423, Jan. 2003.

80. L. E. van Vlerken, Z. Duan, M. V. Seiden, and M. M. Amiji, "Modulation of intracellular ceramide using polymeric nanoparticles to overcome multidrug resistance in cancer," *Cancer Research*, vol. 67, no. 10, pp. 4843–4850, May 2007, doi: 10.1158/0008-5472.CAN-06-1648.

81. H. Devalapally, Z. Duan, M. V. Seiden, and M. M. Amiji, "Modulation of drug resistance in ovarian adenocarcinoma by enhancing intracellular ceramide using tamoxifen-loaded biodegradable polymeric nanoparticles," *Clinical Cancer Research*, vol. 14, no. 10, pp. 3193–3203, May 2008, doi: 10.1158/1078-0432.CCR-07-4973.

82. R. Gaspar and R. Duncan, "Polymeric carriers: Preclinical safety and the regulatory implications for design and development of polymer therapeutics," *Advanced Drug Delivery Reviews*, vol. 61, no. 13, pp. 1220–1231, Nov. 2009, doi: 10.1016/j.addr.2009.06.003.

83. V. Sainz *et al.*, "Regulatory aspects on nanomedicines," *Biochemical Biophysical Research Communication*, vol. 468, no. 3, pp. 504–510, Dec. 2015, doi: 10.1016/j.bbrc.2015.08.023.

84. S. Tinkle *et al.*, "Nanomedicines: Addressing the scientific and regulatory gap," *Annals of the New York Academy of Sciences*, vol. 1313, no. 1, pp. 35–56, 2014, doi: 10.1111/nyas.12403.

85. R. Diab, C. Jaafar-Maalej, H. Fessi, and P. Maincent, "Engineered nanoparticulate drug delivery systems: The next frontier for oral administration?" *American Association of Pharmaceutical Scientist Journal*, vol. 14, no. 4, pp. 688–702, Dec. 2012, doi: 10.1208/s12248-012-9377-y.

86. A. Hafner, J. Lovrić, G. P. Lakoš, and I. Pepić, "Nanotherapeutics in the EU: An overview on current state and future directions," *International Journal of Nanomedicine*, vol. 9, pp. 1005–1023, 2014, doi: 10.2147/IJN.S55359.

87. R. Gaspar, "Regulatory issues surrounding nanomedicines: Setting the scene for the next generation of nanopharmaceuticals," *Nanomedicine*, vol. 2, no. 2, pp. 143–147, Apr. 2007, doi: 10.2217/17435889.2.2.143.

88. R. Karnik *et al.*, "Microfluidic platform for controlled synthesis of polymeric nanoparticles," *Nano Letters*, vol. 8, no. 9, pp. 2906–2912, Sep. 2008, doi: 10.1021/nl801736q.

89. H. Y. Kim, S. H. Kim, M. J. Choi, S. G. Min, and H. S. Kwak, "The effect of high pressure–low temperature treatment on physicochemical properties in milk," *Journal of Dairy Science*, vol. 91, no. 11, pp. 4176–4182, Nov. 2008, doi: 10.3168/jds.2007-0883.

90. M. Rhee, P. M. Valencia, M. I. Rodriguez, R. Langer, O. C. Farokhzad, and R. Karnik, "Synthesis of size-tunable polymeric nanoparticles enabled by 3D hydrodynamic flow focusing in single-layer microchannels," *Advanced Material*, vol. 23, no. 12, pp. H79–83, Mar. 2011, doi: 10.1002/adma.201004333.

91. J. Xu, D. H. C. Wong, J. D. Byrne, K. Chen, C. Bowerman, and J. M. DeSimone, "Future of the particle replication in nonwetting templates (PRINT) technology," *Angewandte Chemie International Edition in English*, vol. 52, no. 26, pp. 6580–6589, Jun. 2013, doi: 10.1002/anie.201209145.

92. J. P. Rolland, B. W. Maynor, L. E. Euliss, A. E. Exner, G. M. Denison, and J. M. DeSimone, "Direct fabrication and harvesting of monodisperse, shape-specific nanobiomaterials," *Journal of the American Chemical Society*, vol. 127, no. 28, pp. 10096–10100, Jul. 2005, doi: 10.1021/ja051977c.

93. A. Albanese, A. K. Lam, E. A. Sykes, J. V. Rocheleau, and W. C. W. Chan, "Tumour-on-a-chip provides an optical window into nanoparticle tissue transport," *Nature Communication*, vol. 4, p. 2718, 2013, doi: 10.1038/ncomms3718.

94. D. Huh, B. D. Matthews, A. Mammoto, M. Montoya-Zavala, H. Y. Hsin, and D. E. Ingber, "Reconstituting organ-level lung functions on a chip," *Science*, vol. 328, no. 5986, pp. 1662–1668, Jun. 2010, doi: 10.1126/science.1188302.

95. S. Y. C. Choi, D. Lin, P. W. Gout, C. C. Collins, Y. Xu, and Y. Wang, "Lessons from patient-derived xenografts for better in vitro modeling of human cancer," *Advance Drug Delivery Reviews*, vol. 79–80, pp. 222–237, Dec. 2014, doi: 10.1016/j.addr.2014.09.009.

96. N. E. Sharpless and R. A. Depinho, "The mighty mouse: Genetically engineered mouse models in cancer drug development," *Nature Review Drug Discovery*, vol. 5, no. 9, pp. 741–754, Sep. 2006, doi: 10.1038/nrd2110.

97. D. Lin *et al.*, "High fidelity patient-derived xenografts for accelerating prostate cancer discovery and drug development," *Cancer Research*, vol. 74, no. 4, pp. 1272–1283, Feb. 2014, doi: 10.1158/0008-5472. CAN-13-2921-T.

98. A. Rongvaux *et al.*, "Development and function of human innate immune cells in a humanized mouse model," *Nature Biotechnology*, vol. 32, no. 4, pp. 364–372, Apr. 2014, doi: 10.1038/nbt.2858.

99. G. K. Hubbard *et al.*, "Combined MYC activation and Pten loss are sufficient to create genomic instability and lethal metastatic prostate cancer," *Cancer Research*, vol. 76, no. 2, pp. 283–292, Jan. 2016, doi: 10.1158/0008-5472.CAN-14-3280.

100. J.-M. Lim *et al.*, "Ultra-high throughput synthesis of nanoparticles with homogeneous size distribution using a coaxial turbulent jet mixer," *ACS Nano*, vol. 8, no. 6, pp. 6056–6065, Jun. 2014, doi: 10.1021/nn501371n.

16 Mangroves as an Alternative Source of Anticancer Drug Leads
Current Evidence and Future Prospects

Vasantha Kavunkal Hridya, Kokkuvayil Vasu Radhakrishnan, Nainarpandian Chandrasekar, and Thadiyan Parambil Ijinu

CONTENTS

16.1 Introduction 275
    16.1.1 Cancer Chemotherapy and the Role of Marine Natural Products......................... 275
    16.1.2 Distribution and Diversity of Mangrove Ecosystem 276
    16.1.3 Medicinal Potential of Mangrove Species.. 277
16.2 Anticancer Phytochemical Leads from Mangroves 277
16.3 Mangrove Potential in Overcoming Drug Resistance 279
16.4 Conclusion and Future Prospects 280
16.5 References... 280

16.1 INTRODUCTION

One of the utmost reasons of mortality, cancer is responsible for nearly 7.6 million deaths worldwide and is expected to reach 13.1 million by 2030 [1]. Several forms of cancers are still intractable despite the number of available treatments. Hence, the current research is still needed to develop more anticancer therapeutic sources and leads. Herein, natural products represent the promising chemical diversity of compounds that serves as a lead for many therapeutic tools. Recently, many approved drugs from natural sources have been developed, which include everolimus, temsirolimus, ixabepilone, trabectedin, and romidepsin, in the field of oncology [1]. Natural compounds play an important role in reversing drug resistance by modulating signaling transduction or target gene expressions through various pharmacological mechanisms [2]. The compounds produced from the marine resources have also demonstrated benefits in the fight against human cancer by either stopping the proliferation of cancerous cells or by promoting apoptosis in cancerous cell lines [3]. In this regard, an effective anticancer therapy with lesser side effects depends on the research in bioactive phytochemicals as they affect cancer cells without affecting normal cells. Natural resources, especially phytochemicals, are pleiotropic in their action and target carcinogenesis by hindering the growth of cancer cells without any side effect; thus, these bio-reservoirs are the suitable source for anticancer drug development progress to consider [4].

16.1.1 CANCER CHEMOTHERAPY AND THE ROLE OF MARINE NATURAL PRODUCTS

History reveals that marine natural products have been used for medicinal purposes in maritime countries for the treatment of many diseases, such as fever, wounds, and gout. Marine flora and

DOI: 10.1201/b23311-16

fauna, including algae, sponges, seaweed, bacteria, fungi, mangroves, diatoms, corals, and ascidians, make up the majority of the total maritime biomass. They provide an incredible lead for the identification of new anticancer compounds and are taxonomically varied with their pharmacological core active unique chemical fingerprints [3]. Natural marine products have recently been identified as the most important source of bioactive components and therapeutic candidates. Additionally, a number of biological features of these compounds have been shown, including antibacterial, antioxidant, and antitumor capabilities. Despite the vast resources enriched with promising compounds, marine biomasses are mostly uncharted for their anticancer properties. Recent advances in the isolation, identification, and characterization of numerous anticancer chemicals from marine sources have led to experimental human trials. The anticancer capabilities of many marine-derived compounds from a variety of flora and fauna, as well as their mechanisms of action, are all included in this review. We also attempted to explain the rise in the use of these compounds for the treatment of malignant growths.

Thus, the largely productive marine floras offer a great possibility for new anticancer drug leads because of their chemically unique nature and biologically active properties. The marine halophytic plants from the mangrove ecosystem are an example of one such lead due to their high levels of phytochemicals, their ethnobotanical relevance, their pharmacological mechanism of action, and particularly their exceptional capacity to flourish in challenging environmental settings [5–7]. For a long it has been documented as mangroves are the huge source of rich in bioactivity and chemical diversity, providing the basis for finding novel entities that serve as pilot initiation ideas for promising therapeutic discoveries.

16.1.2 DISTRIBUTION AND DIVERSITY OF MANGROVE ECOSYSTEM

The definition of a mangrove species is based on its habitat and morphological specialization [8]. Mangroves cover 15.2 million hectares worldwide, predominantly in the intertidal zones of tropical and subtropical coastal regions, and are present on all continents except Antarctica [9]. Spalding et al. [10] estimated that mangroves occupy 18,100,000 ha worldwide, but Giri et al. [9] and Hamilton and Casey later revised their estimate to 13,776,000 ha and 8,349,500 ha, respectively [11]. Mangrove species are defined by their environment and morphological specialization. They are found in 118 different nations, with South and Central America and Asia accounting for the majority of their distribution [8]. Aside from those locations, mangroves can also be found in Australia and New Zealand, East and South Africa, West and Central Africa, South Asia, North and Central America, Southeast Asia, the Pacific Ocean, the Middle East, and East Asia [10].

There are 84 true mangrove species in the world, divided into 12 varieties, 24 genera, and 16 families, of which 14 are semi-mangroves. These species are physiologically a group adapted to the tropics, and they can survive in diverse salinities alongside other plants, like ferns, trees, and shrubs. Indian mangroves comprise 46 real mangrove species, which are divided into 14 families and 22 genera. The Avicenniaceae, Bombacaceae, Combretaceae, Maliaceae, Myrtaceae, Myrsinaceae, Pellicieraceae, Plumbaginaceae, Rhizophoraceae, Rubiaceae, and Sonneratiaceae families include the real mangrove species. The Acanthaceae, Euphorbiaceae, Lythraceae, Palmae, and Sterculiaceae families are among the semi-mangrove plant species that are seen during erratic high tides [12].

Mangroves are a key element in preserving the water quality and shoreline stability by controlling nutrient and sediment circulation and dynamics in estuary waters [13–15]. This is due to their exceptional capacity to grow in difficult environmental conditions, including varying tidal surges and water salinity, as well as their ability to act as a bridging ecosystem between marine and freshwater. Aerial standing roots, a unique physiological mechanism for gas exchange and salt exclusion, and viviparous reproduction are only a few of the distinctive characteristics that set mangrove communities apart from terrestrial plants taxonomically [8, 16]. Mangroves are special in that they promote the biogeochemistry of carbon in coastal ocean environments [17], and tropical mangrove forests account for 30–40% of terrestrial net primary production [18]. As a result, they are crucial to the global carbon cycle. Over time, researchers have begun to recognize the distinctive properties of mangrove plants' ecology and broader ecosystem, as well as their physiological and structural traits. Mangroves are regarded as the source of novel natural and biological resources and carry a specific biochemical action in their bionetwork system. They are abundant sources of flavonoids, alkaloids, tannins, and phenolic compounds, which are new bioactive substances.

16.1.3 MEDICINAL POTENTIAL OF MANGROVE SPECIES

Because of their distinct physiology and shape, mangroves have been employed throughout the world for ethnomedical purposes. Mangroves have long been recognized to be incredibly helpful and successful and active against a wide range of illnesses. They are also good providers of bio-active chemicals, such as antioxidant, antidiarrheal, anti-inflammation, antidiabetic, antimalarial, and anticancer compounds [19]. Due to the presence of specific genes that allow them to survive in harsh environments and the fact that they are known to generate a diverse range of secondary metabolites, they offer a huge amount of uncommon secondary metabolites. In order to explore the scientific potential of mangroves to be used as new natural resources, the Arabs [20] created a rich pharmacopoeia from a variety of distinct mangrove species. Thus, mangroves are crucial to the development of new drugs and drug leads, as well as their safe and efficient delivery against a variety of diseases, regardless of the categories in which they fall. Despite representing a distinct eco-system with significant biological resources, the mangroves' therapeutic benefits have not yet been fully explored [21, 22]. Numerous research efforts have been focused on the therapeutic potential of mangrove plants, including their antibacterial [23–26], antifungal [27], antiviral [28–31], antidiabetic [32, 33], antimalarial [34], and anti-inflammatory [35, 36], and antioxidative [37] activities. An outline of anticancer activities of mangroves and mangrove associates has been summarized in this section with special attention on the current drug-resistant scenario.

16.2 ANTICANCER PHYTOCHEMICAL LEADS FROM MANGROVES

Since the 1960s, efforts have been made to demonstrate the potential therapeutic effects of marine resources through the development of 1- -arabinofuranosylcytosine (Ara-C), the first anticancer drug produced from marine sources for the management of acute myeloid leukemia (AML) [38]. Despite having a slightly larger quantity of diverse bioactive heterogenous chemicals than other marine resources, which is a distinctive feature of mangrove species, mangroves are regrettably the least studied marine resources [19, 39]. Several bioactive substances with anticancer potential have been identified from mangrove plant species (**Table- 1**). Because of the newly discovered isolated alkaloids, tropine and brugine, which are effective against sarcoma 180 and Lewis lung carcinoma as well, Loder and Russell [40] referred to the mangrove plants *Bruguiera sexangula* and *Bruguiera exaristata* (Rhizophoraceae) as "tumour inhibitory plants." The mangrove *Xylocarpus granatum* contains tetranor triterpenoids (xylogranatins A–D), which have been found and charac-terized as anticancer alkaloid chemicals and have been shown to be cytotoxic against a number of cancer cell lines [41]. The same plant also yields the limonoids granaxylocarpins A and B, which are cytotoxic to P-388 leukemia cells [41]. With moderate IC50 values, three types of cells—AGS, MDA-MB-231, and MCF-7—were found to be cytotoxic by seven recognized phytochemicals, including patriscabratine, tetracosane, and five flavonoids of *Acrostichum aureum* [42]. Mei et al. detailed a total of 19 compounds [43–45] from *A. aureum* belonging to different natural compound classes, which include mainly alkaloids, terpenoids, flavonoids, and these compounds are reported to active against various cancer cell lines [46, 47]. The cardiac glycoside 17βH-neriifolin, isolated from *Cerbera odollam*, showed potent anticancer activity against treated and non-treated SKOV-3 cells. These findings revealed the mechanism of action of 17β-H-neriifolin involvement in apoptosis of four proteins: vimentin, pyruvate kinase muscle, heterogeneous nuclear ribonucleoprotein A1, and transgelin [48]. Two new cardenolides and 17β-neriifolin have been reported from the root

TABLE 16.1

Anticancer Compounds from Mangrove Plants

S. No.	Name of Plants	Compounds	Anticancer Activity
1	*Acanthus ebracteatus*	Crude extract	Cervical cancer growth and angiogenesis in a CaSki-cell transplant model in mice [61]
		Bisoxazolinone	
		Methyl apigenin 7-O-β-D- glucuronate flavone glycosides	EAC bearing murine model [62]
2	*Acrostichum aureum*	Tetracosane	(HT-29)- colorectal adenocarcinoma cells against colon cancer
			Breast ductal carcinoma (MDA-MB-231) cells, and Gastric adenocarcinoma (AGS) [46]
		Quercetin-3-O-β-D-glucoside	Gastric adenocarcinoma (AGS)
		Quercetin-3-O-α-L-rhamnoside	Breast ductal carcinoma (MDA-MB-231),
		Quercetin-3-O- β-D-glucosyl-(6 → 1)-α-L-rhamnoside	Estrogen-independent breast cancer (MCF-7) cells [42]
		Quercetin-3-O-α-L-rhamnosyl-7-O-β-D-glucoside (2S,3S)-sulfated pterosin C	
		Kaempferol	Gastric adenocarcinoma (AGS) cells [47]
		Patriscabratine	
3	*Avicennia germinans*	3-Chlorodeoxylapachol	KB human cancer cells in the murine hollow fiber antitumor model [63]
4	*Avicennia officinalis*	Betulinic acid	Human leukemic cell line HL-60 cells [64]
		Triterpene	EAC (Ehrlich ascites carcinoma) cells [65]
5	*Avicennia marina*	Stenocarpoquinone B	K562 (human chronic myeloid leukemia) ell lines [66]
6	*Bruguiera sexangula*	Isobutyric benzoic acid	Lexis lung carcinoma
7	*Bruguiera exaristata*	Tropine	Sarcoma 180 [40]
		Brugine	
8	*Cerbera odollam*	17b-Neriifolin,	KB human oral cancer cells and SKOV-3 ovarian cancer cells [67]
		2′-O-acetyl cerleaside A	
		Cerberin	BC human breast cancer cells and NCI-H187 human lung cancer cells [48, 68]
9	*Cerbera manghas*	17β-Neriifolin	KB human oral cancer cells and Col2 human colon cancer cells [69]
		Cardenolides	
		Deacetyltanghinin and tanghinin	NCI-H187 human lung cancer cells BC
		7,8-Dehydrocerberin	human breast cancer cells [49, 68]
10	*Ceriops decandra*	Quinine	Buccal pouch carcinogenesis and malignant ulcers [70]
11	*Ceriops tagal*	Tagalsins B, C, D, E, F, G, H, W, 9, 10	HCT-8, BGC-823, Bel-7402, A2780, and A549 cell lines
			Hematologic cancer (human T-cell leukemia) [71]
12	*Heritiera fomes*	Methanol extract	EAC (Ehrlich ascites carcinoma) [72]
13	*Phoenix paludosa*	Chloroform	High toxicity against MCF-7, MDA-MB-231,
		Methanol extract	SK-BR-3, and ACHN cell lines [51]
14	*Rhizophora apiculata*	Methanol extract	B16F10 melanoma cells [36]
			A549 lung cancer cells [58]
			Metastatic lung cancer in C57BL/6 Mice, [54]
		Tannins	HepG2 cancer cells [59]
15	*Sonneratia apetala*	Mitomycin C	Diabetes and cancer [23]

TABLE 16.1 *(Continued)*
Anticancer Compounds from Mangrove Plants

S. No.	Name of Plants	Compounds	Anticancer Activity
16	*Sonneratia ovata*	Sonnercerebroside	NCI-H460 (human lung cancer)
		(+)-Dehydrodiconiferyl alcohol	HeLa (human epithelial carcinoma)
			PHF (primary human fibroblast)
			MCF-7 (human breast cancer) [50]
17	*Xylocarpus granatum*	Catechin	[73]
		Epicatechin	
		Xylogranatins A, B, C, D	Human lung carcinoma (A-549) cell lines and murine leukemia cells (P-388) [74]
		Granaxylocarpins A, B	(A-549) human lung carcinoma cell lines and (P-388) murine leukaemia cells [74]
		Xylogranatumine A–F	Tumor cell, A549 [75]
		Photogedunin	CaCo-2 colon cancer cell line [76]
		Xylomexicanin	KT cells and human breast carcinoma [77]
		Gedunin	CaCo-2 colon cancer cell line [76, 78]
		Limnoids	A-549 tumor and P-388 and cell lines [74]
18	*Xylocarpus moluccensis*	Xylomolins	Human triple-negative breast cancer cells (MD-MBA-231) [60]

extract of another mangrove species, *Cerbera manghas*, and the compounds show antiproliferative activity against Col2 human colon cancer cells and antiestrogenic activity against Ishikawa adeno-carcinoma cells [49].

The cancer cell lines NCI-H460 (human lung cancer), HeLa (human epithelial carcinoma), PHF (primary human fibroblast), and MCF-7 (human breast cancer) have all been shown to be actively cytotoxic to *Sonneratia ovata* compounds isolated [50]. The extract-level cytotoxicity studies of *Phoenix palu-dosa* to breast cancer cells (MDA-MB-231 triple negative), MCF-7 (estrogen receptor positive) breast cancer cells, and SKBR-3 (Her 2 negative) reports also exhibited a promising result by Samarakoon et al. [51]. However, despite having comparatively considerable fractions of varied bioactive contents, which is a distinctive characteristic of mangrove species, the research uses and approaches of mangrove species are currently the least investigated among all possible natural therapies [19, 39, 52, 53].

The antimetastatic potential, which is the main task in cancer treatment, of *Rhizophora apiculata* methanolic extract causes a reduction of growth in lung metastasis and tumor cells in a lung metastasis model caused by B16F-10 melanoma in C57BL/6 mice [54]. Traditional medicine uses the mangrove *R. apiculata* (Rhizophoraceae), which has been shown to have an inhibitory effect on the growth of viral, bacterial, and fungal pathogens [55, 56]. Its extract also has a high content of catechin, flavonoids, anthraquinone, tannins, syringol, and pyroligneous acid [19, 57]. The methanol extract of *R. apiculata* has been shown to effectively suppress growth [58] and trigger death by generating reactive oxygen species (ROS), signaling the mitochondrial membrane potential of human adenocarcinoma A549 lung cancer cells, and producing less harm against normal cells. The tannins isolated from *R. apiculata* demonstrated a significant amount of in vivo cytotoxicity against brine shrimp and in vitro cytotoxicity activity against HepG2 cancer cells as well as acute and chronic toxicity [59]. The limnoid compound xylomolins, from the mangrove plant *Xylocarpus moluccensis* seeds, exhibits a modest anticancer effect against human triple-negative breast cancer cells MD-MBA-231 [60].

16.3 MANGROVE POTENTIAL IN OVERCOMING DRUG RESISTANCE

Natural products create a significant role in the findings of new drug leads for chemotherapy, and they have diverse mechanisms of action to hinder various targets contributed in the progress of drug resistance. The heterogeneity of the tumor may lead to metastasis and drug resistance that resists standard treatments. Ninety percent of cancer-related fatalities are triggered by metastasis or drug resistance [79], yet there are currently no medications that can stop even one stage of the metastatic process. In this aspect, the antimetastatic potential of *Rhizophora apiculata*, which causes a tumor mass reduction in a lung metastasis model triggered by B16F-10 melanoma in C57BL/6 mice [54], is pertinent.

Some of the mangrove-derived secondary metabolites of marine fungi from the China Sea have been reported to have promising activity in treating multidrug-resistant (MDR) cancer patients [80]. Those compounds are an isoflavone analogue and a prostaglandin analogue, instigating efforts to isolate promising candidates for further development as clinically beneficial chemotherapeutic drugs. Phomoxanthone-A (PXA), another potent antitumor compound isolated from mangrove associated endophytes *Phomopsis longicolla*, has been reported as a lead for bioprospecting novel anticancer compounds that are required to defeat drug resistance in cancer cells [81]. PXA represents a tetra hydroxanthone atropisomer that exhibits potent antitumor action against blood cancer cell lines and cisplatin-resistant tumor cell lines with significant IC_{50} values. According to the author's summary, the demonstrated activity of PXA may be caused by the induction of caspase-3-dependent apoptosis, and at the same time, PXA was started to trigger immune cells like murine T-lymphocytes, NK cells, and macrophages, which may support in overcoming resistance in cancer chemotherapy research.

16.4   CONCLUSION AND FUTURE PROSPECTS

The research to develop clinical drug leads from mangroves is limited despite the fact that it has significant reported bioactive components [82] against diseases like cancer [83], rheumatism, free radical accumulation [84], inflammatory diseases, diabetes [85], and hepatitis [86]. The study on mangroves has already attracted global attention in terms of conservation aspects. The mangrove ecosystem environment has recently emerged as one of the most interesting hotspots for the bioprospecting of natural product resources. Their halophytic growing circumstances enable some unique bioactive molecules to thrive owing to their biochemical specialization, which may result in the production of numerous herbal and semisynthetic drugs that can reduce the capacity for antibiotic resistance. The reported/isolated compounds' distinctive structures, which can stimulate the activity of membrane transport proteins, are a major factor in the characteristics of MDR and the inhibition of apoptotic pathways in cancer cells.

Marine-derived natural compounds, particularly those of mangrove origin, can be used to frame advanced research on MDR therapies in the years to come. As a result, the mangroves may provide as an alternative source of natural anticancer pharmaceutical leads.

16.5   REFERENCES

1. Mangal, M., Sagar, P., Singh, H., Raghava, G. P., and Agarwal, S. M., 2012. NPACT: Naturally occurring plant-based anti-cancer compound-activity target database. *Nucleic Acids Res*, 41: D1124–D1129. doi: 10.1093/nar/gks1047.
2. Liu, C. M., Su, M. Q., and An, L. J., 2022. Anticancer and overcoming multidrug resistance activities of potential phytochemicals. *Chem Pharm Res.*, 4(1): 1–11.
3. Wali, A. F., Majid, S., and Rasool, S., 2019. Natural products against cancer: Review on phytochemicals from marine sources in preventing cancer. *Saudi Pharm J*, 27(6): 767–777. doi: 10.1016/j.jsps.2019.04.013.
4. Singh, S., Sharma, B., Kanwar, S. S., and Kumar, A., 2016. Lead phytochemicals for anticancer drug development. *Front. Plant Sci*, 7(1667): 1–13.
5. Vannucci, M., 2000. What is so special about mangroves. *Braz J Biol*, 61: 599–603.
6. Dissanayake, N., and Chandrasekara, U., 2014. Effects of mangrove zonation and the physicochemical parameters of soil on the distribution of macrobenthic fauna in kadolkele mangrove forest, a tropical mangrove forest in Sri Lanka. *Adv Ecol Res*, 2014: 1–13.

7. Mahmud, I., Islam, M. K., Saha, S., Barman, A. K., Rahman, M. M., Anisuzzman, M., Rahman, T., Al-Nahain, A., Jahan, R., and Rahmatullah, M., 2014. Pharmacological and ethnomedicinal overview of *Heritiera fomes*: Future prospects. *Int Sch Res Not*, 2014(938543): 1–12.

8. Tomlinson, P. B., 1986. Structural biology. *The Botany of Mangroves*. Cambridge University Press, New York, New York, USA. Chapt.6: 78–84.

9. Giri, C., Ochieng, E., Tieszen, L. L., Zhu, Z., Singh, A., Loveland, T., Masek, J., and Duke, N., 2011. Status and distribution of mangrove forests of the world using earth observation satellite data. *Glob. Ecol. Biogeogr.*, 20: 154–159.

10. Spalding, M., Blasco, F., and Field, C. (Eds), 1997. *World Mangrove Atlas*. The International Society for Mangrove Ecosystems. Okinawa, Japan, 43–171.

11. Hamilton, S. E., and Casey, D., 2016. Creation of a high spatio-temporal resolution global database of continuous mangrove forest cover for the 21st century (CGMFC-21). *Global Ecol. Biogeogr.*, 25: 729–738.

12. Wu, J., Xiao, Q., Xu, J., Li, M. Y., Pan, J. Y., and Yang, M. H., 2008. Natural products from true mangrove flora: Source, chemistry and bioactivities. *Nat. Prod. Rep.*, 25(5): 955–981.

13. Twilley, R. R., 1988. Coupling of mangroves to the productivity of estuarine and coastal waters. *Coastal-Offshore Ecosystem Interactions*. Springer, Berlin, Heidelberg, 22: 155–180.

14. Alongi, D. M., and Sasekumar, A., 1993. Benthic communities. Coastal and estuarine studies. *Tropical Mangrove Ecosystems*, 41: 137–137.

15. Twilley, R. R., Snedaker, S. C., Ya ez-Arancibia, A., and Medina, E., 1996. Biodiversity and ecosystem processes in tropical estuaries: Perspectives of mangrove ecosystems. *Functional Roles of Biodiversity: A Global Perspective*, 55(13): 327–370.

16. Kathiresan, K., and Bingham, B. L., 2001. Biology of mangroves and mangrove ecosystems. *Adv. Mar. Biol.*, 40: 81–251.

17. Twilley, R. R., Chen, R. H., and Hargis, T., 1992. Carbon sinks in mangroves and their implications to carbon budget of tropical coastal ecosystems. *Wat. Air and Soil Poll.*, 64(1): 265–288.

18. Clark, D. A., Brown, S., Kicklighter, D. W., Chambers, J. Q., Thomlinson, J. R., Ni, J., and Holland, E. A., 2001. Net primary production in tropical forests: An evaluation and synthesis of existing field data. *Ecol Appl*, 11(2): 371–384.

19. Bandaranayake, W. M., 2002. Bioactivities, bioactive compounds and chemical constituents of mangrove plants. *Wetland. Ecol. Manag.*, 10: 421–452.

20. Bandaranayake, W. M., 1998. Traditional and medicinal uses of mangroves. *Mangroves and Salt Marshes*, 2: 133–148.

21. Thatoi, H., Samantaray, D., and Das, S. K., 2016a. The genus Avicennia, a pioneer group of dominant mangrove plant species with potential medicinal values: A review. *Front Life Sci*, 9: 267–291.

22. Thatoi, P., Kerry, R. G., Gouda, S, Das, G., Pramanik, K., Thatoi, H., and Patra, J. K., 2016b. Photo-mediated green synthesis of silver and zinc oxide nanoparticles using aqueous extracts of two mangrove plant species, *Heritiera fomes* and *Sonneratia apetala* and investigation of their biomedical applications. *J Photochem Photobiol*, 163: 311–318.

23. Patra, J. K., Das, S. K., and Thatoi, H., 2014. Phytochemical profiling and bioactivity of a mangrove plant, *Sonneratia apetala* from Odisha coast of India. *Chin J Integr Med*, 21: 274–285.

24. Sukumaran, S., Kiruba, S., Mahesh, M., Nisha, S. R., Miller, P. Z., Ben, C. P., and Jeeva, S., 2011. Phytochemical constituents and antibacterial efficacy of the flowers of *Peltophorum pterocarpum* (DC.) Baker ex Heyne. *Asian Pac J Trop Med*, 4: 735–738.

25. Valentin, B., Agnel, D., Franco, J., Merin, M., Geena, M. J., and Elsa, L. J., Thangaraj, M., 2012. Anticancer and antimicrobial activity of mangrove derived fungi *Hypocrea lixii* VB1. *Chin J Nat Med*, 10(1): 77–80.

26. Saad, S., Taher, M., Susanti, D., Qaralleh, H., Binti, N. A., and Rahim, A., 2011. Antimicrobial activity of mangrove plant (*Lumnitzera littorea*). *Asian Pac J Trop Med*, 4: 523–525.

27. Fardin, K. M., Maria, C., and Marx, Y., 2015. Antifungal potential of *Avicennia schaueriana* Stapf & Leech. (Acanthaceae) against Cladosporium and Colletotrichum species. *Lett Appl Microbiol*, 61: 50–57.

28. Jia, L., Yu, J., Yang, J., Song, H., Liu, X., Wang, Y., Xu, Y., Zhang, C., Zhong, Y., and Li, Q., 2009. HCV antibody response and genotype distribution in different areas and races of China. *Int J Biol Sci*, 5: 421–427.

29. Li, W., Jiang, Z., Shen, L., Pedpradab, P., Bruhn, T., Wu, J., and Bringmann, G., 2015. Antiviral limonoids including khayanolides from the Trang mangrove plant *Xylocarpus moluccensis*. *J Nat Prod*, 78: 1570–1578.

30. Sudheer, N. S., Rosamma, P., and Singh, I. S. B., 2012. Anti-white spot syndrome virus activity of *Ceriops tagal* aqueous extract in giant tiger shrimp Penaeus monodon. *Arch Virol*, 157: 1665–1675.

31. Xu, D. B., Ye, W. W., Han, Y., Deng, Z. X., and Hong, K., 2014. Natural products from mangrove Actinomycetes. *Mar Drugs*, 12: 2590–2613.

32. Alikunhi, N. M., Kandasamy, K., and Manivannan, S., 2010. Antidiabetic activity of the mangrove species *Ceriops decandra* in alloxan-induced diabetic rats. *J Diabetes*, 2: 97–103.

33. Reza, H., Haq, W. M., Das, A. K., Rahman, S., Jahan, R., and Rahmatullah, M., 2011. Anti-hyperglycemic and antinociceptive activity of methanol lead and stem extract of *Nypa fruticans* Wrumb. *Pak J Pharm Sci*, 24: 485–488.

34. Hridya, V. K., Prince Godson S, Chandrasekar, N., and Kumaresan, S., 2021. The Antimalarial Potential and Phytochemical Composition of Mangroves from Southeast India: An In vitro Study. *J Aquat Biol Fish*, 9(S1): 29–34.

35. De-Faria, F. M., Almeida, A. C. A., and Luiz, F., 2012. Mechanisms of action underlying the gastric antiulcer activity of the *Rhizophora mangle* L. *J Ethnopharmacol*, 139: 234–243.

36. Prabhu, V. V., and Guruvayoorappan, C., 2012. Anti-inflammatory and antitumor activity of the marine mangrove *Rhizophora apiculate*. *J Immunotoxicol*, 9: 341–352.

37. Kim, K. J., Kim, M. A., and Jung, J. H., 2008. Antitumor and antioxidant activity of protocatechualde-hyde produced from Streptomyces lincolnensis M-20. *Arch Pharm Res*, 31: 1572–1577.

38. Robak, T., and Wierzbowska, A., 2009. Current and emerging therapies for acute myeloid leukemia. *Clin Ther*, 31: 2349–2370.

39. Boopathy, N. S., and Kathiresan, K., 2010. Anticancer drugs from marine flora: An overview. *J. Oncol.*, 2010(214186): 1–18.

40. Loder, J. W., and Russell, G. B., 1969. Tumour inhibitory plants. The alkaloids from *Bruguiera sexangula* and *Bruguiera exaristata* (Rhizophoraceae). *Aust. J. Chem.*, 22(6): 1271–1275.

41. Yin, S., Fan, C. Q., Wang, X. N., Lin, L. P., Ding, J., and Yue, J. M., 2006. Xylogranatins A– D: Novel Tetranortriterpenoids with an unusual 9, 10-s eco scaffold from Marine Mangrove *Xylocarpus granatum*. *Org. Lett.*, 8(21): 4935–4938.

42. Uddin, S. J., Grice, D., and Tiralongo, E., 2012. Evaluation of cytotoxic activity of patriscabratine, tetracosane and various flavonoids isolated from the Bangladeshi medicinal plant *Acrostichum aureum*. *Pharm. Biol.*, 50(10): 1276–1280.

43. Mei, W., Zeng, Y., Ding, Z., and Dai, H., 2006. Isolation and identification from mangrove plant of the chemical constituents from mangrove plant *Acrostichum aureum*. *Zhongguo Yaowu Huaxue Zazhi*, 16(1): 46–48.

44. Nobutoshi, T., Takao, M., Yasuhisa, S., Chiu Ming, C., Gomez, P., and Luis, D., 1981. Chemical and chemotaxonomical studies of ferns. XXXVII. Chemical studies on the constituents of Costa Rican fern. *Chem Pharm Bull*, 29: 3455–3463.

45. Sultana, S., Ilyas, M., and Shaida, W. A., 1986. Chemical investigation of *Acrostichum aureum* Linn. *J Indian Chem Soc*, 63: 1074–1075.

46. Uddin, S. J., Jason, T. L., Beattie, K. D., Grice, I. D., and Tiralongo, E., 2011. (2 S, 3 S)-Sulfated pterosin C, a cytotoxic sesquiterpene from the Bangladeshi mangrove fern *Acrostichum aureum*. *J. Nat. Prod.*, 74(9): 2010–2013.

47. Uddin, S. J., Bettadapura, J., Guillon, P., Darren Grice, I., Mahalingam, S., and Tiralongo, E., 2013. *In-vitro* antiviral activity of a novel phthalic acid ester derivative isolated from the Bangladeshi mangrove fern *Acrostichum aureum*. *J Antivir Antiretrovir*, 5(6): 139–144.

48. Syarifah, M. M. S., Nurhanan, M. Y., Haffiz, J. M., Ilham, A. M., Getha, K., Asiah, O., Norhayati, I., Sahira, H. L., and Suryani, S. A., 2011. Potential anticancer compound from *cerbera odollam*. *J. Trop. For. Sci.*, 23: 89–96.

49. Cheenpracha, S., Karalai, C., Rat-A-Pa, Y., Ponglimanont, C., and Chantrapromma, K., 2004. New cytotoxic cardenolide glycoside from the seeds of *Cerbera manghas*. *Chem. Pharm. Bull.*, 52(8): 1023–1025.

50. Nguyen T.-H.-T., Pham H.-V.-T., Pham N.-K.-T., Quach N.-D.-P., Pudhom K., Hansen P. E., and Nguyen K.-P.-P., 2015. Chemical constituents from *Sonneratia ovata* Backer and their in vitro cytotoxicity and acetylcholinesterase inhibitory activities. *Bioorg. Med. Chem. Lett.*, 25(11): 2366–2371.

51. Samarakoon, S. R., Shanmuganathan, C., Ediriweera, M. K., Tennekoon, K. H., Piyathilaka, P., Thabrew, I., and de Silva, E. D., 2016a. *In vitro* cytotoxic and antioxidant activity of leaf extracts of mangrove plant, *Phoenix paludosa* Roxb. *Trop J Pharm Res*, 15: 127–132.

52. Debbab, A., Aly, A. H., Lin, W. H., and Proksch, P., 2010. Bioactive compounds from marine bacteria and fungi. *Microb. Biotechnol*, 3: 544–563.

53. Valli, S., Suvathi, S. S., Aysha, O. S., Nirmala, P., Vinoth, K. P., and Reena, A., 2012. Antimicrobial potential of Actinomycetes species isolated from marine environment. *Asian Pac. J. Trop. Biomed.*, 2(6), 1: 469–473.

54. Prabhu, V. V., and Guruvayoorappan, C., 2013. Inhibition of metastatic lung cancer in C57BL/6 mice by marine mangrove *Rhizophora apiculata. Asian Pac J Cancer Prev*, 14(3): 1833–1840.

55. Premanathan, M., Arakaki, R., and Izumi, H., 1999. Antiviral properties of a mangrove plant, Rhizophora apiculata Blume, against human immunodeficiency virus. *Antiviral Res*, 44: 113–122.

56. Antony, J. J., Sivalingam, P., and Siva, D., 2011. Comparative evaluation of antibacterial activity of silver nanoparticles synthesized using *Rhizophora apiculata* and glucose. *Colloids Surf B: Biointerfaces*, 88: 134–140.

57. Binh, P. T., Chien, N. V., Trung, N. Q., Thong, V. H., Tuyen, N. V., and Thao, N. P., 2020. Phenolic constituents from the stem barks of *Rhizophora apiculata* Blume. *Vietnam J. Sci. Technol.*, 58(5): 517–525.

58. Ramalingam, V., and Rajaram, R., 2018. Enhanced antimicrobial, antioxidant and anticancer activity of *Rhizophora apiculata*: An experimental report. *3 Biotech*, 8(200): 1–13. doi: 10.1007/s13205-018-1222-2.

59. Hong, L. S., Ibrahim, D., and Kassim, J., 2011. Assessment of *in vivo* and *in vitro* cytotoxic activity of hydrolysable tannin extracted from *Rhizophora apiculata* barks. *World J Microbiol Biotechnol*, 27: 2737–2740.

60. Zhang, J., Li, W., Dai, Y., Shen, L., and Wu, J., 2018. Twenty-nine new limonoids with skeletal diversity from the mangrove plant. *Xylocarpus moluccensis.* Mar Drugs, 16(1): 38. doi: 10.3390/md16010038.

61. Mahasiripanth, T., Hokputsa, S., Niruthisard, S., Bhattarakosol, P., and Patumraj, S., 2012. Effects of *Acanthus ebracteatus* Vahl on tumor angiogenesis and on tumor growth in nude mice implanted with cervical cancer. *Cancer Manag Res.*, 4: 269–279. doi: 10.2147/CMAR.S33596.

62. Chakraborty, T., Bhuniya, D., Chatterjee, M., Rahaman, M., Singh, D., and Chatterjee, B. N., 2007. *Acanthus ilicifolius* plant extracts prevent DNA alterations in a transplantable Ehrlich ascites carcinoma-bearing murine model. *World J Gastroenterol*, 13: 6538–6548.

63. Jones, W. P., Lobo-Echeverri T, Mi Q, Chai, H., Lee, D., Soejarto, D. D., Cordell, G. A., and Pezzuto, J. M., Swanson, S. M., and Kinghorn, A. D., 2005. Antitumour activity of 3-chlorodeoxylapachol, a naphthoquinone from *Avicennia germinans* collected from an experimental plot in southern Florida. *J Pharm Pharmacol*, 57(9): 1101–1108. doi: 10.1211/jpp.57.9.0005.

64. Karami, L., Majd, A., Mehrabian, S., Nabiuni, M., Salehi, M., and Irian, S., 2012. Antimutagenic and anticancer effects of Avicennia marina leaf extract on Salmonella typhimurium TA100 bacterium and human promyelocytic leukaemia HL-60 cells. *Sci. Asia.*, 38: 349–355.

65. Sumithra, M., Anbu, J., Nithya, S., and Ravichandiran, V., 2011. Anticancer activity of Methanolic leaves extract of *Avicennia officinalis* on Ehrlich ascitis carcinoma cell lines in Rodents. *Int. J. Pharmtech. Res.*, 3(3): 1290–1292.

66. Han, Li, Huang, Xueshi, Dahse, Hans-Martin, Moellmann, Ute, Fu, Hongzheng, Grabley, Susanne, Sattler, Isabel, and Lin, Wenhan, 2007. Unusual naphthoquinone derivatives from the twigs of *Avicennia marina. J. Nat. Prod.*, 70(6): 923–927. doi: 10.1021/np060587g.

67. Laphookhieo, S., Cheenpracha, S., Karalai, C., Chan-trapromma, S., Ratapaa, Y., Ponglimanont, C., and Chan-trapromma, K., 2004. Cytotoxic cardenolide glycoside from the seeds of *Cerbera odollam. Phytochemistry*, 65(4): 507–510.

68. Chan, E. W., Wong, S. K., and Chan, H. T., 2016. Apocynaceae species with antiproliferative and/or antiplasmodial properties: A review of ten genera. *J Integr Med*, 14: 269–284. doi: 10.1016/s2095-4964 (16)60261-3.

69. Chang, L. C., Gills, J. J., Bhat, K. P. L., Luyengi, L., Farnsworth, N. R., Pezzuto, J. M., and Kinghorn, A. D., 2000. Activity-guided isolation of constituents of Cerbera manghas with antiproliferative and antiestrogenic activities. *Bioorg. Med. Chem. Lett.*, 10: 2431–2434.

70. Boopathy, N. S., Kathiresan, K., Manivannan, S., and Jeon, Y. J., 2011. Effect of mangrove tea extract from *Ceriops decandra* (Griff.) Ding Hou. on salivary bacterial flora of DMBA induced Hamster buccal pouch carcinoma. *Indian. J. Microbiol.*, 51(3): 338–344.

71. Yang, Y., Zhang, Y., Liu, D., Li-Weber, M., Shao, B., and Lin, W., 2015. Dolabrane-type diterpenes from the mangrove plant *Ceriops tagal* with antitumor activities. *Fitoterapia*, 103: 277–282.

72. Patra, J. K., and Thatoi, H., 2013. Anticancer activity and chromatography characterization of methanol extract of *Heritiera fomes* Buch. Ham., a mangrove plant from Bhitarkanika, India. *Orient Pharm Exp Med.*, 13(2): 133–142.

73. Das, S. K., Samantaray, D., and Thatoi, H., 2014. Ethnomedicinal, antimicrobial and antidiarrhoeal studies on the mangrove plants of the genus xylocarpus: A mini review. *J Bioanal Biomed*, S12: 004. doi: 10.4172/1948-593X.S12-004.

74. Yin, S., Wang, X. N., Fan, C. Q., Lin, L. P., Ding, J., and Yue, J. M., 2007. Limonoids from the seeds of the marine mangrove *Xylocarpus granatum. J. Nat. Prod.,* 70(4): 682–685.

75. Zhou, Z. F., Taglialatela-Scafati, O., Liu, H. L., Gu, Y. C., Kong, L. Y., and Guo, Y. W., 2014. Apotirucallane protolimonoids from the Chinese mangrove Xylocarpus granatum Koenig. *Fitoterapia,* 97: 192–197.

76. Uddin, S. J., Nahar, L., Shilpi, J. A., Shoeb, M., Borkowski, T., Gibbons, S., Middleton, M., Byres, M., and Sarker, S. D., 2007. Gedunin, a limonoid from *Xylocarpus granatum,* inhibits the growth of CaCo-2 colon cancer cell line *in vitro. Phytother Res,* 21(8): 757–761.

77. Li-Ru, S., Mei, D., Bao-Wei, Y., Dong, G., Man-Li, Z., Qing-Wen, S., Chang-Hong, H., Hiromasa, K., Nobuo, S., and Bin, C., 2009. "Xylomexicanins A and B, New Δ14,15-mexicanolides from seeds of the chinese mangrove *Xylocarpus granatum. Zeitschrift für Naturforschung* C, 64(1–2): 37–42.

78. Sahai, R., Bhattacharjee, A., and Shukla, V. N., 2020. Gedunin isolated from the mangrove plant *Xylocarpus granatum* exerts its anti-proliferative activity in ovarian cancer cells through G2/M-phase arrest and oxidative stress-mediated intrinsic apoptosis. *Apoptosis,* 25: 481–499.

79. Gupta, G. P., and Massague, J., 2002. Cancer metastasis: Building a framework. *Cell,* 127: 679–695.

80. Tao, L. Y., Zhang, J. Y., and Liang, Y. J., 2010. Anticancer effect and structure-activity analysis of marine products isolated from metabolites of mangrove fungi in the South China Sea. *Mar Drugs,* 8(4): 1094–1105.

81. Frank, M., Niemann, H., Böhler, P., Stork, B., Wesselborg, S., Lin, W., and Proksch, P., 2015. Phomoxanthone A—from mangrove forests to anticancer therapy. *Curr. Med. Chem.,* 22(30): 3523–3532.

82. Sayantani, M., Nabanita, N., and Punarbasu, C., 2021. A review on potential bioactive phytochemicals for novel therapeutic applications with special emphasis on mangrove species. *Phytomed. Plus,* 1(4): 2667–0313. doi: 10.1016/j.phyplu.2021.100107.

83. Kerry, R. G., Pradhan, P., Das, G., Gouda, S., Swamy, M. K., and Patra, J. K., 2018. Anticancer potential of mangrove plants: Neglected plant species of the marine ecosystem. In: Akhtar, M., Swamy, M. (eds) *Anticancer plants: Properties and Application.* Springer, Singapore, 1: 303–325. doi: 10.1007/978-981-10-8548-2_13.

84. Okla, M. K., Alamri, S. A., Alatar, A. A., Hegazy, A. K., Al-Ghamdi, A. A., Ajarem, J. S., Faisal, M., Abdel-Salam, E. M., Ali, H. M., Salem, M. Z., and Abdel-Maksoud, M. A., 2019. Antioxidant, hypogly-cemic, and neurobehavioral effects of a leaf extract of Avicennia marina on autoimmune diabetic mice. *Evid. Based Complementary Altern. Med.,* 2019(1263260): 1–8. doi: 10.1155/2019/1263260.

85. Sachithanandam, V., Lalitha, P., Parthiban, A., Mageswaran, T., Manmadhan, K., and Sridhar, R., 2019. A review on antidiabetic properties of indian mangrove plants with reference to island ecosystem. *Evid. Based Complementary Altern. Med.,* 2019(4305148): 1–21. doi: 10.1155/2019/4305148.

86. Ravikumar, S., and Gnanadesigan, M., 2011. Hepatoprotective and antioxidant activity of a mangrove plant *Lumnitzera racemosa. Asian Pac J Trop Biomed.,* 1(5): 348–352. doi: 10.1016/S2221-1691(11)60078-6.

Index

Note: Page numbers in *italics* indicate figures and those in **bold** indicate tables.

1'-acetoxychavicol acetate (ACA) and sodium butyrate (SB), 196
2B3'–101, **153**
3,3'-diindolylmethane (DIM), 236–238
4',6-diamidino-2-phenylindole staining, 185
5,4'-dihydroxy-6,7,8,3'-tetramethoxyflavone, 185
5-flourouracil (5-FU)
  apigenin and, 198
  curcumin analog pentagamavunon-1 (PGV-1) and, 197
5-methyl phenazine-1-carboxylic acid, 185
6-shogaol, 66, **104**, **195**

A

AB1–008, **153**
Abraxane, **105**, 110, 115, 119, **119**, 134, 149, 153, 155, 251
acankoreanogein (E13-1) or impressic acid (E12-1), gemcitabine (GEM) and, 198
Acanthopanax trifoliatus, 199
Acanthus ebracteatus, **278**
access in cancer prevention and treatment, 23–24
Acridine orange/ethidium bromide, 185
Acronychia baueri, **186**
Acrostichum aureum, 277, **278**
active targeting, 114, *114*, *245*, 245–246, 265
Adenosma bracteosum, 185
African basil, **193**
African birch, **187**
African teak, **192**
African yellow wood, **191**
Ageratum conyzoides, **186**
albumin-based NPs, 10
albumin-bound paclitaxel NPs, 110, 115, **119**, 155
Alchornea cordifolia, **186**
alkaloids, 88, **105**, 185
allicin, **104**
Allium, 5, 88, **104**, **186**, 232
Allium sativum, 66, **104**, **186**
allyl isothiocyanate (AITC), **228**, 235, *236*
Aloe barbadensis, **186**
α-pinene from bay leaf, 69
Alpinia galangal, **132**, 196
Alstonia boonei, **186**
alvocidib, **105**
amruta, **192**
Anacardium occidentale, **187**
Andrographis paniculata, **187**
anethole, **103**
angiogenesis, 171–172
angiogenesis inhibitors, 89
animal origin, 86
aniseed, 64
Annona atemoya, **187**
Annona muricata, **187**
Anogeissus leiocarpus, **187**
antioxidant effect, 169–170

antioxidant responsive element (ARE), 235
antioxidants in prevention of cancer, 65–66
  selenium, 66
  vitamin C, 65
  vitamin D3, 65
  vitamin E, 65–66
antiproliferative compounds, 90
apigenin
  and 5-flourouracil, 198
  in controlling epigenetic changes and mi RNAs involved in carcinogenesis and cancer prevention, 173–174
apoptosis
  cell cycle arrest and, 171
  inducers, 90
arhar dal, **188**
arsenic trioxide (ATO), genistein and, 197
artar root, **191**
Artemisia annua, 6, **104**, 196
artemisinin (art)
  in preclinical trials for cancer treatment, **104**
  and resveratrol, 185, 196
ashwagandha and winter cherry, **195**
Asian ginger, **103**
Astragalus membranaceus, **187**
Atractylis lancea, **187**
Avicennia germinans, **278**
Avicennia marina, **278**
Avicennia officinalis, **278**
Ayurveda, 76, 85, 185, 199
Azadirachta indica, **104**, **188**

B

bambara, **187**
Basela alba, 185
basic food from a natural source, 90
beans, 89, **103**, 173
belotecan, 217
benzyl isothiocyanate (BITC), 91, 101, **103**, **228**
berberine
  evaluated in clinical trials on various cancers, **105**
  and rapamycin, 197–198
berries, 4, 29, 68–69, 89, 90, 101, 131, **132**
beta-carotene, 5, 43, **228**, *229*
betulinic acid (BA), ginsenoside Rh2 (G-Rh2) and, 196
bhumyamalaki, **193**
bhunimba, **187**
biliary tract cancer, **132**
billygoat weed, **186**
bioactive phytochemicals in cancer therapy and deciphering the genomic puzzle, 168–169
bitter leaf, **195**
black cumin, 66
bladder cancer
  effect of phytochemicals along with therapeutic agents on, **132**

phytochemical-/nutraceutical-based nano-formulation
   used for the prevention of, **250**
blood cancers
   plant-derived substances having anticancer activity
      against, **186–193**, **195**
   polymer-based nano-formulation used for the prevention
      of, **248**
Boerhaavia diffusa, **188**
Boletus edulis, 86, 90
Bonati, 185
bone cancer
   carbon-based nano-formulation used for the prevention
      of, **247**
Boswellia serrata, 29, **188**
boundary tree, **193**
Brahmamanduki, **189**
brain tumors
   liposomal and polymeric NPS in various stages of
      clinical trials for, **153**
   plant-derived substances having anticancer activity
      against, **192**
Brassica phytochemicals, 175, 227–238, *see also* indole-3-
      carbinol (I3C)
   anticarcinogenic properties of, **228**
   beta-carotene, 228, *229*
   Caesar experiment with, 230
   concentration regime, 234–235
   cruciferous vegetables, 233
   direct and indirect interaction, 234
   diversity of, 230
   glucobrassicin, *229*, 235
   integration of collective data, 235
   integrators of metabolic pathway, 234
   interacting with carcinogens, 230–231
   introduction, 227–230
   lutein, 228, *229*
   molecular framework of the human body and its
      response to cancer and, 231–232
   organosulfur compounds, 232–233
   phenylisothiocyanate, *230*
   sulphorafane, **228**, 233–235, 236
   zeaxanthin, 228, *229*
breast cancer
   anticarcinogenic properties of *Brassica* in, **228**
   carbon-based nano-formulation used for the prevention
      of, **247**
   effect of phytochemicals along with therapeutic agents
      on, **132**
   liposomal and polymeric NPS in various stages of
      clinical trials for, **153**
   liposome-based nano-formulation used for the
      prevention of, **249**
   LPHNPs that have been nano-formulated and tested
      in, **249**
   nano-formulations in commercial use or in clinical
      trials for, **119**
   nano-formulations in detection and treatment of, 116
   natural polyphenol for, **28**
   phytochemical-based nano-formulations under clinical
      trials for, **105**
   phytochemical-/nutraceutical-based nano-formulation
      used for the prevention of, **250**
   plant-derived substances having anticancer activity
      against, **187–193**, **195**

polymer-based nano-formulation used for the prevention
   of, **248**
statistics of cases diagnosed in 2020, 243, *244*
vitamin E and, **43**
Western lifestyle and, 23
brimstone tree, **192**
broccoli, 4, 6, 7, 29, 101, 131, 172, 184, **228**, 232, 233–235,
      236
bronchial cancer, 117–118
Brucea javanica, **132**
Bruguiera exaristata, 277, **278**
Bruguiera sexangula, 277, **278**
bush banana, **195**
Byron bay acronychia, **186**

C

Caelyx, 118, **119**, **153**
Caesar, Julius, 230
Caesar experiment with *Brassica*, 230
Cajanus cajan, **188**
calcium-induced fusion method, 149
Calotropis procera, 185
Camellia sinensis, 3, **105**, **188**
Camptotheca acuminata, 58, 207, 208
camptothecin (CPT), 185, 207–217
   belotecan, 217
   biosynthesis of, 208–209, *209*
   camptothecin-bearing plants, 208
   chemistry of, 209, *210*
   chemotherapeutic potential of, and its derivatives,
      215–217
   under clinical trials for cancer treatment, **105**
   conclusion and prospects, 217
   introduction, 207–208
   irinotecan, 216
   synthesis of, 210–212, *211*, *212*
   synthesis of camptothecin derivatives, 212–215,
      213–214, *215*
   topotecan, 215–216
cancer
   carbon-based nano-formulation used for the prevention
      of different types of, 247, **247**
   causes of, *98*, 98–99
   definition of, 184
   genetic correlation of, 167–168
   growth of, 243, *244*
   molecular framework of the human body and its
      response to, 231–232
   nanotechnology in cancer treatment, 244 (*see also*
      nano-formulations in cancer detection and treatment)
   risk factors for, 244
   statistics of cancer cases diagnosed in 2020, 243, *244*
   traditional cancer therapy options for, 244
   types and symptoms of, *99*, 99–100
   use of term, 98
cancer, pathophysiology of, 79–82
   cancer types, *80*
   carcinogenesis, 80–81
   causative agents, 76, *76*
   diagnosis of cancer, 81, *82*
   stage of cancer, 81
   treatment of cancer, 81, *83*
   tumor types, categorizing, 79–80

cancer chemoprevention
  based on nano-formulations, 246–251
  Brassica phytochemicals in, 227–238
  fat-soluble vitamins in, 39–49
  functional foods in, 75–92
  mangroves as an alternative source of anticancer drug leads, 275–280
  medicinal chemistry, pharmacodynamics, and pharmacokinetics of camptothecin and its derivatives in, 207–217
  nano-formulation-based approaches for, 109–120
  nano-formulations in cancer detection and treatment, 243–252
  nano-formulations of chemotherapeutic drugs, 143–158
  natural compounds and, 172
  natural-product-based nano-formulations, potential of, 129–136
  natural products and nano-formulations in, 3–13
  nutrigenomics in, 165–176
  obstacles in utilizing nanomedicine for cancer management, 259–269
  overview, 2–3
  phytochemicals in, 97–106
  polyphenols in, 21–29
  synergistic effects of natural products in cancer treatment, 183–201
  targeting cancer stem cells by natural products for, 57–70
cancer prevention and treatment
  challenges in, 22–24
  growing inequality and societal costs of, 22
  nano-formulations in, 243–252
  naturally occcurring polyphenols for, 26–27
  Western lifestyle and limited access, 23–24
cancer stem cells (CSCs)
  cell signaling pathways and, 59–64
  natural products used in prevention of cancers by acting on, 65–70
cancer stem cells, targeting by natural products for chemoprevention, 57–70
  cancer overview, 58–59
  cell signaling pathways and, 59–64
  chemical and radiation treatment, harmful effects of, 64
  conclusion and prospects, 70
  CSC (cancer stem cell) model, 59
  introduction, 58
  natural products used in prevention of cancers by acting on CSCs, 65–70
  natural products which are used in the prevention of different diseases, 64–65
cancer tissues, nano-formulations for targeting, 114–115, *115*
  active targeting, 114, *114*
  passive targeting, *114*, 115
cancer treatment, phytochemicals in, 102–105
  clinically studied anticancer phytochemicals, 102, 105, **105**
  preclinically studied anticancer phytochemicals, 102, **104**
cancer types, *80*
Cang zhu, **187**
capsaicin, **103**
Capsicum, **103**
carbohydrate-based system, 10–11

carbon-based NPs (CBNs) for drug delivery and cancer treatment, 246–247, **247**
  carbon nanotubes, 247
  graphene, 247
carbon nanotubes (CNTs), *133*, 135, 247, **247**
carbon NPs, 150
carcinogenesis, 80–81
  role of reactive oxygen species in development of, 24
Carica papaya, **189**
carotenoids, 4–5, 88
  evaluated in clinical trials on various cancers, **105**
Cassia occidentalis, **189**
catechins
  in cancer prevention, **103**
  and doxorubicin (DOX), 198
Catharanthus roseus, 58, **132**
causative agents of cancer, 76, *76*
caveolae-mediated endocytosis, 155
Cedrus deodara, **189**
cell cycle arrest, apoptosis and, 171
cell signaling pathways and CSCs, 59–64
  FAK signaling pathway, 63
  Hh signaling pathway, 60, 62
  JAK-STAT signaling pathway, 63
  NF-κB signaling pathway, 62
  Notch pathway, 60
  PI3K/AKT/mTOR pathways, 62
  PPAR signaling pathways, 63–64
  regulation and crosstalk between different signaling pathways, 64
  TGF/SMAD signaling pathway, 63
  wingless/integrated (Wnt) pathway, 60, *61*
cellular pathway
  exploration of, 153–155
  modes of NPs entry into the cell, 153–155, *154*
  techniques for studying cellular interactions of NPs, 155
Centella asiatica, **189**
Cerbera manghas, **278**, 279
Cerbera odollam, 277, **278**
Ceriops decandra, **278**
Ceriops tagal, **278**
cervical cancer
  natural polyphenol for, **28**
  phytochemical-/nutraceutical-based nano-formulation used for the prevention of, **250**
  plant-derived substances having anticancer activity against, **187–195**
Chaetomium globosum, 185
challenges associated with nano-formulated drugs, 156–157
Charak Samhita, 185
chemical treatment, harmful effects of, 64
chemoprevention, *see* cancer chemoprevention
chemosensitization, 237–238
chemotherapeutic drugs
  clinical applications of different nanoformulated, 152–153
  co-delivery of, 156
  marine natural products in, 275–276
  nano-formulations of, 143–158
chickpeas, **103**
chili pepper, 67
chin dok diao, **191**
Chinese medicine, 199, 207

Christmas bush, **186**
Chromolaena odorata, **190**
cisplatin, quercetin and, 198
Citrus aurantifolia, **190**
classification of cancer chemopreventive and
    therapeutically used functional foods, 82–92, *84*
    based on chemical class of phytoconstituents, 86–89, *87*
    based on mechanism of action, 89–90
    based on origin, 83–86
    based on processing methods, 90–92
clathrin- and caveolae-independent endocytosis, 155
clathrin-mediated endocytosis (CME), 155
clinical applications of different nanoformulated
    chemotherapeutic drugs, 152–153
clinically studied anticancer phytochemicals, 102, 105, **105**
clinical trials
    for cancer treatment, phytochemical-based nano-
        formulations under, **105**
    of different nanodrugs, advancement in, 152–153, **153**
    on various cancers, phytochemicals evaluated in, **105**
clove, 64, 68, 196
co-delivery of chemotherapeutic drugs, 156
coffee, 196
coffee senna, **189**
colloids affecting the EPR effect in tumor cells, properties
    of, 263–265
    charge of NPs, 263–264
    effect of EPR of NM in cancerous cells in humans,
        264–265
    shape of NPs, 264
    size of NPs, 263
colon cancer
    anticarcinogenic properties of *Brassica* in, **228**
    carbon-based nano-formulation used for the prevention
        of, **247**
    effect of phytochemicals along with therapeutic agents
        on, **132**
    natural polyphenol for, **28**
    phytochemical-/nutraceutical-based nano-formulation
        used for the prevention of, **250**
    plant-derived substances having anticancer activity
        against, **186–195**, **189**
colorectal cancer
    liposomal and polymeric NPS in various stages of
        clinical trials for, **153**
    nano-formulations in detection and treatment of,
        116–117
    natural polyphenol for, **28**
    plant-derived substances having anticancer activity
        against, **188**, **191**, **194**
    statistics of cases diagnosed in 2020, 243, *244*
complement proteins with nanomedicines, activation of, 261
conventional chemotherapy
    cytotoxicity of, 145
    delivery of, 145, **146**
    drawbacks of, 145
    mode of action, 145
    multidrug resistance of, 145
    selectivity of, 145
    solubility of, 145
convergent growth method, 152
Coptidis rhizoma, 198
CPT, *see* camptothecin (CPT)
CPX-1, **153**

CRLX101, **153**
CRLX301, **105**
Croton zambesicus, **190**
cruciferous vegetables, 6, 29, 77, 85, 90, 92, 100–101, **103**,
    172, 184, 230, 233, 234, 235, 236, 238
crucifers, 175
Cryptolepis sanguinolenta, **190**
Cryptotheca crypta, 58
CSC (cancer stem cell) model, 59
cucurbitaceous plants, **103**
cucurbitacinB, **103**
Curcuma longa, **104, 132, 190**, 196
Curcuma mutabilis rhizome, 185
curcumin, 4, 100
    analog pentagamavunon-1 (PGV-1) and 5-flourouracil
        (5-FU), 197
    under clinical trials for cancer treatment, **105**
    in controlling epigenetic changes and mi RNAs involved
        in carcinogenesis and cancer prevention, 174
    in different types of cancers, 66
    evaluated in clinical trials on various cancers, **105**
    in preclinical trials for cancer treatment, **104**
    and resveratrol, 185
    and vinorelbine, 196–197
cyclodextrins, 10–11
cytochrome P450 (CYP), 231, 234–235, 237
cytotoxicity of conventional chemotherapy, 145

D

delivery of conventional chemotherapy, 145, **146**
dendrimers, 9, *133*, 134
    characteristics and synthesis routes of, 151–152
    convergent growth method, 152
    divergent growth method, 152
    nano-formulations for cancer chemoprevention, 250
    synthesis methods, 151, *151*
deodar cedar, **189**
Derris scandens, **190**
design of nano-formulations, 110–113
    lipid-based particles, *111*, 112, **112**
    nanocrystalline particles, *111*, **112**, 113
    non-polymeric particles, 110–112, *111*, **112**
    polymer-based particles, 110, *111*, **112**
    schematic representation of, *111*
    types of nano-formulations and physicochemical
        properties, **112**, 113
devdar, **189**
DHA, *see* dihydroartemisinin (DHA)
diagnosis of cancer, 81, *82*
diallyl sulfide, 66, 184, **228**, 232
dietary phytochemicals as chemopreventives, 27–29
    examples of, 29
    medicinal plants, 29
    overview, 27, 28–29
dietary signals in nutrigenomics, 168
diet of cancer patients, nutrigenomics to personalize, 167
dihydroartemisinin (DHA), 6
dimerized condensates, 236–238
dioscin, 89
diosgenin, **103**
direct microinjection, 155
diterpenoid, **105**
divergent growth method, 152

DNA
 aptamers, 266
 cleavage, 24
 damage, 2, 25, 39, 66, 81, 83, 89, 135, 145, 170, 197
 dendrimers in gene delivery, 9
 fragmentation, 47
 methylation, 25, 44, 167, 172
 methylation inducers, 90
 oxidative stress-induced DNA damage, 2
 repair, 46, 64, 167, 176, **193**, 238
 ROS-induced DNA instability, 2
 synthesis, **189**, 197
 topoisomerase, 145, **192**, 197, 208, 215, 217
docetaxel, 89, **105**
Doxil, 118, **119**, **153**, 154, 248, 251, 265, 266
doxorubicin (DOX), catechins and, 198
dragon's blood tree, **191**
drug delivery and cancer treatment
 carbon-based NPs for, 246–247, **247**
 conventional chemotherapy, 145, **146**
 metal nano-formulations for, 250, 251
 targeted delivery of NM by nanocarriers to the tumor
   site, 265–266

E

Eclipta alba, 185
EGCG, *see* epigallocatechin gallate (EGCG)
eicosapentaenoic acid (EPA) on antiproliferative action of
   anticancer drugs, 198
electrophile responsive element (EpRE), 170, 231–232, 235
electroporation, 155
ellagic acid (EA), 27
 and quercetin with resveratrol, 197
emulsion polymerization, 150
Enantia chlorantha, **191**
endometrial cancer, **105**
enhanced food, 91
enhanced permeation and retention (EPR) effect
 colloids affecting, in tumor cells, 263–265
 of NM in cancerous cells in humans, 264–265
epidermal growth factor receptors (EGFRs), 2, 64, 67, 114,
   117, **153**, 175, 266
epigallocatechin gallate (EGCG)
 in controlling epigenetic changes and mi RNAs involved
   in carcinogenesis and cancer prevention, 174
 evaluated in clinical trials on various cancers, **105**
epigenetic changes
 mi RNAs involved in carcinogenesis and cancer
   prevention, controlling, 172–175, *173*
 natural products in controlling, 172–175, *173*
Erthyrina suberosa, 29
Escherichia coli, 70
esophageal cancer, **195**
Euphorbia hirta, 29
Euphorbia neriifolia, 185

F

Fagara zanthoxyloides, **191**
FAK signaling pathway, 63
fat-soluble vitamins, 39–49
 ability to act on a variety of cancerous cells and
   different organs in the human body, **43**

conclusion, 49
current scenario of cancer, 40
introduction, 39–40
mechanism of vitamins in cancer prevention, 40–43
vitamins exploited for cancer prevention, 43–49
fennel, **103**
fenugreek, 64, 67–68, **103**
ferulic acid, 27
flavones, 4–5
 carotenoids, 4–5
 evaluated in clinical trials on various cancers, **105**
 lycopene, 5
 quercetin, 4
flavonoids, 26–27, 185
 basic structure of, *22*
 evaluated in clinical trials on various cancers, **105**
 in mint, 67
food products as preventive treatment for cancer, studies
   on, 66–70
 α-pinene from bay leaf, 69
 black cumin, 66
 chili pepper, 67
 cloves, 68
 curcumin in different types of cancers, 66
 fenugreek, 67–68
 flavonoids in mint, 67
 garlic, 66–67
 ginger, 66
 green tea, 69
 koenimbin from curry leaves, 68
 lycopene from tomato, 69
 piperine in pepper, 67
 probiotics for prevention of cancer, use of, 69–70
 quinines, 68
 resveratrol, 68–69
 saffron, 67
 soybean, 69
fortified food, 91
free radicals, 89–90
freeze-thaw method, 148
fruits, 89
functional foods, 75–92
 cancer and its pathophysiology, 79–82
 classification of cancer chemopreventive and
   therapeutically used, 82–92, *84*
 health benefits and market trend, 79
 introduction, 76–77
 mechanism of action, classification based on, 89–90
 origin, classification based on, 83–86
 overview, 77–79
 phytoconstituents, classification based on chemical class
   of, 86–89, *87*
 processing methods, classification based on, 90–92
 regulations and legislation, 77–78
 summary, 92
 technological challenges, 78–79
 types of, *78*

G

gallic acid, 27
gambogic acid (GA) and proteasome inhibitor MS132 or
   MG262, 198
Garcinia hanburyi, 198

garden thyme, **194**
garlic, 29, 66–67, **103**, **186**
garlic mustard, **103**
gastric cancer
  anticarcinogenic properties of *Brassica* in, **228**
  liposomal and polymeric NPS in various stages of
    clinical trials for, **153**
  liposome-based nano-formulation used for the
    prevention of, **249**
  nano-formulations in commercial use or in clinical
    trials for, **119**
  natural polyphenol for, **28**
  plant-derived substances having anticancer activity
    against, **190**
gemcitabine (GEM) and impressic acid (E12–1) or
    acankoreanogein (E13–1), 198
genes
  genetic correlation of cancer, 167–168
  major concepts of interaction between nutrients and,
    166, *166*
  natural phytochemicals targeting, in process of
    carcinogenesis, 169–172
genistein, 6, 101
  and arsenic trioxide (ATO), 197
  in controlling epigenetic changes and mi RNAs involved
    in carcinogenesis and cancer prevention, 174–175
  evaluated in clinical trials on various cancers, **105**
  thearubigin and, 197
genomic puzzle, deciphering, 168–169
genomic stability, effects of dietary phytochemicals on, 170
giant Mexican sunflower, **195**
giloya, **194**
ginger, 66, **195**
gingerol, **104**
ginseng, **193**
ginsenoside Rh2 (G-Rh2) and betulinic acid (BA), 196
Glochidion zeylanicum, **191**
glucobrassicin, *229*, 235
glucosinolates (GSLs), 100, 130, 131, 172, 228, 231, 232,
    234, 235–236
Glycyrrhiza glabra, **191**
goatweed, **194**
God's tree, **186**
Goniothalamus macrophyllus, **191**
government rules and regulations on the manufacturing
    and commercialization of nanomedicines, 267
grapes, 4, 24, 68–69, 89, 131, **132**, 184, 197
graphene, 247, **247**
green tea, 3, 25, 29, 69, **103**, **132**, 184, **188**
guava, 68, 100, **193**
guggul, **188**
Gynandropis pentaphylla, 29

H

harbor, **195**
haridra and haldi, **190**
Harungana madagascariensis, **191**
HCC (hepatocellular carcinoma) (HepG2), 185
HCT-116, 185
head and neck cancer
  anticarcinogenic properties of *Brassica* in, **228**
  liposomal and polymeric NPS in various stages of
    clinical trials for, **153**

nano-formulations in detection and treatment of, 118
herbal approaches in the treatment of cancer, 184–185
  examples of herbs used in, 185
  list of plant-derived substances having anticancer
    activity against various types of cancer with their
    possible mode of action, **186–195**
  natural herb-synthetic drug interaction, 185, 196–199
herb-drug synergistic effect, potential, 197–199
  apigenin and 5-flourouracil (5-FU), 198
  berberine and rapamycin, 197–198
  catechins and doxorubicin (DOX), 198
  eicosapentaenoic acid (EPA) on antiproliferative action
    of anticancer drugs, 198
  gambogic acid (GA) and proteasome inhibitor MS132 or
    MG262, 198
  gemcitabine (GEM) and impressic acid (E12-1) or
    acankoreanogein (E13-1), 198
  genistein and arsenic trioxide (ATO), 197
  quercetin and cisplatin, 198
herb-herb synergistic effect, potential, 185, 196–197
  1′-acetoxychavicol acetate (ACA) and sodium butyrate
    (SB), 196
  artemisinin (art) and resveratrol, 185, 196
  clove, oregano, thyme, walnuts, and coffee, 196
  curcumin analog pentagamavunon-1 (PGV-1) and
    5-flourouracil (5-FU), 197
  curcumin and resveratrol, 185
  curcumin and vinorelbine, 196–197
  ellagic acid (EA) and quercetin with resveratrol, 197
  ginsenoside Rh2 (G-Rh2) and betulinic acid (BA), 196
  matrine and resveratrol, 196
  thearubigin and genistein, 197
Herceptin, 119, 184, 237
Heritiera fomes, **278**
Hh signaling pathway, 60, 62
Himalayan cedar, **189**
hog creeper, **190**
hole formation, 155
holy basil, **132**
hTERT (human telomerase reverse transcriptase), **43**, 45
humanized monoclonal antibody, 116, **119**
hydroxybenzoic acid, 27
hydroxycinnamic acid, 27

I

I3C, *see* indole-3-carbinol (I3C)
impressic acid (E12-1) or acankoreanogein (E13-1),
    gemcitabine (GEM) and, 198
Indian aloe, **186**
indole-3-carbinol (I3C), 7, 89, 101, 172, *173*, 175, 227, **228**,
    229
  3,3′-diindolylmethane, 236–238
  chemistry of, 235–238
  chemosensitization's companionship, 237–238
  condensate, 236
  in controlling epigenetic changes and mi RNAs involved
    in carcinogenesis and cancer prevention, 175
  diversity within, 236–237
  mode of action by, 238
inequality of cancer prevention, 22
inflammatory mediators, 170–171
inorganic NPs
  carbon NPs, 150

characteristics and synthesis routes of, 150–151
mesoporous silica NPs, 150–151
irinotecan, 216
isoflavone
  in cancer prevention, **103**
  evaluated in clinical trials on various cancers, **105**
isolated and purified preparation of active phytochemical
  from food constituents (herbal/product/dosage form),
  91–92
isothiocyanates (ITCs), 6, 28, 29, 79, 100–101, 172–173,
  173, 184, 228, 231, 232, 234, 235, 236, *236*
  in controlling epigenetic changes and mi RNAs involved
  in carcinogenesis and cancer prevention, 172, 173

J

JAK-STAT signaling pathway, 63
Japanese Kampo, 199
jaramla, **193**
jewel vine, **190**

K

Kadcyla, 118–119, **119**, **153**
kaempferol, *23*, 25, **28**, 169, 172, **186**, **189**, **190**, 228, **228**,
  236, 238, **278**
kalgur, **192**
kalmegha, **187**
katuka, **193**
Key lime, **190**
Khaya senegalensis, **191**
kidney cancer
  anticarcinogenic properties of *Brassica* in, **228**
  plant-derived substances having anticancer activity
  against, **192**, **193**
koenimbin from curry leaves, 68
kutki, **193**

L

L9-NC, **105**
labdane diterpenoid,(E)-14,15-epoxylabda-8(17),12-dien-
  16-al (Cm epoxide), 185
Lactobacillus, 70, 86
Languas galangal, 196
laryngeal cancer, **187**
L-asparaginase, 89
L-BLP25, **119**
leaves, 89
LE-DT/ATI-1123, **105**
leukemia
  effect of phytochemicals along with therapeutic agents
  on, **132**
  liposome-based nano-formulation used for the
  prevention of, **249**
  nano-formulations in commercial use or in clinical
  trials for, **119**
  phytochemicals evaluated in clinical trials on, **105**
  plant-derived substances having anticancer activity
  against, **187**, **189**, **192**
  polymer-based nano-formulation used for the prevention
  of, **248**
  role of other phytochemicals in prevention of, **103**
licorice, **191**

licorice weed, **194**
lipid-based NPs (LBNPs), *111*, 112, **112**, 115, 118, **119**, 260
lipid-polymer hybrid NPs (LPHNPs), 248, 249, **249**
Lipocurc, **105**
Lipoplatin, 119, **119**, 153, **153**
liposomal formulations, 117, 118, **119**
liposomal vaccine, **119**
liposome-based formulations for cancer prevention,
  248, **249**
liposomes, 131, 133, *133*
  calcium-induced fusion method, 149
  characteristics and synthesis routes of, 147–149
  classification of, 147
  freeze-thaw method, 148
  reverse-phase evaporation method, 148
  schematic representation of, *148*
  sonication method, 148
  synthesis methods, 148
  synthesis routes of, 147–149
  thin-film method, 148
Lipotecan, **153**
Lipusu, **105**, 153, **249**
liver cancer
  anticarcinogenic properties of *Brassica* in, **228**
  effect of phytochemicals along with therapeutic agents
  on, **132**
  liposomal and polymeric NPS in various stages of
  clinical trials for, **153**
  nano-formulations in detection and treatment of, 118
  natural polyphenol for, **28**
  phytochemical-/nutraceutical-based nano-formulation
  used for the prevention of, **250**
  plant-derived substances having anticancer activity
  against, **186–195**
  statistics of cases diagnosed in 2020, 243, *244*
Lophira alata, **191**
lung cancer
  anticarcinogenic properties of *Brassica* in, **228**
  effect of phytochemicals along with therapeutic agents
  on, **132**
  liposomal and polymeric NPS in various stages of
  clinical trials for, **153**
  liposome-based nano-formulation used for the
  prevention of, **249**
  LPHNPs that have been nano-formulated and tested in,
  249
  nano-formulations in commercial use or in clinical
  trials for, **119**
  nano-formulations in detection and treatment of,
  117–118
  natural polyphenol for, **28**
  phytochemical-based nano-formulations under clinical
  trials for, **105**
  phytochemical-/nutraceutical-based nano-formulation
  used for the prevention of, **250**
  plant-derived substances having anticancer activity
  against, **186–191**, **193**, **195**
  polymer-based nano-formulation used for the prevention
  of, **248**
  quantum-dot-based nano-formulation used for the
  prevention of, **251**
  statistics of cases diagnosed in 2020, 243, *244*
Lupron Depot, **153**
lutein, 228, *229*

lycopene, 5, 100
  evaluated in clinical trials on various cancers, **105**
  from tomato, 69
lymphoma, **192**, **193**

M

macrophage phagocytic system (MPS), nanomedicine
    uptake and clearance by, 261–262
mamaphal, **187**
Mangifera indica, **192**
mango, **192**
mangroves as an alternative source of anticancer drug
    leads, 275–280
  anticancer compounds from mangrove plants, **278–279**
  anticancer phytochemical leads from, 277, 279
  conclusion and future prospects, 280
  distribution and diversity of mangrove ecosystem,
    276–277
  introduction, 275–277
  mangrove potential in overcoming drug resistance,
    279–280
  marine natural products, 275–276
  medicinal potential of mangrove species, 277
MAPK (mitogen-activated protein kinase), 66, 67, 68, 69,
    101, **103**, **132**, 169, 170, 176, 233, 234
Mappia foetida, **192**
marine natural products in cancer chemotherapy,
    275–276
Marqibo, 118, **119**, 133
matrine and resveratrol, 196
matrix metalloproteinase inhibitors, 90
mechanism of action, classification of cancer
    chemopreventive and therapeutically used functional
    foods based on, 89–90
  angiogenesis inhibitors, 89
  antiproliferative compounds, 90
  apoptosis inducers, 90
  DNA methylation inducers, 90
  free radicals and ROS inhibitors, 89–90
  matrix metalloproteinase inhibitors, 90
  metastasis inhibitors, 89
medicinal plants in chemoprevention of cancer, 29
mesoporous silica NPs (MSNs), 150–151
metal nano-formulations for drug delivery and cancer
    treatment, 250, 251
metastasis inhibitors, 89
micellar NPs, **105**, 119, **119**, 245, **248**
microbial origin, 86
micropinocytosis, 154
Milicia excelsa, **192**
Millettia pinnata, **185**
mi RNAs involved in carcinogenesis and cancer
    prevention, natural products in controlling,
    172–175, *173*
mode of action, in conventional chemotherapy, 145
mohaguno, **190**
Mongolian milkvetch, **187**
monoterpene and triterpenoids, 5
Morinda lucida, **192**
Moringa oleifera, **185**
mRNA, 25–26, *41*, **45**, 45–46, 65, 174, 198, 233
mulberries, **132**
multi-biopolymer systems, 11

multidrug resistance (MDR)
  of conventional chemotherapy, 145
  mangrove potential in overcoming, 279–280
  by NPs, eliminating, 266–267

N

nanocrystalline particles, *111*, **112**, 113
nanoemulsions, 9–10
nanofibers, 9
nano-formulations, 7–13
  advantages over conventional therapeutic formulations,
    113
  approaches for chemoprevention, 109–120
  based on natural products, challenges in producing,
    11–12
  carbon nanotubes, *133*, 135
  in chemoprevention in various cancer types, 115–118
  of chemotherapeutic drugs, 143–158
  clinical insights, 118–119
  in commercial use or in clinical trials for
    chemoprevention, **119**
  conclusions and future perspectives/prospects, 12–13,
    120, 136
  dendrimers, *133*, 134
  design, 110–113
  introduction to approaches to, 109–110
  liposomes, 131, 133, *133*
  nanomicelles, *133*, 135–136
  natural products and, 7–13
  NPs, *133*, 133–134
  opportunities with, 243–252
  overview, 7–8
  physicochemical properties of, **112**, 113
  phytochemicals and their anticancer effect, **132**
  properties of, having phytochemicals and nanospecies,
    133
  schematic representation of, 111
  targeting cancer tissues by, 114–115, *115*
  types of, 8–11, *11*, **112**, 131, 133–136
nano-formulations, cancer chemoprevention based on,
    246–251
  carbon-based NPs for drug delivery and cancer
    treatment, 246–247
  dendrimer-based nano-formulations for cancer
    chemoprevention, 250
  lipid-polymer hybrid nano-formulations for cancer
    therapy, 248, 249, **249**
  liposome-based formulations for cancer prevention,
    248, **249**
  metal nano-formulations for drug delivery and cancer
    treatment, 250, 251
  nutraceuticals and phytochemical-based nano-
    formulations for cancer treatment, 249–250, **250**
  polymeric nano-formulations for cancer therapy, 247,
    248, **248**
  quantum dot nano-formulations for cancer prevention,
    251, **251**
nano-formulations in cancer detection and treatment,
    243–252, *see also* nano-formulations, cancer
    chemoprevention based on
  breast cancer, 116
  colorectal cancer, 116–117
  conclusion and future prospects, 251, 252

head and neck carcinoma, 118
introduction, 243–244
liver cancer, 118
lung and bronchial cancer, 117–118
opportunities and challenges, 245–246
oral squamous cell carcinoma, 118
pancreatic cancer, 115
prostate cancer, 118
skin cancer, 118
targeting, active and passive, *245*, 245–246
in various cancer types, 115–118
nano-formulations of chemotherapeutic drugs, 143–158
advantages of nano-formulation-based drug delivery
over conventional dug delivery, **146**
cellular pathway, exploration of, 153–155
characteristics and synthesis routes of different
nano-formulations, 147–152
clinical applications of different nanoformulated
chemotherapeutic drugs, 152–153
conclusion, 157–158
drawbacks of conventional chemotherapy, 145
exploration of cellular pathway, 153–155
introduction, 144
nano-co-delivery for cancer treatment, recent advances
in, 155–156
nano-formulations as delivery systems for
chemotherapeutic drugs, 146
recent advances in nano-co-delivery for cancer
treatment, 155–156
toxicity issues, challenges, and regulations associated
with nano-formulated drugs, 156–157
nanoliposomes, 8
nanomedicine for cancer management, obstacles in
utilizing, 259–269
activation of complement proteins with nanomedicines,
261
challenges associated with the translation of NM from
laboratory to clinic, 268–269
conclusions, 269
eliminating multidrug resistance by NPs, 266–267
government rules and regulations on the manufacturing
and commercialization of nanomedicines, 267
introduction, 259–260
limitations crossing the physiological barriers of body,
260–262
nanomedicine uptake and clearance by macrophage
phagocytic system, 261–262
properties of colloids affecting the EPR effect in tumor
cells, 263–265
targeted delivery of NM by nanocarriers to the tumor
site, 265–266
tumor environment and vasculature as a barrier to
delivery of NPs to the tumor site, 262–263
nanomicelles, *133*, 135–136
nanoparticles (NPs), *133*, 133–134
charge of, 263–264
shape of, 264
size of, 263
nanoparticles (NPs) entry into the cell, modes of, 153–155,
154
additional entryway processes, 155
caveolae-mediated endocytosis, 155
clathrin- and caveolae-independent endocytosis, 155
clathrin-mediated endocytosis, 155

micropinocytosis, 154
phagocytosis, 154
Nanotax, **153**
nanotechnology for chemoprevention, emergence of, 7
narkya, **192**
natural compounds, chemoprevention and, 172
natural herb–synthetic drug interaction, 185, 196–199
potential herb-drug synergistic effect, 197–199
potential herb-herb synergistic effect, 185, 196–197
natural phytochemicals targeting genes in process of
carcinogenesis, 169–172
angiogenesis, 171–172
antioxidant effect, 169–170
apoptosis and cell cycle arrest, 171
effects of dietary phytochemicals on genomic stability
and non-genotoxicity, 170
inflammatory mediators, 170–171
natural products, 3–13
antioxidants in prevention of cancer, 65–66
challenges associated with using as chemopreventive
agents, 131
challenges in producing anticancer nanoformulations
based on, 11–12
conclusion and future perspective, 12–13
in controlling epigenetic changes and mi RNAs
involved in carcinogenesis and cancer prevention,
172–175, *173*
emergence of nanotechnology for chemoprevention, 7
future prospects of, in the treatment of cancer, 200
introduction, 129–130
nano-formulations, 7–13
overview, 3
plant products with anticancer properties, 130–131
potential of, 129–136
shortcomings of, 7
studies on food products as preventive treatment for
cancer, 66–70
synergistic effects of, in cancer treatment, 183–201
targeting cancer stem cells by, 57–70
types of, 3–7
used in prevention of cancers by acting on CSCs, 65–70
which are used in the prevention of different diseases,
64–65
Navelbine/NanoVNB, **105**
NCI-H460, 185
neem, **188**
Newbouldia laevis, **193**
NF-κB, 6, 62, 100, **132**, 172, *173*, 174, 175, 199, 233, 237
nimbolide, **104**
NK 105, **105**
non-genotoxicity, effects of dietary phytochemicals on,
170
non-polymeric NPs, 110–112, *111*, **112**
Notch pathway, 60
NPs, *see* nanoparticles (NPs)
Nrf2, **132**, 196, 231, 233, 235
nutraceuticals and phytochemical-based nano-
formulations for cancer treatment, 249–250, **250**
nutrient, major concepts of interaction between gene and,
166, *166*
nutrigenomics in cancer chemoprevention, 165–176
bioactive phytochemicals in cancer therapy and
deciphering the genomic puzzle, 168–169
cancer and its genetic correlation, 167–168

chemoprevention and natural compounds, 172
conclusion and future perspective, 176
controlling epigenetic changes and mi RNAs involved
    in carcinogenesis and cancer prevention, 172–175,
    173
crosstalk, 175–176
dietary signals in nutrigenomics, 168
elementary principles of, 166
introduction, 165–167
major concepts of interaction between nutrient and
    gene, 166, *166*
natural phytochemicals targeting the key genes in
    process of carcinogenesis, 169–172
to personalize diet of cancer patients, 167

O

Ocimum gratissimum, **193**
Oldenlandia diffusa, **104**
onion, **132**
Onivyde, 115, **119**
oral cancer
    effect of phytochemicals along with therapeutic agents
        on, **132**
    nano-formulation-based chemoprevention in, 118
oregano, 196
organosulfur compounds (OSC), 88, 232–233
    enzymatic interaction, 232
    molecular insights of brassicaceae metabolites,
        232–233
    in preclinical trials for cancer treatment, **104**
origin, classification of cancer chemopreventive and
        therapeutically used functional foods based on,
        83–86
    animal origin, 86
    microbial origin, 86
    plant origin, 83–86, *85*
    probable mechanisms shown by phytochemicals in
        cancer chemoprevention, 83, *84*
ovarian cancer
    effect of phytochemicals along with therapeutic agents
        on, **132**
    liposomal and polymeric NPS in various stages of
        clinical trials for, **153**
    liposome-based nano-formulation used for the
        prevention of, **249**
    phytochemical-based nano-formulations under clinical
        trials for, **105**
    plant-derived substances having anticancer activity
        against, **189**, **192**, **193**, **194**
    polymer-based nano-formulation used for the prevention
        of, **248**

P

paclitaxel, 89, 185
    under clinical trials for cancer treatment, **105**
    evaluated in clinical trials on various cancers, **105**
Panax ginseng, **193**, 196
pancreatic cancer
    nano-formulation-based chemoprevention in, 115
    nano-formulations in commercial use or in clinical
        trials for, **119**
    nano-formulations in detection and treatment of, 115

phytochemical-based nano-formulations under clinical
    trials for, **105**
phytochemical-/nutraceutical-based nano-formulation
    used for the prevention of, **250**
plant-derived substances having anticancer activity
    against, **186**, **189**, **190**, **193**, **194**
polymer-based nano-formulation used for the prevention
    of, **248**
papaya, **189**
papaya seeds, **103**
Paris polyphylla, **132**
passive diffusion, 155
passive targeting, *114*, 115, *245*, 245–246, 265
Peaderia foetida, 29
peanuts, 4, 68–69, 101, 131, **132**
PEGylated liposomes, 8, 116, **119**, 147, 265
PEGylated polylactic glycolic acid (PLGA) NPs, 149
pegylated silica hybrid nanomaterials, **250**
PEITC, **103**
penawar hitam, **191**
peritoneal neoplasms, **153**
phagocytosis, 154
phase I and phase II enzymes, 231–232
Phellinus linteus, 86
phenol, **104**
phenolic acids, 27
phenolics, 87
phenyl isothiocyanates (PITC), *230*, 235, *236*
phenylpropanoid, **104**
Phoenix paludosa, **278**, 279
Phomopsis longicolla, 280
Phyllanthus amarus, **193**
physiological barriers of body, limitations crossing,
    260–262
    activation of complement proteins with nanomedicines,
        261
    nanomedicine uptake and clearance by macrophage
        phagocytic system, 261–262
phytochemicals, 97–106
    in cancer prevention, 100–102
    in cancer treatment, 102–105
    conclusion and future prospects, 105–106
    cytotoxic effects, 25
    effects on genomic stability and non-genotoxicity, 170
    hemolytic effects, 25
    introduction, 97–98
    isolated and purified preparation of active, from food
        constituents (herbal/product/dosage form), 91–92
    modulating signaling molecules, 25–26
    nano-formulations for cancer treatment, nutraceuticals
        and, 249–250, **250**
phytochemicals in cancer prevention, 100–102, **103**
    curcumin, 100
    genistein, 101
    isothiocyanates (ITCs), 100–101
    lycopene, 100
    other phytochemicals in, 102
    probable mechanisms shown by, 83, *84*
    resveratrol, 101–102
phytoconstituents, classification of cancer
    chemopreventive and therapeutically used functional
    foods based on, 86–89, *87*
    alkaloids, 88
    carotenoids, 88

organosulfur compounds, 88
other miscellaneous compounds, 88–89
phenolics, 87
quinones, 88
terpenes and triterpenoids, 88
phytopolyphenol, **104**
phytosterols, 89
PI3K/AKT/mTOR pathways, 62
picrorhiza, **193**
Picrorhiza kurroa, **193**
Picrorrhiza kurroa, 29
pilu oil, **103**
pines, 4
pink trumpet, **194**
piperine in pepper, 67
piperlongumine, **103**
plant origin, 83–86, *85*
plant roots, 89
PMs, *see* polymeric micelles (PMs)
podophyllotoxin, 185
Podophyllum peltatum, 58, 185
poison devil's-pepper, **194**
Polygonum cuspidatum, 4, 101, **104**, 185
polymer-drug conjugates, 9
polymeric micelles (PMs), 8
polymeric nano-formulations for cancer therapy, 247, 248, **248**
polymeric NPs (PNPs), 110, *111*, **112**, 116, 117, 118, 248, 269
  characteristics and synthesis routes of, 149–150
  emulsion polymerization, 150
  methods of polymeric NPs synthesis, 149
  physicochemical properties of, **112**, 113
  schematic representation of, *111*
  solvent evaporation, 149
  synthesis methods, 149
  in various stages of clinical trials, **153**
polypeptide NPs, 10
polyphenols, 3–4, 21–29, 185
  anticancer properties, **28**
  challenges in cancer prevention and treatment, 22–24
  classifications of, 22, *23*
  curcumin (bis-α, β-unsaturated β-diketone), 4
  dietary phytochemicals as chemopreventives, 27–29
  epigallocatechin-3-gallate (EGCG), 3–4
  evaluated in clinical trials on various cancers, **105**
  flavonoids, 26–27
  introduction, 21–22
  naturally occcurring, for cancer prevention and treatment, 26–27, **28**
  nature of, 22
  phenolic acids, 27
  phytochemicals modulating signaling molecules, 25–26
  in preclinical trials for cancer treatment, **104**
  as pro-oxidants, 24–25
  reactive oxygen species in development of carcinogenesis, role of, 24
  resveratrol and trans-resveratrol (3,5,4'-trihydroxy-trans-stilbene), 4
  role and mechanism of, in cancer chemoprevention, 24–26
pomegranate, 29, 89, **194**
PPAR signaling pathways, 63–64
preclinically studied anticancer phytochemicals, 102, **104**

preclinical trials for cancer treatment, phytochemicals in, **104**
probiotics for prevention of cancer, use of, 69–70
processed food, 90–91
processing methods, classification of cancer chemopreventive and therapeutically used functional foods based on, 90–92
  basic food from a natural source, 90
  enhanced food, 91
  fortified food, 91
  isolated and purified preparation of active phytochemical from food constituents (herbal/product/dosage form), 91–92
  processed food, 90–91
pro-oxidants, polyphenols as, 24–25
prostate cancer
  anticarcinogenic properties of *Brassica* in, **228**
  liposomal and polymeric NPS in various stages of clinical trials for, **153**
  nano-formulations in detection and treatment of, 118
  natural polyphenol for, 28
  phytochemical-/nutraceutical-based nano-formulation used for the prevention of, **250**
  plant-derived substances having anticancer activity against, **186**, **188–195**
  polymer-based nano-formulation used for the prevention of, **248**
  quantum-dot-based nano-formulation used for the prevention of, **251**
  statistics of cases diagnosed in 2020, 243, *244*
proteasome inhibitor MS132 or MG262, gambogic acid (GA) and, 198
protein-based systems, 10
Pseudomonas aeruginosa, 70, 234
Pseudomonas putida, 185
Psidium guajava, **193**
punarnava, **188**
Punica granatum, **194**

Q

QDs, *see* quantum dots (QDs)
quantum dots (QDs), 9, 251, **251**
quercetin, 4, *23*, 25, 27, **28**, 29, 67, 89, **132**, 169, 170, 172, 175, **189**, **190**, 197, 198–199, 228, **228**, *236*, 249, **250**, **278**
  and cisplatin, 198
  in controlling epigenetic changes and mi RNAs involved in carcinogenesis and cancer prevention, 175
  with resveratrol, ellagic acid (EA) and, 197
quinines, 68
quinones, 88, 185

R

radiation treatment, harmful effects of, 64
Rapamune, 119, **119**
rapamycin, berberine and, 197–198
Rauwolfia vomitoria, **194**
reactive oxygen species in development of carcinogenesis, role of, 24
red ironwood tree, **191**

regulation and crosstalk between different signaling
     pathways, 64
regulation associated with nano-formulated drugs,
     156–157
renal cancer
     liposomal and polymeric NPS in various stages of
          clinical trials for, **153**
     nano-formulations in commercial use or in clinical
          trials for, **119**
resveratrol, 68–69, 101–102
     artemisinin (art) and, 185, 196
     in controlling epigenetic changes and mi RNAs involved
          in carcinogenesis and cancer prevention, 174
     curcumin and, 185
     evaluated in clinical trials on various cancers, **105**
     matrine and, 196
     in preclinical trials for cancer treatment, **104**
     and trans-resveratrol (3,5,4'-trihydroxy-trans-stilbene), 4
reverse-phase evaporation method, 148
R-Gossypol acetic acid, **105**
Rhizophora apiculata, **278**, 279, 280
RNA
     micro (miRNA), 26, 174
     noncoding, 26, 45, 167, 174
     synthesis, 208
roots of long pepper, **103**
roots of wild yam, **103**
rosemary, **103**, 184
ROS inhibitors, 89–90

S

saffron, 67
Salvia miltiorrhiza, **132**
saponins, 185
Scoparia dulcis, **194**
selectivity of conventional chemotherapy, 145
selenium, 66
Senegal mahogany, **191**
sesquiterpene lactone, **104**
shallaki, **188**
shiny-leaf prickly ash, **195**
Siam weed, **190**
signaling molecules, phytochemicals modulating, 25–26
skin cancer
     effect of phytochemicals along with therapeutic agents
          on, **132**
     nano-formulations in detection and treatment of, 118
     phytochemical-/nutraceutical-based nano-formulation
          used for the prevention of, **250**
     plant-derived substances having anticancer activity
          against, **188**, **191**, **193**
     statistics of cases diagnosed in 2020, 243, *244*
societal costs of cancer prevention, 22
sodium butyrate (SB), 1'-acetoxychavicol acetate (ACA)
     and, 196
solid lipid NPs (SLNs), 112, 115, 118
solubility of conventional chemotherapy, 145
sonication method, 148
Sonneratia apetala, **278**
Sonneratia ovata, 279, **279**
Sophora flavescens, 196
soybean, 6, 29, 69, **103**, 184
stage of cancer, 81

stilbenoid, **105**
stomach cancer
     effect of phytochemicals along with therapeutic agents
          on, **132**
     plant-derived substances having anticancer activity
          against, **187**, **188**, **192**, **194**
     statistics of cases diagnosed in 2020, 243, *244*
stone breaker, **193**
streptavidin-conjugated superparamagnetic NPs, 116, **119**
*Streptomyces hygroscopicu*s, 86, 197
studying cellular interactions of NPs, techniques for, 155
sulfur compounds, 5–6
sulphorafane (SFN), 6, 175, 184, **228**, 233–235, 236
superparamagnetic NPs, 116–117, **119**
synergistic effects of natural products in cancer treatment,
     183–201
     conclusion, 200
     current status of synergistic approach of natural
          products in the treatment of cancer, 199
     future prospects of natural products in the treatment of
          cancer, 200
     herbal approaches in the treatment of cancer, 184–185
     introduction, 184
     natural herb–synthetic drug interaction, 185, 196–199
synthesis methods
     dendrimers, 151, *151*
     liposomes, 148
     polymeric NPs, 149
synthesis routes
     dendrimers, 151–152
     inorganic NPs, 150–151
     polymeric NPs, 149–150
SYP-0709, **105**

T

Tabebuia avellanedae, **194**
tamoxifen, 10, 25, 76, 82, 116, 237–238, 267
tannins, 185
targeted delivery of NM by nanocarriers to the tumor site,
     265–266
     active targeting of NMS, 265
     passive targeting of NMS, 265
     triggered release of NMS, 265–266
targeting
     active, 114, *114*, *245*, 245–246
     passive, *114*, 115, *245*, 245–246
targeting cancer stem cells (CSCs) by natural products for
     chemoprevention, 57–70
     cancer overview, 58–59
     cell signaling pathways and, 59–64
     chemical and radiation treatment, harmful effects of, 64
     conclusion and prospects, 70
     CSC (cancer stem cell) model, 59
     introduction, 58
     natural products used in prevention of cancers by acting
          on CSCs, 65–70
     natural products which are used in the prevention of
          different diseases, 64–65
taxanes, 89
Taxus brevifolia, 58, **105**
terpenes, 88
TGF/SMAD signaling pathway, 63
thearubigin and genistein, 197

thin-film method, 148
thyme, 196
Thymus vulgaris, **194**
Tinospora cordifolia, **194**
Tithonia diversifolia, **195**
tocopherols (TOCs), 46–48
tocotrienols (TTs), 46–48
  cell cycle arrest caused by, 46–47
  effect of, on cancer stem cells, 47–48
  induction of autophagy and apoptosis by, 47
  inhibition of invasion, metastasis, and angiogenesis by, 47
  overview, 46
TOCs, *see* tocopherols (TOCs)
topotecan (TPT), 215–216
toxicity issues associated with nano-formulated drugs, 156–157
treatment of cancer, 81, *83*
Tricholoma matsutake, 86, 90
triggered release of NMS, 265–266
triterpenes, **104**, 185
triterpenoids, 88, **104**
TTs, *see* tocotrienols (TTs)
tumor environment and vasculature as a barrier to delivery of NPs to the tumor site, 262–263
tumor necrosis factor (TNF), 25, **103**, 233, 235
tumor necrosis factor alpha (TNF-α), 66, 170–171, *173*
tumor types, categorizing, 79–80
tupbel, **190**
turmeric, 64, 66, 70, 89, 90, 100, **105**, 174, 184

U

umbrella cheese tree, **191**
ursolic acid, 185
  in cancer prevention, **103**
  in preclinical trials for cancer treatment, **104**
Uvaria chamae, **195**

V

variegated laurel, **190**
VDR, *see* vitamin D receptor (VDR)
VDREs, *see* vitamin D response elements (VDREs)
Vernonia amygdalina, **195**
vincristine sulfate, **105**
vinorelbine, curcumin and, 196–197
vinorelbine tartrate, **105**, 197–198
vitamin A, 43–45
  ability to act on a variety of cancerous cells and different organs in the human body, **43**
  cancer and, 44–45
  overview, 43
vitamin C
  antioxidants in prevention of cancer, 65–66
  mechanism of, in cancer prevention, 42, 43
  vitamin K co-administration of, 48–49
vitamin D, 45–46
  ability to act on a variety of cancerous cells and different organs in the human body, **43**

cell cycle arrest triggered by, 41, *42*
effect of, on TGF-β, 46
mechanism of, in cancer prevention, 40–41, *41*
molecular mechanism of, 40, *41*
overview, 45
in regulating DNA repair proteins, 46
tumor progression and, 45
vitamin D receptor, 40, *41*, 45, 46
vitamin D response elements, 40, *41*
vitamin D3, 65
vitamin D receptor (VDR), 40, *41*, 45, 46
vitamin D response elements (VDREs), 40, *41*
vitamin E, 46–48, 65–66
  ability to act on a variety of cancerous cells and different organs in the human body, **43**
  antioxidants in prevention of cancer, 65–66
  cell cycle arrest caused by tocotrienols, 46–47
  effect of tocotrienols on cancer stem cells, 47–48
  induction of autophagy and apoptosis by tocotrienols, 47
  inhibition of invasion, metastasis, and angiogenesis by tocotrienols, 47
  overview, 46
vitamin K, 48–49
  mechanism of, in cancer prevention, 42, 43
  overview, 48
  vitamin C co-administration of, 48–49
vitamins exploited for cancer prevention, 43–49
  vitamin A, 43–45
  vitamin C, 42, 43, 48–49
  vitamin D, 45–46
  vitamin E, 46–48
  vitamin K, 48–49

W

walnuts, 196
watercress, **132**
Western lifestyle, in cancer prevention and treatment, 23–24
wild jujube, **195**
wingless/integrated (Wnt) pathway, 60, *61*
Withania somnifera, **195**
Women's Healthy Eating and Living (WHEL), 238

X

xanthmicrol, 185
Xue song, **189**
Xylocarpus granatum, 277, **279**
Xylocarpus moluccensis, 279, **279**

Z

Zanthoxylum nitidum, **195**
zeaxanthin, 228, *229*
zerumbone, **103**
Zingiber officinale, 66, **104**, **195**
Ziziphus nummularia, **195**

For Product Safety Concerns and Information please contact our EU
representative GPSR@taylorandfrancis.com
Taylor & Francis Verlag GmbH, Kaufingerstraße 24, 80331 München, Germany

www.ingramcontent.com/pod-product-compliance
Lightning Source LLC
Chambersburg PA
CBHW080932220326
41598CB00034B/5759